Stem Cell Biology and Regenerative Medicine

Series editor

Kursad Turksen, Ph.D.
e-mail: kursadturksen@gmail.com

AF786540

Our understanding of stem cells has grown rapidly over the last decade. While the apparently tremendous therapeutic potential of stem cells has not yet been realized, their routine use in regeneration and restoration of tissue and organ function is greatly anticipated. To this end, many investigators continue to push the boundaries in areas such as the reprogramming, the stem cell niche, nanotechnology, biomimetics and 3D bioprinting, to name just a few. The objective of the volumes in the Stem Cell Biology and Regenerative Medicine series is to capture and consolidate these developments in a timely way. Each volume is thought-provoking in identifying problems, offering solutions, and providing ideas to excite further innovation in the stem cell and regenerative medicine fields.

More information about this series at http://www.springer.com/series/7896

Vladimir A. Botchkarev • Sarah E. Millar
Editors

Epigenetic Regulation of Skin Development and Regeneration

Humana Press

Editors
Vladimir A. Botchkarev
Centre for Skin Sciences
University of Bradford
Bradford, UK

Department of Dermatology
Boston University School of Medicine
Boston, MA, USA

Sarah E. Millar
Department of Dermatology,
Perelman School of Medicine
University of Pennsylvania
Philadelphia, PA, USA

Department of Cell and Developmental
Biology, Perelman School of Medicine
University of Pennsylvania
Philadelphia, PA, USA

Department of Anatomy and Cell Biology,
School of Dental Medicine
University of Pennsylvania
Philadelphia, PA, USA

ISSN 2196-8985 ISSN 2196-8993 (electronic)
Stem Cell Biology and Regenerative Medicine
ISBN 978-3-030-13237-8 ISBN 978-3-319-16769-5 (eBook)
https://doi.org/10.1007/978-3-319-16769-5

© Springer International Publishing AG, part of Springer Nature 2018
Softcover re-print of the Hardcover 1st edition 2018
This work is subject to copyright. All rights are reserved by the Publisher, whether the whole or part of the material is concerned, specifically the rights of translation, reprinting, reuse of illustrations, recitation, broadcasting, reproduction on microfilms or in any other physical way, and transmission or information storage and retrieval, electronic adaptation, computer software, or by similar or dissimilar methodology now known or hereafter developed.
The use of general descriptive names, registered names, trademarks, service marks, etc. in this publication does not imply, even in the absence of a specific statement, that such names are exempt from the relevant protective laws and regulations and therefore free for general use.
The publisher, the authors and the editors are safe to assume that the advice and information in this book are believed to be true and accurate at the date of publication. Neither the publisher nor the authors or the editors give a warranty, express or implied, with respect to the material contained herein or for any errors or omissions that may have been made. The publisher remains neutral with regard to jurisdictional claims in published maps and institutional affiliations.

Printed on acid-free paper

This Humana Press imprint is published by the registered company Springer International Publishing AG part of Springer Nature.
The registered company address is: Gewerbestrasse 11, 6330 Cham, Switzerland

Preface

Living organisms are protected from the external environment by an integument that forms the outer body surface and permits both survival and adaptation to environmental challenges. In mammals, the integument is formed by the skin and its appendages, including feathers, hairs, and glands, and displays a high degree of evolutionary, anatomical and environmentally dependent-variability.

Mammalian skin is one of the largest organs of the body. It serves a number of critical roles, including protecting the organism from external insults, maintaining body temperature and water balance, and transmitting sensory information. To efficiently fulfill these complex functions, the skin has developed the capacity to provide a high degree of plasticity in response to changing environmental conditions, while maintaining its structural integrity. In postnatal life, the epithelial outer layer of the skin, known as the epidermis, continuously regenerates due to the ability of epithelial stem cells in its basal layer to supply progeny capable of forming all of the epidermal cell layers and generating a functional barrier. Unlike the continuously regenerating epidermis, hair follicles undergo cyclical transitions between active growth, involution, and resting phases. After skin injury, stem cells in both epithelial and underlying mesenchymal components of the skin contribute to skin regeneration, supplying progeny that repair the wounded area.

During the last two decades, tremendous progress has been made in understanding the molecular mechanisms that underlie skin development, regeneration, and both rare and common diseases. Genome-wide association studies and comparative genome analyses have provided invaluable insights into the physiological role of genetic information. Nevertheless, how the genomes of diverse populations of epithelial and mesenchymal skin cells are organized beyond their linear sequence, and the mechanisms regulating lineage-specific responsiveness of distinct genomic regions to external signals in healthy and diseased skin, remain to be clarified.

It is now widely accepted that in addition to their regulation by signaling/transcription factor-mediated mechanisms, lineage-specific gene expression programs are also controlled epigenetically by covalent DNA and histone modifications, as well as via higher-order chromatin remodeling and topological arrangement of genes and their distal regulatory elements in 3D nuclear space. Epigenetic mechanisms play important roles in controlling cellular functions in living organisms and are considered to be a driving force of phenotypic plasticity and evolutionary adaptation. Variability in epigenetic status helps explain the relationships between an individual's genetic background and the effects of the environment on susceptibility to different diseases.

As an important and accessible source of epithelial, mesenchymal, pigmentary, and neuronal stem cells, the skin serves as an excellent model for studying how extrinsic signals coordinate gene expression by directing the activity and distribution of distinct epigenetic regulators and orchestrating the execution of lineage-specific gene expression programs and their adaptation to environmental cues.

This volume presents and summarizes recent major findings that shed light on the roles of critical components of the epigenetic regulatory machinery in the control of skin development and regeneration. Chapter 1 outlines how signaling/transcription factor-mediated and epigenetic mechanisms operate in concert to regulate skin development and regeneration, and highlights the role of the cell nucleus as a command center integrating signals received from the external environment and transforming them into distinct transcriptional outcomes.

Chapter 2 focuses on the importance of DNA methylation as a keeper of epigenetic memory in the control of skin development and physiological regeneration. Chapters 3, 4, and 5 discuss the impact of distinct post-translational histone modifications and their corresponding epigenetic regulators, including Polycomb and Trithorax genes and histone deacetylases, in the control of transcriptional silencing and activation in epithelial cells of the developing and adult skin. In Chapter 6, the role of ATP-dependent chromatin remodeling in the control of gene expression in the epidermis is discussed.

Chapters 7 and 8 highlight the roles of noncoding and micro-RNAs in regulating keratinocyte differentiation, while Chapter 9 is devoted to the emerging roles of RNA modifications in the control of epithelial stem cell activity.

Chapter 10 discusses the mechanisms coordinating three-dimensional organization of epidermal genes and their regulatory elements (enhancers) in the nucleus, while Chapter 11 describes the role of the nuclear lamina in transmitting signals from the external environment to the genome and in controlling lineage-specific differentiation programs in normal skin. Finally, Chapter 12 summarizes recent data on the functions of distinct components of the epigenetic machinery in skin regeneration after injury and during wound healing.

This collection of work offers a brief introduction to this exciting and rapidly developing area of research and provides readers with an understanding of the experimental underpinnings of current models that will aid in critical evaluation of new literature in the field. The exhilarating pace of discovery will undoubtedly ensure that significant new developments and unexpected findings will be revealed before this book is widely available.

In summary, we believe that this volume provides a useful introduction to skin epigenetics for many categories of researchers. Our hope is that this work will serve as a platform and inspiration for future research in this field that is necessary to bridge the gap between our knowledge of basic epigenetic mechanisms and clinical practice. Progress in this direction will ultimately permit the development of novel approaches for modulating the epigenome and epitranscriptome to protect the skin against aging and environmental stressors, as well as in the treatment of skin disorders.

Bradford, UK Vladimir A. Botchkarev
Philadelphia, PA, USA Sarah E. Millar

Contents

Contributors

Bogi Andersen Departments of Biological Chemistry and Medicine, University of California, Irvine, CA, USA

Vladimir A. Botchkarev Centre for Skin Sciences, University of Bradford, Bradford, UK

Department of Dermatology, Boston University School of Medicine, Boston, MA, USA

Natalia V. Botchkareva Centre for Skin Sciences, University of Bradford, Bradford, UK

Donna M. Brennan-Crispi Department of Dermatology, Perelman School of Medicine, University of Pennsylvania, Philadelphia, PA, USA

Howard Y. Chang Program in Epithelial Biology, Stanford University School of Medicine, Stanford, CA, USA

Cheng-Ming Chuong Department of Pathology, Keck School of Medicine, University of Southern California, Los Angeles, CA, USA

Integrative Stem Cell Center, China Medical University Hospital, China Medical University, Taichung, Taiwan

Idan Cohen Department of Cell, Developmental, and Regenerative Biology, Black Family Stem Cell Institute, Icahn School of Medicine at Mount Sinai, NY, New York, USA

Katherine L. Dauber-Decker Department of Cell, Developmental, and Regenerative Biology, Black Family Stem Cell Institute, Icahn School of Medicine at Mount Sinai, NY, New York, USA

Elena Ezhkova Department of Cell, Developmental, and Regenerative Biology, Black Family Stem Cell Institute, Icahn School of Medicine at Mount Sinai, NY, New York, USA

Michael Y. Fessing Centre for Skin Sciences, University of Bradford, Bradford, UK

Michaela Frye University of Cambridge, Department of Genetics, Cambridge, UK

Gitali Ganguli-Indra Department of Pharmaceutical Sciences, College of Pharmacy, Oregon State University, Corvallis, OR, USA

Molecular Cell Biology Program, Oregon State University, Corvallis, OR, USA

Knight Cancer Institute, Oregon Health & Science University (OHSU), Portland, OR, USA

Cristina de Guzman Strong Division of Dermatology, Department of Medicine, Washington University in St. Louis School of Medicine, St. Louis, MO, USA

Center for Pharmacogenomics, Department of Medicine, Washington University in St. Louis School of Medicine, St. Louis, MO, USA

Center for the Study of Itch, Washington University in St. Louis School of Medicine, St. Louis, MO, USA

Valerie Horsley Department of Dermatology, Yale School of Medicine, New Haven, CT, USA

Department of Molecular, Cell and Developmental Biology, Yale University, New Haven, CT, USA

Arup K. Indra Department of Pharmaceutical Sciences, College of Pharmacy, Oregon State University, Corvallis, OR, USA

Molecular Cell Biology Program, Oregon State University, Corvallis, OR, USA

Knight Cancer Institute, Oregon Health & Science University (OHSU), Portland, OR, USA

Department of Biochemistry and Biophysics, Oregon State University, Corvallis, OR, USA

Linus Pauling Institute, Oregon State University, Corvallis, OR, USA

Departments of Dermatology, OHSU, Portland, OR, USA

Megan C. King Department of Cell Biology, Yale School of Medicine, New Haven, CT, USA

Rachel Herndon Klein Departments of Biological Chemistry and Medicine, University of California, Irvine, CA, USA

Ya-Chen Liang Department of Pathology, Keck School of Medicine, University of Southern California, Los Angeles, CA, USA

Integrative Stem Cell Center, China Medical University Hospital, China Medical University, Taichung, Taiwan

Andrei N. Mardaryev Centre for Skin Sciences, School of Chemistry and Biosciences, Faculty of Life Science, University of Bradford, Bradford, UK

Sarah E. Millar Department of Dermatology, Perelman School of Medicine, University of Pennsylvania, Philadelphia, PA, USA

Department of Cell and Developmental Biology, Perelman School of Medicine, University of Pennsylvania, Philadelphia, PA, USA

Department of Anatomy and Cell Biology, School of Dental Medicine, University of Pennsylvania, Philadelphia, PA, USA

Inez Y. Oh Division of Dermatology, Department of Medicine, Washington University in St. Louis School of Medicine, St. Louis, MO, USA

Center for Pharmacogenomics, Department of Medicine, Washington University in St. Louis School of Medicine, St. Louis, MO, USA

Center for the Study of Itch, Washington University in St. Louis School of Medicine, St. Louis, MO, USA

Abdulrahim A. Sajini University of Cambridge, Department of Genetics, Cambridge, UK

Department of Biomedical Engineering, Khalifa University of Science and Technology, Abu Dhabi, United Arab Emirates

Andrei A. Sharov Department of Dermatology, Boston University School of Medicine, Boston, MA, USA

Rachel M. Stewart Department of Cell Biology, Yale School of Medicine, New Haven, CT, USA

Kevin C. Wang Program in Epithelial Biology, Stanford University School of Medicine, Stanford, CA, USA

Veterans Affairs Palo Alto Healthcare System, Palo Alto, CA, USA

Randall Widelitz Department of Pathology, Keck School of Medicine, University of Southern California, Los Angeles, CA, USA

Rui Yi Department of Molecular, Cellular and Developmental Biology, University of Colorado, Boulder, Denver, CO, USA

Chapter 1
All Roads Go to the Nucleus: Integration of Signaling/Transcription Factor-Mediated and Epigenetic Regulatory Mechanisms in the Control of Skin Development and Regeneration

Vladimir A. Botchkarev, Andrey A. Sharov, and Michael Y. Fessing

Abbreviations

KC keratinocyte

1.1 Introduction

Living organisms interact with the external world via their integument or outermost body surface that transmits signals between the external environment and internal milieu and provides conditions for survival. The complexity of the integument's organization varies greatly among the species and depends on their position in the "evolutional tree", as well as on the environment in which they live [1]. In mammals, the integument is formed by the skin and its appendages, which can include hairs, nails and a variety of glands. These structures show a large degree of evolutionary, anatomical, and environmentally-dependent variability [2].

V. A. Botchkarev (✉)
Centre for Skin Sciences, University of Bradford, Bradford, UK

Department of Dermatology, Boston University School of Medicine, Boston, MA, USA
e-mail: v.a.botchkarev@bradford.ac.uk; vladbotc@bu.edu

A. A. Sharov
Department of Dermatology, Boston University School of Medicine, Boston, MA, USA

M. Y. Fessing
Centre for Skin Sciences, University of Bradford, Bradford, UK

© Springer International Publishing AG, part of Springer Nature 2018

V. A. Botchkarev, S. E. Millar (eds.), *Epigenetic Regulation of Skin Development and Regeneration*, Stem Cell Biology and Regenerative Medicine,
https://doi.org/10.1007/978-3-319-16769-5_1

Mammalian skin fulfills a large number of functions, the most important of which are the protection of the organism against environmental (physical, chemical, biological) stressors; maintenance of body temperature and water balance; and a transmission of sensory information. As one of the largest organs in mammals, the skin also possesses endocrine and immune functions and represents an important source of the growth factors, cytokines and hormones [3, 4]. In contributing to visual appearance, the skin is critically important for establishing contacts with the external environment, psycho-social communications and mimicry [1].

To efficiently fulfill its complex functions, the skin develops as an organ that is capable of both maintaining its structural integrity, and providing a high degree of plasticity in response to changing environmental conditions. The skin is formed from two embryologically distinct tissue compartments: the outer layer or epidermis, which arises from the surface ectoderm; and the inner layer or dermis, which has mesenchymal or mesodermal developmental origins that vary according to anatomical location [5–7]. Molecular interactions between the two skin layers, as well as between cells migrating into or invading the skin, such as endothelial cells, neurons, melanocytes, and immune cells, play fundamental roles in development of the skin as an organ that can operate efficiently as an interface between the organism and its environment.

Skin morphogenesis is a complex process characterized by precise spatial and temporal coordination of several developmental programs, leading to the formation of the epidermis, dermis, cutaneous appendages, skin microvasculature, innervation, immune and pigmentary system [5, 8–10]. In postnatal life, the epidermis continuously regenerates due to the ability of epithelial stem cells to supply progeny that form all of the epidermal cell layers; by contrast, hair follicles undergo cyclical transitions between active growth (anagen), involution (catagen) and resting (telogen) phases [11, 12]. After skin injury, both epidermal and hair follicle stem cells contribute to skin regeneration, supplying progeny that form the epithelial layer covering the wound area [13].

Over the last two decades, significant progress has been achieved in defining the molecular organization of the skin and delineating signaling pathways that are involved in the development of each skin component. Fundamental discoveries on the roles of keratin intermediate filaments, collagens, adhesion molecules, growth factors and their signaling components, and transcriptional regulators in the maintenance of structural skin integrity have provided a molecular basis for the development of novel therapeutic approaches to treat many skin diseases [14–19].

Molecular control of skin development and regeneration is based on integration of at least three distinct layers of regulation:

1. *Extracellular* mechanisms including secreted regulatory molecules (e.g., growth factors and their modulators/inhibitors), components of extracellular matrix and adhesion molecules that help to establish proper cell-cell and/or cell-matrix signaling and communication [20–24].

2. *Intracellular (cytoplasmic)* mechanisms that include regulatory molecules capable of modulating the activity of signaling pathways and transcription factors that transmit signals from cytoplasm to the nucleus, as well as microRNA/mRNA interactions that modulate transcriptional outcomes [22, 24–29].
3. *Intranuclear* regulatory mechanisms including the activity of lineage-specific transcription factors; and epigenetic machinery that controls covalent DNA and histone modifications, higher-order chromatin remodeling, and nuclear compartmentalization of specific sets of genes, enhancer elements and transcriptional machinery. Together, these mechanisms modulate the responsiveness of the distinct genomic regions to external signals [30–32].

Increasing evidence indicates that signaling/transcription factor-mediated and epigenetic regulatory mechanisms operate in concert to control gene expression in skin epithelial cells. The genetic program of epithelial cell differentiation in the skin includes stage- and lineage-specific activation of gene expression at keratinocyte-specific gene loci including the Epidermal Differentiation Complex [EDC], and Keratin type I/II loci, that are located on mouse chromosomes 3, 11 and 15, respectively [33, 34]. In developing and regenerating skin, key signaling pathways including Wnt, BMP/TGF-beta, EDA/EDAR/NFκB, EGF/FGF/MAPK, Hedgehog, HIF, Notch, and SCF/c-kit, lineage-specific transcription factors such as p63 and MITF, and epigenetic regulators including DNA- and histone modifying enzymes, higher-order chromatin remodelers, long non-coding (Lnc) RNAs, and microRNAs (miRNAs), regulate gene expression at these and other loci in a highly integrated manner [30, 35–37].

Epigenetic mechanisms play an important role in the control of cellular functions in living organisms and are considered to be a driving force of phenotypic plasticity and evolutionary adaptation [38]. Variability in epigenetic status helps to explain the relationships between an individual's genetic background and the effects of the environment on susceptibility to different diseases [39]. As an important source of epithelial, mesenchymal, pigmentary and neuronal stem cells, the skin serves as an excellent model to study how extrinsic signals coordinate gene expression by directing the activity and distribution of distinct epigenetic regulators to orchestrate the execution of lineage-specific gene expression programs and their adaptation to environmental cues.

In this Chapter, we describe how signaling/transcription factor-mediated and epigenetic mechanisms operate in concert to regulate skin development and regeneration. We review the mechanisms regulating the development of the epidermis and its appendages and emphasize the role of the cell nucleus as a command center integrating the signals received from external environment and transforming them into distinct transcriptional outcomes. We also summarize existing data on how specific epigenetic signatures confer responsiveness to extrinsic cues at discrete regions of the keratinocyte genome.

1.2 Development of the Epidermis and Skin Appendages: Transition from Single-Layered Epithelium to Multicellular Mini-Organs

Epidermal Development Development of the epidermis occurs as a multi-focused program to create a self-renewing layer of cells that is capable of establishing an effective barrier to prevent the influx of harmful agents from the outside and limit water loss from the inside, as well as to sense and generate appropriate adaptive tissue responses to environmental challenges. This program is achieved by tightly coordinated interactions in the epidermal cellular ensemble that includes keratinocytes, melanocytes, Langerhans cells, gamma-delta-T-cells, Merkel cells and intra-epidermal nerve fibers.

In mouse embryos, the epidermal-specific gene expression program is initiated at about day 8.5 of embryonic development (E8.5), when cells of the embryonic ectoderm expressing epithelial keratins K8 and K18 become committed to an epidermal fate and begin to express the key epidermal transcription factor p63. This is followed at E9.5 by onset of expression of the basal epidermal keratins K5/K14 [25, 26, 40] (Fig. 1.1a). Between E9.5 and E12.5, K5/K14-positive epidermal progenitors expand laterally via symmetric cell division to cover the rapidly growing embryo [41, 42]. Nuclei in basal epidermal progenitor cells are oriented vertically and contain multiple nucleoli indicating their active metabolic state [42–44]. These

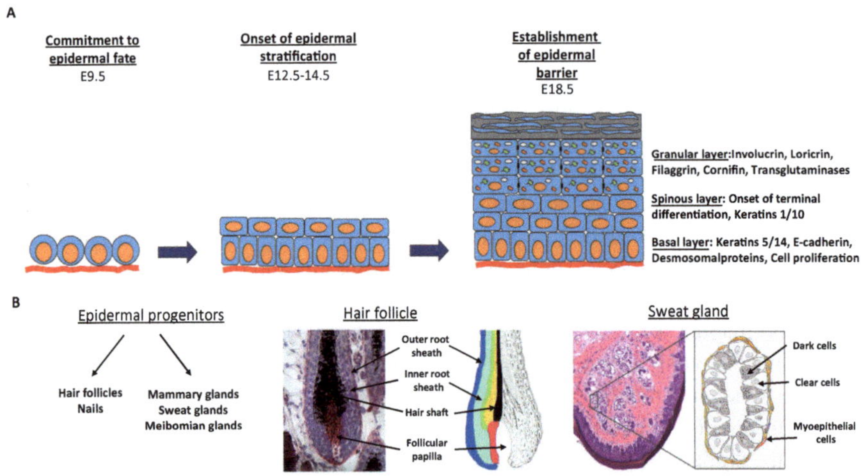

Fig. 1.1 Development of the epidermis and skin appendages. (**a**) Schematic showing major stages of epidermal development in the mouse embryo and selected markers expressed in distinct epidermal layers. (**b**) Schematic showing the fates of epidermal progenitor cells differentiating into distinct types of skin appendage (left). Distinct cell lineages occurring as a result of differentiation of epidermal progenitor cells into hair follicle keratinocytes or sweat gland epithelial cells, as well as images illustrating morphology of the hair follicle (central panel) and sweat gland (right panel) are shown

basal cells retain their multi-potent properties, and at later stages of skin development give rise to several lineage-committed progenitor populations that form the periderm; epidermal appendages such as mammary glands, hair follicles, sweat glands and nail; and specialized sensory Merkel cells (see below).

During this period, epidermal progenitor cells also establish adhesive connections with each other via formation of desmosomes, and connect to the underlying basement membrane via hemidesmosomes [45]. Cytoplasmic K5/K14 intermediate filaments provide mechanical support to epidermal progenitor cells, and connect desmosomes and hemidesmosomes to the cell nucleus via interactions with proteins such as nesprins and plectin that are located at the outer nuclear membrane [45]. Keratin filaments also participate in regulation of cell signaling, translation and protein synthesis, providing a scaffold for many processes associated with epidermal stratification and barrier formation [46, 47].

The onset of epidermal stratification occurs at about E12.5 of mouse embryogenesis, when basal keratinocytes undergo asymmetric cell division to form an intermediate epidermal layer between the basal layer and the periderm [48] (Fig. 1.1a). The intermediate layer subsequently develops into the spinous layer [40], in which nuclei are oriented mostly horizontally. Spinous layer cells are smaller than basal epidermal cells, and display fewer nucleoli [42–44]. At E13.5-E14.5, intermediate/spinous layer cells occasionally proliferate, but terminate expression of K5/K14, instead beginning to express the suprabasal keratins K1/K10 [40]. K1/K10 play an important roles in maintaining the mechanical integrity of the cytoplasm and nucleus, and in positioning desmosomes in keratinocytes of the intermediate layer [49]. These cells are also connected to each other via adherens junctions, tight junctions and gap junctions [45, 50]. Cells in the upper spinous layer start to accumulate lamellar granules that are enriched for lipids and lipid-processing enzymes, essential components of the epidermal barrier [51].

By day E14.5-E15.5 of mouse embryonic development the granule-containing keratinocytes of the upper spinous layer form the granular epidermal layer [52], that displays smaller and flattened nuclei. Nucleoli in these cells are fused and concentrated at the nuclear interior, while heterochromatin becomes spread all over the nucleus, suggesting cessation of metabolic activity [43, 44]. Cells of the granular epidermal layer remain connected to each other via tight junctions and desmosomes [45]. A characteristic feature of these cells is the accumulation of keratohyalin granules, that are composed of keratin intermediate filaments and proteins encoded by genes that lie within the Epidermal Differentiation Complex (EDC) locus located on chromosome 3 [53]. These proteins, including filaggrin, involucrin, loricirin, trychohyalin, small proline-rich and late cornified cell envelope proteins, promote aggregation of keratin filaments into thick bundles. After transglutaminase-mediated cross-linking, these proteins complex with intracellular lipids released from the lamellar granules to form a scaffold just beneath the plasma membrane [51]. Formation of the cornified layer is the final step in the differentiation of epidermal cells. In mouse embryos, this takes place between E17.5-E18.5 [52] (Fig. 1.1a). Cornification represents a unique form of cell death and includes formation of the cytoplasmic cornified envelope, a submembranous structure in which intermediate filaments, proteins, lipids, fragments of the degraded organelles and nuclear remnants are cross-linked [45]. The process of cornified envelope formation is regulated by enzymes including Caspase-14 that is involved in pro(fillagrin) processing, and transglutaminases that regulate protein cross-linking [45, 51, 54]. Transglutaminases are also involved in cross-linking of proteins with desmosomes to form corneodesmosomes. These structures play key roles in attaching the cornified envelope to the plasma membrane and in binding cornified cells together [45, 51]. Cornified cells are surrounded by a continuous extracellular lipid matrix, forming a multilayered barrier structure that prevents penetration of harmful organisms and chemicals and protects the organism from water loss [53]. The uppermost cornified cells are shed continuously from the epidermal surface and are replaced by terminally differentiating granular layer cells. This process is driven by gradual degradation of the corneo-desmosomes in the upper corneal layers [53].

The epidermis of newborn mice is composed of a basal layer, one to two layers of spinous cells, three or four layers of granular cells, and about twenty layers of cornified cells [55]. This relatively thick epidermis is maintained in postnatal life only in foot-pad and plantar skin, while in hairy skin the epidermis becomes progressively thinner from about P10. By the time mice reach weaning age at approximately P21, their hair follicles are synchronized at the telogen stage of the hair follicle growth cycle, and the interfollicular epidermis consists of only a few keratinocyte layers [55, 56]. Several mechanisms contribute to the decrease in epidermal thickness of postnatal hairy skin, including lower rates of epidermal proliferation, increased keratinocyte apoptosis, and phagocytosis of cell fragments by neighboring keratinocytes [55, 57, 58].

Molecular Control of Epidermal Development, Self-Renewal and Differentiation Establishment of a functional epidermal barrier is one of the major goals of the epidermal differentiation program, which includes a tightly regulated progression of keratinocyte proliferation, terminal differentiation, apoptosis and shedding. This program is regulated at several levels including extracellular ligand/receptor signaling pathways, lineage-specific transcription factors, and epigenetic regulators. Together, these mechanisms establish a sequential and well-coordinated process of transformation of keratinocyte progenitor cells from a proliferative, metabolically active state to a fully keratinized cellular end product of terminal differentiation.

The transcription factor p63 operates as master regulator of this program, inducing expression of multiple sets of genes important for epidermal differentiation, including those encoding essential components of the cytoskeleton, such as keratins 5 and 14; cell adhesion molecules including P-cadherin, integrin-α3, Perp, and dystonin; cell matrix regulators such as Fras-1; the transcription factors AP-2γ, IKKα, and IRF6; and the epigenetic regulators Satb1, Brg1, Cbx4 [25, 30, 59–61].

Genetic ablation of p63 in mice results in failure of stratification of the epidermis and other squamous epithelia, lack of the formation of epidermal appendages including hair follicles, sweat glands, and teeth, and severe abnormalities in the development of limbs and external genitalia [62–64]. In line with the phenotypes of mouse mutants, heterozygous mutations in the human p63 gene underlie several ectodermal dysplasia syndromes that are characterized by abnormalities in the development of digits, teeth, hairs, nails and sweat glands (reviewed in [65, 66]).

p63 encodes two classes of isoforms, TAp63 isoforms and ΔNp63 isoforms, which are characterized, respectively, by the presence or absence for an N-terminal transactivation domain. Each isoform class consists of C-terminal variants known as α, β and γ. The TAp63 and ΔNp63 isoforms show distinct, although partially overlapping roles in the control of epidermal differentiation and stratification [25, 59]. ΔNp63α dominates in the epidermis compared with TAp63 isoforms, is strongly expressed in basal epidermal keratinocytes and is markedly downregulated in the spinous layer [67]. ΔNp63α plays a major role in mediating the effects of p63 in epidermal development, whereas TAp63 protects keratinocytes from senescence and suppress tumorigenesis in postnatal epidermis [67, 68].

Cell proliferation in the epidermis is an essential prerequisite of its self-renewal capacity and is normally restricted to the basal epidermal keratinocytes [58]. Approximately 35–45% of basal epidermal cells in newborn mice are positive for the proliferation marker Ki-67; this decreases to approximately 20–25% in adult mice [57, 69]. Keratinocyte proliferation is stringently controlled by autocrine, juxtacrine and paracrine mechanisms. These involve stimulatory and inhibitory signals arising from epidermal cells, intra-epidermal nerve endings, underlying dermal cells, and immune cells, as well as circulating factors such as hormones (reviewed in [22]).

Keratinocytes produce and secrete multiple ligands including epidermal growth factor (EGF), nerve growth factor (NGF), granulocyte-macrophage colony stimulating factor (GM-CSF) and endothelins that interact with their corresponding cell membrane receptors and stimulate cell proliferation in an autocrine manner [70]. Growth factor receptors on the surface of basal epidermal cells interact with adhesion molecules and other receptors; for example, signaling through EGFR can be activated by G-protein-coupled receptors that promote shedding of EGFR ligands from the cell surface by activation of metalloproteinases, whereas integrins that connect basal epidermal cells with the underlying basement membrane trigger ligand-independent EGFR activation [45, 70]. Keratinocyte proliferation is also stimulated in a paracrine manner by ligands of the fibroblast growth factor (FGF), hepatocyte growth factor (HGF) and insulin growth factor (IGF) families that are predominantly secreted by dermal cells in the skin (reviewed in [22, 71, 72]. In addition, neuropeptides, such as substance P, CGRP, and VIP released from sensory nerve endings in the epidermis or dermis, and POMC-derived peptides, such as beta-endorphin that are released from melanocytes, stimulate keratinocyte proliferation (reviewed in [73, 74]).

Cell proliferation in the basal epidermis is positively regulated by several transcription factors, including c-Myc and ΔNp63 [26, 75]. c-Myc stimulates keratinocyte proliferation by controlling expression of the cell cycle regulators Cdk4 and CdkN2B, while ΔNp63 maintains the progenitor status of basal epidermal keratinocytes by stimulating expression of the FGF receptors Fgfr2/3, as well as by directly repressing the expression of anti-proliferative target genes, including 14-3-3σ, p16/

Ink4a, p19/Arf and p21 [75–78]. A number of epigenetic regulators including DNA methyltransferase DNMT1, histone deacetylases HDAC1/2, and the Polycomb components Cbx4, Bmi1, and Ezh1/2, stimulate proliferation of basal epidermal progenitors via repression of cell cycle inhibitory genes (reviewed in [30]). p63 interacts with HDAC1/2 at the *p16/Ink4a* promoter to repress gene transcription, suggesting that HDAC1/2 plays an important role in mediating repressive p63 functions [79].

Stimulatory effects on cell proliferation in the epidermis are counter-balanced by numerous inhibitory mechanisms that involve a variety of signaling molecules and transcriptional and epigenetic regulators. Signaling molecules that inhibit keratinocyte proliferation include transforming growth factor-beta/bone morphogenetic protein (TGF-β/BMP) and Notch ligands, vitamin D3 and interferon-gamma, which interact with their corresponding receptors on keratinocytes [22]. TGF-β/BMP and Notch pathways suppress proliferation in part by stimulating the expression of cyclin-dependent kinase inhibitor p21, which in turn inhibits the expression of cell cycle-associated genes and mediates keratinocyte exit from the cell cycle [80–83].

Transcription factors inhibit proliferation in the basal epidermis via diverse mechanisms. For instance, the transcription factor AP-2α inhibits keratinocyte proliferation via repression of EGFR [84], whereas ΔNp63α, besides its repressive effects on the expression of cell cycle inhibitors (see above), stimulates expression of IKKα, an important component of the NF-κB signaling pathway which is required for cell cycle withdrawal [85]. ΔNp63α also positively regulates expression of the transcription factor IRF6, which, in turn, induces proteasome-mediated ΔNp63α degradation and exit of keratinocytes from the cell cycle [86]. Expression of p63 is inhibited by miR-203 expressed in suprabasal epidermal cells, which provides a negative regulatory loop allowing keratinocytes to exit the cell cycle and limiting their proliferative capacity [87].

Terminal differentiation of keratinocytes is initiated by asymmetric cell division, which causes one or more daughter cells to lose adhesion to the basement membrane and move into the suprabasal layer [26, 45]. Asymmetric cell division of basal cells is controlled by the LGN, NuMA and dynactin (Dctn1) proteins, which are concentrated in the apex of keratinocytes during mitosis [42, 88]. Notch signaling operates downstream of these proteins [88], and, similar to its role in intestinal stem cells, serves as a target for protein kinase aPKC, a component of the Par complex that interacts with the mitotic spindle machinery to align spindle orientation along the apico-basal axis [89]. aPKC activity enhances Delta/Notch signaling in daughter cells and induces terminal differentiation [89].

Signaling through Notch receptors in suprabasal epidermal cells is induced by Notch ligands expressed in basal cells and helps to initiate an *early differentiation program* that stimulates expression of keratin 1 (K1), a marker of spinous layer keratinocytes [90, 91]. Notch signaling operates in parallel with AP-2α/γ transcription factors, which, by stimulating the expression of C/EBPα/β transcription factors, also mediate the switch in expression from basal (K5/K14) to suprabasal (K1/K10) keratins [40, 91]. This process is additionally promoted by the transcription factor

FOXN1 [92]. p63 contributes to this process indirectly, via TAp63α-mediated induction of AP-2γ expression [93].

The transition of keratinocytes from the spinous to granular layer is associated with onset of expression of *late differentiation* genes that both contribute to the formation of the cornified cell envelope (for example, EDC genes and transglutaminase genes), and regulate lipid synthesis and promote formation of the lipid-containing lamellar bodies [94]. The expression of EDC genes encoding the filaggrin, involucrin, loricirin, trychohyalin, small proline-rich and late cornified cell envelope proteins is controlled by numerous transcription factors including AP1, AP2, ARNT, FOXN1, GATA3, Grainyhead-like 3, KLF4, NRF2, m-OVO, PPAR-α, Sp1, Sp3, and TALE homeobox factors in a distinct and partially overlapping manner (reviewed in [95, 96]). Many of these factors operate as both stimulators and repressors of transcription, depending on the target gene and on their interactions with specifics sets of epigenetic regulators to form distinct transcriptional complexes.The coordinated involvement of these factors helps to fine-tune expression of different terminal-differentiation-associated genes during epidermal barrier formation. Some of these transcription factors, such as KLF4, also operate in concert with corticosteroids and regulate expression of common target genes in keratinocytes [97].

p63 contributes to the control of terminal keratinocyte differentiation indirectly, via modulating the expression of the transcription factor ZNF750, which, in turn, positively regulates expression of Klf4 and some of its target genes [98]. In addition, p63 regulates expression of the higher-order chromatin remodeler SATB1 and promotes establishment of a specific three-dimensional chromatin structure at the central EDC domain, which is required for coordinated regulation of gene expression [69]. The ATP-dependent chromatin remodelers Brg1 and Mi-2β, and the histone demethylase JMJD3, also act to enhance expression of terminal differentiation-associated genes involved in epidermal barrier formation [99–101].

Transcription of EDC genes is modulated by a number of inhibitory factors including the epigenetic regulators DNA methyltransferase 1, HDAC1/2, and components of the Polycomb complex (Cbx4, Ezh1/2 and Bmi1), which promote formation of a repressive chromatin structure and inhibit premature activation of terminal differentiation-associated genes [102–106]. The Polycomb component Cbx4 also acts to inhibit expression of non-keratinocyte lineage (neuronal) genes in the epidermis [61]. In addition, epigenetic regulators can control expression of transcription factors in keratinocytes via formation of active or repressive local chromatin structures at their promoter regions; for instance, p63 expression in keratinocytes is inhibited by the histone methyltransferase Setd8 and the histone deacetylase HDAC1/2 [79, 107].

Cellular and Molecular Mechanisms Regulating the Development of Skin Appendages In mice, formation of distinct types of epidermal appendages is initiated at different time-points in embryogenesis: mammary gland morphogenesis begins at E10.5, vibrissa follicles are initiated at E12.5, pelage hair follicles are induced between E14.5 and E18.5, while development of the claws and sweat glands begins at E14.5 and E16.5, respectively [108–111].

The fates of ectodermal cells and their capacity to differentiate into distinct types of appendage is linked to their topology and depends on interactions with the underlying mesoderm or mesenchyme [2, 112, 113]. Progenitor cells in the surface ectoderm differentiate into keratinized cells that serve as essential components of the epidermis, hairs and nails, and also form the epithelial compartments of ectodermal glands including mammary, sweat, and sebaceous glands, where subsets of these cells assume a secretory phenotype (Fig. 1.1b) [113, 114]. Gene expression programs activated in keratinizing and secretory epithelial cells are markedly different [115, 116]. It has been proposed that ectodermal progenitors are intrinsically programmed to form keratinized differentiation products such as hair shaft and nail, while their differentiation towards a glandular/secretory phenotype requires inhibition of this basic program [113, 117].

Several lines of evidence support this model. For instance, in mice hair follicles are present almost all over the body surface, while ectodermal glands develop in restricted areas of the skin: mammary glands form only in the ventro-lateral part of the trunk; sweat glands develop in the foot pads; and Meibomian glands form only in the eyelid area. In those skin areas that show presence of both principal types of appendages, such as mammary glands and hair follicles in the ventral skin, the processes of the their initiation during embryogenesis are separated by at least 3 days, thus suggesting that, at least in mice, appendageal glandular and keratinizing programs are not initiated simultaneously in the same anatomical regions [113, 117]. Furthermore, heterotypic epidermal-dermal recombination studies [113, 118] showed that the ability of mesenchyme to induce glandular differentiation is much more restricted both spatially and temporarily than the mesenchymal capacity to promote hair follicle cell fate [119]. Finally, in those skin areas that show presence of only glands (for example, foot pads) and where hair follicle cell fate is intrinsically inhibited, hair follicle development can be re-activated by injection of follicular papilla cells or via overexpression of the BMP antagonist Noggin [120, 121].

Despite the remarkable differences in the morphology and functions of the distinct ectodermal appendages, their development occurs through several common stages (Fig. 1.2):

(i). **Pre-placode/placode stage**: characterized by a specific molecular signature of a group of epidermal progenitors that are committed to form the distinct type of epidermal appendage. Focal thickening of the epidermis is a characteristic feature of placode formation and occurs via centripetal migration of the basal epidermal cells and their compaction [122]. This stage is also characterized by changes in the spatial organization of the mesenchymal cells underlying the placode.

(ii). **Bud stage**: characterized by downgrowth of the epithelial cells into the mesenchyme, which undergoes significant remodeling in areas closely opposed to the invading bud.

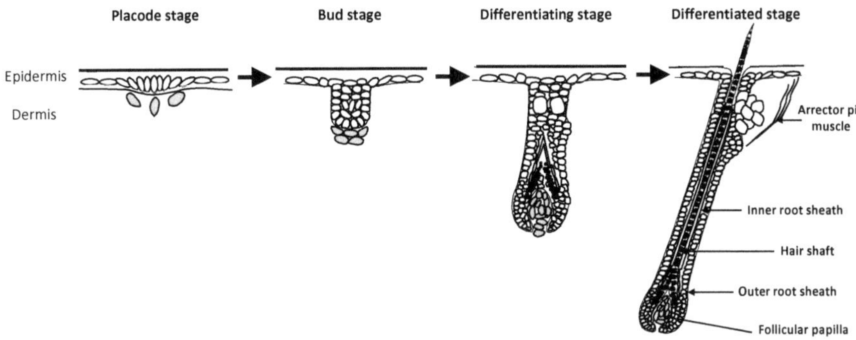

Fig. 1.2 Stages of skin appendage development. Schematic drawings illustrating the stages of hair follicle morphogenesis. Distinct compartments in the fully mature hair follicle are shown (right panel) [110]

(iii). **Differentiating stage**: includes several sub-stages characterized by the development of distinct appendage-specific morphological features, such as branching into the fat pad mesenchyme in developing mammary glands, formation of the hair bulb engulfing the dermal papilla in the hair follicle, or coiling of the epithelial portion of the developing sweat gland.

(iv). **Fully differentiated stage:** characterized by complete differentiation of the appendage-specific cell lineages and full functional capacity of skin appendages, such as production of the hair shaft, sebum, milk or sweat [109–112, 123, 124].

Despite many differences in the molecular controls of the development of distinct skin appendages, regulation of the initial stage of their development includes key common mechanisms. These appear to include Turing's reaction-diffusion interactions between the activators and inhibitors that specify sites of the appendage formation [20, 125, 126]. Similarly to neural tube induction, the initiation stages of mammary gland and hair follicle development occur via interactions between Wnt and BMP ligands and their secreted antagonists that establish gradients of activity of the corresponding signaling pathways in both embryonic ectoderm and mesenchyme [125, 127–129]. The Wnt signaling pathway promotes placode formation, while BMP signaling inhibits this process and, together with the Wnt inhibitors Dkk1/2/4 promotes inter-placode cell fate in epidermal progenitors [125, 127–129]. Inhibition of BMP signaling by BMP antagonists such as Noggin enhances formation of hair follicle placodes, and Noggin overexpression induces ectopic hair follicle formation in mouse footpads and eyelids at expense of formation of sweat or Meibomian glands, respectively [121, 128, 130, 131].

Fibroblast Growth Factor (FGF) and Ectodysplasin receptor (EDAR) signaling pathways operate as essential stimulators of the formation of both glandular and follicular epidermal placodes (reviewed in [112, 132]). The FGF and EDAR pathways promote placode fate at least in part by inhibiting BMP signaling [133, 134]. EDAR also signals via the NF-κB pathway to stimulate expression of BMP antagonists such as Ctgf and Follistatin [135]. During hair follicle placode formation, EDAR serves as a downstream effector of Wnt signaling, which is initially activated in pre-placodes independently of EDAR/NF-κB activity, while EDAR signaling at later stages is required to refine the pattern of Wnt/β-catenin activity by stimulating expression of Wnt10b in placode progenitor cells [136]. In addition, EDAR signaling promotes hair follicle placode formation by enhancing expression of Sonic hedgehog (Shh) [137], which operates as a potent stimulator of keratinocyte proliferation and lateral expansion of developing hair follicles [138, 139].

Recent data reveal that hair follicles are specified when mesenchymal BMP signaling is blocked, permitting Shh production, while sweat glands are specified by mesenchymal-derived BMPs and FGFs that signal to epithelial buds and suppress epithelial-derived Shh production [140]. Similarly, Shh signaling is inactive in mammary gland placodes [141]. Thus, the presence or absence of Shh activity provides an important molecular predictor of hair follicle vesus glandular fate.

Hair follicle and sweat gland placodes also show differences in expression of a number of transcription factors that contribute to cell fate specification and organ development: hair follicle placodes express LHX2, MSX2, and SOX9 transcription factors, while sweat gland placodes are characterized by their low expression [140, 142, 143]. At later stages of development, hair follicles and sweat glands also show appendage-specific differences in the transcriptome signatures underlying the production of keratinized hair fiber or sweat secretion [115, 116].

The hair shaft differentiation program is characterized by expression of hair-specific keratins (Krt31-37, Krt81-86) and keratin-associated proteins in hair progenitor cells [144, 145], while these genes are not expressed in sweat gland epithelia [116]. The hair shaft-specific program is regulated by coordinated involvement of BMP, EDAR, FGF, Hedgehog, IGF, Notch and Wnt signaling pathways, as well as by a number of transcription factors including DLX3, FOXN1, HOXC13, KROX20, and MSX2 (reviewed in [12, 24]). These transcription factors regulate target genes involved in the control of cell proliferation and differentiation of hair matrix progenitors, including direct regulation of hair keratin or keratin-associated protein genes by FOXN1, LEF1 and HOXC13 [12, 24].

Differentiation of the inner root sheath in the hair follicle that provides mechanical support for growing hair fiber is regulated by mechanisms distinct from those that promote hair shaft formation [11]. Epidermal growth factor receptor and its ligand TGF-α, as well as enzymes involved in the TGF-α ectodomain shedding (TNFα-converting enzyme and Lysophosphatidic acid producing enzyme PA-PLAα1α) prevent premature keratinization of inner root sheath cells, and their deficiency leads to the formation of curly hair [146–149]. The BMP and Notch signaling pathways, and the transcription factors CUTL1, DLX3, GATA3, and MSX2 are also involved in the control of inner root sheath differentiation [150–155].

In sweat gland epithelial cells, a FOXA1-BEST2 cascade regulates sodium/potassium/chloride exchange and Ca^{2+}, which are required for sweat production [156]. In addition, the FOXC1 transcription factor plays an important role in the control of sweat gland-specific differentiation by inhibiting expression of epidermal keratinocyte-specific genes, such as Sprr2a [157].

1.3 Distinct Classes of Skin Stem Cells as Sources for Postnatal Skin Regeneration

During early postnatal development, distinct populations of epithelial, melanocyte, and mesenchymal stem cells establish their niches in micro-anatomical skin compartments including the epidermis, skin appendages and dermis [158–162]. Each population of skin stem cells derives from distinct classes of progenitor cells that enter into a relatively quiescent state shortly after formation of the corresponding skin compartments or structures and contribute to their maintenance and/or self-renewal during postnatal life [163, 164]. Skin stem cells contribute to skin regeneration after injury, and are capable of generating multiple classes of progenitor cells upon removal from the skin and culture *ex vivo* [165, 166].

Epithelial stem cells in the skin are localized in the epidermis and in skin appendages. In the *epidermis*, the basal cell population includes slow-cycling Keratin 5/14+ stem cells that express high levels of α2 and β1 integrins, and a more rapidly replicating population of Keratin 5-negative committed progenitor cells displaying intermediate levels of α2 and β1 integrins [167]. Both populations contribute to epidermal self-renewal and differentiate into suprabasal cells via asymmetric cell division in a stochastic manner [167]. However, only the stem cell population is capable of actively suppling progenitor cells during epidermal injury, while contribution of the committed progenitor cells to wound repair appears to be minimal [167].

In the *hair follicle*, several populations of epithelial stem cells exist in the bulge, isthmus and infundibulum [159]. The hair follicle bulge is formed during the first few days of postnatal development and contains a population of rarely cycling keratinocytes expressing CD34, Keratin 15, NFATc1, TCF3, SOX9 and LHX2 [142, 143, 168–172]. Bulge cells also express nephronectin, which is deposited into the hair follicle connective tissue sheath where it serves as a receptor for α8β1 integrin, inducing mesenchymal cells to form the insertion site for the arrector pili muscle [163].

In parallel to the specification of bulge stem cells in early postnatal hair follicles, niches are established for CD34/Keratin 15-negative stem cell populations in the upper regions of the follicle. These include LGR6- and MTS24-expressing stem cells in the isthmus, and an LRIG1+ stem cell population in the infundibulum [173–175]. BLIMP1-expressing cells become specified at the sebaceous gland mouth in early postnatal follicles and are postulated to serve as a source of stem cells for the sebaceous gland [176]. During progression of hair follicles through the catagen regression phase (P17-P20 in mouse dorsal skin), an LGR5+ stem cell population that expresses P-cadherin, LHX2, SOX9, and TCF3 is specified in the secondary hair germ [177, 178].

The distinct populations of follicular epithelial stem cells differentially contribute to physiological epidermal and hair follicle regeneration. During activation of the hair follicle growth phase (anagen), LGR5+/P-cadherin+ cells of the secondary hair germ participate in the initial phase of hair follicle regeneration, while CD34+/

Keratin 15+ bulge cells maintain later stages of anagen [177–181]. LGR6+ cells located in the isthmus only infrequently contribute to hair follicle regeneration and more actively supply progenitor cells for the interfollicular epidermis and sebaceous glands [175]. LRIG1+ stem cells located in the infundibulum are involved in maintenance of this hair follicle compartment, and can also contribute to regeneration of the epidermis [174, 182].

In adult epithelial stem cells in the skin, actively transcribed genes that maintain stemness display H3K4me3 and H3K79me2 histone marks, while repressed non-epidermal genes and genes activated during cell differentiation are enriched by H3K27me3 [183]. Stem cell transition towards hair follicle-specific differentiation is accompanied by loss of H3K27me3 and appearance of H3K4me3 and H3K79me2 marks in the chromatin of those genes (Lef1, Bmp4, Wnt5a, Msx1, etc.) that become active, whereas the chromatin of key stemness genes (e.g. Cd34, Sox9, Nfatc1) that become repressed in transient-amplifying cells shows loss of H3K4me3/H3K79me2 and appearance of H3K27me3 [183]. Interestingly, the chromatin of cell cycle-associated genes that are repressed in stem cells does not contain H3K27me3 and instead is decorated with H3K4me3, while their activation in transient amplifying cells is accompanied, in addition to H3K4me3, by the appearance of H3K79me2 [183]. These data suggest that complex regulatory mechanisms are involved in the control of repression of cell cycle-associated genes in stem cells; these may include the activity of anti-proliferative signaling pathways, and lack of pro-proliferative signaling molecules.

Recent data demonstrate that levels of both active (H3K4me3) and repressive (H3K9/K27me3) methylation are reduced in hair follicle stem cells during the telogen resting stage of the hair cycle, which corresponds to G0 quiescence [184]. Inhibition of BMP signaling in vivo promoted histone H3 methylation followed by cell proliferation onset and hair cycle progression. Similarly, addition of BMP4 caused reduced levels of histone methylases and increased transcription of demethylase mRNAs in cultured skin epithelial cells. Thus, BMP signaling couples hair follicle stem cell quiescence with reduced H3 K4/K9/K27me3 levels [184].

While skin stem cells have generally restricted fates in homeostasis, during wound induced skin regeneration essentially all of these populations are mobilized from the hair follicles to supply progenitor cells to the epidermis [143, 175, 181, 185, 186]. The switch between hair follicle versus epidermal cell fate in bulge stem cells during wound healing is regulated by the transcription factor LHX2, which promotes differentiation of bulge progenitors towards an epidermal phenotype by stimulating Sox9 and inhibiting Lgr5 expression [186].

Distinct populations of epithelial stem cells are also established in developing *mammary and sweat glands* [116, 187]. In mammary glands, two lineage-restricted stem cell populations expressing either Keratins 5/14 or Keratins 8/18 contribute to maintenance of the myoepithelial and luminal cells, respectively [116, 187]. In sweat glands, myoepithelial stem cells express Keratins 5/14 and smooth muscle actin and can generate both myoepithelial and luminal cells, while luminal stem cells express Keratins 18/19 and 15 and appear to be unipotent [116]. Additionally, a stem cell population expressing Keratins 5/14/18 and residing in the sweat gland duct contributes to regeneration of the epidermis in close vicinity to the sweat gland duct orifice after injury [116].

A melanocyte stem cell population is established in mouse skin within the first few days of postnatal development [164]. Melanocyte stem cells are located in the bulge of developing hair follicles, express dopachrome tautomerase (Trp2), and contribute to regeneration of the hair follicle pigmentary unit during hair cycle [188–190]. Melanocyte stem cells are located close to the epithelial stem cells in the bulge and secondary hair germ, and their attachment to extracellular matrix is maintained via Col17a1 produced by epithelial stem cells in the niche [191]. TGF-beta signaling promotes a quiescent state in melanocyte stem cells via suppression of MITF expression [192], while Wnt and SCF/c-kit signaling activate the stem cells and promote their proliferation during anagen [188, 193]. Following epidermal injury or UVB irradiation, melanocyte stem cells are capable of migrating into the epidermis in an MC1R-dependent manner [194].

Mesenchymal stem cells are located in distinct compartments of the skin including the dermal papilla of the hair follicle, and the hair follicle connective tissue sheath [165, 195, 196]. Dermal papilla fibroblasts possess hair-inductive properties in culture and after transplantation into non-hair bearing skin [120, 197, 198]. Interestingly, these cells are also capable of generating progenitor cells for erythroid and myeloid cell lineages *in vitro* and can reconstitute a hematopoetic system *in vivo* [199]. Furthermore, SOX2+ dermal papilla cells of the hair follicles are capable of generating neurons and Schwann cells in culture and possess a higher capacity for reprogramming towardsinduced pluripotent stem cells than non-follicular dermal fibroblasts [198, 200–202]. SOX2+ dermal papilla cells also contribute to regeneration of the dermis during wound healing [198].

Mesenchymal cells residing in the hair follicle connective tissue sheath can also generate non-follicular cells including adipocyte and osteoblast lineages and mast cells *in vitro* [203, 204]. During active hair growth *in vivo*, progenitor cells residing in the connective tissue sheath migrate into the dermal papilla and are involved in maintenance of anagen [205]. These cells also serve as a source of dermal progenitors for repair of connective tissue during wound healing [206, 207].

1.4 The Nucleus as a Hub Transforming Signaling Information into Lineage- and Differentiation Stage-Specific Transcriptional Outcomes

The cell nucleus integrates signals from the extracellular space and transforms them into specific gene expression programs that underly cell differentiation programs, and permit cells to adapt to external stimuli and changes in the microenvironment. After entering the nucleus, transcription factors interact with DNA in a sequence-specific fashion to either increase or decrease gene transcription [208]. Interaction of transcription factors with DNA is a complex process dependent on the chromatin status (active or repressed) at the target gene promoter. Certain transcription factors serve as pioneer factors that possess the ability to target DNA in nucleosomes and elicit chromatin opening, allowing other transcription factors to bind DNA and regulate gene expression [209]. Recent data reveal that transcription factors are also capable of binding non-coding genomic regions, regulating the activity of enhancer elements and enhancer-promoter interactions in 3D nuclear space [210].

The cell nucleus is a highly complex organelle that consists of the nuclear membrane, individual chromosomes occupying distinct territories, and a number of nuclear bodies such as nucleoli, Cajal bodies, promyelocytic leukaemia (PML) bodies, nuclear speckles, and Polycomb bodies. Together, these components facilitate execution of gene expression programs and other nuclear functions [211–215] (Fig. 1.3).

During terminal differentiation of keratinocytes in the epidermis and hair follicle, the nucleus undergoes programmed transformation from a highly active status, associated with execution of the genetic programs for epidermal barrier or hair fiber formation, to fully a inactive condition, and finally is lost as cells keratinize to form the cornified epidermal layer or hair shaft. In epidermal keratinocytes, the transition from proliferation to differentiation is accompanied by marked remodeling of the 3D nuclear organization and micro-anatomy including changes in the spatial arrangement of lineage-specific genes, nuclear bodies and heterochromatin [216].

Nuclear Envelope The nucleus is separated from the cytoplasm by the nuclear envelope that consists of outer and inner membranes, nuclear pore complexes, and a lamina located beneath the inner nuclear membrane (reviewed in [217–219] (Fig. 1.3a). The nuclear envelope provides sites for anchoring cytoskeletal components and chromatin to the outer and inner nuclear membranes, respectively, and plays crucial roles in regulating mechanical stability and positioning of the nucleus, nucleo-cytoplasmic transport, chromatin organization, and gene expression [218]. The nuclear envelope is completely disintegrated during mitosis, and is formed again at interphase, when it re-establishes its functions in separating the nucleus and cytoplasm [218].

The outer and inner membranes of the nuclear envelope are comprised of over one hundred proteins that fall into at least four distinct functional groups:

(i). Nucleoporins form the nuclear pore complexes, the sites of bidirectional exchange of proteins, RNA and ribonucleoprotein complexes between the nucleus and the cytoplasm [218]. Certain nucleoporins, such as Sec13 and Nup98, are also capable of modulating the expression of developmentally-regulated genes, suggesting that they contribute to transcription control [220, 221].

(ii). Proteins localized at the outer nuclear membrane, such as Nesprins, interact directly with distinct components of the cytoskeleton including actin, dynein, and plectin, and contribute to regulation of nuclear positioning in the cytoplasm, which is essential for cell polarity and migration, among other functions. (Fig. 1.3a). These proteins also interact with SUN1/2 proteins located at

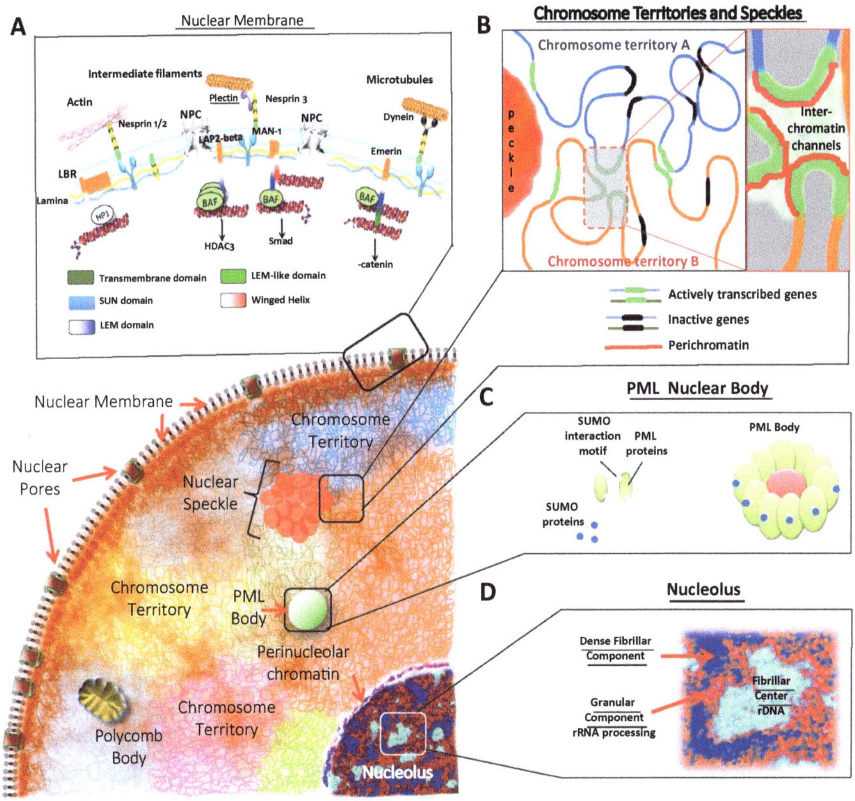

Fig. 1.3 Micro-anatomy of the cell nucleus. The nucleus is surrounded by a nuclear envelope that consists of outer and inner membranes, nuclear pore complexes, and a lamina located beneath the inner nuclear membrane (**a**). Chromosomes occupy distinct territories, in which different chromatin domains are permeated by interchromatin channels connected with a network of larger channels and lacunas separating the chromosomes and harboring a number of nuclear bodies including speckles (**b**). PML bodies consist of an outer shell enriched by PML protein, and an inner component containing poly-SUMO2/3 chains and other proteins (**c**). The nucleolus is the largest nuclear body and consists of fibrillar centers, the sites of location of rDNA, surrounded by electron-dense fibrillar and granular components (**d**)

the inner membrane, forming "bridges" that link the outer and inner membranes and establishing physical connections between the cytoskeleton and chromatin. These functions are thought to be important for regulation of higher-order chromatin remodeling, transcription, DNA replication and DNA repair [218]. In addition, Nesprin-2 forms complexes with α-catenin, emerin and β-catenin, regulating the bio-availability of the latter and thus modulating activity of the Wnt signaling pathway [222].

(iii). Proteins localized at the inner nuclear membrane interact with lamins and chromatin and play roles in the control of chromatin organization and gene expression. These proteins include lamin B receptor, lamina-associated polypeptides 1/2 or LAP1/2, emerin, MAN1, and SUN1/2. LAP1/2, emerin and MAN1 share a common LEM domain that binds the chromatin protein BAF, which in turn interacts with histones and tethers chromatin to the nuclear lamina [219]. MAN1 can also interact with DNA directly, while LAP2 binds HDAC3 and lamin B receptor binds heterochromatin protein 1a, thus forming a complex network involved in anchoring chromatin to the nuclear lamina [218]. Some of these factors can also interact with distinct components of the Wnt, Tgf-beta/BMP and Rb signaling pathways, modulating their activity (see below) [219] (Fig. 1.3a).

(iv). Lamins are intermediate filaments that form an inter-connected meshwork (lamina) underlying the inner nuclear membrane. The lamina is critical for nuclear stability and regulation of gene expression. Lamins A and B interact with proteins located at the inner nuclear membrane (see above), as well as with numerous other proteins and play important roles in the control of DNA replication, transcription and chromatin organization [223].

The expression of lamins and lamin-associated proteins shows developmental stage- and tissue-specific variability: undifferentiated mouse and human embryonic stem cells express lamins B1/B2, while expression of lamin A/C is activated during cell differentiation [224]. In mouse skin, lamins A/C show broad patterns of expression in the basal and suprabasal layers of the epidermis, hair follicle, sebaceous gland and dermis [225]. In the testis, ablation of the B1 lamin LamDm(o) results in detachment of testis-specific gene clusters from the nuclear envelope and their selective transcriptional up-regulation in somatic cells, thus suggesting involvement of lamins in the control of tissue-specific gene expression programs [226]. In line with this, the nuclear envelope-associated proteins Emerin and MAN1 regulate signal transduction by interacting with β-catenin and R-Smads and interfering with the activity of Wnt and Tgf-beta/Bmp signaling pathways during development and tissue regeneration [227, 228].

The biological relevance of lamins and lamin-associated proteins is also evident from a number of human diseases, known as "laminopathies", that are linked to mutations in the genes encoding these proteins [229]. Most laminopathies are caused by mutations in the LMNA gene, encoding lamin A/C, and result in diverse pathological conditions including muscular dystrophy, lipodystrophy, neuropathy, and progeroid syndromes [230]. Prelamin Hutchinson-Gilford progeria syndrome is caused by point mutation in LMNA that results in expression of a truncated prelamin A (progerin) protein and accelerated aging [229]. Cells from patients with progeria syndrome show abnormal nuclear shapes, altered chromatin structure and increased DNA damage [231].

Progerin transcripts are expressed at low levels in normal human skin of all ages, while protein is accumulated in dermal fibroblasts and selected terminally differentiated keratinocytes of aged individuals [232]. In human mesenchymal stem cells, progerin activates downstream effectors of the Notch signaling pathway and alters their molecular signature and differentiation potential [233]. Transgenic mice expressing progerin in the epidermis under the control of a Keratin 14 promoter show abnormalities in the morphology of keratinocyte nuclei, including nuclear envelope lobulation [234]. However, epidermal morphology, dynamics of the hair follicle cycling and wound healing were apparently normal in these mice [234]. In contrast to K14-progerin mice, inducible overexpression of the human mutant LMNA gene under the control of a Keratin 5 promoter results in the development of foci of epidermal hyperproliferation, hair thinning, sebaceous gland hypoplasia and dermal fibrosis [235]. These data suggest that a proper balance between different lamin A isoforms is required for the control of gene expression programs in keratinocytes.

Recent data reveal that in mice lacking the p63 transcription factor, epidermal progenitor cells of the developing skin display alterations in nuclear shape accompanied by a marked decrease in expression of several nuclear envelope-associated components including Lamin B1, Lamin A/C, SUN1, Nesprin-3, and Plectin, compared with controls [236]. Furthermore, ChIP-qPCR assays revealed enrichment of p63 at the Sun1, Syne3 and Plec promoters, suggesting these as direct p63 targets. Alterations in nuclear shape and in the expression of nuclear envelope-associated proteins were accompanied by altered distribution patterns of the repressive histone marks H3K27me3, H3K9me3 and heterochromatin protein 1- alpha in p63-null keratinocytes. These changes were also associated with downregulation of transcriptional activity and relocation of keratinocyte-specific gene loci away from sites of active transcription and towards heterochromatin-enriched repressive nuclear compartments in p63-null cells. These data demonstrate functional links between nuclear envelope organization, chromatin architecture and gene expression in keratinocytes and suggest nuclear envelope-associated genes as important targets mediating p63-regulated gene expression programs in the epidermis [236].

The **nucleolus** is the largest nuclear body, and serves as a site for ribosomal DNA (rDNA) transcription, rRNA maturation and ribosome production, as well as fulfilling a number of non-canonical functions [237–239]. Ribosomal RNA genes are clustered in the genome at nucleolar organizer regions (NORs) located on human chromosomes 13–15, 21 and 22 and on mouse chromosomes 12, 15, 16, 17, 18 and 19 (reviewed in [238]). The nucleolus is formed around the NOR-bearing chromosomes at the end of telophase. At this point in the cell cycle, rDNA begins to be transcribed by RNA polymerase I resulting in appearance of the primary transcripts, which are subsequently processed into mature 18S, 5.8S and 28S rRNA [237, 238].

At the ultrastructural level, nucleoli consist of fibrillar centers, whererDNA is localized, surrounded by electron-dense fibrillar and granular components [240–242] (Fig. 1.3d). Transcription of rRNA occurs at the fibrillarin-enriched border between the fibrillar centers and dense fibrillar components. rRNA is further processed at these sites, and is prepared for subsequent assembly into pre-ribosomal subunits in the granular component of the nucleoli [215].

Epidermal keratinocytes contain several nucleoli, as determined by immunofluorescent visualization of the nucleolus marker nucleophosmin B23. During terminal keratinocyte differentiation in human epidermis, the shape of the nucleolus changes, and its size and the quantities of the granular component decrease in granular layer compared with basal and spinous layer cells [243, 244]. These alterations are associated with cessation of protein synthesis and are accompanied by accumulation of heterochromatin around nucleoli in terminally-differentiated cells [244].

Together with its essential role in ribosome production, the nucleolus fulfills multiple non-canonical functions (reviewed in [239, 242]). Nucleolus-associated domains (NADs) contain over 700 genes, and, in addition to rDNA, include several other gene families including zinc-finger (ZNF), olfactory receptor family, immunoglobulin, defensin and T-cell receptor genes [245]. NADs are also enriched by multiple repressive histone marks, while markers of active chromatin are depleted here, suggesting that these domains may represent sites of gene silencing [245].

The nucleolus also serves as a site of accumulation of over four thousand proteins. The majority of these are implicated in the control of cellular functions beyond ribosome production, such as stress response, cell cycle control, viral infection, and apoptosis [239]. For instance, the nucleolus plays an important role in regulating the levels and activity of p53 protein, which accumulates in the intra-nucleolar cavities and is activated in response to nucleolar stress and reduction of ribosomal biogenesis [239]. Some nucleolar proteins, such as Rpl11, can bind MDM2 and promote p53 stabilization, cell cycle arrest and apoptosis [246]. In keratinocytes, the nucleolus contains retinoblastoma protein, which is markedly downregulated after UV irradiation, contributing to the control of cell proliferation [247, 248]. The nucleolus also serves as a site of concentration of papilloma viruses in epidermal keratinocytes [249, 250].

The nucleolus is surrounded by perinucleolar heterochromatin, a compact compartment containing satellite DNA and silent rDNA domains [215]. Perinucleolar heterochromatin represents the sites of gene silencing for many chromosomes in the nucleus [215]. The periphery of the nucleolus contains the perinucleolar compartment, a subnuclear body enriched by RNA-binding proteins and pol III RNA [251]. The perinucleolar body is associated with malignancy both in vitro and in vivo and its presence positively correlates with metastatic capacity, making it a potential cancer marker [251].

Nuclear speckles have irregular shapes and can be visualized as Interchromatin Granule Clusters (IGC) under electron microscopy. Nuclear speckles can be visualized under light microscopy by immunostaining with antibody to the SC-35 antigen, and are located in regions of the nucleoplasm that contain little or no DNA [252]. Nuclear speckles contain important constituents of the pre-mRNA processing machinery, such as polyadenylation and splicing factors including small nuclear ribonuclear proteins (snRNPs), as well as poly-A^+ RNA and other splicing-related proteins [253]. Many of these factors are either recruited to transcription sites from the speckles or are involved in mRNA processing within the speckles [252, 253]. Normally, 25–50 speckles with size from a half to several micrometers can be observed in interphase nuclei, occupying approximately 5–10% of the nucleoplasmic volume [253].

Speckles are frequently seen close to actively transcribed genes (Fig. 1.3b): for instance, in erythroid cells highly expressed genes show a tendency to cluster around speckles [254]. Speckles are present in epidermal keratinocytes, and the Epidermal Differentiation Complex locus, which contains a large number of genes activated during terminal differentiation, is frequently associated with speckles. It was recently demonstrated that the protein SON operates as a scaffolding factor for components of the RNA processing machinery, and its depletion results in severe disorganization of snRNP and splicing factors in nuclear speckles [255]. However, the significance of other proteins found to be enriched in speckles, including transcription factors (such as hepatocyte nuclear factors 1α/4α) and chromatin-remodeling factors (such as BAF53A/BAF57), for organization of the RNA processing machinery and expression of highly active genes associated with speckles remains to be defined [256].

Cajal and Histone Locus Bodies The Cajal Body and Histone Locus Body are nuclear sub-compartments, recognized by immunostaining for coilin and containing important components of the RNA processing machinery [215, 257]. Cajal Bodies are enriched in proteins and RNAs that control processing of small nuclear RNPs (snRNPs) and show association with genes encoding snRNAs [257, 258]. The Histone Locus Body is associated with genes encoding histones and plays a role in the processing of histone pre-mRNAs [257]. Cajal Bodies are very numerous during early stages of embryogenesis, while their numbers decrease at later developmental stages. In the developing zebrafish epidermis two or three Cajal Bodies are detected in each cell [259].

In addition to their role in RNA processing, Cajal Bodies are proposed to serve as sites of telomerase maturation, storage and assembly in human cells [215, 257, 260]. In cancer cells, telomerase-containing Cajal Bodies are associated with telomeres during S-phase of the cell cycle, indicating a possible role in the control of telomere elongation [260, 261]. Studies with Fluorescence Recovery after Photobleaching (FRAP) revealed that many components of Cajal Bodies, such as coilin, fibrillarin and snRNPs, show rapid exchange between the nucleoplasm and Cajal bodies, suggesting their possible roles in nuclear metabolism [257, 258]. Thus, Cajal Bodies represent interesting and dynamic nuclear structures, but their precise roles in the control of RNA processing and telomere metabolism in keratinocytes requires further investigation.

Promyelocytic Leukaemia (PML) Bodies PML bodies are nuclear domains of about 0.1–1 μm in diameter, containing several dozens of proteins including PML protein [262]. PML protein is a product of the *PML* gene that shows chromosomal translocation in acute promyelocytic leukemia and plays a key role in organizing PML bodies, which are absent in $PML^{-/-}$ cells [263]. PML protein constitutes a scaffold component, interacting with a large number of other proteins present in the PML bodies. In addition, PML possesses SUMO ligase activity and is itself sumoylated by SUMO1/2/3 ligases [264] (Fig. 1.3c).

High-resolution confocal microscopy revealed that PML bodies consist of an outer spherical shell stabilized by interactions between PML and Sp100, and an inner component enriched by poly-SUMO2/3 chains and other proteins [265]. Proteins that are present in PML bodies can be divided into two groups: "transient" factors that accumulate upon distinct conditions such as DNA damage, environmental stress, and viral infection, and "constitutive" proteins that are localized in PML bodies rather permanently [266].

PML bodies are involved in the control of many key cellular processes such as storage and post-translational modification of proteins, and regulation of transcription and chromatin organization, and additionally contribute to cellular senescence, DNA damage/apoptosis and responses to viral infections [262]. It is not clear whether there is any heterogeneity in the involvement of PML bodies in the control of such diverse functions, which may depend on their position relative to other nuclear compartments, or on their distinct biochemical composition [264].

PML bodies are increased in number and size in an ATM- and ATR-dependent manner during the DNA damage response, and co-localize with sites of single-stranded DNA and DNA repair in the nucleus [267]. In addition, PML bodies contain many components of the DNA repair machinery, such as ATR and CHK2 kinases, and, after DNA damage, accumulate ATM, WRN helicase, BRCA1 and H2AX [267]. PML bodies are also involved in the control of apoptosis: PML protein promotes p53 acetylation and phosphorylation and is also capable of binding to and inhibiting MDM2 [264]. PML bodies serve as storage sites for DAXX protein, which sensitizes cells to Fas-dependent apoptosis and can participate in numerous cellular functions as a mediator of protein interactions [266].

PML bodies are present in keratinocytes, and transgenic mice expressing a PML/RARalpha fusion gene under the control of a human MRP8 promoter recapitulated the phenotype of acute promyelocytic leukemia and developed multiple squamous papillomas [268]. Conversely, PML overexpression under the control of a K5 promoter resulted in hair loss, increased resistance of mice to chemically-induced carcinogenesis, and premature senescence of cultured keratinocytes associated with upregulation of p16 and Rb, but not p19 and p53 [269]. Thus, PML protein operates as tumor suppressor in keratinocytes, and local interruption of PML and RARalpha signaling in the skin, together with systemic retinoid deficiency, promotes formation of epidermal tumors in a Ras-independent manner [268].

Chromosomes and Chromosomal Territories Chromosomes are the largest units of genome organization, occupying distinct territories in the interphase nucleus [270–272] (Fig. 1.3). DNA is compacted in chromosomes up to several thousand fold and is organized into a DNA-protein complex (chromatin) that allows the genome to be transcribed, replicated and repaired [35, 273, 274]. Each chromosome contains a centromer (pericentromeric chromatin enriched in α-satellite repetitive sequences); chromosome arms containing gene-rich and gene-poor domains enriched in GC- and AT-sequences, respectively, and visualized as light and dark bands by Gimsa staining; and telomeres [275]. Chromosomes can be visualized by the three-dimensional fluorescence hybridization (3D-FISH) technique with specific paints that allow their positions in the nucleus to be defined [276, 277].

The term "chromosomal territory" was first introduced by Theodor Boveri in 1909 (reviewed in [278, 279]). Research in Thomas Cremer's laboratory performed during the last three decades has brought tremendous progress in our understanding of the spatial organization of genes and chromosomes in the interphase nucleus (for reviews, see [271, 272, 276]). Confocal microscopic analyses of tissue sections or isolated cells using whole chromosome 3D-FISH probes demonstrated that in interphase nuclei the relative positioning of chromosomes within the 3D nuclear space is not random and depends on many factors including cell type, differentiation stage, chromosome size and gene-rich or gene-poor status [280]. Data obtained from mouse skin *in situ* show that in basal epidermal keratinocytes, chromosome 3 harboring the Epidermal Differentiation Complex (EDC) locus is always located at the nuclear periphery: its positioning does not change during embryonic or post-natal development, or during terminal differentiation and keratinocyte transition to the spinous and granular epidermal layers [60, 69, 216]. However, chromosomes 11 and 15, harboring Keratin type I and type II loci, respectively, occupy predominantly central positions in keratinocyte nuclei [30].

In the interphase nucleus, positioning of chromosomes is controlled through several mechanisms that include interactions between specialized lamina-associated domains (LADs) and the nuclear lamina, as well as through association of chromosomes bearing nucleolar-organizing region domains with nucleoli (reviewed in [212, 237, 281, 282]). Distinct chromosomes may be arranged in the nuclei of differentiated cells in a cell lineage-specific manner, which may explain the increased frequency of translocations between distinct chromosomal parts in the corresponding tumors [283–286]. However, it is unclear whether genes from neighboring chromosomes may share common regulatory mechanisms required for their transcription [271].

The introduction of super-resolution confocal microscopy has allowed improved resolution of fluorescence images up to 20–100 nm and has served as an important next step in analyses of nuclear architecture [272, 287]. Super-resolution confocal microscopy revealed that each chromosome territory resembles a sponge-like structure and consists of chromatin domains permeated by interchromatin channels, and connected with a network of larger channels and lacunas separating distinct chromosomes and harboring a number of nuclear bodies [288]. Inter-chromatin channels serve as a reservoir for macromolecular complexes, transcription factors, regulators of splicing, replication, and repair, as well as for exporting mRNA-containing ribonucleoprotein complexes [272]. The network of interchromatin channels starts at nuclear pores and expands throughout the nuclear space, while chromatin domains in each territory are separated from the interchromatin channels by a 100–200-nm layer of decondensed chromatin, called perichromatin and enriched by nascent DNA and RNA, RNA polymerase II (RNA Pol II), and by the H3K4me3 histone modification which is specific for transcriptionally active chromatin [272, 288].

These observations were further developed into a model that suggests the presence of active and inactive nuclear compartments inside each chromosome territory that harbor transcriptionally active or inactive genes, respectively [272] (Fig. 1.3b). This model also suggests a large degree of flexibility in the positioning of distinct chromatin domains inside each chromosome territory, which is in line with the observation that some gene loci, such as IFN-γ and T_H2 cytokine loci in T_H lymphocytes, globin genes in erythroid cells, and the Nanog locus in iPS cells, can change

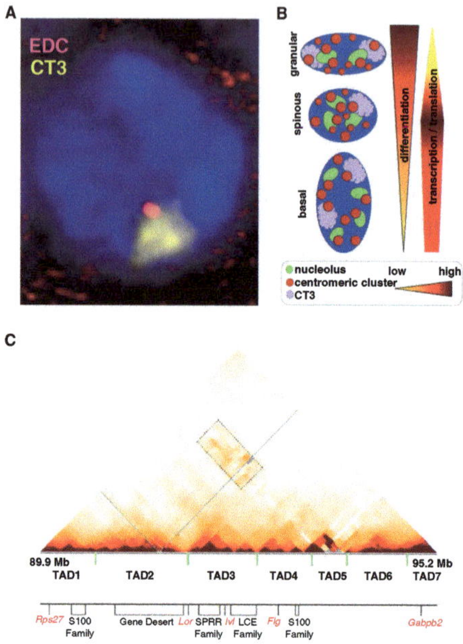

Fig. 1.4 Changes in spatial organization of the keratinocyte nucleus during epidermal development and differentiation. (**a**) 3D-FISH image of the nucleus of a murine basal epidermal keratinocyte showing chromosome territory 3 (CT3, yellow) with the EDC locus located at the internal part of the CT3 (red) (courtesy of I. Malashchuk). (**b**) Schematic illustrating remodeling of 3D nuclear organization during terminal keratinocyte differentiation in the epidermis [216]. (**c**) Organization of TADs at the EDC locus and neighboring regions in mouse keratinocytes. A heatmap representing 5C data after normalization and binning (bin size 150 kb, step size 15 kb) is shown. The positions of TAD border midpoints (average for the midpoints calculated based on the insulation index analysis in two replicates independently) are identified by green lines under the heatmaps

their positioning relative to other loci or to chromosomal territories associated with either gene activation or silencing [60, 289].

During epidermal morphogenesis and differentiation of basal epidermal progenitor cells, the lineage-specific EDC locus shows marked remodeling of its higher-order chromatin structure and relocates away from the peripheral part of the chromosomal territory 3 towards its internal part. This repositioning is associated with an increase in the transcriptional activity of genes involved in the control of terminal keratinocyte differentiation and epidermal barrier formation [60]. Such developmentally-regulated relocation of the EDC towards the nuclear interior is a keratinocyte-specific event, which does not occur in dermal cells, and it is maintained during adulthood despite the many cycles of cell division that occur in this rapidly proliferating and self-renewing epithelial tissue [60] (Fig. 1.4a).

These data are generally consistent with previous observations showing looping out from chromosome territory 1 of the EDC locus in cultured human keratinocytes, which suggests that positioning of this genomic domain within the nucleus is quite flexible [290]. Developmentally regulated relocation of the EDC locus into the nuclear interior is associated with an increase in the number of SC-35 nuclear speckles present within the vicinity of the EDC, suggesting that this nuclear compartment may provide a permissive environment for efficient transcription and maintenance of high expression levels of genes activated during keratinocyte differentiation [60]. However, the impact of distinct speckle components in the control of gene expression within the EDC and other keratinocyte-specific gene loci remains to be further determined.

Systematic analyses of the remodeling of nuclear architecture during terminal keratinocyte differentiation in the mouse epidermis has demonstrated that terminally differentiated keratinocytes show marked differences in micro-anatomical organization of the nucleus compared to basal epidermal cells, including: (i) decrease in nuclear volume; (ii) decrease in expression of markers of transcriptionally-active chromatin; (iii) internalization and decrease in the number of nucleoli; (iv) increased numbers of pericentromeric heterochromatic clusters; and (v) increased frequency of associations between nucleoli, pericentromeric clusters and chromosomal territory 3 [216]. These changes are likely to contribute to the global changes in the transcriptional landscape in terminally differentiating keratinocytes and transition of the keratinocyte nucleus from a metabolically active status to an inactive condition [216]. These data also suggest the nucleoli and pericentromeric clusters as important elements of the nuclear architecture which may control the local transcriptional micro-environment of distinct chromatin domains by modulating chromosome tethering, positioning, folding and/or orientation (Fig. 1.4b).

Spatial proximity of the genes and chromosomes in the nucleus plays an important role in the occurrence of chromosomal translocations during neoplastic transformation: neighboring chromosomes show higher frequencies of translocations compared to distal chromosomes, and translocations are formed predominantly between proximal chromosome breaks [291]. In a subset of basal cell carcinoma, the *SHH* gene is translocated between chromosomes 7 and Y, which may contribute to its abnormal activation in the absence of the *PTCH1* and *SMO* mutations [292]. Thus, it appears to be important to carefully dissect how topological organization of the genome in keratinocytes is changed in pathological skin conditions including epidermal tumors, and disorders of epidermal differentiation such as psoriasis, and how such changes may contribute to alterations in the transcriptional landscape that contribute to these diseases.

Three-Dimensional Genome Organization and Enhancer-Promoter Interactions Chromosome conformation capture technologies (3C and its variations 4C, 5C and Hi-C) were developed by Job Dekker and his laboratory [293] and are based on formaldehyde-mediated cross-linking of closely located chromatin domains and multi-protein complexes followed by DNA digestion with restriction enzymes and ligation at high dilution to facilitate the formation of intra-molecular but not inter-molecular products [294, 295]. These techniques have allowed investigator to define chromatin interactions between two distinct genomic sites (3C or "one-versus-one") or between a genomic site of interest and the genome globally (4C or "one-versus-all"), as well as assessing the complex interactions within a distinct genomic locus (5C or "many-versus-many") or global interactions within the whole genome (Hi-C or "all-versus-all") [296].

Hi-C analyses of global chromatin interactions revealed that genes and chromatin domains from the same chromosomes show a higher frequency of interaction with each other than with genes from other chromosomes, confirming the presence of chromosome territories at the molecular level [297]. Furthermore, these analyses demonstrated the existence of at least two types of sub-chromosomal compartments, which respectively segregate actively transcribed and transcriptionally silenced chromatin domains [297]. Such sub-chromosomal compartments were subsequently confirmed using 3D structural illumination microscopy that revealed presence of active and inactive sub-chromosomal compartments enriched either by the elongating form of PolII and H3K4me3 or by H3K9me3 histone modifications, respectively [298].

Most importantly, Hi-C analyses also revealed the existence of another level of chromatin folding and the presence of Topologically Associating Domains (TADs) on each interphase chromosome. TADs range in size from several hundred Kb to 1–2 Mb [299], and are characterized by much higher interaction frequencies between the distinct elements within a TAD (intra-TAD interactions) compared to the interactions between different TADs (inter-TAD interactions) [299]. Interestingly, the borders of TADs are conserved between humans and mice and are not altered during cell differentiation, while TADs are lost within the inactive X-chromosome, and during mitosis [300–302].

TAD borders in the mammalian genome are enriched in binding sites for a number of architectural proteins including CTCF and cohesin [301, 303]. CTCF and cohesin binding sites also exist within TADs, and CTCF is involved in organizing smaller-sized (100–200 kb) intra-TAD chromatin loops [304] and in mediating enhancer-promoter contacts [299]. SATB1 is another chromatin architectural protein that binds specialized DNA regions with an ATC-sequence context and folds chromatin into loops involving tissue-specific gene loci, including at the T_{H2}-cytokine, MHC class I, and globin loci [305–307]. SATB1 also targets chromatin remodelers and transcription factors to specific gene targets and plays a unique role in the execution of lineage-specific gene expression programs by integrating high-order chromatin organization with the regulation of gene expression [307, 308].

5C technology has been applied to characterize the spatial chromatin interaction network of the EDC locus and its flanking regions in epidermal keratinocytes. These experiments showed that the EDC locus is organized into four different TADs displaying distinct chromatin interaction networks and spatial compartmentalization patterns based on their gene-rich or gene-poor status (Fig. 1.4c) [309]. Comparison of ChIP-seq and 5C data revealed that the spatial chromatin interactome of gene-rich TADs at the EDC locus in keratinocytes involves extensive intra- and inter-TAD networks connecting gene promoters and enhancers. These promoter-enhancer interaction sites are enriched for binding of the chromatin architectural proteins CTCF and Rad21, and the chromatin remodeler Brg1. In contrast to gene-rich TADs, gene-poor TADs show preferential spatial contacts with each other, do not contain active enhancers, and display decreased binding of CTCF, Rad21 and Brg1. Thus, spatial interactions involving gene promoters and enhancers at the multi-TAD EDC locus in skin epithelial cells are not restricted by the TAD boundaries and involve, together with intra-TAD interactions, extensive contacts between different gene-rich TADs, forming a framework for lineage-specific transcription [309].

Data from chromosome conformation capture analyses have helped to identify enhancer-promoter interactions as a major driving force facilitating execution of lineage-specific differentiation programs [310]. Enhancers are mostly present in non-coding regions of the genome, and may be located far from their target genes or even on different chromosomes [311]. In normal differentiating cells, interactions between gene promoters and their enhancers occur via chromatin looping and are critical for the execution of lineage-specific gene expression programs [310, 311]. For example, in keratinocytes, an epidermal-specific regulatory enhancer designated as 923 is present within the EDC locus and interacts with multiple EDC gene promoters; a subset of these interactions is regulated by the AP-1 transcription factor [312]. Furthermore, calcium stimulation in differentiating keratinocytes results in increased physical proximity of the enhancer and the promoter regions of the peptidylarginine deiminase 3 gene that controls filaggrin processing [313]. However, the roles of CTCF, cohesin, SATB1 and other chromatin architectural proteins in regulation of enhancer-promoter interactions during establishment and maintenance of the epidermal differentiation program in keratinocytes remain to be clarified.

1.5 Integration of Signaling/Transcription Factor-Mediated and Epigenetic Regulatory Mechanisms in the Control of Keratinocyte Differentiation and Stem Cell Activity

The program of epidermal differentiation and barrier formation is tightly controlled by a number of transcription regulators including p63, c-Myc, AP-2, KLF4, GRLH3, PPAR-alpha, and m-Ovo [26, 314–316]. In keratinocytes, these transcription factors regulate a large number of genes that encode distinct adhesion/signaling molecules, transcription factors, and cell cycle-associated proteins, as well as tissue-specific proteins, such as keratins, involucrin, and loricrin (reviewed in [316, 317]. Transcription factors operate in concert with various classes of epigenetic regulators to precisely control gene activation and silencing [37, 318]. Cross-talk between transcription factors and distinct groups of epigenetic regulators in the nucleus is mediated by diverse mechanisms, including the following (Fig. 1.5):

1. Transcription factors directly regulate expression of distinct components of epigenetic machinery to achieve proper execution of gene expression programs. For example, the p63 transcription factor, a master regulator of epidermal development, controls the expression of at least three important components of the epigenetic machinery: the ATP-dependent chromatin remodeler Brg1, the genome organizer Satb1, and the Polycomb group member Cbx4 [60, 69, 319]. Brg1 and Satb1 are direct targets of p63, and promote chromatin mobility/higher order folding and coordinated gene activity at the keratinocyte-specific EDC locus during terminal differentiation in the epidermis, while Cbx4 maintains

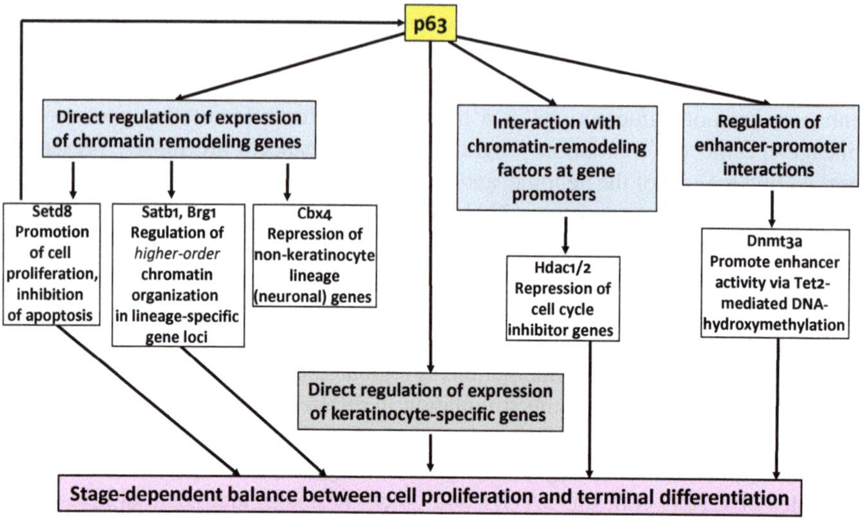

Fig. 1.5 Model illustrating cross-talk between p63 transcription factor and epigenetic regulators in the control of gene expression in keratinocytes. In addition to direct regulation of tissue-specific genes, p63 transcription factor directly regulates expression of distinct epigenetic regulators, interacts with epigenetic regulators at gene promoters, and regulates enhancer-promoter interactions. Setd8 histone methyltransferase positively regulates p63 gene expression

epithelial identity by repressing non-keratinocyte (neuronal and mesodermal) genes in keratinocytes [60, 61, 69]. These findings are consistent with recent ChIP-seq data demonstrating that p63 targets over 60 distinct components of the epigenetic machinery in keratinocytes including Dnmt3a, Dnmt3b, Hdac9, Jarid2, Ring1, and Suv39H1 [320].

2. Epigenetic regulators play important roles in controlling the expression of transcription factors. For instance, p63 expression in keratinocytes is regulated by the histone methyltrasferase Setd8, which, in turn, is a target of the c-Myc transcription factor and mediates the effects of c-Myc on epidermal differentiation [107]. c-Myc is an important regulator of epidermal homeostasis and promotes keratinocyte proliferation and differentiation into epidermal and sebaceous gland lineages [321–323]. In addition to direct involvement in the control of expression of epidermal-specific genes, c-Myc also exhibits potent activity as a modulator of local chromatin structure: it regulates expression of the histone acetyltransferase Gcn5 and promotes histone acetylation [324, 325].
3. Transcription factors directly interact with chromatin remodeling factors and target them to gene promoters: for instance, c-Myc targets histone acetyltransferases Gcn5 to chromatin and promotes global euchromatisation [325], and NFATc1 transcription factor forms a complex with the ATP-dependent chromatin remodeler Brg1 and stimulates formation of DNAse1 hypersensitive sites and recruitment of other transcription factors to target genes [326, 327]. Similar mechanisms involving interactions of pioneer transcription factors, such as

FOXA1 and GATA1, with ATP-dependent chromatin remodeling factors play important roles in opening closed chromatin regions and making them accessible to secondary factors [328].

Transcription factors can also operate as transcriptional repressors by interacting with HDACs and targeting repressive complexes to gene promoters. For instance, the transcriptional repressor NFX1 interacts with mSin3A/HDAC and recruits the repression complex to the hTERT promoter in keratinocytes, while p53 inhibits expression of its target genes during the DNA damage response by interacting with the Sin3B/HDAC complex and targeting it to promoters [329, 330]. HDAC1/2 also serve as important epigenetic regulators by interacting with p63 and mediating its repression of the *p16/Ink4a* gene [79].

4. Transcription factors regulate expression of microRNAs and RNA-modifying enzymes that establish feedback loops balancing gene expression programs at the post-transcriptional level. For instance, BMP signaling inhibits expression of miR21 in keratinocytes; miR21, in turn, is capable of inhibiting Smad7 activity at posttranscriptional level, thus providing a regulatory loop that limits BMP signaling [331, 332]. As another example, c-Myc regulates expression of the RNA methyltransferase Misu, which modulates c-Myc's effects on cell proliferation and differentiation [333, 334].
5. Transcription factors are capable of binding enhancer elements and regulating enhancer-promoter interactions. Recent data revealed that in human keratinocytes, approximately 50% of p63's binding sites display the H3K27ac histone modification that is specific for active enhancers [335]. Similar data were obtained using mouse keratinocytes [320], suggesting that p63-mediated regulation of the epidermal differentiation program is far more complex than previously appreciated and includes the control of enhancer-promoter interactions at p63 target genes [336]. This view is consistent with recent data demonstrating the role of the transcription factor SOX9 in assembling super-enhancers in hair follicle bulge stem cells [337], as well as with substantial reorganization of super-enhancer profiles in squamous cell carcinomas compared to normal keratinocytes [37].

Thus, chromatin remodeling factors operate as an integral part of the genetic programs governed by transcription factors, and modulate their effects on local and global chromatin structure. Below, we review data demonstrating interactions between transcription factors mediating the effects of the Wnt and BMP signaling pathways and the chromatin remodeling machinery in hair follicle epithelial stem cells during their quiescence and hair cycle-associated activation in adult skin. Wnt and BMP pathways act antagonistically in bulge stem cells during quiescence, while in differentiating hair matrix keratinocytes these signals promote execution of lineage-specific differentiation programs towards hair shaft- or inner root sheath-specific lineages, respectively [338–341].

The **canonical Wnt signaling pathway** operates via its downstream effector β-catenin, which acts as a bipartite transcription co-factor for the LEF1, TCF1, TCF3 and TCF4 DNA-binding proteins [342]. The transcriptional outcomes of Wnt signaling (activation or repression) depend on the combination of proteins bound to the promoters of target genes. In quiescent hair follicle stem cells, TCF3, TCF4 and transcriptional repressor TLE (Groucho) bind coordinately with histone deacetylase HDAC1 and transcriptionally repress Wnt target genes [343]. However, when TCF3/TCF4 levels are decreased or levels of β-catenin are increased, the repressive activities of TCF3/TCF4/TLE/HADC1 are outweighed by those of the Wnt/β--catenin/LEF1 complex, which promotes transcription of β-catenin target genes, leading to stem cell activation [343].

Analyses of direct Wnt target genes in hair follicle epithelial stem cells revealed that Cbx8, a component of the Polycomb repressive complex, is strongly upregulated upon β-catenin ablation, which prevents stem cell activation [343]. These findings correspond well with previous data demonstrating that stem cell transition towards hair follicle-specific differentiation is accompanied by loss of the repressive H3K27me3 mark, and suggest that downregulation of Polycomb activities might play a role in β-catenin-mediated stem cell activation and hair follicle growth.

The **BMP/Smad signaling pathway** exerts inhibitory effects on the activity of bulge epithelial stem cells and promotes differentiation of inner root sheath progenitors in the hair matrix via binding of Smad1/Smad4 or Smad5/Smad4 complexes to DNA [338–341]. Quiescent hair follicle stem cells show reduced levels of both active and repressive histone H3 methylation marks (H3K4me3 and H3K9/K27me3, respectively), and BMP4 treatment reduces histone methylases and increases demethylases mRNA levels in cultured skin epithelial cells [184]. Interestingly, comparison of these data with ChIP-seq analyses of Smad1/5 target genes in hair follicle stem cells did not reveal the histone methyltransferases Ezh2, Suv39h2 and Setd1b among the direct pSmad1/5 targets [344]. These data suggest that other mechanisms including indirect effects of the BMP/Smad pathway on the activities of histone methylation enzymes need to be considered to substantiate the links between the BMP signaling and low histone methylation status in quiescent stem cells.

ChIP-seq analyses of Smad1/5 target genes in quiescent hair follicle stem cells and differentiating transient amplifying hair matrix cells demonstrated an enrichment of the genes showing both H3K4me3 and H3K27me3 histone modifications that correspond to the bivalent or poised genes [344]. Because entry of hair follicles into the anagen growth stage is accompanied by both gene activation and silencing, these data suggest possible involvement of BMP/Smad signaling in control of the transitional chromatin state in hair follicle stem cells and their differentiating progeny [344].

ChiP-seq analyses of pSmad1/5 targets revealed that many epigenetic regulators including Cbx4, Cbx8, Dnmt3a, Hdac5, Hdac7, Kdm6b, Mll1, and Smarcd2/3 are potential direct transcriptional targets of BMP signaling in hair follicle stem cells and differentiating hair matrix cells [344]. These epigenetic regulators may contribute to the opening or closing of distinct chromatin regions to allow for transcriptional regulation during inner root sheath-specific differentiation of hair follicle stem cells [344].

Over 50% of pSmad1/5 binding sites in hair follicle stem cells are located outside gene promoters and are associated with enhancer elements [344]. These data are consistent with ChIP-seq analyses of p63 target genes, which demonstrated that in human and mouse keratinocytes approximately 50% of p63's binding sites are co-localized with the H3K27ac histone modification that is specific for active enhancers [320, 335]. Together with data showing that in human epidermal stem cells p63 interacts with Dnmt3a to maintain high levels of DNA hydroxymethylation at the center of enhancers in a Tet2-dependent manner [345], binding of pSmad1/5 to enhancers suggests that these factors control enhancer activities and gene regulation by modulating enhancer/promoter interactions. However, it remains to be determined which coactivators and corepressors interact with pSmad1/5 [346] in hair follicle stem cells and their differentiating progeny to modulate chromatin structure and enhancer activity.

Spatial chromatin interactions in the nucleus involving gene promoters and distal regulatory elements located in the non-coding genomic domains are thought to provide a major force that drives evolution of the mammalian genome [311]. Furthermore, genome-wide association studies (GWAS) have demonstrated that many human diseases are associated with single nucleotide polymorphisms (SNPs) in intergenic regions, suggesting that these mutations might perturb normal gene expression programs by affecting the activity of distal gene regulatory elements [347]. The global chromatin landscape and spatial arrangements between genes and their regulatory elements are also substantially re-organized in malignant cells, and are functionally important for their growth [308, 312, 348]. Thus, further research aimed at establishing enhancer-promoter regulatory network maps in healthy and diseased skin is important for a full understanding of human skin evolution and both genetic and sporadic skin diseases.

1.6 Conclusions

In summary, the current state of research shows that cross-talk between signaling/transcription factor-mediated and epigenetic mechanisms is of key importance for the execution of tissue-specific programs of gene activation and silencing in the skin during development, regeneration and adaptation to environmental factors. However, additional efforts are required to fully understand how the keratinocyte genome is organized at multiple levels to achieve a proper balance between cell proliferation and differentiation, as well as to interpret macro-/micro-environmental changes and to respond to diverse stimuli.

Recent advances in pharmaco-epigenomics have resulted in the development of a number of molecules that are capable of modulating distinct epigenetic regulatory mechanisms, and some of these are already FDA approved, or are being tested in clinical trials (reviewed in [349]). In combination with modulators of signaling pathway activity, epigenetic drugs may prove useful in the management of skin disorders. Future research in this area will help to bridge the gap between our current knowledge of signaling/transcription factor-mediated and epigenetic mechanisms and potential applications of signaling pathway and epigenetic modulators, and has the potential to identify a new generation of therapeutic agents for treatment of skin disorders, protection of the skin against environmental stressors, and combating the effects of skin aging.

References

1. Chuong CM, Nickoloff BJ, Elias PM, Goldsmith LA, Macher E, Maderson PA, et al. What is the 'true' function of skin? Exp Dermatol 2002;11(2):159–187. Epub 2002/05/08. doi: 112 [pii]. PubMed PMID: 11994143.
2. Chuong CM, editor. Molecular basis of epithelial appendage morphogenesis. Austin: Landes Biosciences; 1998.
3. Slominski A, Wortsman J. Neuroendocrinology of the skin. Endocrine Rev. 2000;21:457–87.
4. Kupper TS, Fuhlbrigge RC. Immune surveillance in the skin: mechanisms and clinical consequences. Nat Rev Immunol. 2004;4:211–22.
5. Sengel P. Morphogenesis of the skin. In: Abercrombie M, Newth DR, Torrey JG, editors. Developmental and cell biology series. Cambridge: Cambridge University Press; 1976.
6. Gilbert SF. Developmental biology. Sunderland: Sinauer Associates, Inc.; 2003. 784 p
7. Dhouailly D, Olivera-Martinez I, Fliniaux I, Missier S, Viallet JP, Thelu J. Skin field formation: morphogenetic events. Int J Dev Biol 2004;48(2–3):85–91. doi:https://doi.org/10.1387/ijdb.031753dd. Epub 2004/07/24. 031753dd [pii]. PubMed PMID: 15272373.
8. Loomis CA. Development and morphogenesis of the skin. Adv Dermatol 2001;17:183–210. Epub 2002/01/05. PubMed PMID: 11758116.
9. Fuchs E, Raghavan S. Getting under the skin of epidermal morphogenesis. Nat Rev Genet. 2002;3:199–209.
10. Chu DH, Haake AR, Hobrook K, Loomis CA. The structure and development of skin. In: Freedberg IM, Eisen AZ, Wolf K, Austen KF, Goldsmith LA, Katz SI, editors. Fitzpatrick's dermatology in general medicine, vol. 1. New York: McGraw-Hill; 2003. p. 58–88.
11. Stenn KS, Paus R. Controls of hair follicle cycling. Physiol Rev 2001;81(1):449–494. Epub 2001/01/12. PubMed PMID: 11152763.
12. Schneider MR, Schmidt-Ullrich R, Paus R. The hair follicle as a dynamic miniorgan. Curr Biol 2009;19(3):R132–R142. doi: https://doi.org/10.1016/j.cub.2008.12.005. Epub 2009/02/13. S0960-9822(08)01626-6 [pii]. PubMed. PMID: 19211055.
13. Arwert EN, Hoste E, Watt FM. Epithelial stem cells, wound healing and cancer. Nat Rev Cancer 2012;12(3):170–180. doi:https://doi.org/10.1038/nrc3217. PubMed PMID: 22362215.
14. Cotsarelis G, Millar SE. Towards a molecular understanding of hair loss and its treatment. Trends Mol Med. 2001;7:293–301.
15. Uitto J, Pulkkinen L, Ringpfeil F. Progress in molecular genetics of heritable skin diseases: the paradigms of epidermolysis bullosa and pseudoxanthoma elasticum. J Invest Dermatol Symp Proc. 2002;7:6–16.
16. Irvine AD, McLean WH. The molecular genetics of the genodermatoses: progress to date and future directions. Br J Dermatol. 2003;148:1–13.
17. Lane EB, McLean WH. Keratins and skin disorders. J Pathol 2004;204(4):355–366. doi:https://doi.org/10.1002/path.1643. Epub 2004/10/21. PubMed PMID: 15495218.
18. O'Shaughnessy RFCA. Inherited disorders of the skin in human and mouse: from development to differentiation. Int J Dev Biol. 2004;48:171–9.
19. Schmidt-Ullrich R, Paus R. Molecular principles of hair follicle induction and morphogenesis. BioEssays. 2005;27:247–61.

20. Maini PK, Baker RE, Chuong CM. Developmental biology. The Turing model comes of molecular age. Science. 2006;314(5804):1397–8. doi:https://doi.org/10.1126/science.1136396 314/5804/1397 [pii]. Epub 2006/12/02. PubMed PMID: 17138885.
21. Chuong CM, Richardson MK. Pattern formation today. Int J Dev Biol. 2009;53(5–6):653–8. doi:https://doi.org/10.1387/ijdb.082594cc. Epub 2009/06/27. 082594cc [pii]. PubMed PMID: 19557673; PubMed Central PMCID: PMCPMC2874132.
22. Shirakata Y. Regulation of epidermal keratinocytes by growth factors. J Dermatol Sci 2010;59(2):73–80. https://doi.org/10.1016/j.jdermsci.2010.05.002. Epub 2010/06/24. doi: S0923-1811(10)00163-5 [pii]. PubMed PMID: 20570492.
23. Chuong CM, Randall VA, Widelitz RB, Wu P, Jiang TX. Physiological regeneration of skin appendages and implications for regenerative medicine. Physiology (Bethesda) 2012;27(2):61–72. doi: https://doi.org/10.1152/physiol.00028.2011.27/2/61 [pii]. Epub 2012/04/17. PubMed PMID: 22505663.
24. Lee J, Tumbar T. Hairy tale of signaling in hair follicle development and cycling. Semin Cell Dev Biol. 2012;23(8):906–16. doi:https://doi.org/10.1016/j.semcdb.2012.08.003. Epub 2012/09/04. S1084-9521(12)00144-9 [pii]. PubMed PMID: 22939761; PubMed Central PMCID: PMCPMC3496046.
25. Koster MI, Roop DR. Mechanisms regulating epithelial stratification. Annu Rev Cell Dev Biol 2007;23:93–113. doi:https://doi.org/10.1146/annurev.cellbio.23.090506.123357. Epub 2007/05/11. PubMed PMID: 17489688.
26. Blanpain C, Fuchs E. Epidermal homeostasis: a balancing act of stem cells in the skin. Nat Rev Mol Cell Biol. 2009;10(3):207–17. doi: https://doi.org/10.1038/nrm2636. nrm2636 [pii]. Epub 2009/02/12. PubMed PMID: 19209183; PubMed Central PMCID: PMCPMC2760218.
27. Ning MS, Andl T. Control by a hair's breadth: the role of microRNAs in the skin. Cell Mol Life Sci. 2012. doi:https://doi.org/10.1007/s00018-012-1117-z. Epub 2012/09/18. PubMed PMID: 22983383; PubMed Central PMCID: PMCPMC3548021.
28. Botchkareva NV. MicroRNA/mRNA regulatory networks in the control of skin development and regeneration. Cell Cycle 2012;11(3):468–474. doi: https://doi.org/10.4161/cc.11.3.19058. Epub 2012/01/21. 19058 [pii]. PubMed PMID: 22262186.
29. Yi R, Fuchs E. A miR image of stem cells and their lineages. Curr Top Dev Biol 2012;99:175–199. doi: https://doi.org/10.1016/B978-0-12-387038-4.00007-0. .Epub 2012/03/01. B978-0-12-387038-4.00007-0 [pii]. PubMed PMID: 22365739.
30. Botchkarev VA, Gdula MR, Mardaryev AN, Sharov AA, Fessing MY. Epigenetic regulation of gene expression in keratinocytes. J Invest Dermatol 2012;132(11):2505–2521. doi: https://doi.org/10.1038/jid.2012.182. Epub 2012/07/06. jid2012182 [pii]. PubMed PMID: 22763788.
31. Frye M, Benitah SA. Chromatin regulators in mammalian epidermis. Semin Cell Dev Biol 2012;23(8):897–905. doi: https://doi.org/10.1016/j.semcdb.2012.08.009. Epub 2012/09/05. S1084-9521(12)00150-4 [pii]. PubMed PMID: 22944592.
32. Zhang J, Bardot E, Ezhkova E. Epigenetic regulation of skin: focus on the Polycomb complex. Cell Mol Life Sci 2012;69(13):2161–2172. doi: https://doi.org/10.1007/s00018-012-0920-x. Epub 2012/02/09. PubMed PMID: 22314499.
33. Martin N, Patel S, Segre JA. Long-range comparison of human and mouse Sprr loci to identify conserved noncoding sequences involved in coordinate regulation. Genome Res. 2004;14(12):2430–8.
34. Magin TM, Vijayaraj P, Leube RE. Structural and regulatory functions of keratins. Exp Cell Res 2007;313(10):2021–2032. doi:https://doi.org/10.1016/j.yexcr.2007.03.005. S0014-4827(07)00108-5 [pii]. Epub 2007/04/17. PubMed PMID: 17434482.
35. Ho L, Crabtree GR. Chromatin remodelling during development. Nature. 2010;463(7280):474–84.
36. Perdigoto CN, Valdes VJ, Bardot ES, Ezhkova E. Epigenetic regulation of epidermal differentiation. Cold Spring Harb Perspect Med. 2014;4(2). doi:https://doi.org/10.1101/cshperspect.a015263. a015263 [pii]. 4/2/a015263 [pii]. Epub 2014/02/05. PubMed PMID: 24492849.

37. Adam RC, Fuchs E. The Yin and Yang of chromatin dynamics in stem cell fate selection. Trends Genet 2016;32(2):89–100. doi: https://doi.org/10.1016/j.tig.2015.11.002. PubMed PMID: 26689127; PubMed Central PMCID: PMCPMC4733603.
38. Feinberg AP. Phenotypic plasticity and the epigenetics of human disease. Nature 2007;447(7143):433–440. doi:https://doi.org/10.1038/nature05919. nature05919 [pii]. Epub 2007/05/25. PubMed PMID: 17522677.
39. Feinberg AP. The epigenetic basis of common human disease. Trans Am Clin Climatol Assoc. 2013;124:84–93. PubMed PMID: 23874013; PubMed Central PMCID: PMCPMC3715917.
40. Byrne C, Tainsky M, Fuchs E. Programming gene expression in developing epidermis. Development 1994;120(9):2369–2383. Epub 1994/09/01. PubMed PMID: 7525178.
41. Weiss LW, Zelickson AS. Embryology of the epidermis: ultrastructural aspects. 1. Formation and early development in the mouse with mammalian comparisons. Acta Derm Venereol 1975;55(3):161–168. Epub 1975/01/01. PubMed PMID: 50684.
42. Lechler T, Fuchs E. Asymmetric cell divisions promote stratification and differentiation of mammalian skin. Nature. 2005;437(7056):275–80. doi: https://doi.org/10.1038/nature03922. Epub 2005/08/12. nature03922 [pii]. PubMed PMID: 16094321; PubMed Central PMCID: PMCPMC1399371.
43. Rowden G. Ultrastructural studies of keratinized epithelia of the mouse. III. Determination of the volumes of nuclei and cytoplasm of cells in murine epidermis. J Invest Dermatol 1975;64(1):1–3. Epub 1975/01/01. PubMed PMID: 1110303.
44. Gdula MR, Poterlowicz K, Mardaryev AN, Sharov AA, Peng Y, Fessing MY, et al. Remodelling of three-dimensional organization of the nulceus during terminal keratinocyte differentiation in the epidermis. 2012.
45. Simpson CL, Patel DM, Green KJ. Deconstructing the skin: cytoarchitectural determinants of epidermal morphogenesis. Nat Rev Mol Cell Biol. 2011;12(9):565–80. doi: https://doi.org/10.1038/nrm3175. Epub 2011/08/24. nrm3175 [pii]. PubMed PMID: 21860392; PubMed Central PMCID: PMCPMC3280198.
46. Gu LH, Coulombe PA. Keratin function in skin epithelia: a broadening palette with surprising shades. Curr Opin Cell Biol 2007;19(1):13–23. doi:https://doi.org/10.1016/j.ceb.2006.12.007. Epub 2006/12/21. S0955-0674(06)00188-8 [pii]. PubMed PMID: 17178453.
47. Kim S, Coulombe PA. Emerging role for the cytoskeleton as an organizer and regulator of translation. Nat Rev Mol Cell Biol 2010;11(1):75–81. doi:https://doi.org/10.1038/nrm2818. Epub 2009/12/23. nrm2818 [pii]. PubMed PMID: 20027187.
48. Poulson ND, Lechler T. Asymmetric cell divisions in the epidermis. Int Rev Cell Mol Biol 2012;295:199–232. doi:https://doi.org/10.1016/B978-0-12-394306-4.00012-5. Epub 2012/03/28. B978-0-12-394306-4.00012-5 [pii]. PubMed PMID: 22449491.
49. Wallace L, Roberts-Thompson L, Reichelt J. Deletion of K1/K10 does not impair epidermal stratification but affects desmosomal structure and nuclear integrity. J Cell Sci 2012;125(Pt 7):1750–1758. doi:https://doi.org/10.1242/jcs.097139. jcs.097139 [pii]. Epub 2012/03/01. PubMed PMID: 22375063.
50. Weiss LW, Zelickson AS. Embryology of the epidermis: ultrastructural aspects. II. Period of differentiation in the mouse with mammalian comparisons. Acta Derm Venereol 1975;55(5):321–329. Epub 1975/01/01. PubMed PMID: 52965.
51. Candi E, Schmidt R, Melino G. The cornified envelope: a model of cell death in the skin. Nat Rev Mol Cell Biol 2005;6(4):328–340. doi: https://doi.org/10.1038/nrm1619. nrm1619 [pii]. Epub 2005/04/02. PubMed PMID: 15803139.
52. Byrne C, Hardman M, Nield K. Covering the limb – formation of the integument. J Anat. 2003;202(1):113–23. Epub 2003/02/18. PubMed PMID: 12587926; PubMed Central PMCID: PMCPMC1571060.
53. Chu DH, Haake AR, Holbrook K, Loomis CA. The structure and development of skin. In: Freedberg I, editor. Dermatology in general medicine. New York: McGraw-Hill; 2007. p. 58–98.

54. Denecker G, Ovaere P, Vandenabeele P, Declercq W. Caspase-14 reveals its secrets. J Cell Biol. 2008;180(3):451–8. doi:https://doi.org/10.1083/jcb.200709098. jcb.200709098 [pii]. Epub 2008/02/06. PubMed PMID: 18250198; PubMed Central PMCID: PMCPMC2234247.
55. Weiss LW, Zelickson AS. Embryology of the epidermis: ultrastructural aspects. III. Maturation and primary appearance of dendritic cells in the mouse with mammalian comparisons. Acta Derm Venereol 1975;55(6):431–442. Epub 1975/01/01. PubMed PMID: 55035.
56. Peters EM, Botchkarev VA, Muller-Rover S, Moll I, Rice FL, Paus R. Developmental timing of hair follicle and dorsal skin innervation in mice. J Comp Neurol 2002;448(1):28–52. doi: https://doi.org/10.1002/cne.10212. Epub 2002/05/16. PubMed PMID: 12012374.
57. Botchkarev VA, Metz M, Botchkareva NV, Welker P, Lommatzsch M, Renz H, et al. Brain-derived neurotrophic factor, neurotrophin-3, and neurotrophin-4 act as "epitheliotrophins" in murine skin. Lab Investig 1999;79(5):557–572. Epub 1999/05/20. PubMed PMID: 10334567.
58. Magerl M, Tobin DJ, Muller-Rover S, Hagen E, Lindner G, McKay IA, et al. Patterns of proliferation and apoptosis during murine hair follicle morphogenesis. J Invest Dermatol 2001;116(6):947–955. doi:https://doi.org/10.1046/j.0022-202x.2001.01368.x. jid1368 [pii]. Epub 2001/06/16. PubMed PMID: 11407986.
59. Vanbokhoven H, Melino G, Candi E, Declercq W. p63, a story of mice and men. J Invest Dermatol 2011;131(6):1196–1207. doi:https://doi.org/10.1038/jid.2011.84. jid201184 [pii]. Epub 2011/04/08. PubMed PMID: 21471985.
60. Mardaryev AN, Gdula MR, Yarker JL, Emelianov VN, Poterlowicz K, Sharov AA, et al. p63 and Brg1 control developmentally regulated higher-order chromatin remodelling at the epidermal differentiation complex locus in epidermal progenitor cells. Development. 2014;141(1):101–11. doi: https://doi.org/10.1242/dev.103200. Epub 2013/12/19. 141/1/101 [pii]. PubMed PMID: 24346698; PubMed Central PMCID: PMCPMC3865752.
61. Mardaryev AN, Liu B, Rapisarda V, Poterlowicz K, Malashchuk I, Rudolf J, et al. Cbx4 maintains the epithelial lineage identity and cell proliferation in the developing stratified epithelium. J Cell Biol 2016;212(1):77–89. doi: https://doi.org/10.1083/jcb.201506065. PubMed PMID: 26711500; PubMed Central PMCID: PMCPMC4700479.
62. Mills AA, Zheng B, Wang XJ, Vogel H, Roop DR, Bradley A. p63 is a p53 homologue required for limb and epidermal morphogenesis. Nature. 1999;398(6729):708–13.
63. Yang A, Schweitzer R, Sun D, Kaghad M, Walker N, Bronson RT, et al. p63 is essential for regenerative proliferation in limb, craniofacial and epithelial development. Nature 1999;398(6729):714–718. doi: https://doi.org/10.1038/19539. Epub 1999/05/05. PubMed PMID: 10227294.
64. Ince TA, Cviko AP, Quade BJ, Yang A, McKeon FD, Mutter GL, et al. p63 Coordinates anogenital modeling and epithelial cell differentiation in the developing female urogenital tract. Am J Pathol. 2002;161(4):1111–7. doi:https://doi.org/10.1016/S0002-9440(10)64387-8. Epub 2002/10/09. S0002-9440(10)64387-8 [pii]. PubMed PMID: 12368184; PubMed Central PMCID: PMCPMC1867285.
65. Koster MI. p63 in skin development and ectodermal dysplasias. J Invest Dermatol. 2010;130(10):2352–8. doi: https://doi.org/10.1038/jid.2010.119. jid2010119 [pii]. Epub 2010/05/07. PubMed PMID: 20445549; PubMed Central PMCID: PMCPMC2919658.
66. Rinne T, Brunner HG, van Bokhoven H. p63-associated disorders. Cell Cycle 2007;6(3):262–268. Epub 2007/01/17. doi: 3796 [pii]. PubMed PMID: 17224651.
67. Romano RA, Smalley K, Magraw C, Serna VA, Kurita T, Raghavan S, et al. DeltaNp63 knockout mice reveal its indispensable role as a master regulator of epithelial development and differentiation. Development. 2012;139(4):772–82. doi: https://doi.org/10.1242/dev.071191. 139/4/772 [pii]. Epub 2012/01/26. PubMed PMID: 22274697; PubMed Central PMCID: PMCPMC3265062.
68. Guo X, Keyes WM, Papazoglu C, Zuber J, Li W, Lowe SW, et al. TAp63 induces senescence and suppresses tumorigenesis in vivo. Nat Cell Biol. 2009;11(12):1451–7. doi: https://doi.org/10.1038/ncb1988. ncb1988 [pii]. Epub 2009/11/10. PubMed PMID: 19898465; PubMed Central PMCID: PMCPMC2920298.

69. Fessing MY, Mardaryev AN, Gdula MR, Sharov AA, Sharova TY, Rapisarda V, et al. p63 regulates Satb1 to control tissue-specific chromatin remodeling during development of the epidermis. J Cell Biol. 2011;194(6):825–39. doi: https://doi.org/10.1083/jcb.201101148. jcb.201101148 [pii]. Epub 2011/09/21. PubMed PMID: 21930775; PubMed Central PMCID: PMCPMC3207288.
70. Pastore S, Mascia F, Mariani V, Girolomoni G. The epidermal growth factor receptor system in skin repair and inflammation. J Invest Dermatol 2008;128(6):1365–1374. doi:https://doi.org/10.1038/sj.jid.5701184. 5701184 [pii]. Epub 2007/12/01. PubMed PMID: 18049451.
71. Braun S, auf dem Keller U, Steiling H, Werner S. Fibroblast growth factors in epithelial repair and cytoprotection. Philos Trans R Soc Lond B Biol Sci. 2004;359(1445):753–7. doi:https://doi.org/10.1098/rstb.2004.1464. Epub 2004/08/06. B8J00UNA7V7V04PB [pii]. PubMed PMID: 15293802; PubMed Central PMCID: PMCPMC1693362.
72. Botchkarev VA, Yaar M, Peters EM, Raychaudhuri SP, Botchkareva NV, Marconi A, et al. Neurotrophins in skin biology and pathology. J Invest Dermatol 2006;126(8):1719–1727. doi:https://doi.org/10.1038/sj.jid.5700270. 5700270 [pii]. Epub 2006/07/18. PubMed PMID: 16845411.
73. Paus R, Peters EM, Eichmuller S, Botchkarev VA. Neural mechanisms of hair growth control. J Investig Dermatol Symp Proc. 1997;2(1):61–8.
74. Scholzen T, Armstrong CA, Bunnett NW, Luger TA, Olerud JE, Ansel JC. Neuropeptides in the skin: interactions between the neuroendocrine and the skin immune systems. Exp Dermatol 1998;7(2–3):81–96. Epub 1998/05/16. PubMed PMID: 9583747.
75. Watt FM, Frye M, Benitah SA. MYC in mammalian epidermis: how can an oncogene stimulate differentiation? Nat Rev Cancer. 2008;8(3):234–42. doi: https://doi.org/10.1038/nrc2328. nrc2328 [pii]. Epub 2008/02/23. PubMed PMID: 18292777; PubMed Central PMCID: PMCPMC2494614.
76. Ferone G, Thomason HA, Antonini D, De Rosa L, Hu B, Gemei M, et al. Mutant p63 causes defective expansion of ectodermal progenitor cells and impaired FGF signalling in AEC syndrome. EMBO Mol Med. 2012;4(3):192–205. doi: https://doi.org/10.1002/emmm.201100199. Epub 2012/01/17. PubMed PMID: 22247000; PubMed Central PMCID: PMCPMC3376849.
77. Su X, Cho MS, Gi YJ, Ayanga BA, Sherr CJ, Flores ER. Rescue of key features of the p63-null epithelial phenotype by inactivation of Ink4a and Arf. EMBO J. 2009;28(13):1904–15. doi: https://doi.org/10.1038/emboj.2009.151. emboj2009151 [pii]. Epub 2009/06/06. PubMed PMID: 19494829; PubMed Central PMCID: PMCPMC2711186.
78. Westfall MD, Mays DJ, Sniezek JC, Pietenpol JA. The Delta Np63 alpha phosphoprotein binds the p21 and 14-3-3 sigma promoters in vivo and has transcriptional repressor activity that is reduced by Hay-Wells syndrome-derived mutations. Mol Cell Biol. 2003;23(7):2264–76. Epub 2003/03/18. PubMed PMID: 12640112; PubMed Central PMCID: PMCPMC150720.
79. LeBoeuf M, Terrell A, Trivedi S, Sinha S, Epstein JA, Olson EN, et al. Hdac1 and Hdac2 act redundantly to control p63 and p53 functions in epidermal progenitor cells. Dev Cell. 2010;19(6):807–18. doi: https://doi.org/10.1016/j.devcel.2010.10.015. S1534-5807(10)00468-5 [pii]. Epub 2010/11/26. PubMed PMID: 21093383; PubMed Central PMCID: PMCPMC3003338.
80. Dotto GP. p21(WAF1/Cip1): more than a break to the cell cycle? Biochim Biophys Acta 2000;1471(1):M43–M56. Epub 2000/09/01. doi: S0304-419X(00)00019-6 [pii]. PubMed PMID: 10967424.
81. Watt FM, Estrach S, Ambler CA. Epidermal Notch signalling: differentiation, cancer and adhesion. Curr Opin Cell Biol. 2008;20(2):171–9. doi: https://doi.org/10.1016/j.ceb.2008.01.010. S0955-0674(08)00024-0 [pii]. Epub 2008/03/18. PubMed PMID: 18342499; PubMed Central PMCID: PMCPMC2324124.
82. Dotto GP. Crosstalk of Notch with p53 and p63 in cancer growth control. Nat Rev Cancer 2009;9(8):587–595. doi: https://doi.org/10.1038/nrc2675. nrc2675 [pii]. Epub 2009/07/18. PubMed PMID: 19609265.

83. Ferrandiz N, Caraballo JM, Garcia-Gutierrez L, Devgan V, Rodriguez-Paredes M, Lafita MC, et al. p21 as a transcriptional co-repressor of S-phase and mitotic control genes. PLoS One. 2012;7(5):e37759. doi: https://doi.org/10.1371/journal.pone.0037759. Epub 2012/06/05. PONE-D-11-22619 [pii]. PubMed PMID: 22662213; PubMed Central PMCID: PMCPMC3360621.
84. Wang X, Bolotin D, Chu DH, Polak L, Williams T, Fuchs E. AP-2alpha: a regulator of EGF receptor signaling and proliferation in skin epidermis. J Cell Biol. 2006;172(3):409–21. doi: https://doi.org/10.1083/jcb.200510002. jcb.200510002 [pii]. Epub 2006/02/02. PubMed PMID: 16449191; PubMed Central PMCID: PMCPMC2063650.
85. Marinari B, Ballaro C, Koster MI, Giustizieri ML, Moretti F, Crosti F, et al. IKKalpha is a p63 transcriptional target involved in the pathogenesis of ectodermal dysplasias. J Invest Dermatol 2009;129(1):60–69. doi: https://doi.org/10.1038/jid.2008.202. jid2008202 [pii]. Epub 2008/07/18. PubMed PMID: 18633439.
86. Moretti F, Marinari B, Lo Iacono N, Botti E, Giunta A, Spallone G, et al. A regulatory feedback loop involving p63 and IRF6 links the pathogenesis of 2 genetically different human ectodermal dysplasias. J Clin Invest. 2010;120(5):1570–7. doi: https://doi.org/10.1172/JCI40267. 40267 [pii]. Epub 2010/04/29. PubMed PMID: 20424325; PubMed Central PMCID: PMCPMC2860936.
87. Yi R, Poy MN, Stoffel M, Fuchs E. A skin microRNA promotes differentiation by repressing 'stemness'. Nature 2008;452(7184):225–229. doi:https://doi.org/10.1038/nature06642. Epub 2008/03/04. doi: nature06642 [pii]. PubMed PMID: 18311128.
88. Williams SE, Beronja S, Pasolli HA, Fuchs E. Asymmetric cell divisions promote Notch-dependent epidermal differentiation. Nature. 2011;470(7334):353–8. doi: https://doi.org/10.1038/nature09793. nature09793 [pii]. Epub 2011/02/19. PubMed PMID: 21331036; PubMed Central PMCID: PMCPMC3077085.
89. Goulas S, Conder R, Knoblich JA. The par complex and integrins direct asymmetric cell division in adult intestinal stem cells. Cell Stem Cell 2012;11(4):529–540. doi: https://doi.org/10.1016/j.stem.2012.06.017. S1934-5909(12)00478-X [pii]. Epub 2012/10/09. PubMed PMID: 23040479.
90. Blanpain C, Lowry WE, Pasolli HA, Fuchs E. Canonical notch signaling functions as a commitment switch in the epidermal lineage. Genes Dev. 2006;20(21):3022–35. doi: https://doi.org/10.1101/gad.1477606. 20/21/3022 [pii]. Epub 2006/11/03. PubMed PMID: 17079689; PubMed Central PMCID: PMCPMC1620020.
91. Wang X, Pasolli HA, Williams T, Fuchs E. AP-2 factors act in concert with Notch to orchestrate terminal differentiation in skin epidermis. J Cell Biol. 2008;183(1):37–48. doi:https://doi.org/10.1083/jcb.200804030. jcb.200804030 [pii]. Epub 2008/10/01. PubMed PMID: 18824566; PubMed Central PMCID: PMCPMC2557040.
92. Baxter RM, Brissette JL. Role of the nude gene in epithelial terminal differentiation. J Invest Dermatol 2002;118(2):303–309. doi:https://doi.org/10.1046/j.0022-202x.2001.01662.x. 1662 [pii]. Epub 2002/02/14. PubMed PMID: 11841548.
93. Koster MI, Kim S, Huang J, Williams T, Roop DR. TAp63alpha induces AP-2gamma as an early event in epidermal morphogenesis. Dev Biol 2006;289(1):253–261. doi:https://doi.org/10.1016/j.ydbio.2005.10.041. Epub 2005/12/06. S0012-1606(05)00769-4 [pii]. PubMed PMID: 16324689.
94. Kalinin AE, Kajava AV, Steinert PM. Epithelial barrier function: assembly and structural features of the cornified cell envelope. BioEssays 2002;24(9):789–800. doi:https://doi.org/10.1002/bies.10144. Epub 2002/09/05. PubMed PMID: 12210515.
95. Brown SJ, Tilli CM, Jackson B, Avilion AA, MacLeod MC, Maltais LJ, et al. Rodent Lce gene clusters; new nomenclature, gene organization, and divergence of human and rodent genes. J Invest Dermatol 2007;127(7):1782–1786. doi:https://doi.org/10.1038/sj.jid.5700792. 5700792 [pii]. Epub 2007/04/06. PubMed PMID: 17410201.
96. Kypriotou M, Huber M, Hohl D. The human epidermal differentiation complex: cornified envelope precursors, S100 proteins and the 'fused genes' family. Exp Dermatol 2012;21(9):643–

649. doi: https://doi.org/10.1111/j.1600-0625.2012.01472.x. Epub 2012/04/18. PubMed PMID: 22507538.

97. Patel S, Xi ZF, Seo EY, McGaughey D, Segre JA. Klf4 and corticosteroids activate an overlapping set of transcriptional targets to accelerate in utero epidermal barrier acquisition. Proc Natl Acad Sci U S A. 2006;103(49):18668–73. doi: https://doi.org/10.1073/pnas.0608658103. 0608658103 [pii]. Epub 2006/11/30. PubMed PMID: 17130451; PubMed Central PMCID: PMCPMC1693720.
98. Sen GL, Boxer LD, Webster DE, Bussat RT, Qu K, Zarnegar BJ, et al. ZNF750 is a p63 target gene that induces KLF4 to drive terminal epidermal differentiation. Dev Cell. 2012;22(3):669–77. doi:https://doi.org/10.1016/j.devcel.2011.12.001. S1534-5807(11)00567-3 [pii]. Epub 2012/03/01. PubMed PMID: 22364861; PubMed Central PMCID: PMCPMC3306457.
99. Indra AK, Dupe V, Bornert JM, Messaddeq N, Yaniv M, Mark M, et al. Temporally controlled targeted somatic mutagenesis in embryonic surface ectoderm and fetal epidermal keratinocytes unveils two distinct developmental functions of BRG1 in limb morphogenesis and skin barrier formation. Development. 2005;132(20):4533–44. doi:https://doi.org/10.1242/dev.02019. 132/20/4533 [pii]. Epub 2005/09/30. PubMed PMID: 16192310.
100. Kashiwagi M, Morgan BA, Georgopoulos K. The chromatin remodeler Mi-2beta is required for establishment of the basal epidermis and normal differentiation of its progeny. Development 2007;134(8):1571–1582. doi: https://doi.org/10.1242/dev.001750. dev.001750 [pii]. Epub 2007/03/16. PubMed PMID: 17360773.
101. Sen GL, Webster DE, Barragan DI, Chang HY, Khavari PA. Control of differentiation in a self-renewing mammalian tissue by the histone demethylase JMJD3. Genes Dev. 2008;22(14):1865–70. doi: https://doi.org/10.1101/gad.1673508. 22/14/1865 [pii]. Epub 2008/07/17. PubMed PMID: 18628393; PubMed Central PMCID: PMCPMC2492733.
102. Sen GL, Reuter JA, Webster DE, Zhu L, Khavari PA. DNMT1 maintains progenitor function in self-renewing somatic tissue. Nature. 2010;463(7280):563–7. doi: https://doi.org/10.1038/nature08683. nature08683 [pii]. Epub 2010/01/19. PubMed PMID: 20081831; PubMed Central PMCID: PMCPMC3050546.
103. Ezhkova E, Lien WH, Stokes N, Pasolli HA, Silva JM, Fuchs E. EZH1 and EZH2 cogovern histone H3K27 trimethylation and are essential for hair follicle homeostasis and wound repair. Genes Dev. 2011;25(5):485–98. doi:https://doi.org/10.1101/gad.2019811. Epub 2011/02/15. gad.2019811 [pii]. PubMed PMID: 21317239; PubMed Central PMCID: PMCPMC3049289.
104. Ezhkova E, Pasolli HA, Parker JS, Stokes N, Su IH, Hannon G, et al. Ezh2 orchestrates gene expression for the stepwise differentiation of tissue-specific stem cells. Cell. 2009;136(6):1122–35. doi:https://doi.org/10.1016/j.cell.2008.12.043. S0092-8674(09)00006-3 [pii]. Epub 2009/03/24. PubMed PMID: 19303854; PubMed Central PMCID: PMCPMC2716120.
105. Mejetta S, Morey L, Pascual G, Kuebler B, Mysliwiec MR, Lee Y, et al. Jarid2 regulates mouse epidermal stem cell activation and differentiation. EMBO J. 2011;30(17):3635–46. doi: https://doi.org/10.1038/emboj.2011.265. emboj2011265 [pii]. Epub 2011/08/04. PubMed PMID: 21811233; PubMed Central PMCID: PMCPMC3181486.
106. Luis NM, Morey L, Mejetta S, Pascual G, Janich P, Kuebler B, et al. Regulation of human epidermal stem cell proliferation and senescence requires polycomb- dependent and -independent functions of Cbx4. Cell Stem Cell 2011;9(3):233–246. doi:https://doi.org/10.1016/j.stem.2011.07.013. S1934-5909(11)00343-2 [pii]. Epub 2011/09/03. PubMed PMID: 21885019.
107. Driskell I, Oda H, Blanco S, Nascimento E, Humphreys P, Frye M. The histone methyltransferase Setd8 acts in concert with c-Myc and is required to maintain skin. EMBO J. 2012;31(3):616–29. doi: https://doi.org/10.1038/emboj.2011.421. emboj2011421 [pii]. Epub 2011/11/26. PubMed PMID: 22117221; PubMed Central PMCID: PMCPMC3273381.
108. Van Exan RJ, Hardy MH. A spatial relationship between innervation and the early differentiation of vibrissa follicles in the embryonic mouse. J Anat. 1980;131(Pt 4):643–56. Epub 1980/12/01. PubMed PMID: 7216903; PubMed Central PMCID: PMCPMC1233217.

109. Veltmaat JM, Mailleux AA, Thiery JP, Bellusci S. Mouse embryonic mammogenesis as a model for the molecular regulation of pattern formation. Differentiation 2003;71(1):1–17. doi:https://doi.org/10.1046/j.1432-0436.2003.700601.x. S0301-4681(09)60262-0 [pii]. Epub 2003/02/01. PubMed PMID: 12558599.
110. Paus R, Muller-Rover S, Van Der Veen C, Maurer M, Eichmuller S, Ling G, et al. A comprehensive guide for the recognition and classification of distinct stages of hair follicle morphogenesis. J Invest Dermatol 1999;113(4):523–532. doi: https://doi.org/10.1046/j.1523--1747.1999.00740.x. Epub 1999/10/03. jid740 [pii]. PubMed PMID: 10504436.
111. Kunisada M, Cui CY, Piao Y, Ko MS, Schlessinger D. Requirement for Shh and Fox family genes at different stages in sweat gland development. Hum Mol Genet. 2009;18(10):1769–78. doi: https://doi.org/10.1093/hmg/ddp089. ddp089 [pii]. Epub 2009/03/10. PubMed PMID: 19270025; PubMed Central PMCID: PMCPMC2671986.
112. Mikkola ML, Millar SE. The mammary bud as a skin appendage: unique and shared aspects of development. J Mammary Gland Biol Neoplasia 2006;11(3–4):187–203. doi: https://doi.org/10.1007/s10911-006-9029-x. Epub 2006/11/18. PubMed PMID: 17111222.
113. Dhouailly D. A new scenario for the evolutionary origin of hair, feather, and avian scales. J Anat. 2009;214(4):587–606. doi: https://doi.org/10.1111/j.1469-7580.2008.01041.x. JOA1041 [pii]. Epub 2009/05/09. PubMed PMID: 19422430; PubMed Central PMCID: PMCPMC2736124.
114. Widelitz RB, Baker RE, Plikus M, Lin CM, Maini PK, Paus R, et al. Distinct mechanisms underlie pattern formation in the skin and skin appendages. Birth Defects Res C Embryo Today 2006;78(3):280–291. doi: https://doi.org/10.1002/bdrc.20075. Epub 2006/10/25. PubMed PMID: 17061271.
115. Rendl M, Lewis L, Fuchs E. Molecular dissection of mesenchymal-epithelial interactions in the hair follicle. PLoS Biol. 2005;3(11):e331. doi: https://doi.org/10.1371/journal.pbio.0030331. 05-PLBI-RA-0332R1 [pii]. Epub 2005/09/16. PubMed PMID: 16162033; PubMed Central PMCID: PMCPMC1216328.
116. Lu CP, Polak L, Rocha AS, Pasolli HA, Chen SC, Sharma N, et al. Identification of stem cell populations in sweat glands and ducts reveals roles in homeostasis and wound repair. Cell. 2012;150(1):136–50. doi: https://doi.org/10.1016/j.cell.2012.04.045. S0092-8674(12)00647-2 [pii]. Epub 2012/07/10. PubMed PMID: 22770217; PubMed Central PMCID: PMCPMC3423199.
117. Oftedal OT, Dhouailly D. Evo-devo of the mammary gland. J Mammary Gland Biol Neoplasia 2013;18(2):105–120. doi: https://doi.org/10.1007/s10911-013-9290-8. Epub 2013/05/18. PubMed PMID: 23681303.
118. Sengel P. Pattern formation in skin development. Int J Dev Biol 1990;34(1):33–50. Epub 1990/03/01. PubMed PMID: 2203463.
119. Viallet JP, Dhouailly D. Retinoic acid and mouse skin morphogenesis. II. Role of epidermal competence in hair glandular metaplasia. Dev Biol 1994;166(1):277–288. doi: https://doi.org/10.1006/dbio.1994.1314. S0012-1606(84)71314-5 [pii]. Epub 1994/11/01. PubMed PMID: 7958452.
120. Jahoda CA, Horne KA, Oliver RF. Induction of hair growth by implantation of cultured dermal papilla cells. Nature 1984;311(5986):560–562. Epub 1984/10/11. PubMed PMID: 6482967.
121. Plikus M, Wang WP, Liu J, Wang X, Jiang TX, Chuong CM. Morpho-regulation of ectodermal organs: integument pathology and phenotypic variations in K14-Noggin engineered mice through modulation of bone morphogenic protein pathway. Am J Pathol. 2004;164(3):1099–114. doi: https://doi.org/10.1016/S0002-9440(10)63197-5. S0002-9440(10)63197-5 [pii]. Epub 2004/02/26. PubMed PMID: 14982863; PubMed Central PMCID: PMCPMC1614723.
122. Ahtiainen L, Lefebvre S, Lindfors PH, Renvoise E, Shirokova V, Vartiainen MK, et al. Directional cell migration, but not proliferation, drives hair placode morphogenesis. Dev Cell 2014;28(5):588–602. doi: https://doi.org/10.1016/j.devcel.2014.02.003. PubMed PMID: 24636260.

123. Duverger O, Morasso MI. Role of homeobox genes in the patterning, specification, and differentiation of ectodermal appendages in mammals. J Cell Physiol. 2008;216(2):337–46. doi: https://doi.org/10.1002/jcp.21491. Epub 2008/05/07. PubMed PMID: 18459147; PubMed Central PMCID: PMCPMC2561923.
124. Cowin P, Wysolmerski J. Molecular mechanisms guiding embryonic mammary gland development. Cold Spring Harb Perspect Biol. 2010;2(6):a003251. doi: https://doi.org/10.1101/cshperspect.a003251. Epub 2010/05/21. cshperspect.a003251 [pii]. PubMed PMID: 20484386; PubMed Central PMCID: PMCPMC2869520.
125. Sick S, Reinker S, Timmer J, Schlake T. WNT and DKK determine hair follicle spacing through a reaction-diffusion mechanism. Science 2006;314(5804):1447–1450. doi:https://doi.org/10.1126/science.1130088. 1130088 [pii]. Epub 2006/11/04. PubMed PMID: 17082421.
126. Kondo S, Miura T. Reaction-diffusion model as a framework for understanding biological pattern formation. Science 2010;329(5999):1616–1620. doi:https://doi.org/10.1126/science.1179047. 329/5999/1616 [pii]. Epub 2010/10/12. PubMed PMID: 20929839.
127. Jiang TX, Jung HS, Widelitz RB, Chuong CM. Self-organization of periodic patterns by dissociated feather mesenchymal cells and the regulation of size, number and spacing of primordia. Development 1999;126(22):4997–5009. Epub 1999/10/26. PubMed PMID: 10529418.
128. Botchkarev VA, Botchkareva NV, Roth W, Nakamura M, Chen LH, Herzog W, et al. Noggin is a mesenchymally derived stimulator of hair-follicle induction. Nat Cell Biol 1999;1(3):158–164. doi: https://doi.org/10.1038/11078. Epub 1999/11/13. PubMed PMID: 10559902.
129. Andl T, Reddy ST, Gaddapara T, Millar SE. WNT signals are required for the initiation of hair follicle development. Dev Cell 2002;2(5):643–653. Epub 2002/05/23. doi: S1534580702001673 [pii]. PubMed PMID: 12015971.
130. Botchkarev VA, Botchkareva NV, Sharov AA, Funa K, Huber O, Gilchrest BA. Modulation of BMP signaling by noggin is required for induction of the secondary (nontylotrich) hair follicles. J Invest Dermatol 2002;118(1):3–10. doi:https://doi.org/10.1046/j.1523--1747.2002.01645.x. 1645 [pii]. Epub 2002/02/20. PubMed PMID: 11851869.
131. Huang J, Dattilo LK, Rajagopal R, Liu Y, Kaartinen V, Mishina Y, et al. FGF-regulated BMP signaling is required for eyelid closure and to specify conjunctival epithelial cell fate. Development. 2009;136(10):1741–50. doi: https://doi.org/10.1242/dev.034082. dev.034082 [pii]. Epub 2009/04/17. PubMed PMID: 19369394; PubMed Central PMCID: PMCPMC2673764.
132. Mikkola ML. Genetic basis of skin appendage development. Semin Cell Dev Biol. 2007;18:225–36.
133. Eivers E, Fuentealba LC, De Robertis EM. Integrating positional information at the level of Smad1/5/8. Curr Opin Genet Dev. 2008;18(4):304–10. Epub 2008/07/02. doi: https://doi.org/10.1016/j.gde.2008.06.001. S0959-437X(08)00069-5 [pii]. PubMed PMID: 18590818; PubMed Central PMCID: PMCPMC3033225.
134. Plouhinec JL, Zakin L, De Robertis EM. Systems control of BMP morphogen flow in vertebrate embryos. Curr Opin Genet Dev. 2011;21(6):696–703. Epub 2011/09/23. doi: https://doi.org/10.1016/j.gde.2011.09.001. S0959-437X(11)00136-5 [pii]. PubMed PMID: 21937218; PubMed Central PMCID: PMCPMC3224208.
135. Mou C, Jackson B, Schneider P, Overbeek PA, Headon DJ. Generation of the primary hair follicle pattern. Proc Natl Acad Sci U S A. 2006;103(24):9075–80. doi: https://doi.org/10.1073/pnas.0600825103. Epub 2006/06/14. 0600825103 [pii]. PubMed PMID: 16769906; PubMed Central PMCID: PMCPMC1482568.
136. Zhang Y, Tomann P, Andl T, Gallant NM, Huelsken J, Jerchow B, et al. Reciprocal requirements for EDA/EDAR/NF-kappaB and Wnt/beta-catenin signaling pathways in hair follicle induction. Dev Cell. 2009;17(1):49–61. doi: https://doi.org/10.1016/j.devcel.2009.05.011. S1534-5807(09)00213-5 [pii]. Epub 2009/07/22. PubMed PMID: 19619491; PubMed Central PMCID: PMCPMC2859042.

137. Pummila M, Fliniaux I, Jaatinen R, James MJ, Laurikkala J, Schneider P, et al. Ectodysplasin has a dual role in ectodermal organogenesis: inhibition of Bmp activity and induction of Shh expression. Development. 2007;134(1):117–25.
138. Chiang C, Swan RZ, Grachtchouk M, Bolinger M, Litingtung Y, Robertson EK, et al. Essential role for Sonic hedgehog during hair follicle morphogenesis. Dev Biol. 1999;205(1):1–9.
139. St_Jacques B, Dassule HR, Karavanova I, Botchkarev VA, Li J, Danielian PS, et al. Sonic hedgehog signaling is essential for hair development. Curr Biol. 1998;8(19):1058–68.
140. Lu CP, Polak L, Keyes BE, Fuchs E. Spatiotemporal antagonism in mesenchymal-epithelial signaling in sweat versus hair fate decision. Science. 2016;354(6319). doi: https://doi.org/10.1126/science.aah6102. PubMed PMID: 28008008; PubMed Central PMCID: PMCPMC5333576.
141. Lee MY, Sun L, Veltmaat JM. Hedgehog and Gli signaling in embryonic mammary gland development. J Mammary Gland Biol Neoplasia. 2013;18(2):133–8. doi: https://doi.org/10.1007/s10911-013-9291-7. Epub 2013/05/17. PubMed PMID: 23677624; PubMed Central PMCID: PMCPMC3691482.
142. Rhee H, Polak L, Fuchs E. Lhx2 maintains stem cell character in hair follicles. Science. 2006;312:1946–99.
143. Nowak JA, Polak L, Pasolli HA, Fuchs E. Hair follicle stem cells are specified and function in early skin morphogenesis. Cell Stem Cell. 2008;3(1):33–43. doi: https://doi.org/10.1016/j.stem.2008.05.009. S1934-5909(08)00223-3 [pii]. Epub 2008/07/03. PubMed PMID: 18593557; PubMed Central PMCID: PMCPMC2877596.
144. Langbein L, Schweizer J. Keratins of the human hair follicle. Int Rev Cytol 2005;243:1–78. doi:https://doi.org/10.1016/S0074-7696(05)43001-6. S0074769605430016 [pii]. Epub 2005/03/31. PubMed PMID: 15797458.
145. Rogers MA, Langbein L, Praetzel-Wunder S, Winter H, Schweizer J. Human hair keratin-associated proteins (KAPs). Int Rev Cytol 2006;251:209–263. doi:https://doi.org/10.1016/S0074-7696(06)51006-X. S0074-7696(06)51006-X [pii]. Epub 2006/08/31. PubMed PMID: 16939781.
146. Luetteke NC, Qiu TH, Peiffer RL, Oliver P, Smithies O, Lee DC. TGF alpha deficiency results in hair follicle and eye abnormalities in targeted and waved-1 mice. Cell 1993;73(2):263–278. Epub 1993/04/23. doi: 0092-8674(93)90228-I [pii]. PubMed PMID: 8477445.
147. Luetteke NC, Phillips HK, Qiu TH, Copeland NG, Earp HS, Jenkins NA, et al. The mouse waved-2 phenotype results from a point mutation in the EGF receptor tyrosine kinase. Genes Dev 1994;8(4):399–413. Epub 1994/02/15. PubMed PMID: 8125255.
148. Peschon JJ, Slack JL, Reddy P, Stocking KL, Sunnarborg SW, Lee DC, et al. An essential role for ectodomain shedding in mammalian development. Science 1998;282(5392):1281–1284. Epub 1998/11/13. PubMed PMID: 9812885.
149. Inoue A, Arima N, Ishiguro J, Prestwich GD, Arai H, Aoki J. LPA-producing enzyme PA-PLA(1)alpha regulates hair follicle development by modulating EGFR signalling. EMBO J. 2011;30(20):4248–60. doi: https://doi.org/10.1038/emboj.2011.296. emboj2011296 [pii]. Epub 2011/08/23. PubMed PMID: 21857648; PubMed Central PMCID: PMCPMC3199385.
150. Ellis T, Gambardella L, Horcher M, Tschanz S, Capol J, Bertram P, et al. The transcriptional repressor CDP (Cutl1) is essential for epithelial cell differentiation of the lung and the hair follicle. Genes & Development. 2001;15(17):2307–19. doi: https://doi.org/10.1101/gad.200101. Epub 2001/09/07. PubMed PMID: 11544187; PubMed Central PMCID: PMCPMC312776.
151. Kobielak K, Pasolli HA, Alonso L, Polak L, Fuchs E. Defining BMP functions in the hair follicle by conditional ablation of BMP receptor IA. J Cell Biol. 2003;163:609–23.
152. Kaufman CK, Zhou P, Pasolli HA, Rendl M, Bolotin D, Lim KC, et al. GATA-3: an unexpected regulator of cell lineage determination in skin. Genes Dev. 2003;17(17):2108–22. doi: https://doi.org/10.1101/gad.1115203. Epub 2003/08/19. 1115203 [pii]. PubMed PMID: 12923059; PubMed Central PMCID: PMCPMC196453.

153. Hwang J, Mehrani T, Millar SE, Morasso MI. Dlx3 is a crucial regulator of hair follicle differentiation and cycling. Development. 2008;135(18):3149–59. doi: https://doi.org/10.1242/dev.022202. dev.022202 [pii]. Epub 2008/08/08. PubMed PMID: 18684741; PubMed Central PMCID: PMCPMC2707782.
154. Ma L, Liu J, Wu T, Plikus M, Jiang TX, Bi Q, et al. 'Cyclic alopecia' in Msx2 mutants: defects in hair cycling and hair shaft differentiation. Development 2003;130(2):379–389. Epub 2002/12/06. PubMed PMID: 12466204.
155. Satokata I, Ma L, Ohshima H, Bei M, Woo I, Nishizawa K, et al. Msx2 deficiency in mice causes pleiotropic defects in bone growth and ectodermal organ formation. Nat Genet 2000;24(4):391–395. doi: https://doi.org/10.1038/74231. Epub 2000/03/31. PubMed PMID: 10742104.
156. Cui CY, Childress V, Piao Y, Michel M, Johnson AA, Kunisada M, et al. Forkhead transcription factor FoxA1 regulates sweat secretion through Bestrophin 2 anion channel and Na-K-Cl cotransporter 1. Proc Natl Acad Sci USA. 2012;109(4):1199–203. doi: https://doi.org/10.1073/pnas.1117213109. 1117213109 [pii]. Epub 2012/01/10. PubMed PMID: 22223659; PubMed Central PMCID: PMCPMC3268268.
157. Cui CY, Ishii R, Campbell DP, Michel M, Piao Y, Kume T, et al. Foxc1 Ablated Mice Are Anhidrotic and Recapitulate Features of Human Miliaria Sweat Retention Disorder. J Invest Dermatol. 2017;137(1):38–45. doi: https://doi.org/10.1016/j.jid.2016.08.012. PubMed PMID: 27592801; PubMed Central PMCID: PMCPMC5183533.
158. Cotsarelis G. Epithelial stem cells: a folliculocentric view. J Invest Dermatol 2006;126(7):1459–1468. doi:https://doi.org/10.1038/sj.jid.5700376. 5700376 [pii]. Epub 2006/06/17. PubMed PMID: 16778814.
159. Goldstein J, Horsley V. Home sweet home: skin stem cell niches. Cell Mol Life Sci. 2012;69(15):2573–82. doi: https://doi.org/10.1007/s00018-012-0943-3. Epub 2012/03/14. PubMed PMID: 22410738; PubMed Central PMCID: PMCPMC3449145.
160. Jaks V, Kasper M, Toftgard R. The hair follicle-a stem cell zoo. Exp Cell Res 2010;316(8):1422–1428. doi: https://doi.org/10.1016/j.yexcr.2010.03.014. S0014-4827(10)00121-7 [pii]. Epub 2010/03/27. PubMed PMID: 20338163..
161. Beck B, Blanpain C. Mechanisms regulating epidermal stem cells. EMBO J. 2012;31(9):2067–75. doi:https://doi.org/10.1038/emboj.2012.67. emboj201267 [pii]. Epub 2012/03/22. PubMed PMID: 22433839; PubMed Central PMCID: PMCPMC3343465.
162. Van Keymeulen A, Blanpain C. Tracing epithelial stem cells during development, homeostasis, and repair. J Cell Biol. 2012;197(5):575–84. doi: https://doi.org/10.1083/jcb.201201041. jcb.201201041 [pii]. Epub 2012/05/30. PubMed PMID: 22641343; PubMed Central PMCID: PMCPMC3365500.
163. Fujiwara H, Ferreira M, Donati G, Marciano DK, Linton JM, Sato Y, et al. The basement membrane of hair follicle stem cells is a muscle cell niche. Cell. 2011;144(4):577–89. doi: https://doi.org/10.1016/j.cell.2011.01.014. Epub 2011/02/22. S0092-8674(11)00015-8 [pii]. PubMed PMID: 21335239; PubMed Central PMCID: PMCPMC3056115.
164. Osawa M, Egawa G, Mak SS, Moriyama M, Freter R, Yonetani S, et al. Molecular characterization of melanocyte stem cells in their niche. Development. 2005;132(24):5589–99. doi: https://doi.org/10.1242/dev.02161. 132/24/5589 [pii]. Epub 2005/11/30. PubMed PMID: 16314490.
165. Hunt DP, Jahoda C, Chandran S. Multipotent skin-derived precursors: from biology to clinical translation. Curr Opin Biotechnol 2009;20(5):522–530. doi:https://doi.org/10.1016/j.copbio.2009.10.004. S0958-1669(09)00132-3 [pii]. Epub 2009/11/10. PubMed PMID: 19896826.
166. Fuchs E, Chen T. A matter of life and death: self-renewal in stem cells. EMBO Rep. 2013;14(1):39–48. doi: https://doi.org/10.1038/embor.2012.197. embor2012197 [pii]. Epub 2012/12/12. PubMed PMID: 23229591; PubMed Central PMCID: PMCPMC3537149.
167. Mascre G, Dekoninck S, Drogat B, Youssef KK, Brohee S, Sotiropoulou PA, et al. Distinct contribution of stem and progenitor cells to epidermal maintenance. Nature

2012;489(7415):257–262. doi: https://doi.org/10.1038/nature11393. nature11393 [pii]. Epub 2012/09/04. PubMed PMID: 22940863.

168. Cotsarelis G, Sun TT, Lavker RM. Label-retaining cells reside in the bulge area of pilosebaceous unit: implications for follicular stem cells, hair cycle, and skin carcinogenesis. Cell 1990;61(7):1329–1337. doi: 0092-8674(90)90696-C [pii]. Epub 1990/06/29. PubMed PMID: 2364430.
169. Morris RJ, Liu Y, Marles L, Yang Z, Trempus C, Li S, et al. Capturing and profiling adult hair follicle stem cells. Nat Biotechnol 2004;22(4):411–417. doi: https://doi.org/10.1038/nbt950. nbt950 [pii]. Epub 2004/03/17. PubMed PMID: 15024388.
170. Tumbar T, Guasch G, Greco V, Blanpain C, Lowry WE, Rendl M, et al. Defining the epithelial stem cell niche in skin. Science. 2004;303(5656):359–63. doi: https://doi.org/10.1126/science.1092436. 1092436 [pii]. Epub 2003/12/13. PubMed PMID: 14671312; PubMed Central PMCID: PMCPMC2405920.
171. Horsley V, Aliprantis AO, Polak L, Glimcher LH, Fuchs E. NFATc1 balances quiescence and proliferation of skin stem cells. Cell. 2008;132(2):299–310. doi: https://doi.org/10.1016/j.cell.2007.11.047. S0092-8674(07)01614-5 [pii]. Epub 2008/02/05. PubMed PMID: 18243104; PubMed Central PMCID: PMCPMC2546702.
172. Nguyen H, Rendl M, Fuchs E. Tcf3 governs stem cell features and represses cell fate determination in skin. Cell 2006;127(1):171–183. doi:https://doi.org/10.1016/j.cell.2006.07.036. S0092-8674(06)01162-7 [pii]. Epub 2006/10/05. PubMed PMID: 17018284.
173. Nijhof JG, Braun KM, Giangreco A, van Pelt C, Kawamoto H, Boyd RL, et al. The cell-surface marker MTS24 identifies a novel population of follicular keratinocytes with characteristics of progenitor cells. Development 2006;133(15):3027–3037. doi: https://doi.org/10.1242/dev.02443. dev.02443 [pii]. Epub 2006/07/05. PubMed PMID: 16818453.
174. Jensen KB, Collins CA, Nascimento E, Tan DW, Frye M, Itami S, et al. Lrig1 expression defines a distinct multipotent stem cell population in mammalian epidermis. Cell Stem Cell. 2009;4(5):427–39. doi: https://doi.org/10.1016/j.stem.2009.04.014. S1934-5909(09)00168-4 [pii]. Epub 2009/05/12. PubMed PMID: 19427292; PubMed Central PMCID: PMCPMC2698066.
175. Snippert HJ, Haegebarth A, Kasper M, Jaks V, van Es JH, Barker N, et al. Lgr6 marks stem cells in the hair follicle that generate all cell lineages of the skin. Science 2010;327(5971):1385–1389. doi: https://doi.org/10.1126/science.1184733.327/5971/1385 [pii]. Epub 2010/03/13. PubMed PMID: 20223988.
176. Horsley V, O'Carroll D, Tooze R, Ohinata Y, Saitou M, Obukhanych T, et al. Blimp1 defines a progenitor population that governs cellular input to the sebaceous gland. Cell. 2006;126(3):597–609. doi: https://doi.org/10.1016/j.cell.2006.06.048. S0092-8674(06)00971-8 [pii]. Epub 2006/08/12. PubMed PMID: 16901790; PubMed Central PMCID: PMCPMC2424190.
177. Jaks V, Barker N, Kasper M, van Es JH, Snippert HJ, Clevers H, et al. Lgr5 marks cycling, yet long-lived, hair follicle stem cells. Nat Genet 2008;40(11):1291–1299. doi: https://doi.org/10.1038/ng.239. ng.239 [pii]. Epub 2008/10/14. PubMed PMID: 18849992.
178. Greco V, Chen T, Rendl M, Schober M, Pasolli HA, Stokes N, et al. A two-step mechanism for stem cell activation during hair regeneration. Cell Stem Cell. 2009;4(2):155–69. doi: https://doi.org/10.1016/j.stem.2008.12.009. S1934-5909(08)00626-7 [pii]. Epub 2009/02/10. PubMed PMID: 19200804; PubMed Central PMCID: PMCPMC2668200.
179. Panteleyev AA, Jahoda CA, Christiano AM. Hair follicle predetermination. J Cell Sci 2001;114(Pt 19):3419–3431. PubMed PMID: 21538920..
180. Oshima H, Rochat A, Kedzia C, Kobayashi K, Barrandon Y. Morphogenesis and renewal of hair follicles from adult multipotent stem cells. Cell. 2001;104(2):233–45.
181. Ito M, Liu Y, Yang Z, Nguyen J, Liang F, Morris RJ, et al. Stem cells in the hair follicle bulge contribute to wound repair but not to homeostasis of the epidermis. Nat Med 2005;11(12):1351–1354. doi:https://doi.org/10.1038/nm1328. nm1328 [pii]. Epub 2005/11/17. PubMed PMID: 16288281.

182. Page ME, Lombard P, Ng F, Gottgens B, Jensen KB. The epidermis comprises autonomous compartments maintained by distinct stem cell populations. Cell Stem Cell. 2013. doi: https://doi.org/10.1016/j.stem.2013.07.010. S1934-5909(13)00318-4 [pii]. Epub 2013/08/21. PubMed PMID: 23954751.
183. Lien WH, Guo X, Polak L, Lawton LN, Young RA, Zheng D, et al. Genome-wide maps of histone modifications unwind in vivo chromatin states of the hair follicle lineage. Cell Stem Cell. 2011;9(3):219–32. doi: https://doi.org/10.1016/j.stem.2011.07.015. S1934-5909(11)00378-X [pii]. Epub 2011/09/03. PubMed PMID: 21885018; PubMed Central PMCID: PMCPMC3166618.
184. Lee J, Kang S, Lilja KC, Colletier KJ, Scheitz CJ, Zhang YV, et al. Signalling couples hair follicle stem cell quiescence with reduced histone H3 K4/K9/K27me3 for proper tissue homeostasis. Nat Commun. 2016;7:11278. doi: https://doi.org/10.1038/ncomms11278. PubMed PMID: 27080563; PubMed Central PMCID: PMCPMC4835553.
185. Kasper M, Jaks V, Are A, Bergstrom A, Schwager A, Svard J, et al. Wounding enhances epidermal tumorigenesis by recruiting hair follicle keratinocytes. Proc Natl Acad Sci U S A. 2011;108(10):4099–104. doi: https://doi.org/10.1073/pnas.1014489108. 1014489108 [pii]. Epub 2011/02/16. PubMed PMID: 21321199; PubMed Central PMCID: PMCPMC3054016.
186. Mardaryev AN, Meier N, Poterlowicz K, Sharov AA, Sharova TY, Ahmed MI, et al. Lhx2 differentially regulates Sox9, Tcf4 and Lgr5 in hair follicle stem cells to promote epidermal regeneration after injury. Development 2011;138(22):4843–4852. doi: https://doi.org/10.1242/dev.070284.138/22/4843 [pii]. Epub 2011/10/27. PubMed PMID: 22028024.
187. Van Keymeulen A, Rocha AS, Ousset M, Beck B, Bouvencourt G, Rock J, et al. Distinct stem cells contribute to mammary gland development and maintenance. Nature 2011;479(7372):189–193. doi:https://doi.org/10.1038/nature10573. nature10573 [pii]. Epub 2011/10/11. PubMed PMID: 21983963.
188. Botchkareva NV, Khlgatian M, Longley BJ, Botchkarev VA, Gilchrest BA. SCF/c-kit signaling is required for cyclic regeneration of the hair pigmentation unit. FASEB J 2001;15(3):645–658. doi: https://doi.org/10.1096/fj.00-0368com.15/3/645. [pii]. Epub 2001/03/22. PubMed PMID: 11259383.
189. Nishimura EK, Jordan SA, Oshima H, Yoshida H, Osawa M, Moriyama M, et al. Dominant role of the niche in melanocyte stem-cell fate determination. Nature 2002;416(6883):854–860. doi: https://doi.org/10.1038/416854a. 416854a [pii]. Epub 2002/04/27. PubMed PMID: 11976685.
190. Botchkareva NV, Botchkarev VA, Gilchrest BA. Fate of melanocytes during development of the hair follicle pigmentary unit. J Investig Dermatol Symp Proc 2003;8(1):76–79. doi: https://doi.org/10.1046/j.1523-1747.2003.12176.x. Epub 2003/08/05. PubMed PMID: 12894999.
191. Tanimura S, Tadokoro Y, Inomata K, Binh NT, Nishie W, Yamazaki S, et al. Hair follicle stem cells provide a functional niche for melanocyte stem cells. Cell Stem Cell 2011;8(2):177–187. doi: https://doi.org/10.1016/j.stem.2010.11.029. S1934-5909(10)00651-X [pii]. Epub 2011/02/08. PubMed PMID: 21295274.
192. Nishimura EK, Suzuki M, Igras V, Du J, Lonning S, Miyachi Y, et al. Key roles for transforming growth factor beta in melanocyte stem cell maintenance. Cell Stem Cell. 2010;6(2):130–40. doi: https://doi.org/10.1016/j.stem.2009.12.010. S1934-5909(09)00633-X [pii]. Epub 2010/02/11. PubMed PMID: 20144786; PubMed Central PMCID: PMCPMC3437996.
193. Rabbani P, Takeo M, Chou W, Myung P, Bosenberg M, Chin L, et al. Coordinated activation of Wnt in epithelial and melanocyte stem cells initiates pigmented hair regeneration. Cell 2011;145(6):941–955. doi:https://doi.org/10.1016/j.cell.2011.05.004. S0092-8674(11)00530-7 [pii]. Epub 2011/06/15. PubMed PMID: 21663796.
194. Chou WC, Takeo M, Rabbani P, Hu H, Lee W, Chung YR, et al. Direct migration of follicular melanocyte stem cells to the epidermis after wounding or UVB irradiation is dependent on Mc1r signaling. Nat Med 2013;19(7):924–929. doi: https://doi.org/10.1038/nm.3194. nm.3194 [pii]. Epub 2013/06/12. PubMed PMID: 23749232.

195. Richardson GD, Arnott EC, Whitehouse CJ, Lawrence CM, Reynolds AJ, Hole N, et al. Plasticity of rodent and human hair follicle dermal cells: implications for cell therapy and tissue engineering. J Investig Dermatol Symp Proc 2005;10(3):180–183. doi:https://doi.org/10.1111/j.1087-0024.2005.10101.x. Epub 2005/12/31. PubMed PMID: 16382659.
196. Waters JM, Richardson GD, Jahoda CA. Hair follicle stem cells. Semin Cell Dev Biol 2007;18(2):245–254. doi: https://doi.org/10.1016/j.semcdb.2007.02.003. S1084-9521(07)00036-5 [pii]. Epub 2007/05/08. PubMed PMID: 17481931.
197. Reynolds AJ, Lawrence C, Cserhalmi-Friedman PB, Christiano AM, Jahoda CA. Trans-gender induction of hair follicles. Nature 1999;402(6757):33–34. doi:https://doi.org/10.1038/46938. Epub 1999/11/26. PubMed PMID: 10573414.
198. Biernaskie J, Paris M, Morozova O, Fagan BM, Marra M, Pevny L, et al. SKPs derive from hair follicle precursors and exhibit properties of adult dermal stem cells. Cell Stem Cell. 2009;5(6):610–23. doi: https://doi.org/10.1016/j.stem.2009.10.019. S1934-5909(09)00573-6 [pii]. Epub 2009/12/03. PubMed PMID: 19951689; PubMed Central PMCID: PMCPMC2828150.
199. Lako M, Armstrong L, Cairns PM, Harris S, Hole N, Jahoda CA. Hair follicle dermal cells repopulate the mouse haematopoietic system. J Cell Sci 2002;115(Pt 20):3967–3974. Epub 2002/09/24. PubMed PMID: 12244134.
200. Fernandes KJ, Kobayashi NR, Gallagher CJ, Barnabe-Heider F, Aumont A, Kaplan DR, et al. Analysis of the neurogenic potential of multipotent skin-derived precursors. Exp Neurol 2006;201(1):32–48. doi:https://doi.org/10.1016/j.expneurol.2006.03.018. Epub 2006/05/09. S0014-4886(06)00205-6 [pii]. PubMed PMID: 16678161.
201. McKenzie IA, Biernaskie J, Toma JG, Midha R, Miller FD. Skin-derived precursors generate myelinating Schwann cells for the injured and dysmyelinated nervous system. J Neurosci. 2006;26(24):6651–60. doi: https://doi.org/10.1523/JNEUROSCI.1007-06.2006. 26/24/6651 [pii]. Epub 2006/06/16. PubMed PMID: 16775154.
202. Tsai SY, Bouwman BA, Ang YS, Kim SJ, Lee DF, Lemischka IR, et al. Single transcription factor reprogramming of hair follicle dermal papilla cells to induced pluripotent stem cells. Stem Cells 2011;29(6):964–971. doi: https://doi.org/10.1002/stem.649. Epub 2011/05/13. PubMed PMID: 21563278.
203. Jahoda CA, Whitehouse J, Reynolds AJ, Hole N. Hair follicle dermal cells differentiate into adipogenic and osteogenic lineages. Exp Dermatol 2003;12(6):849–859. Epub 2004/01/13. PubMed PMID: 14714566.
204. Kumamoto T, Shalhevet D, Matsue H, Mummert ME, Ward BR, Jester JV, et al. Hair follicles serve as local reservoirs of skin mast cell precursors. Blood 2003;102(5):1654–1660. doi:https://doi.org/10.1182/blood-2003-02-0449. 2003-02-0449 [pii]. Epub 2003/05/10. PubMed PMID: 12738661.
205. Elliott K, Stephenson TJ, Messenger AG. Differences in hair follicle dermal papilla volume are due to extracellular matrix volume and cell number: implications for the control of hair follicle size and androgen responses. J Invest Dermatol 1999;113(6):873–877. doi:https://doi.org/10.1046/j.1523-1747.1999.00797.x. jid797 [pii]. Epub 1999/12/14. PubMed PMID: 10594724.
206. Jahoda CA, Reynolds AJ. Hair follicle dermal sheath cells: unsung participants in wound healing. Lancet 2001;358(9291):1445–1448. doi:https://doi.org/10.1016/S0140-6736(01)06532-1. S0140-6736(01)06532-1 [pii]. Epub 2001/11/14. PubMed PMID: 11705511.
207. Gharzi A, Reynolds AJ, Jahoda CA. Plasticity of hair follicle dermal cells in wound healing and induction. Exp Dermatol 2003;12(2):126–136. Epub 2003/04/19. doi: 106 [pii]. PubMed PMID: 12702140.
208. MacQuarrie KL, Fong AP, Morse RH, Tapscott SJ. Genome-wide transcription factor binding: beyond direct target regulation. Trends Genet. 2011;27(4):141–8. doi: https://doi.org/10.1016/j.tig.2011.01.001. PubMed PMID: 21295369; PubMed Central PMCID: PMCPMC3068217.

209. Zaret KS, Mango SE. Pioneer transcription factors, chromatin dynamics, and cell fate control. Curr Opin Genet Dev. 2016;37:76–81. doi: https://doi.org/10.1016/j.gde.2015.12.003. PubMed PMID: 26826681; PubMed Central PMCID: PMCPMC4914445.
210. Hnisz D, Shrinivas K, Young RA, Chakraborty AK, Sharp PA. A Phase Separation Model for Transcriptional Control. Cell. 2017;169(1):13–23. doi: https://doi.org/10.1016/j.cell.2017.02.007. PubMed PMID: 28340338; PubMed Central PMCID: PMCPMC5432200.
211. Lanctot C, Cheutin T, Cremer M, Cavalli G, Cremer T. Dynamic genome architecture in the nuclear space: regulation of gene expression in three dimensions. Nat Rev Genet. 2007;8(2):104–15.
212. Misteli T. Beyond the sequence: cellular organization of genome function. Cell 2007;128(4):787–800. doi:https://doi.org/10.1016/j.cell.2007.01.028. Epub 2007/02/27. S0092-8674(07)00126-2 [pii]. PubMed PMID: 17320514.
213. Hubner MR, Spector DL. Chromatin dynamics. Annu Rev Biophys. 2010;39:471–89. Epub 2010/05/14. doi: https://doi.org/10.1146/annurev.biophys.093008.131348. PubMed PMID: 20462379; PubMed Central PMCID: PMCPMC2894465.
214. Pederson T. The nucleus introduced. Cold Spring Harb Perspect Biol. 2011;3(5). doi: https://doi.org/10.1101/cshperspect.a000521. a000521 [pii]. cshperspect.a000521 [pii]. Epub 2010/07/28. PubMed PMID: 20660024; PubMed Central PMCID: PMCPMC3101846.
215. Mao YS, Zhang B, Spector DL. Biogenesis and function of nuclear bodies. Trends Genet 2011;27(8):295–306. doi:https://doi.org/10.1016/j.tig.2011.05.006. Epub 2011/06/18. S0168-9525(11)00076-X [pii]. PubMed PMID: 21680045; PubMed Central PMCID: PMC3144265.
216. Gdula MR, Poterlowicz K, Mardaryev AN, Sharov AA, Peng Y, Fessing MY, et al. Remodelling of three-dimensional Organization of the Nucleus during Terminal Keratinocyte Differentiation in the epidermis. J Invest Dermatol. 2013. doi:https://doi.org/10.1038/jid.2013.66. jid201366 [pii]. Epub 2013/02/15. PubMed PMID: 23407401.
217. Dechat T, Adam SA, Taimen P, Shimi T, Goldman RD. Nuclear lamins. Cold Spring Harb Perspect Biol 2010;2(11):a000547. doi:https://doi.org/10.1101/cshperspect.a000547. Epub 2010/09/10. cshperspect.a000547 [pii]. PubMed PMID: 20826548.
218. Hetzer MW. The nuclear envelope. Cold Spring Harb Perspect Biol. 2010;2(3):a000539. doi: https://doi.org/10.1101/cshperspect.a000539. Epub 2010/03/20. PubMed PMID: 20300205; PubMed Central PMCID: PMCPMC2829960.
219. Wilson KL, Foisner R. Lamin-binding Proteins. Cold Spring Harb Perspect Biol 2010;2(4):a000554. doi:https://doi.org/10.1101/cshperspect.a000554. Epub 2010/05/11. cshperspect.a000554 [pii]. PubMed PMID: 20452940.
220. Capelson M, Liang Y, Schulte R, Mair W, Wagner U, Hetzer MW. Chromatin-bound nuclear pore components regulate gene expression in higher eukaryotes. Cell. 2010;140(3):372–83. doi: https://doi.org/10.1016/j.cell.2009.12.054. S0092–8674(09)01681-X [pii]. Epub 2010/02/11. PubMed PMID: 20144761; PubMed Central PMCID: PMCPMC2821818.
221. Hoelz A, Debler EW, Blobel G. The structure of the nuclear pore complex. Annu Rev Biochem 2011;80:613–643. doi: https://doi.org/10.1146/annurev-biochem-060109-151030. Epub 2011/04/19. PubMed PMID: 21495847.
222. Neumann S, Schneider M, Daugherty RL, Gottardi CJ, Eming SA, Beijer A, et al. Nesprin-2 interacts with {alpha}-catenin and regulates Wnt signaling at the nuclear envelope. J Biol Chem. 2010;285(45):34932–8. doi: https://doi.org/10.1074/jbc.M110.119651. M110.119651 [pii]. Epub 2010/08/31. PubMed PMID: 20801886; PubMed Central PMCID: PMCPMC2966107.
223. Kubben N, Voncken JW, Demmers J, Calis C, van Almen G, Pinto Y, et al. Identification of differential protein interactors of lamin A and progerin. Nucleus. 2010;1(6):513–25. doi: https://doi.org/10.4161/nucl.1.6.13512. Epub 2011/02/18. PubMed PMID: 21327095; PubMed Central PMCID: PMCPMC3027055.
224. Constantinescu D, Gray HL, Sammak PJ, Schatten GP, Csoka AB. Lamin A/C expression is a marker of mouse and human embryonic stem cell differentiation. Stem Cells 2006;24(1):177–

185. doi:https://doi.org/10.1634/stemcells.2004-0159. Epub 2005/09/24. 2004-0159 [pii]. PubMed PMID: 16179429.
225. Hanif M, Rosengardten Y, Sagelius H, Rozell B, Eriksson M. Differential expression of A-type and B-type lamins during hair cycling. PLoS One. 2009;4(1):e4114. doi: https://doi.org/10.1371/journal.pone.0004114. Epub 2009/01/06. PubMed PMID: 19122810; PubMed Central PMCID: PMCPMC2606029.
226. Shevelyov YY, Lavrov SA, Mikhaylova LM, Nurminsky ID, Kulathinal RJ, Egorova KS, et al. The B-type lamin is required for somatic repression of testis-specific gene clusters. Proc Natl Acad Sci U S A. 2009;106(9):3282–7. doi:https://doi.org/10.1073/pnas.0811933106. Epub 2009/02/17. 0811933106 [pii]. PubMed PMID: 19218438; PubMed Central PMCID: PMCPMC2651240.
227. Pan D, Estevez-Salmeron LD, Stroschein SL, Zhu X, He J, Zhou S, et al. The integral inner nuclear membrane protein MAN1 physically interacts with the R-Smad proteins to repress signaling by the transforming growth factor-{beta} superfamily of cytokines. J Biol Chem 2005;280(16):15992–16001. doi:https://doi.org/10.1074/jbc.M411234200. Epub 2005/01/14. M411234200 [pii]. PubMed PMID: 15647271.
228. Markiewicz E, Tilgner K, Barker N, van de Wetering M, Clevers H, Dorobek M, et al. The inner nuclear membrane protein emerin regulates beta-catenin activity by restricting its accumulation in the nucleus. EMBO J. 2006;25(14):3275–85. doi: https://doi.org/10.1038/sj.emboj.7601230. Epub 2006/07/22. 7601230 [pii]. PubMed PMID: 16858403; PubMed Central PMCID: PMCPMC1523183.
229. Worman HJ, Fong LG, Muchir A, Young SG. Laminopathies and the long strange trip from basic cell biology to therapy. J Clin Invest. 2009;119(7):1825–36. doi: https://doi.org/10.1172/JCI37679. 37679 [pii]. Epub 2009/07/10. PubMed PMID: 19587457; PubMed Central PMCID: PMCPMC2701866.
230. Maraldi NM, Capanni C, Cenni V, Fini M, Lattanzi G. Laminopathies and Lamin-associated signaling pathways. J Cell Biochem 2011;112(4):979–992. doi: https://doi.org/10.1002/jcb.22992. Epub 2011/03/15. PubMed PMID: 21400569.
231. Scaffidi P, Misteli T. Reversal of the cellular phenotype in the premature aging disease Hutchinson-Gilford progeria syndrome. Nat Med. 2005;11(4):440–5. doi:https://doi.org/10.1038/nm1204. Epub 2005/03/08. nm1204 [pii]. PubMed PMID: 15750600; PubMed Central PMCID: PMCPMC1351119.
232. McClintock D, Ratner D, Lokuge M, Owens DM, Gordon LB, Collins FS, et al. The mutant form of lamin A that causes Hutchinson-Gilford progeria is a biomarker of cellular aging in human skin. PLoS One. 2007;2(12):e1269. doi: https://doi.org/10.1371/journal.pone.0001269. Epub 2007/12/07. PubMed PMID: 18060063; PubMed Central PMCID: PMCPMC2092390.
233. Scaffidi P, Misteli T. Lamin A-dependent misregulation of adult stem cells associated with accelerated ageing. Nat Cell Biol. 2008;10(4):452–9. doi:https://doi.org/10.1038/ncb1708. Epub 2008/03/04. ncb1708 [pii]. PubMed PMID: 18311132; PubMed Central PMCID: PMCPMC2396576.
234. Wang Y, Panteleyev AA, Owens DM, Djabali K, Stewart CL, Worman HJ. Epidermal expression of the truncated prelamin A causing Hutchinson-Gilford progeria syndrome: effects on keratinocytes, hair and skin. Hum Mol Genet. 2008;17(15):2357–69. doi: https://doi.org/10.1093/hmg/ddn136. Epub 2008/04/30. ddn136 [pii]. PubMed PMID: 18442998; PubMed Central PMCID: PMCPMC2733813.
235. Sagelius H, Rosengardten Y, Hanif M, Erdos MR, Rozell B, Collins FS, et al. Targeted transgenic expression of the mutation causing Hutchinson-Gilford progeria syndrome leads to proliferative and degenerative epidermal disease. J Cell Sci 2008;121(Pt 7):969–978. doi:https://doi.org/10.1242/jcs.022913. Epub 2008/03/13. jcs.022913 [pii]. PubMed PMID: 18334552.
236. Rapisarda V, Malashchuk I, Asamaowei IE, Poterlowicz K, Fessing MY, Sharov AA, et al. p63 transcription factor regulates nuclear shape and expression of nuclear envelope-associated genes in epidermal keratinocytes. J Invest Dermatol 2017. doi: https://doi.org/10.1016/j.jid.2017.05.013. PubMed PMID: 28595999.

237. McKeown PC, Shaw PJ. Chromatin: linking structure and function in the nucleolus. Chromosoma 2009;118(1):11–23. doi: https://doi.org/10.1007/s00412-008-0184-2. Epub 2008/10/18. PubMed PMID: 18925405.
238. Bartova E, Horakova AH, Uhlirova R, Raska I, Galiova G, Orlova D, et al. Structure and epigenetics of nucleoli in comparison with non-nucleolar compartments. J Histochem Cytochem. 2010;58(5):391–403. doi: https://doi.org/10.1369/jhc.2009.955435. Epub 2009/12/23. jhc.2009.955435 [pii]. PubMed PMID: 20026667; PubMed Central PMCID: PMCPMC2857811.
239. Boulon S, Westman BJ, Hutten S, Boisvert FM, Lamond AI. The nucleolus under stress. Mol Cell. 2010;40(2):216–27. doi: https://doi.org/10.1016/j.molcel.2010.09.024. Epub 2010/10/23. S1097-2765(10)00752-5 [pii]. PubMed PMID: 20965417; PubMed Central PMCID: PMCPMC2987465.
240. Hernandez-Verdun D. Nucleolus: from structure to dynamics. Histochem Cell Biol 2006;125(1–2):127–137. doi: https://doi.org/10.1007/s00418-005-0046-4. Epub 2005/12/06. PubMed PMID: 16328431.
241. Boisvert FM, van Koningsbruggen S, Navascues J, Lamond AI. The multifunctional nucleolus. Nat Rev Mol Cell Biol 2007;8(7):574–585. doi:https://doi.org/10.1038/nrm2184. Epub 2007/05/24. nrm2184 [pii]. PubMed PMID: 17519961.
242. Cisterna B, Biggiogera M. Ribosome biogenesis: from structure to dynamics. Int Rev Cell Mol Biol 2010;284:67–111. doi:https://doi.org/10.1016/S1937-6448(10)84002-X. Epub 2010/09/30. doi: S1937-6448(10)84002-X [pii]. PubMed PMID: 20875629.
243. Karasek J, Smetana K, Hrdlicka A, Dubinin I, Hornak O, Ohelert W. Nuclear and nucleolar ultrastructure during the late stages of normal human keratinocyte maturation. Br J Dermatol 1972;86(6):601–607. Epub 1972/06/01. PubMed PMID: 5045951.
244. Tsuji T, Cox AJ. Ultrastructural studies of nuclei and nucleoli in psoriatic epidermis compared to those in normal epidermis. J Invest Dermatol. 1977;69:205–10.
245. Nemeth A, Conesa A, Santoyo-Lopez J, Medina I, Montaner D, Peterfia B, et al. Initial genomics of the human nucleolus. PLoS Genet 2010;6(3):e1000889. doi:https://doi.org/10.1371/journal.pgen.1000889. Epub 2010/04/03. PubMed PMID: 20361057; PubMed Central PMCID: PMC2845662.
246. Sasaki M, Kawahara K, Nishio M, Mimori K, Kogo R, Hamada K, et al. Regulation of the MDM2-P53 pathway and tumor growth by PICT1 via nucleolar RPL11. Nat Med 2011;17(8):944–951. doi: https://doi.org/10.1038/nm.2392. nm.2392 [pii]. Epub 2011/08/02. PubMed PMID: 21804542.
247. Edwards MJ, Thomas RC. Protein phosphatase type 1-dependent dephosphorylation of the retinoblastoma tumor suppressor protein in ultraviolet-irradiated human skin and keratinocytes. J Invest Dermatol 2000;115(1):88–94. doi:https://doi.org/10.1046/j.1523--1747.2000.00026.x. Epub 2000/07/08. jid026 [pii]. PubMed PMID: 10886513.
248. Kahlina K, Goren I, Pfeilschifter J, Frank S. p68 DEAD box RNA helicase expression in keratinocytes. Regulation, nucleolar localization, and functional connection to proliferation and vascular endothelial growth factor gene expression. J Biol Chem 2004;279(43):44872–44882. doi: https://doi.org/10.1074/jbc.M402467200. Epub 2004/08/12. M402467200 [pii]. PubMed PMID: 15304501.
249. Nicholls PK, Doorbar J, Moore RA, Peh W, Anderson DM, Stanley MA. Detection of viral DNA and E4 protein in basal keratinocytes of experimental canine oral papillomavirus lesions. Virology 2001;284(1):82–98. doi:https://doi.org/10.1006/viro.2001.0868. S0042-6822(01)90868-2 [pii]. Epub 2001/05/16. PubMed PMID: 11352670.
250. Bohl J, Hull B, Vande Pol SB. Cooperative transformation and coexpression of bovine papillomavirus type 1 E5 and E7 proteins. J Virol. 2001;75(1):513–21. doi: https://doi.org/10.1128/JVI.75.1.513-521.2001. Epub 2000/12/19. PubMed PMID: 11119620; PubMed Central PMCID: PMCPMC113944.
251. Pollock C, Huang S. The perinucleolar compartment. Cold Spring Harb Perspect Biol. 2010;2(2):a000679. doi: https://doi.org/10.1101/cshperspect.a000679. PubMed PMID: 20182614; Epub 2010/02/26. PubMed Central PMCID: PMCPMC2828281.

252. Hall LL, Smith KP, Byron M, Lawrence JB. Molecular anatomy of a speckle. Anat Rec A Discov Mol Cell Evol Biol. 2006;288((7):664–75.
253. Spector DL, Lamond AI. Nuclear speckles. Cold Spring Harb Perspect Biol. 2011;3(2). doi: https://doi.org/10.1101/cshperspect.a000646. a000646 [pii]. Epub 2010/10/12. cshperspect. a000646 [pii]. PubMed PMID: 20926517.
254. Brown JM, Green J, das Neves RP, Wallace HA, Smith AJ, Hughes J, et al. Association between active genes occurs at nuclear speckles and is modulated by chromatin environment. J Cell Biol 2008;182(6):1083–1097. doi: https://doi.org/10.1083/jcb.200803174. Epub 2008/09/24. jcb.200803174 [pii]. PubMed PMID: 18809724.; PubMed Central PMCID: PMC2542471.
255. Sharma A, Takata H, Shibahara K, Bubulya A, Bubulya PA. Son is essential for nuclear speckle organization and cell cycle progression. Mol Biol Cell. 2010;21(4):650–63. doi: https://doi.org/10.1091/mbc.E09-02-0126. E09-02-0126 [pii]. Epub 2010/01/08. PubMed PMID: 20053686; PubMed Central PMCID: PMCPMC2820428.
256. Saitoh N, Spahr CS, Patterson SD, Bubulya P, Neuwald AF, Spector DL. Proteomic analysis of interchromatin granule clusters. Mol Biol Cell. 2004;15(8):3876–90. doi: https://doi.org/10.1091/mbc.E04-03-0253. E04-03-0253 [pii]. Epub 2004/06/01. PubMed PMID: 15169873; PubMed Central PMCID: PMCPMC491843.
257. Nizami Z, Deryusheva S, Gall JG. The Cajal body and histone locus body. Cold Spring Harb Perspect Biol 2010;2(7):a000653. doi:https://doi.org/10.1101/cshperspect.a000653. Epub 2010/05/28. cshperspect.a000653 [pii]. PubMed PMID: 20504965.
258. Dundr M, Hebert MD, Karpova TS, Stanek D, Xu H, Shpargel KB, et al. In vivo kinetics of Cajal body components. J Cell Biol. 2004;164(6):831–42. doi: https://doi.org/10.1083/jcb.200311121. jcb.200311121 [pii]. Epub 2004/03/17. PubMed PMID: 15024031; PubMed Central PMCID: PMCPMC1630494.
259. Strzelecka M, Oates AC, Neugebauer KM. Dynamic control of Cajal body number during zebrafish embryogenesis. Nucleus. 2010;1(1):96–108. doi: https://doi.org/10.4161/nucl.1.1.10680. Epub 2011/02/18. PubMed PMID: 21327108; PubMed Central PMCID: PMCPMC3035118.
260. Tomlinson RL, Abreu EB, Ziegler T, Ly H, Counter CM, Terns RM, et al. Telomerase reverse transcriptase is required for the localization of telomerase RNA to cajal bodies and telomeres in human cancer cells. Mol Biol Cell. 2008;19(9):3793–800. doi: https://doi.org/10.1091/mbc.E08-02-0184. E08-02-0184 [pii]. Epub 2008/06/20. PubMed PMID: 18562689; PubMed Central PMCID: PMCPMC2526684.
261. Zhu Y, Tomlinson RL, Lukowiak AA, Terns RM, Terns MP. Telomerase RNA accumulates in Cajal bodies in human cancer cells. Mol Biol Cell. 2004;15(1):81–90. Epub 2003/10/07. doi: https://doi.org/10.1091/mbc.E03-07-0525. E03-07-0525 [pii]. PubMed PMID: 14528011; PubMed Central PMCID: PMCPMC307529.
262. Lallemand-Breitenbach V, de The H. PML nuclear bodies. Cold Spring Harb Perspect Biol. 2010;2(5):a000661. doi: https://doi.org/10.1101/cshperspect.a000661. cshperspect. a000661 [pii]. Epub 2010/05/11. PubMed PMID: 20452955; PubMed Central PMCID: PMCPMC2857171.
263. Ishov AM, Sotnikov AG, Negorev D, Vladimirova OV, Neff N, Kamitani T, et al. PML is critical for ND10 formation and recruits the PML-interacting protein daxx to this nuclear structure when modified by SUMO-1. J Cell Biol. 1999;147(2):221–34. Epub 1999/10/20. PubMed PMID: 10525530; PubMed Central PMCID: PMCPMC2174231.
264. Bernardi R, Pandolfi PP. Structure, dynamics and functions of promyelocytic leukaemia nuclear bodies. Nat Rev Mol Cell Biol 2007;8(12):1006–1016. doi:https://doi.org/10.1038/nrm2277. Epub 2007/10/12. nrm2277 [pii]. PubMed PMID: 17928811.
265. Lang M, Jegou T, Chung I, Richter K, Munch S, Udvarhelyi A, et al. Three-dimensional organization of promyelocytic leukemia nuclear bodies. J Cell Sci. 2010;123(Pt 3):392–400. doi: https://doi.org/10.1242/jcs.053496. Epub 2010/02/05. 123/3/392 [pii]. PubMed PMID: 20130140.

266. Lindsay CR, Morozov VM, Ishov AM. PML NBs (ND10) and Daxx: from nuclear structure to protein function. Front Biosci 2008;13:7132–7142. Epub 2008/05/30. doi: 3216 [pii]. PubMed PMID: 18508722.
267. Dellaire G, Bazett-Jones DP. PML nuclear bodies: dynamic sensors of DNA damage and cellular stress. BioEssays 2004;26(9):963–977. Epub 2004/09/08. doi: https://doi.org/10.1002/bies.20089. PubMed PMID: 15351967.
268. Hansen LA, Brown D, Virador V, Tanaka T, Andreola F, Strain K, et al. A PMLRARA transgene results in a retinoid-deficient phenotype associated with enhanced susceptibility to skin tumorigenesis. Cancer Res 2003;63(17):5257–5265. Epub 2003/09/23. PubMed PMID: 14500356.
269. Virador VM, Flores-Obando RE, Berry A, Patel R, Zakhari J, Lo YC, et al. The human promyelocytic leukemia protein is a tumor suppressor for murine skin carcinogenesis. Mol Carcinog. 2009;48(7):599–609. doi: https://doi.org/10.1002/mc.20498. Epub 2008/12/06. PubMed PMID: 19058256; PubMed Central PMCID: PMCPMC2711034.
270. Cremer M, von Hase J, Volm T, Brero A, Kreth G, Walter J, et al. Non-random radial higher-order chromatin arrangements in nuclei of diploid human cells. Chromosom Res 2001;9(7):541–567. Epub 2001/11/28. PubMed PMID: 11721953.
271. Cremer T, Cremer M. Chromosome territories. In: Misteli T, Spector DL, editors. The nucleus. Cold-Spring Harbor: Cold Spring Harbor Laboratory Press; 2011. p. 93–114.
272. Cremer T, Cremer M, Hubner B, Strickfaden H, Smeets D, Popken J, et al. The 4D nucleome: evidence for a dynamic nuclear landscape based on co-aligned active and inactive nuclear compartments. FEBS Lett 2015. doi:https://doi.org/10.1016/j.febslet.2015.05.037. S0014-5793(15)00431-7 [pii]. Epub 2015/06/02. PubMed PMID: 26028501.
273. Hemberger M, Dean W, Reik W. Epigenetic dynamics of stem cells and cell lineage commitment: digging Waddington's canal. Nat Rev Mol Cell Biol 2009;10(8):526–537. Epub 2009 Jul 15.
274. Sequeira-Mendes J, Gutierrez C. Genome architecture: from linear organisation of chromatin to the 3D assembly in the nucleus. Chromosoma 2015. doi: https://doi.org/10.1007/s00412-015-0538-5. Epub 2015/09/04. PubMed PMID: 26330112.
275. Fukui K. Structural analyses of chromosomes and their constituent proteins. Cytogenet Genome Res 2009;124(3–4):215–227. doi:https://doi.org/10.1159/000218127. 000218127 [pii]. Epub 2009/06/27. PubMed PMID: 19556775.
276. Cremer T, Cremer C. Chromosome territories, nuclear architecture and gene regulation in mammalian cells. Nat Rev Genet 2001;2(4):292–301. doi: https://doi.org/10.1038/35066075. 35066075 [pii]. Epub 2001/04/03. PubMed PMID: 11283701.
277. Solovei I, Cremer M. 3D-FISH on cultured cells combined with immunostaining. Methods Mol Biol 2010;659:117–126. doi: https://doi.org/10.1007/978-1-60761-789-1_8. Epub 2010/09/03. PubMed PMID: 20809307.
278. Cremer T, Cremer C. Rise, fall and resurrection of chromosome territories: a historical perspective. Part II. Fall and resurrection of chromosome territories during the 1950s to 1980s. Part III. Chromosome territories and the functional nuclear architecture: experiments and models from the 1990s to the present. Eur J Histochem 2006;50(4):223–272. Epub 2007/01/11. PubMed PMID: 17213034.
279. Cremer T, Cremer C. Rise, fall and resurrection of chromosome territories: a historical perspective. Part I. The rise of chromosome territories. Eur J Histochem 2006;50(3):161–176. Epub 2006/08/22. PubMed PMID: 16920639.
280. Cremer T, Cremer M. Chromosome territories. Cold Spring Harb Perspect Biol 2010;2(3):a003889. doi: https://doi.org/10.1101/cshperspect.a003889. Epub 2010/03/20. PubMed PMID: 20300217; PubMed Central PMCID: PMC2829961.
281. Joffe B, Leonhardt H, Solovei I. Differentiation and large scale spatial organization of the genome. Curr Opin Genet Dev 2010;20(5):562–569. doi:https://doi.org/10.1016/j.gde.2010.05.009. S0959-437X(10)00098-5 [pii]. Epub 2010/06/22. PubMed PMID: 20561778.

282. Kind J, van Steensel B. Stochastic genome-nuclear lamina interactions: modulating roles of Lamin A and BAF. Nucleus. 2014;5(2):124–30. doi: https://doi.org/10.4161/nucl.28825. 28825 [pii]. Epub 2014/04/11. PubMed PMID: 24717229; PubMed Central PMCID: PMCPMC4049918.
283. Roix JJ, McQueen PG, Munson PJ, Parada LA, Misteli T. Spatial proximity of translocation-prone gene loci in human lymphomas. Nat Genet 2003;34(3):287–291. doi:https://doi.org/10.1038/ng1177. ng1177 [pii]. Epub 2003/06/17. PubMed PMID: 12808455.
284. Parada LA, Sotiriou S, Misteli T. Spatial genome organization. Exp Cell Res 2004;296(1):64–70. doi: https://doi.org/10.1016/j.yexcr.2004.03.013. S0014482704001296 [pii]. Epub 2004/05/04. PubMed PMID: 15120995.
285. Brianna Caddle L, Grant JL, Szatkiewicz J, van Hase J, Shirley BJ, Bewersdorf J, et al. Chromosome neighborhood composition determines translocation outcomes after exposure to high-dose radiation in primary cells. Chromosom Res 2007;15(8):1061–1073. doi: https://doi.org/10.1007/s10577-007-1181-7. Epub 2007/12/07. PubMed PMID: 18060570.
286. Khalil A, Grant JL, Caddle LB, Atzema E, Mills KD, Arneodo A. Chromosome territories have a highly nonspherical morphology and nonrandom positioning. Chromosom Res 2007;15(7):899–916. Epub 2007/10/11. doi: https://doi.org/10.1007/s10577-007-1172-8. PubMed PMID: 17926137.
287. Schermelleh L, Heintzmann R, Leonhardt H. A guide to super-resolution fluorescence microscopy. J Cell Biol. 2010;190(2):165–75. doi: https://doi.org/10.1083/jcb.201002018. jcb.201002018 [pii]. Epub 2010/07/21. PubMed PMID: 20643879; PubMed Central PMCID: PMCPMC2918923.
288. Markaki Y, Gunkel M, Schermelleh L, Beichmanis S, Neumann J, Heidemann M, et al. Functional Nuclear Organization of Transcription and DNA Replication: A Topographical Marriage between Chromatin Domains and the Interchromatin Compartment. Cold Spring Harb Symp Quant Biol. 2011. doi: https://doi.org/10.1101/sqb.2010.75.042. sqb.2010.75.042 [pii]. Epub 2011/04/07. PubMed PMID: 21467142.
289. Jost KL, Haase S, Smeets D, Schrode N, Schmiedel JM, Bertulat B, et al. 3D-Image analysis platform monitoring relocation of pluripotency genes during reprogramming. Nucleic Acids Res. 2011;39(17):e113. doi: https://doi.org/10.1093/nar/gkr486. gkr486 [pii]. Epub 2011/06/28. PubMed PMID: 21700670; PubMed Central PMCID: PMCPMC3177216.
290. Williams RR, Broad S, Sheer D, Ragoussis J. Subchromosomal positioning of the epidermal differentiation complex (EDC) in keratinocyte and lymphoblast interphase nuclei. Exp Cell Res 2002;272(2):163–175. doi: https://doi.org/10.1006/excr.2001.5400. S001448270195400X [pii]. Epub 2002/01/05. PubMed PMID: 11777341.
291. Roukos V, Misteli T. The biogenesis of chromosome translocations. Nat Cell Biol 2014;16(4):293–300. doi: https://doi.org/10.1038/ncb2941. ncb2941 [pii]. Epub 2014/04/03. PubMed PMID: 24691255.
292. Gomez-Ospina N, Chang AL, Qu K, Oro AE. Translocation affecting sonic hedgehog genes in basal-cell carcinoma. N Engl J Med 2012;366(23):2233–2234. doi: https://doi.org/10.1056/NEJMc1115123. PubMed PMID: 22670922; PubMed Central PMCID: PMCPMC3839666.
293. Dekker J, Rippe K, Dekker M, Kleckner N. Capturing chromosome conformation. Science 2002;295(5558):1306–1311. doi: https://doi.org/10.1126/science.1067799.295/5558/1306. [pii]. Epub 2002/02/16. PubMed PMID: 11847345.
294. Dekker J, Marti-Renom MA, Mirny LA. Exploring the three-dimensional organization of genomes: interpreting chromatin interaction data. Nat Rev Genet 2013;14(6):390–403. Epub 2013/05/10. doi: https://doi.org/10.1038/nrg3454. nrg3454 [pii]. PubMed PMID: 23657480.
295. Lajoie BR, Dekker J, Kaplan N. The Hitchhiker's guide to Hi-C analysis: practical guidelines. Methods. 2015;72:65–75. doi: https://doi.org/10.1016/j.ymeth.2014.10.031. S1046-2023(14)00358-2 [pii]. Epub 2014/12/03. PubMed PMID: 25448293; PubMed Central PMCID: PMCPMC4347522.
296. de Laat W, Dekker J. 3C-based technologies to study the shapc of the genome. Methods. 2012;58(3):189–91. doi: https://doi.org/10.1016/j.ymeth.2012.11.005. S1046-

2023(12)00291-5 [pii]. Epub 2012/12/04. PubMed PMID: 23199640; PubMed Central PMCID: PMCPMC3874834.
297. Lieberman-Aiden E, van Berkum NL, Williams L, Imakaev M, Ragoczy T, Telling A, et al. Comprehensive mapping of long-range interactions reveals folding principles of the human genome. Science. 2009;326(5950):289–93.
298. Popken J, Brero A, Koehler D, Schmid VJ, Strauss A, Wuensch A, et al. Reprogramming of fibroblast nuclei in cloned bovine embryos involves major structural remodeling with both striking similarities and differences to nuclear phenotypes of in vitro fertilized embryos. Nucleus 2014;5(6):555–589. doi: https://doi.org/10.4161/19491034.2014.979712. Epub 2014/12/09. PubMed PMID: 25482066.
299. Dekker J, Heard E. Structural and functional diversity of Topologically Associating Domains. FEBS Lett. 2015. doi:https://doi.org/10.1016/j.febslet.2015.08.044. S0014-5793(15)00814-5 [pii]. Epub 2015/09/09. PubMed PMID: 26348399.
300. Nora EP, Lajoie BR, Schulz EG, Giorgetti L, Okamoto I, Servant N, et al. Spatial partitioning of the regulatory landscape of the X-inactivation centre. Nature 2012;485(7398):381–385. doi: https://doi.org/10.1038/nature11049. nature11049 [pii]. Epub 2012/04/13. PubMed PMID: 22495304.
301. Dixon JR, Selvaraj S, Yue F, Kim A, Li Y, Shen Y, et al. Topological domains in mammalian genomes identified by analysis of chromatin interactions. Nature. 2012;485(7398):376–80. doi: https://doi.org/10.1038/nature11082. nature11082 [pii]. Epub 2012/04/13. PubMed PMID: 22495300; PubMed Central PMCID: PMCPMC3356448.
302. Naumova N, Imakaev M, Fudenberg G, Zhan Y, Lajoie BR, Mirny LA, et al. Organization of the mitotic chromosome. Science 2013;342(6161):948–953. doi: https://doi.org/10.1126/science.1236083. science.1236083 [pii]. Epub 2013/11/10. PubMed PMID: 24200812.
303. Gomez-Diaz E, Corces VG. Architectural proteins: regulators of 3D genome organization in cell fate. Trends Cell Biol. 2014;24(11):703–11. doi: https://doi.org/10.1016/j.tcb.2014.08.003. S0962-8924(14)00135-4 [pii]. Epub 2014/09/15. PubMed PMID: 25218583; PubMed Central PMCID: PMCPMC4254322.
304. Rao SS, Huntley MH, Durand NC, Stamenova EK, Bochkov ID, Robinson JT, et al. A 3D map of the human genome at kilobase resolution reveals principles of chromatin looping. Cell 2014;159(7):1665–1680. doi:https://doi.org/10.1016/j.cell.2014.11.021. S0092-8674(14)01497-4 [pii]. Epub 2014/12/17. PubMed PMID: 25497547.
305. Cai S, Han HJ, Kohwi-Shigematsu T. Tissue-specific nuclear architecture and gene expression regulated by SATB1. Nat Genet. 2003;34(1):42–51.
306. Cai S, Lee CC, Kohwi-Shigematsu T. SATB1 packages densely looped, transcriptionally active chromatin for coordinated expression of cytokine genes. Nat Genet. 2006;38(11):1278–88. Epub 2006 Oct 22
307. Kohwi-Shigematsu T, Kohwi Y, Takahashi K, Richards HW, Ayers SD, Han HJ, et al. SATB1-mediated functional packaging of chromatin into loops. Methods. 2012;58(3):243–54. doi: https://doi.org/10.1016/j.ymeth.2012.06.019. S1046-2023(12)00154-5 [pii]. Epub 2012/07/12. PubMed PMID: 22782115; PubMed Central PMCID: PMCPMC4029128.
308. Kohwi-Shigematsu T, Poterlowicz K, Ordinario E, Han HJ, Botchkarev VA, Kohwi Y. Genome organizing function of SATB1 in tumor progression. Semin Cancer Biol. 2013;23(2):72–9. doi: https://doi.org/10.1016/j.semcancer.2012.06.009. S1044-579X(12)00101-0 [pii]. Epub 2012/07/10. PubMed PMID: 22771615; PubMed Central PMCID: PMCPMC4048061.
309. Poterlowicz K, Yarker JL, Malashchuk I, Lajoie BR, Mardaryev AN, Gdula MR, Sharov AA, Kohwi-Shigematsu T, Botchkarev VA, Fessing MY. 5C analysis of the Epidermal Differentiation Complex locus reveals distinct chromatin interaction networks between gene-rich and gene-poor TADs in skin epithelial cells. PLoS Genet. 2017;13(9):e1006966
310. Dowen JM, Fan ZP, Hnisz D, Ren G, Abraham BJ, Zhang LN, et al. Control of cell identity genes occurs in insulated neighborhoods in mammalian chromosomes. Cell. 2014;159(2):374–87. doi: https://doi.org/10.1016/j.cell.2014.09.030. S0092-8674(14)01179-9 [pii]. Epub 2014/10/11. PubMed PMID: 25303531; PubMed Central PMCID: PMCPMC4197132.

311. de Laat W, Duboule D. Topology of mammalian developmental enhancers and their regulatory landscapes. Nature 2013;502(7472):499–506. doi: https://doi.org/10.1038/nature12753. nature12753 [pii]. Epub 2013/10/25. PubMed PMID: 24153303.
312. Oh IY, Albea DM, Goodwin ZA, Quiggle AM, Baker BP, Guggisberg AM, et al. Regulation of the dynamic chromatin architecture of the epidermal differentiation complex is mediated by a c-Jun/AP-1-modulated enhancer. J Invest Dermatol 2014. doi: https://doi.org/10.1038/jid.2014.44. jid201444 [pii]. Epub 2014/01/29. PubMed PMID: 24468747.
313. Adoue V, Chavanas S, Coudane F, Mechin MC, Caubet C, Ying S, et al. Long-range enhancer differentially regulated by c-Jun and JunD controls peptidylarginine deiminase-3 gene in keratinocytes. J Mol Biol 2008;384(5):1048–1057. doi: https://doi.org/10.1016/j.jmb.2008.10.019. S0022-2836(08)01287-4 [pii]. Epub 2008/10/28. PubMed PMID: 18952102.
314. Gierman HJ, Indemans MH, Koster J, Goetze S, Seppen J, Geerts D, et al. Domain-wide regulation of gene expression in the human genome. Genome Res. 2007;17(9):1286–95. doi: https://doi.org/10.1101/gr.6276007. gr.6276007 [pii]. Epub 2007/08/19. PubMed PMID: 17693573; PubMed Central PMCID: PMCPMC1950897.
315. Truong AB, Khavari PA. Control of keratinocyte proliferation and differentiation by p63. Cell Cycle 2007;6(3):295–299. Epub 2007/02/01. doi: 3753 [pii]. PubMed PMID: 17264679.
316. Klein RH, Andersen B. Dynamic networking for epidermal differentiation. Dev Cell 2015;32(6):661–662. doi: https://doi.org/10.1016/j.devcel.2015.03.006. PubMed PMID: 25805131.
317. Fuchs E. Epithelial skin biology: three decades of developmental biology, a hundred questions answered and a thousand new ones to address. Curr Top Dev Biol 2016;116:357–374. doi: https://doi.org/10.1016/bs.ctdb.2015.11.033. PubMed PMID: 26970628; PubMed Central PMCID: PMCPMC5026047.
318. Botchkarev VA. Integration of the transcription factor-regulated and epigenetic mechanisms in the control of keratinocyte differentiation. J Investig Dermatol Symp Proc 2015;17(2):30–32. doi: https://doi.org/10.1038/jidsymp.2015.37. PubMed PMID: 26551942; PubMed Central PMCID: PMCPMC5080658.
319. Botchkarev VA, Mardaryev AN. Repressing the keratinocyte genome: how the Polycomb complex subunits operate in concert to control skin and hair follicle development. J Invest Dermatol 2016;136(8):1538–1540. doi: https://doi.org/10.1016/j.jid.2016.04.026. PubMed PMID: 27450498.
320. Sethi I, Gluck C, Zhou H, Buck MJ, Sinha S. Evolutionary re-wiring of p63 and the epigenomic regulatory landscape in keratinocytes and its potential implications on species-specific gene expression and phenotypes. Nucleic Acids Res 2017. doi: https://doi.org/10.1093/nar/gkx416. PubMed PMID: 28505376.
321. Waikel RL, Kawachi Y, Waikel PA, Wang XJ, Roop DR. Deregulated expression of c-Myc depletes epidermal stem cells. Nat Genet 2001;28(2):165–168. doi: https://doi.org/10.1038/88889. 88889 [pii]. Epub 2001/05/31. PubMed PMID: 11381265.
322. Lo Celso C, Berta MA, Braun KM, Frye M, Lyle S, Zouboulis CC, et al. Characterization of bipotential epidermal progenitors derived from human sebaceous gland: contrasting roles of c-Myc and beta-catenin. Stem Cells 2008;26(5):1241–1252. doi: https://doi.org/10.1634/stemcells.2007-0651. 2007-0651 [pii]. Epub 2008/03/01. PubMed PMID: 18308950.
323. Nascimento EM, Cox CL, MacArthur S, Hussain S, Trotter M, Blanco S, et al. The opposing transcriptional functions of Sin3a and c-Myc are required to maintain tissue homeostasis. Nat Cell Biol. 2011;13(12):1395–405. doi: https://doi.org/10.1038/ncb2385. ncb2385 [pii]. Epub 2011/11/22. PubMed PMID: 22101514; PubMed Central PMCID: PMCPMC3242072.
324. Habib T, Park H, Tsang M, de Alboran IM, Nicks A, Wilson L, et al. Myc stimulates B lymphocyte differentiation and amplifies calcium signaling. J Cell Biol. 2007;179(4):717–31. doi: https://doi.org/10.1083/jcb.200704173. Epub 2007/11/14. PubMed PMID: 17998397; PubMed Central PMCID: PMCPMC2080907.

325. Knoepfler PS, Zhang XY, Cheng PF, Gafken PR, McMahon SB, Eisenman RN. Myc influences global chromatin structure. EMBO J. 2006;25(12):2723–34. doi: https://doi.org/10.1038/sj.emboj.7601152. 7601152 [pii]. Epub 2006/05/26. PubMed PMID: 16724113; PubMed Central PMCID: PMCPMC1500848.
326. Wurster AL, Pazin MJ. BRG1-mediated chromatin remodeling regulates differentiation and gene expression of T helper cells. Mol Cell Biol. 2008;28(24):7274–85. doi: https://doi.org/10.1128/MCB.00835-08. MCB.00835-08 [pii]. Epub 2008/10/15. PubMed PMID: 18852284; PubMed Central PMCID: PMCPMC2593447.
327. Pham LV, Tamayo AT, Li C, Bueso-Ramos C, Ford RJ. An epigenetic chromatin remodeling role for NFATc1 in transcriptional regulation of growth and survival genes in diffuse large B-cell lymphomas. Blood. 2010;116(19):3899–906. doi: https://doi.org/10.1182/blood-2009-12-257378. Epub 2010/07/29. PubMed PMID: 20664054; PubMed Central PMCID: PMCPMC2981542.
328. Swinstead EE, Paakinaho V, Presman DM, Hager GL. Pioneer factors and ATP-dependent chromatin remodeling factors interact dynamically: a new perspective: multiple transcription factors can effect chromatin pioneer functions through dynamic interactions with ATP-dependent chromatin remodeling factors. BioEssays 2016;38(11):1150–1157. doi: https://doi.org/10.1002/bies.201600137. PubMed PMID: 27633730.
329. Xu M, Luo W, Elzi DJ, Grandori C, Galloway DA. NFX1 interacts with mSin3A/histone deacetylase to repress hTERT transcription in keratinocytes. Mol Cell Biol. 2008;28(15):4819–28. doi: https://doi.org/10.1128/MCB.01969-07. MCB.01969-07 [pii]. Epub 2008/05/29. PubMed PMID: 18505829; PubMed Central PMCID: PMCPMC2493374.
330. Bansal N, Kadamb R, Mittal S, Vig L, Sharma R, Dwarakanath BS, et al. Tumor suppressor protein p53 recruits human Sin3B/HDAC1 complex for down-regulation of its target promoters in response to genotoxic stress. PLoS One. 2011;6(10):e26156. doi: https://doi.org/10.1371/journal.pone.0026156. PONE-D-11-10406 [pii]. Epub 2011/10/27. PubMed PMID: 22028823; PubMed Central PMCID: PMCPMC3197607.
331. Ahmed MI, Mardaryev AN, Lewis CJ, Sharov AA, Botchkareva NV. MicroRNA-21 is an important downstream component of BMP signalling in epidermal keratinocytes. J Cell Sci. 2011;124(Pt 20):3399–404. doi: https://doi.org/10.1242/jcs.086710. PubMed PMID: 21984808; PubMed Central PMCID: PMCPMC3196856.
332. Lin L, Gan H, Zhang H, Tang W, Sun Y, Tang X, et al. MicroRNA21 inhibits SMAD7 expression through a target sequence in the 3′ untranslated region and inhibits proliferation of renal tubular epithelial cells. Mol Med Rep 2014;10(2):707–712. doi: https://doi.org/10.3892/mmr.2014.2312. PubMed PMID: 24913635.
333. Driskell I, Oda H, Blanco S, Nascimento E, Humphreys P, Frye M. The histone methyltransferase Setd8 acts in concert with c-Myc and is required to maintain skin. EMBO J 2011. doi: https://doi.org/10.1038/emboj.2011.421. emboj2011421 [pii]. Epub 2011/11/26. PubMed PMID: 22117221.
334. Frye M, Watt FM. The RNA methyltransferase Misu (NSun2) mediates Myc-induced proliferation and is upregulated in tumors. Curr Biol 2006;16(10):971–981. doi: https://doi.org/10.1016/j.cub.2006.04.027. S0960-9822(06)01488-6 [pii]. Epub 2006/05/23. PubMed PMID: 16713953.
335. Kouwenhoven EN, Oti M, Niehues H, van Heeringen SJ, Schalkwijk J, Stunnenberg HG, et al. Transcription factor p63 bookmarks and regulates dynamic enhancers during epidermal differentiation. EMBO Rep. 2015;16(7):863–78. doi: https://doi.org/10.15252/embr.201439941. embr.201439941 [pii]. Epub 2015/06/03. PubMed PMID: 26034101; PubMed Central PMCID: PMCPMC4515125.
336. Kouwenhoven EN, van Bokhoven H, Zhou H. Gene regulatory mechanisms orchestrated by p63 in epithelial development and related disorders. Biochim Biophys Acta 2015;1849(6):590–600. doi: https://doi.org/10.1016/j.bbagrm.2015.03.003. S1874-9399(15)00068-1 [pii]. Epub 2015/03/24. PubMed PMID: 25797018.

337. Adam RC, Yang H, Rockowitz S, Larsen SB, Nikolova M, Oristian DS, et al. Pioneer factors govern super-enhancer dynamics in stem cell plasticity and lineage choice. Nature. 2015. doi: https://doi.org/10.1038/nature14289. nature14289 [pii]. Epub 2015/03/25. PubMed PMID: 25799994.
338. Botchkarev VA, Botchkareva NV, Nakamura M, Huber O, Funa K, Lauster R, et al. Noggin is required for induction of hair follicle growth phase in postnatal skin. FASEB J. 2001;15:2205–14.
339. Kandyba E, Leung Y, Chen YB, Widelitz R, Chuong CM, Kobielak K. Competitive balance of intrabulge BMP/Wnt signaling reveals a robust gene network ruling stem cell homeostasis and cyclic activation. Proc Natl Acad Sci U S A 2013;110(4):1351–1356. doi: https://doi.org/10.1073/pnas.1121312110. PubMed PMID: 23292934; PubMed Central PMCID: PMCPMC3557042.
340. Kobielak K, Pasolli HA, Alonso L, Polak L, Fuchs E. Defining BMP functions in the hair follicle by conditional ablation of BMP receptor IA. J Cell Biol 2003;163(3):609–623. doi: https://doi.org/10.1083/jcb.200309042. PubMed PMID: 14610062; PubMed Central PMCID: PMCPMC2173651.
341. Andl T, Ahn K, Kairo A, Chu EY, Wine-Lee L, Reddy ST, et al. Epithelial Bmpr1a regulates differentiation and proliferation in postnatal hair follicles and is essential for tooth development. Development. 2004;131:2257–68.
342. Clevers H, Nusse R. Wnt/beta-catenin signaling and disease. Cell 2012;149(6):1192–1205. doi: https://doi.org/10.1016/j.cell.2012.05.012. PubMed PMID: 22682243.
343. Lien WH, Polak L, Lin M, Lay K, Zheng D, Fuchs E. In vivo transcriptional governance of hair follicle stem cells by canonical Wnt regulators. Nat Cell Biol. 2014;16(2):179–90. doi: https://doi.org/10.1038/ncb2903. PubMed PMID: 24463605; PubMed Central PMCID: PMCPMC3984009.
344. Genander M, Cook PJ, Ramskold D, Keyes BE, Mertz AF, Sandberg R, et al. BMP signaling and its pSMAD1/5 target genes differentially regulate hair follicle stem cell lineages. Cell Stem Cell 2014;15(5):619–633. doi: https://doi.org/10.1016/j.stem.2014.09.009. PubMed PMID: 25312496; PubMed Central PMCID: PMCPMC4276600.
345. Rinaldi L, Datta D, Serrat J, Morey L, Solanas G, Avgustinova A, et al. Dnmt3a and Dnmt3b Associate with Enhancers to Regulate Human Epidermal Stem Cell Homeostasis. Cell Stem Cell 2016;19(4):491–501. doi: https://doi.org/10.1016/j.stem.2016.06.020. PubMed PMID: 27476967.
346. Hill CS. Transcriptional control by the SMADs. Cold Spring Harb Perspect Biol 2016;8(10). doi: https://doi.org/10.1101/cshperspect.a022079. PubMed PMID: 27449814.
347. Maurano MT, Humbert R, Rynes E, Thurman RE, Haugen E, Wang H, et al. Systematic localization of common disease-associated variation in regulatory DNA. Science. 2012;337(6099):1190–5. doi: https://doi.org/10.1126/science.1222794. science.1222794 [pii]. Epub 2012/09/08. PubMed PMID: 22955828; PubMed Central PMCID: PMCPMC3771521.
348. Gondor A. Nuclear architecture and chromatin structure on the path to cancer. Semin Cancer Biol 2013;23(2):63–64. doi: https://doi.org/10.1016/j.semcancer.2013.02.006. S1044-579X(13)00018-7 [pii]. Epub 2013/03/05. PubMed PMID: 23454238.
349. Tough DF, Lewis HD, Rioja I, Lindon MJ, Prinjha RK. Epigenetic pathway targets for the treatment of disease: accelerating progress in the development of pharmacological tools: IUPHAR Review 11. Br J Pharmacol. 2014;171(22):4981–5010. doi: https://doi.org/10.1111/bph.12848. Epub 2014/07/26. PubMed PMID: 25060293; PubMed Central PMCID: PMCPMC4253452.

Chapter 2
DNA Methylation as an Epigenetic Memory Keeper during Skin Development and Regeneration

Ya-Chen Liang, Randall Widelitz, and Cheng-Ming Chuong

Abbreviations

5mC	5-methylcytosine
aHF-SC	active hair follicle stem cells
CpG	cytosine and guanine separated by one phosphate
DMR	differentially methylated region
Dnmt1/3a/3b	DNA methyltransferase 1/3a/3b
HF	hair follicle
IFE	interfollicular epidermis
K14	keratin 14
ORS	outer root sheath
qHF-SC	quiescent hair follicle stem cells
TAC	transient-amplifying matrix cells
Tet	Ten-eleven translocation methylcytosine dioxygenase

Y.-C. Liang · C.-M. Chuong (✉)
Department of Pathology, Keck School of Medicine, University of Southern California, Los Angeles, CA, USA

Integrative Stem Cell Center, China Medical University Hospital, China Medical University, Taichung, Taiwan
e-mail: yachen.liang@usc.edu; cmchuong@usc.edu

R. Widelitz
Department of Pathology, Keck School of Medicine, University of Southern California, Los Angeles, CA, USA

© Springer International Publishing AG, part of Springer Nature 2018

V. A. Botchkarev, S. E. Millar (eds.), *Epigenetic Regulation of Skin Development and Regeneration*, Stem Cell Biology and Regenerative Medicine,
https://doi.org/10.1007/978-3-319-16769-5_2

2.1 Introduction

2.1.1 *Skin as a Model for Epigenetic Regulation of Organ Development and Regeneration*

The skin is the largest organ in the body. It covers the whole body surface and consists of three layers: the epidermis, dermis, and hypodermis or subcutaneous tissue. The skin provides three major functions for individuals: (1) it prevents dehydration and serves as a protective barrier against mechanical impacts and stresses, microbes and chemical elements; (2) it regulates physiological homeostasis such as body temperature and peripheral circulation; (3) the skin also harbors sensory nerve endings that detect and relay changes in the environment.

During embryonic development, the overall framework of the whole skin structure is formed. After birth, the skin is renewed by physiological regeneration that takes place throughout life. The skin also can repair and regenerate in response to injury. The skin's accessibility and life-long regenerative ability make it a model system that is well-suited for the study of organ development and regeneration. Physiologically, the interfollicular epidermis (IFE) undergoes continuous regeneration to replenish the outermost cutaneous layer, whereas the follicular epidermis of the hair follicle (HF) undergoes cyclic regeneration during hair cycling. The skin is located on the body surface and can be induced to regenerate after micro-trauma injuries such as hair plucking or macro-trauma injuries such as wounding. The recent emergence of molecular markers to track the development of specific skin lineages has facilitated regenerative studies enabling us to visualize the step-wise progression of skin specification in a systematic manner. Combined with the properties of easy-access, short-term regenerative cycles, and mature culture and transplantation technologies, the skin has emerged as an excellent tool for fundamental studies of organ development and regeneration.

The skin model is also a reliable system in which to study the indispensable roles of epigenetic regulation in organizing global changes of gene expression during organ development and regeneration. In brief, genes were previously thought to be regulated solely by their local promoter regions. However, it was later found that accessibility of chromatin to the transcriptional machinery is controlled through a number of DNA and histone modifications which affect chromatin micro- and macro-conformations. To date, three major epigenetic mechanisms have been identified: DNA methylation; post-translational modifications of core histone proteins, also called histone modifications; and higher-order chromatin looping. These events occur as cells undergo dramatic cell fate changes such as differentiation, de-differentiation, and trans-differentiation. Hypothetically, epigenetic machineries globally re-organize cell nuclei and shape tissue-specific profiles of gene expression during cellular differentiation. On the other hand, maintenance of epigenetic marks permits inheritance of specific cell identities, also known as epigenetic memory. Among the three major epigenetic modifications, DNA methylation is considered to be relatively stable and serves as an epigenetic memory keeper for cells. During organ development and regeneration, maintaining stem cell identity and controlling commitments to different cell fates are critical. Here we use the skin system as a model and specifically explore roles of DNA methylation in skin development and regeneration.

2.1.2 *Principles of DNA Methylation*

DNA methylation is a biochemical process. In vertebrate animals, some lower eukaryotes and plants, a methyl group (CH_3) can be covalently attached to DNA at the fifth position of a cytosine base in the dinucleotide sequence CpG, forming 5-methylcytosine (5mC). This modification is conducted by the *de novo* methylating enzymes DNA methyltransferase 3a and 3b (Dnmt3a and Dnmt3b), and is maintained by DNA methyltransferase 1 (Dnmt1) when cell divide [29, 40, 47]. Among the three Dnmts, Dnmt1 preferentially methylates hemimethylated DNA during DNA replication, resulting in faithful mitotic inheritance of genomic methylation profiles (reviewed in [12]).

DNA methylation is reversible by either active or passive DNA demethylation pathways. In active Tet-mediated DNA demethylation, the Ten-eleven translocation methylcytosine dioxygenase (Tet) protein family oxidizes 5mC to generate 5-hydroxymethylcytosine (5hmC) and to further form 5-formylcytosine (5fC) and/or 5-carboxylcytosine (5caC), which induces the base excision repair machinery to remove a modified cytosine. On the other hand, in replication-dependent passive DNA demethylation, the absence of maintenance DNA methyltransferase (Dnmt1) or loss of the maintenance enzyme activity results in a gradual loss of DNA methylation after rounds of cellular replication [29, 47].

DNA methylation is one of the most prominently studied epigenetic modifications and is involved in various biological processes including maintenance of cellular memory [46], regulation of gene expression, X-chromosome inactivation [38], silencing of transposons, and genomic imprinting [18, 21].

DNA methylation relies on two molecular actions to repress gene transcription: 1) direct interference with transcription factor binding; and 2) recruitment of methyl-CpG binding proteins and repressor complexes. These two actions on target promoters lead to transcriptional repression of target genes. In addition, although DNA methyltransferases generally methylate DNA at CpG sites, there are regions containing clusters of CpG sequences called CpG islands that are DNA methylation-free [29]. In studying mechanisms of embryonic development, it is of interest to identify regions termed differentially methylated regions (DMRs), which represent DNA sequences in the genome that have different DNA methylation patterns across different cell types [41]. In this chapter, we focus on our current understanding of how DNA methylation participates in embryonic skin development and adult skin regeneration.

2.2 DNA Methylation in Skin Development

2.2.1 Dynamic DNA Methylation in the Early Embryo

Previous studies have shown that DNA methylation is indispensable in development since genetic depletion of *Dnmt1*, *Dnmt3a*, or *Dnmt3b* causes embryonic lethality [30, 40]. During early development, the genome undergoes dynamic changes in DNA methylation [36] as shown in Fig. 2.1 After fertilization, the first global DNA demethylation occurs in both the paternal and maternal genome until the blastocyst stage begins at 3 days of mouse embryonic development (E3.5) [42]. Subsequently, gradual *de novo* methylation takes place, ceasing at E6.5 with the formation of three lineages: the trophoblast, hypoblast and epiblast. Among these, the epiblast is the source of embryonic stem (ES) cells that will give rise to the whole fetus. Around the time of gastrulation after E6.5, the epiblast genome undergoes *de novo* DNA methylation while differentiating into three primary germ layers: endoderm, mesoderm, and ectoderm. The mesoderm will form the future dermis among other tissues while the neural plate and ectoderm will form the epidermis [4, 27].

2.2.2 DNA Methylation Fixes Skin Lineage Commitments during Embryonic Development

DNA methylation plays key roles during the lineage commitment process from ES cells to the three primary germ layers [23]. Isagawa et al. traced genome-wide DNA methylation patterns and showed that most promoter regions remain hypomethylated during ES cell differentiation, while only a small number of gene promoters such as germ cell-specific promoters are *de novo* methylated. These results indicate that DNA methylation is important to fix lineage commitments by repressing the transcription of germ cell-specific genes.

DNA methylation is also crucial in patterning the commitment of skin cells derived from different germ layer origins [34]. After gastrulation, a single layer of surface ectoderm (SE) will develop into the future skin epidermis, and mesoderm-derived dermal fibroblasts will form the skin dermis. Lowdon et al. performed whole-genome bisulfite sequencing (WGBS) and compared global DNA methylation patterns across SE-derived keratinocytes and mesoderm-derived dermal fibroblasts. Their results indicate that lineage specific DMRs of the genome enable expression in one cell lineage while suppressing it in others. For example, the SE-DMR is hypomethylated in skin keratinocytes but hypermethylated in fibroblasts and melanocytes. During skin development, these DMRs serve as an epigenetic switch controlling cell type-specific expression networks to set cells on the right path for their respective fates (Fig. 2.1).

The above studies show that DNA methylation profiles restrict skin lineage commitments from ES cells to germ cell-derived skin cells; however, the factors that

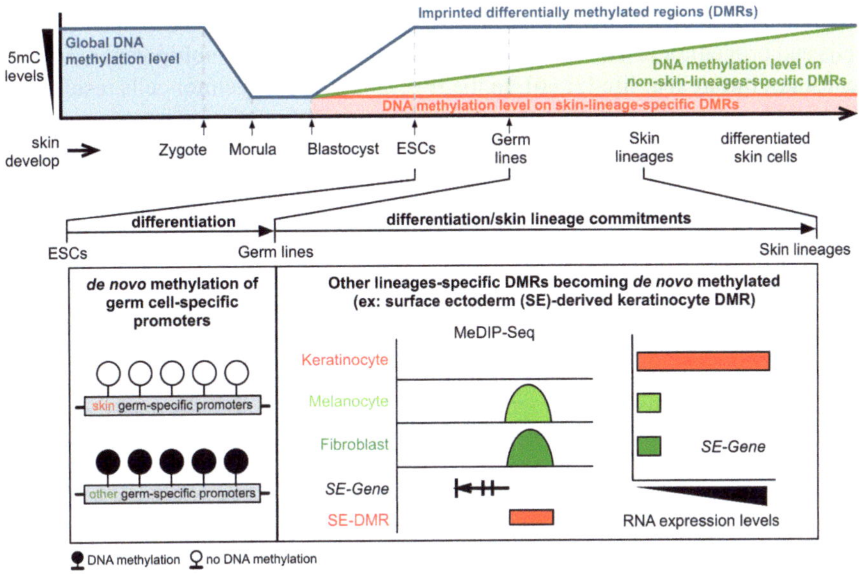

Fig. 2.1 DNA Methylation differentially marks skin lineage and non-skin lineage DMRs during embryonic development. After fertilization, DNA is globally de-methylated. As cells differentiate from the epiblast (embryonic stem cells, ESCs) to ectoderm which gives rise to the future skin epithelium, DNA methylation is gradually re-established on other germ-specific promoters to silence those paths and allow expression from only skin-specific promoters. The skin is composed of cells from different lineages such as keratinocytes, dermal fibroblasts, melanocytes etc. As cells become committed to each skin lineage, differentially methylated regions (DMRs) further restricts gene expression. For example, DMRs in surface ectoderm (SE)-derived keratinocytes remain un-methylated, but these regions are methylated in other skin lineages such as neural crest-derived melanocytes and mesoderm-derived dermal fibroblasts

control the establishment of each skin cell type-specific DMR remain elusive. The roles of DNA methylation at later stages of embryonic skin morphogenesis including stratification of the epidermis and hair follicle development remain to be studied.

2.3 DNA Methylation in Epidermal Renewal and Hair Cycling

2.3.1 Maintenance of DNA Methylation Keeps the Identity of Epidermal Progenitor Cells in the Interfollicular Epidermis

Cell identity describes the ability of a cell to preserve its cell type and properties after division. Epigenetic regulation, especially DNA methylation, is essential to establish and maintain the blueprint of cell identity [7], generating a comprehensive gene expression profile for each specific cell type [2, 6, 33, 39].

Maintaining cell identity is crucial especially in self-renewing cells, such as stem cells and progenitor cells, as these cells have to maintain their multipotent or unipotent states throughout life [3, 46]. In the IFE, epidermal progenitor cells residing in the basal layer are capable of both self-renewal, and differentiation into spinous and granular keratinocytes, which are displaced upward and ultimately terminally differentiate to generate the outermost cutaneous permeability barrier [19].

During this process, *DNMT1* is essential for the epidermal progenitors to maintain their abilities of self-renewal and proper differentiation (Fig. 2.2). As we mentioned above, Dnmt1 mediates faithful mitotic inheritance of genomic methylation profiles. In adult skin, *DNMT1* is expressed in the basal layer of the epidermis and its expression level decreases upon differentiation [31, 45]. Sen et al. used primary human keratinocytes expressing *DNMT1* shRNAs to generate human *DNMT1*-knockdown epidermis in a xenograft model. They showed that decreased *DNMT1* expression causes loss of the epidermal progenitor population and precocious differentiation of the basal layer. Moreover, using *in vitro* differentiation studies and methylated DNA immunoprecipitation (MeDIP) followed by hybridization on a human promoter tiling array, Sen et al. found that DNA methylation sustains the epidermal progenitor state by inducing silencing of the differentiation program, while the DNA demethylation machinery functions to promote differentiation [45]. This study provides insight into how DNA methylation / demethylation coordinate the renewal and differentiation of human epidermal progenitors *in vitro*.

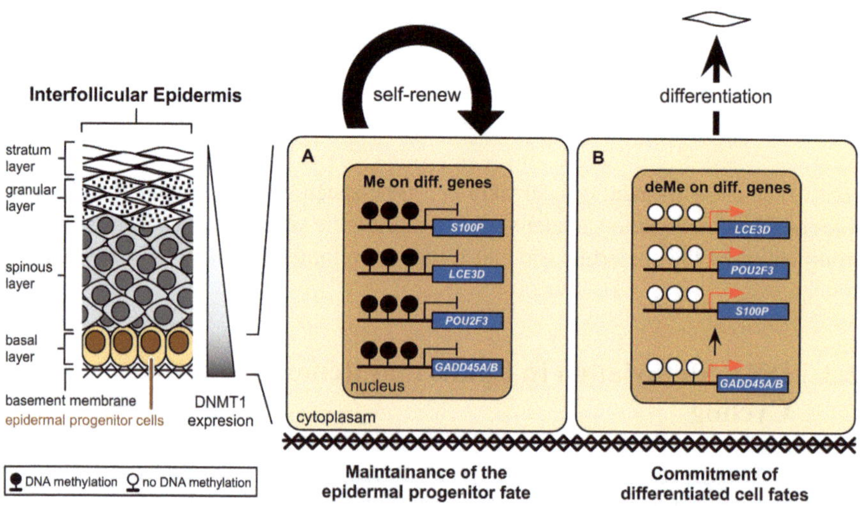

Fig. 2.2 DNA methylation maintains epidermal progenitor cell identity in the interfollicular epidermis (IFE). (**a**) To maintain epidermal progenitor fate, DNA methylation occurs on differentiation gene promoters to repress their gene expression. (**b**) When an epidermal progenitor cell commits to differentiate, it is crucial for early differentiation gene promoters, especially GADD45A and GADD45B, to lose their DNA methylation so other differentiation gene promoters, such as LCE3D, POU2F3, and S100P also can be de-methylated, resulting in up-regulation of their transcription and initiation of cell differentiation

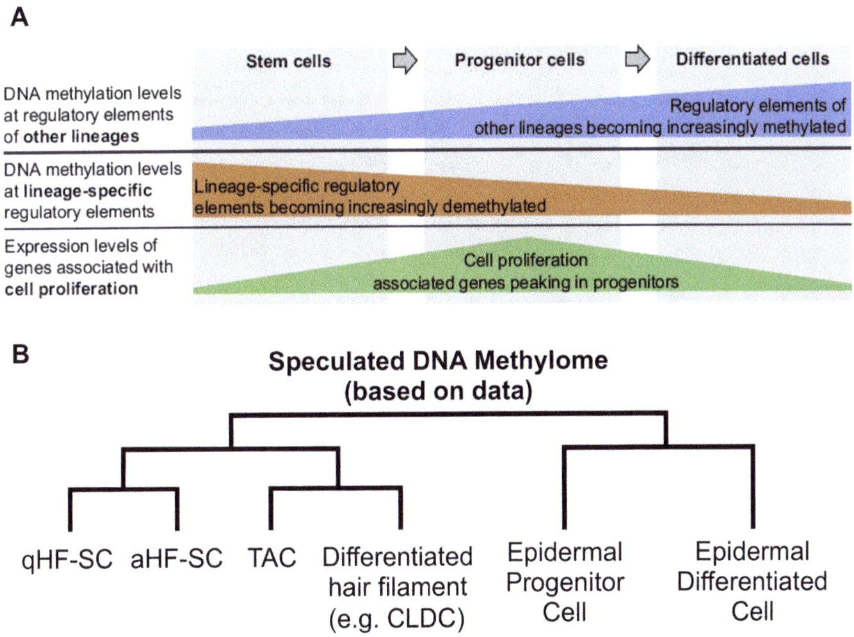

Fig. 2.3 DNA methylation orchestrates hierarchies of skin lineages. (**a**) The illustration summarizes DNA methylation and gene expression levels of regulatory elements in skin and non-skin lineages during differentiation. (**b**) DNA methylation orchestrates skin lineage hierarchies. Bock et al. performed reduced representation bisulfite sequencing to map global DNA methylation profiles across seven skin lineages and found that patterns of DNA methylation reflect strict hierarchies of skin lineages corresponding to their differentiation stages. CLDC, companion layer differentiated cells. This panel is adapted from [6]

However, the results on the roles of Dnmt1 obtained from *in vivo* are different. Li et al. used an *in vivo* approach to study the importance of *Dnmt1* in regulating epidermal progenitor homeostasis in the epidermis and hair follicles. These authors generated mice with a conditional knockout of *Dnmt1* in the *K14*-expressing epidermis, and showed that the mutant mice exhibit uneven thickness, abnormal expression of differentiation markers, and hyperproliferation of the IFE [31]. Because *K14* is expressed in the mitotically active basal layer [1], these results indicated that *Dnmt1* knockout in the basal layer causes dysregulation of the progenitor cells in both self-renewal and differentiation processes. The distinct pathologic phenotypes between the *in vivo* and *in vitro* studies are likely to be due to the different experimental models. In the *in vitro* model, a new homeostasis condition is established using human keratinocytes [45] without feedback signals from adjacent niche or dermis. In the *in vivo* genetic model, compensatory mechanisms may have developed to adjust epidermal homeostasis (Fig. 2.4b). However, both of these studies suggest that *DNMT1* is important to maintain epidermal progenitor functions in the IFE (Fig. 2.4b). Further investigations will be required to reveal the downstream targets of DNMT1 and their biological pathways.

A

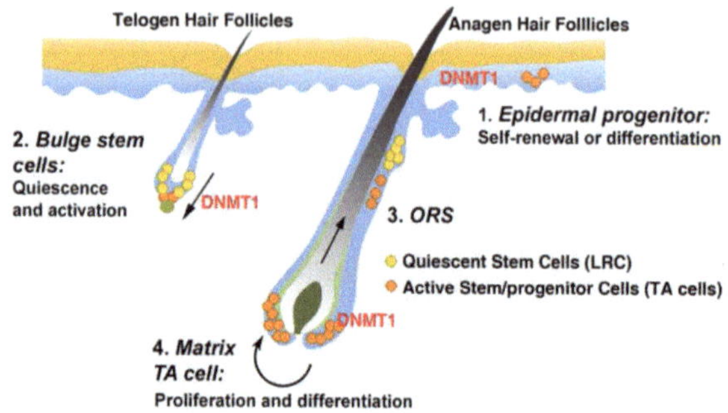

B

Gene name	Symbol	Experimental system	Region	Model used in the study	Phenotypes of the gene depletion under physiological conditions	Reference
DNA methyltransferase 1	*DNMT1*	*in vitro*	IFE	Primary human keratinocytes expressing shRNAs for *DNMT1* gene were used to regenerate human epidermis on immuno-deficient mice	• Exit from the progenitor cell compartment • Premature differentiation of the basal layer • Loss of tissue self-renewal	Sen *et al.* (2010)
	Dnmt1	*in vivo*	IFE HE	Transgenic mice with a conditional knockout of *Dnmt1* in the K14-expressing epidermis	• Alterations in the epidermal thickness • Progressive reduction in hair follicle size, number, and cycling activity, resulting in alopecia • Prolonged telogen with hyperplastic sebaceous glands • Reduced proliferation and increased apoptosis in anagen hair follicles • Delayed activation of hair stem cells after hair plucking	Li *et al.* (2012)
DNA methyltransferase 3 alpha or 3 beta	*DNMT3A* or *DNMT3B*	*in vitro*	IFE	Primary human keratinocytes expressing shRNAs for *DNMT3A or DNMT3B* gene were used to regenerate human epidermis on immuno-deficient mice	• Fewer *DNMT3A*-KD and *DNMT3B*-KD keratinocytes in the basal layer after the 4-week reconstitution • *DNMT3A*-KD and *DNMT3B*-KD keratinocytes mainly located in the differentiated strata after the 4-week reconstitution • *DNMT3A*-KD and *DNMT3B*-KD keratinocytes significantly decreased the clonogenic potential in the colony formation assay	Rinaldi *et al.* (2016)
	Dnmt3a or *Dnmt3b*	*in vivo*	IFE HE	Transgenic mice with a conditional knockout of *Dnmt3a or Dnmt3b* in the K14-expressing epidermis	• No notable epidermal phenotypic differences compared to wild-type littermates at different postnatal ages • Loss of *Dnmt3a* in hair-follicle stem cells didn't cause changes in hair follicle cycling and pelage growth	Rinaldi *et al.* (2017)

*IFE: Interfollicular Epidermis; HE: Hair Epidermis

Fig. 2.4 Role of DNA methylation in the homeostasis of hair follicule (HF) epithelia and inter-follicular epidermis. (**a**) Dnmtl is involved in regulating DNA methylation which maintains homoeostasis of cycling hair follicles (HFs). In telogen, Dnmt1 is expressed in quiescent hair follicle stem cells (qHF-SCs) and the hair germ, whereas in anagen it is expressed in active hair follicle stem cells (aHF-SCs), transient amplifying cells (TACs), and the outer root sheath. Based on the progressive alopecia phenotype in K14 mediated knock down of Dnmtl mice, we propose Dnmt1 is involved in activation of bulge stem cells and maintenance of proliferation of matrix TA cells. Thus, Dnmtl may control hair follicle homoeostasis by directly regulating activation, quiescence, or proliferation states of each cell. (**b**) Dynamic DNA methylation in regulating the homeostasis of inter-follicular epidermis. Here we summarize four different studies, showing the different roles of maintenance methylation (Dnmtl) and de novo methylation (Dnmt3). Different results obtained from in vitro and in vivo studies involving depletion of DNA methylation-associated factors in the skin. The *in vitro* condition might lack epidermal niches and signals from the supporting environment while the *in vivo* condition might be compensated by other factors or mechanisms. Panel A is from [31]

2.3.2 Roles of de novo DNA Methyltransferases in Epidermal Homeostasis

The functions of *de novo* DNA methyltransferases, DNMT3A and DNMT3B, in epidermal homeostasis are also controversial. Rinaldi, et al. used *in vitro* cultured primary human keratinocytes as a model to study the roles of DNMT3A and DNMT3B in epidermis. They found both *DNMT3A* and *DNMT3B* were highly expressed in keratinocyte progenitors. *DNMT3A* is increased during differentiation while *DNMT3B* is diminished at the onset of differentiation. Through omics and virus-mediated shRNA knockdown approaches, they found that DNMT3A and DNMT3B are located at active typical and clustered enhancers (H3K4me1 and H3K27ac) through their H3K36me3-binding domains. Whereas DNMT3A targets specifically at the center, DNMT3B occupied the center and body of their target enhancers. They further demonstrated that DNMT3A works with Tet2 to maintain 5-hmC at the enhancers while DNMT3B promotes 5-mC methylation along the enhancers. Loss of either DNMT3A or DNMT3B decreases enhancer activity. These results suggest that, during keratinocyte differentiation, *de novo* DNA methylation is required to control proper enhancer activity which leads to expression of enhancer RNAs and ectoderm lineage genes.

Using an *in vivo* competition assay, Rinaldi, et al. transplanted *DNMT3A*- or *DNMT3B*-depleted primary human keratinocytes into immunocompromised mice. They found that cells depleted of either *DNMT3A* or *DNMT3B* could not be detected in the basal layers after 4 weeks, and also exhibited impaired colony forming ability in culture. The authors concluded that DNMT3A and DNMT3B are required for homeostasis of human keratinocyte progenitors, whereas DNMT3A is also required for further differentiation [44].

The same group published a subsequent paper claiming that loss of *Dnmt3a* and *Dnmt3b* does not affect epidermal homeostasis but promotes tumorigenesis *in vivo* [43]. In this study, they generated epidermal-specific conditional knockout (cKO) mice by crossing Keratin14 (K14)-Cre-ROSA26-YFP mice with *Dnmt3a* $^{fl/fl}$ or *Dnmt3b* $^{fl/fl}$ mice. They didn't observe any epidermal phenotypic differences in IFE nor in hair follicle cycling, so they concluded that epidermal deletion of *Dnmt3a in vivo* does not affect epidermal homeostasis under physiological conditions. Nevertheless, Rinaldi et al. found that epidermal-depletion of *Dnmt3a* and *Dnmt3b* promoted carcinogen-induced tumorigenesis. Interestingly, they found K14-mediated depletion of *Dnmt3a* resulted in up-regulation of genes involved in lipid metabolism. Their omics results were consistent with these findings and showed that Dnmt3a binds and methylates promoters of genes that regulate cell proliferation and lipid metabolism. Among these genes, the master regulator of lipid metabolism, PPARγ, proved to be controlled by DNA methylation. Deletion of Dnmt3a promoted squamous transformation mediated by activation of PPARγ [43].

Taken together (Fig. 2.4b), the above two works lead to different interpretations regarding the roles of Dnmt3a and Dnmt3b in maintaining epidermal homeostasis under physiological conditions. Similar to the studies of Dnmt1 described above, these reports highlight the different results one may obtain using *in vitro* and *in vivo* models. The *in vitro* condition might lack epidermal niches and signals from the supporting environment while the *in vivo* condition might be compensated by other factors or mechanisms. Thus, the role of *de novo* DNA methylation hasn't been proved *in vivo* remains to be elucidated. Further studies on the molecular targets of Dnmts in specific cell populations will help to clarify their functions.

2.3.3 DNA Methylation Maintains Hair Follicle Homeostasis

At approximately E9 of mouse embryonic development, the surface ectoderm starts to form a stratified epithelium, which eventually develops into the adult IFE. At E14.5, the ectoderm forms periodically arrayed hair placodes, which progress to form primary hair follicles (HFs). Hair placode cells adopt a fate that is clearly distinct from that of the IFE. They progress to form hair follicle stem cells, proliferating transit amplifying cells (TACs), and differentiating cells that produce hair filaments [4, 15]. The roles of DNA methylation during the specification of HF placodes remain to be investigated. The following studies bring the functions of DNA methylation in adult hair follicle regeneration to light.

Unlike the IFE, which can continuously regenerate to replenish the shed cutaneous layer, hair follicles undergo cyclic regeneration throughout life, requiring the period activation of hair follicle stem cells (HF-SCs). In each hair cycle, quiescent HF-SCs (qHF-SCs) residing in the bulge region are stimulated to become active HF-SCs (aHF-SCs). These aHF-SCs interact with the dermal papilla, and give rise to TACs in the matrix, which proliferate and further differentiate into seven hair follicle lineages [19]. *Dnmt1* is essential for this cyclic regeneration of normal hair follicles as epidermal deletion of *Dnmt1* results in progressive hair loss in mice [31]. Li et al. generated K14-Cre Dnmt1$^{fl/fl}$ mice in which *Dnmt1* was conditionally depleted in the epidermal basal layer and in K14-positive hair follicle progenitors. These progenitor cells generate HF-SCs, hair germs, outer root sheath (ORS) as well as TACs [5, 17, 34]. *Dnmt1* was detected in the anagen ORS and hair matrix, and in the telogen hair germ [31]. The results of K14-Cre-mediated deletion of *Dnmt1* implied that Dnmt1 is not only involved in regulating the activation of hair stem cells, but also the proliferation of hair matrix cells and their differentiation into mature hair follicle components.

Interestingly, *in vivo*, the phenotypes caused by Dnmt1 deletion become more severe as the mice age. Hair loss is first apparent at approximately 4-months, and becomes more prominent in aged mice. *Dnmt1* and *Dnmt2* are likely to be involved in activation of HF-SC, and Dnmt1 deficiency leads to a lower probability of successful HF-SC activation. As hairs cycle many times, the probability of successful hair growth becomes progressively lower. Apoptosis also occurs, reducing the size of the hair stem cell population.

In young *K14*-Cre *Dnmt1$^{fl/fl}$* mice, the hair bulge stem cell population does not differ from that found in wild type mice during early anagen stages, but is decreased by 50% in one-year-old mutant mice as determined by long-term label-retention [31]. Since the aging of an individual entails the comprehensive outcome of global cellular senescence which blocks both self-renewal and differentiation processes to keep cells in a quiescent state (van [9]), it is likely that Dnmt1 maintains HF-SC populations by inhibiting cellular senescence and sustaining their identity as occurs in epidermal progenitor cells. This could explain why mice with depletion of *Dnmt1* exhibited prolonged telogen and delayed HF regeneration after hair plucking in old mice. Loss of *Dnmt1* may induce permanent senescence and a gradually decreased HF-SC population as mice age, resulting in insufficient activation of HF-SCs and resultant delayed anagen re-entry. Although there are other mechanisms that might delay HF regeneration, such as a slower morphogenesis process and weaker activation signals generated from the other *Dnmt1*-depleted HF lineages, regulation of Dnmt1 in senescence appears to be the most straightforward explanation for the observed progressive alopecia. However, direct evidence connecting Dnmt1 and cellular senescence in hair follicle lineages remains to be obtained. Currently, we can only conclude that *Dnmt1* is required for normal HF regeneration during aging.

In addition to controlling the onset of HF regeneration, Dnmt1 also regulates the cellular composition of follicular epithelia. One-year-old *K14*-Cre *Dnmt1*$^{fl/fl}$ mice showed high variability in the size of hair follicles and reduced hair fiber thickness, indicating abnormal follicular morphogenesis [31]. In examining how Dnmt1 controls HF morphogenesis, Li et al. further identified reduced proliferation, accumulated DNA damage, and increased apoptosis of TACs in aged anagen HFs. These results demonstrated that, although the basic hair architecture remained similar, the cellular composition of aged mutant HFs was abnormal. Taken together, the role of DNA methylation in follicular epithelia is to maintain the homeostasis of hair follicle lineages (Fig. 2.4a).

A recent study showed that alopecia in aging mice is caused by both intrinsic changes in hair follicle stem cells and extrinsic changes in the stem cell environment [28]. Hair follicle aging is manifest as the miniaturization of hair follicles, loss of dermal papillae, loss of hair bulge stem cells, and eventually loss of hair follicles [35]. K14 mediated-depletion of DNMT 1 leads to a similar progression except for the loss of the dermal papilla. This may reflect that DNMT1 deficient hair stem cells exhibit defective intrinsic properties, leading them to age faster.

2.3.4 DNA Methylation Orchestrates Hierarchies of Skin Lineages

While genetic depletion studies greatly impact functional understanding of *Dnmt1* in skin regeneration, genome-wide DNA methylation profiling further illustrates the underlying regulatory networks. Bock et al. performed reduced representation bisulfite sequencing (RRBS) to map global DNA methylation profiles across seven skin lineages, including qHF-SCs, aHF-SCs, TACs, companion layer differentiated cells, epidermal progenitor cells, and epidermal differentiated cells, *in vivo* [6].

Their results indicated that patterns of DNA methylation reflect strict hierarchies of skin lineages corresponding to their differentiation stages. Moreover, this DNA methylation hierarchy correlates with the transcriptome hierarchy of skin lineages, although the transcriptome hierarchy is looser in distinguishing between-lineage differences, suggesting that DNA methylation functions upstream of gene transcription (Fig. 2.3).

Comparison of genome-wide DNA methylation and gene expression profiles has also identified key regulatory factors. For example, Keratin genes such as *Krt5*, *Krt15*, *Krt27*, and *Krt35* as well as the transcription factors, *Cebpb*, *Gata3*, and *Hoxa5*, have been identified as lineage-specific markers or regulators in skin differentiation. By contrast, *Hoxc6*, *Sox9*, *Tcf7l2*, and *Runx1* have been identified as putative skin stem cell specific genes [6]. However, the precise roles of these factors in either skin stemness or differentiation remain to be further clarified.

2.4 DNA Methylation in Skin Wound Healing

Wound healing is an important topic in regenerative medicine. There are two types of wound healing, *reparative wound healing* by which injured skin is replaced by scar tissue which lacks normal appearance and functions, and *regenerative wound healing* which can re-construct dermis and re-epithelialize epidermis with functional skin appendages [8]. Three concurrent phases are involved in *regenerative wound healing*, including inflammatory, tissue regenerative, and tissue remodeling phases. However, little is known about how DNA methylation participates in these processes.

This topic is currently challenging to study because of the heterogeneity of healing tissues and lack of knowledge of the origins of the cells that contribute to *regenerative wound healing*. During wound re-epithelialization, suprabasal epidermal cells are found to contribute the re-establishment of IFE. However, the origin of the cells that contribute to wound-induced HF neogenesis is still controversial [24]. Without clear identification of the origins and fates of the cells that contribute to HF neogenesis, it is premature to address the functions of DNA methylation in this process.

2.5 Crosstalk between DNA Methylation and Histone Modifications in Skin

A key question in the epigenetic field concerns how different epigenetic modulations coordinate with each other, or crosstalk, to regulate transcription and other chromatin events. DNA methylation factors interact with histone modifiers that deposit marks such as histone H3 lysine 9 tri-methylation (H3K9me3) and histone

H3 lysine 4 methylation, as well as with ATP-dependent nucleosome remodelers [11]. Although no report to date provides direct evidence of crosstalk between DNA methylation and other epigenetic modifications in the skin, we can gain some insights from genetic depletion studies involving epigenetic processes.

Epithelial-specific deletion of *Hdac1* [20], *Setd8* [10], *Ezh1/2* [14], *Brg1* [22], or *Mi-2β* [25] causes abnormalities in epidermal morphogenesis and defects in HF regeneration, suggesting that histone acetylation, histone H4K20 mono-methylation (H4K20me1), histone H3K27 tri-methylation (H3K27me3), ATP-dependent chromatin remodeling, and other epigenetic processes are involved in these processes. Notably, epidermal deletion of *Jarid2* (H3K27me3 demethylase) does not affect epidermal and hair follicle morphogenesis, but delays onset of the first hair follicle growth cycle [37]. Similarly, loss of histone acetylation in *K14-Cre* $Hdac1^{cKO/cKO}$ mice causes a defect in onset of the first hair cycle. Moreover, in *Ezh2*-deficient basal cells *Ink4A/4B* is de-repressed and transcription factor AP1 aberrantly targets terminal differentiation gene promoters, causing cell-cycle arrest and premature epidermal terminal differentiation [13]. These effects are similar to the pathological phenotypes observed when cells are depleted for *DNMT1* [45]. Based on these studies, one might speculate that DNA methylation crosstalks with histone acetylation and H3K27me3 to control the onset of HF regeneration.

In addition, genome-wide profiling of H3K4me3 and H3K27me3 coupled with microarray analysis identified possible key genes during the telogen-anagen transition. *Sox9, Nfatc1, Hoxa7/8, Pknox2, Tbx1,* and *Nfia* were suggested to be possible key regulators in HF regeneration [14]. By comparing the candidate target lists between DNA methylation and other histone modifications, it may be possible to identify novel regulatory factors and delineate possible epigenetic networks during hair regeneration.

2.6 Potential for Future Applications

Abnormal genomic methylation has long been linked to tumorigenesis [32], and DNA methylation contributes to the pathogenesis of many skin disorders, including cancer and chronic inflammation. In skin cancer, hypermethylation of promoter CpG islands is found in many cases [32]. Lauss et al. compared genome-wide DNA methylation profiles of 50 stage IV melanoma cases as well as normal melanocytes, keratinocytes, and dermal fibroblasts, and found promoter hypermethylation of the microphthalmia-associated transcription factor (*MITF*) and its co-regulated differentiation pathway genes in melanoma [26]. Using the concept of cancer-specific DNA hypermethylation, Gao et al. characterized melanoma-specific hypermethylation of genes such as *CLDN11*, *CDH11*, *PPP1R3C*, *MAPK13*, and *GNMT,* which can be used as epigenetic biomarkers for identifying progressive stages of this disease [16].

Zhang et al. conducted MeDIP-Seq to examine whole-genome DNA methylation profiles in clinical samples of psoriasis, a chronic inflammatory skin disorder [48].

Their results indicate abnormal hypermethylated DMRs in several different ontological domains such as the immune system, cell cycle regulation, and apoptosis. These authors also provided evidence that the *PDCD5* and *TIMP2* genes could play important roles in psoriasis. This study provides a classical workflow to identify possible regulatory factors by comparing alterations in DNA methylation patterns in autoimmune diseases. Furthermore, there are several agents in clinical use for cancer treatments that change DNA methylation patterns. Although currently there are no reports describing the use of anti-DNA methylation agents to specifically treat skin cancer, this therapeutic strategy is a possible option to be explored in the future. In summary, aberrant DNA methylation in diseases can serve as a biomarker for diagnosis, a predictor for candidate regulators, and a potential therapeutic target. Profiling of the DNA methylome in skin diseases may therefore be a promising means to develop novel and personalized clinical approaches.

2.7 Concluding Remarks

The skin possesses numerous properties enabling it to serve as a model in studying organ development and regeneration: (1) it undergoes rounds of physiological regeneration; (2) it is experimentally accessible; (3) its lineages are well-characterized; and (4) hair follicles have a relatively short regenerative cycle time. In studies of epigenetic roles during skin development and regeneration, we focused on DNA methylation and discovered that DNA methylation functions as an epigenetic memory keeper. In embryonic development, DNA methylation serves as an epigenetic switch that controls cell type-specific expression networks and sets cells on the right path for their respective fates. In adult skin regeneration, DNA methylation maintains tissue homeostasis by sustaining the identity of epidermal progenitor cells and hair follicle stem cells. The demethylation machinery is carefully coordinated as cells progress toward their specific fates. DNA methylation may also participate in the regulation of cellular senescence during skin regeneration, although detailed mechanisms remain to be further clarified. While studies of DNA methylation in regenerative wound healing are still limited due to a lack of understanding of the key cellular components in this process, we speculate that this mechanism plays key roles in regenerating skin Furthermore, based on current studies of DNA methylation in skin diseases, DNA methylation could serve as a biomarker for cancer diagnosis, a predictor for candidate regulators, and a potential therapeutic target in the future.

Acknowledgments We thank the Chuong lab members for their critical review of the manuscript and discussion. We thank the following funding sources: Dr. C.-M. Chuong and Dr. R.B. Widelitz are supported by National Institute of Arthritis and Musculoskeletal and Skin Diseases grants RO1-AR47364 and -AR60306; Dr. Y.-C. Liang is supported by the Overseas Postdoctoral Program for Frontier Technologies (102-2917-I-564-002-A1), the Ministry of Science and Technology, Taiwan.

References

1. Alam H, Sehgal L, Kundu ST, Dalal SN, Vaidya MM. Novel function of keratins 5 and 14 in proliferation and differentiation of stratified epithelial cells. Mol Biol Cell. 2011;22:4068–78.
2. Álvarez-Errico D, Vento-Tormo R, Sieweke M, Ballestar E. Epigenetic control of myeloid cell differentiation, identity and function. Nat Rev Immunol. 2014;15:7–17.
3. Barrero MJ, Boué S, Izpisúa Belmonte JC. Epigenetic mechanisms that regulate cell identity. Cell Stem Cell. 2010;7:565–70.
4. Benitah SA, Frye M. Stem cells in ectodermal development. J Mol Med. 2012;90:783–90.
5. Blanpain C, Lowry WE, Geoghegan A, Polak L, Fuchs E. Self-renewal, multipotency, and the existence of two cell populations within an epithelial stem cell niche. Cell. 2004;118:635–48.
6. Bock C, Beerman I, Lien WH, Smith ZD, Gu H, Boyle P, Gnirke A, Fuchs E, Rossi DJ, Meissner A. DNA methylation dynamics during in vivo differentiation of blood and skin stem cells. Mol Cell. 2012;47:633–47.
7. Cheedipudi S, Genolet O, Dobreva G. Epigenetic inheritance of cell fates during embryonic development. Front Genet. 2014;5:19.
8. Chueh SC, Lin SJ, Chen CC, Lei M, Wang LM, Widelitz R, Hughes MW, Jiang TX, Chuong CM. Therapeutic strategy for hair regeneration: hair cycle activation, niche environment modulation, wound-induced follicle neogenesis, and stem cell engineering. Expert Opin Biol Ther. 2013;13:377–91.
9. Van Deursen JM. The role of senescent cells in ageing. Nature. 2014;509:439–46.
10. Driskell I, Oda H, Blanco S, Nascimento E, Humphreys P, Frye M. The histone methyltransferase Setd8 acts in concert with c-Myc and is required to maintain skin. EMBO J. 2012;31:616–29.
11. Du J, Johnson LM, Jacobsen SE, Patel DJ. DNA methylation pathways and their crosstalk with histone methylation. Nat Rev Mol Cell Biol. 2015;16:519–32.
12. Edwards JR, Yarychkivska O, Boulard M, Bestor TH. DNA methylation and DNA methyltransferases. Epigenetics Chromatin. 2017;10:23.
13. Ezhkova E, Pasolli HA, Parker JS, Stokes N, Su IH, Hannon G, Tarakhovsky A, Fuchs E. Ezh2 orchestrates gene expression for the stepwise differentiation of tissue-specific stem cells. Cell. 2009;136:1122–35.
14. Ezhkova E, Lien WH, Stokes N, Pasolli HA, Silva JM, Fuchs E. EZH1 and EZH2 cogovern histone H3K27 trimethylation and are essential for hair follicle homeostasis and wound repair. Genes Dev. 2011;25:485–98.
15. Fuchs E. Scratching the surface of skin development. Nature. 2007;445:834–42.
16. Gao L, van den Hurk K, Moerkerk PT, Goeman JJ, Beck S, Gruis NA, van den Oord JJ, Winnepenninckx VJ, van Engeland M, van Doorn R. Promoter CpG island hypermethylation in dysplastic nevus and melanoma: CLDN11 as an epigenetic biomarker for malignancy. J Invest Dermatol. 2014;134:2957–66.
17. Hoi CS, Lee SE, Lu SY, McDermitt DJ, Osorio KM, Piskun CM, Peters RM, Paus R, Tumbar T. Runx1 directly promotes proliferation of hair follicle stem cells and epithelial tumor formation in mouse skin. Mol Cell Biol. 2010;30:2518–36.
18. Howell CY, Bestor TH, Ding F, Latham KE, Mertineit C, Trasler JM, Chaillet JR. Genomic imprinting disrupted by a maternal effect mutation in the Dnmt1 gene. Cell. 2001;104:829–38.
19. Hsu YC, Li L, Fuchs E. Emerging interactions between skin stem cells and their niches. Nat Med. 2014;20:847–56.
20. Hughes MW, Jiang TX, Lin SJ, Leung Y, Kobielak K, Widelitz RB, Chuong CM. Disrupted ectodermal organ morphogenesis in mice with a conditional histone deacetylase 1, 2 deletion in the epidermis. J Invest Dermatol. 2014;134:24–32.
21. Ikeda Y, Nishimura T. The role of DNA methylation in transposable element silencing and genomic imprinting. In: Pontes O, Jin H, editors. Nuclear functions in plant transcription, signaling and development. Springer New York: New York; 2015. p. 13–29.

22. Indra AK, Dupé V, Bornert JM, Messaddeq N, Yaniv M, Mark M, Chambon P, Metzger D. Temporally controlled targeted somatic mutagenesis in embryonic surface ectoderm and fetal epidermal keratinocytes unveils two distinct developmental functions of BRG1 in limb morphogenesis and skin barrier formation. Development. 2005;132:4533–44.
23. Isagawa T, Nagae G, Shiraki N, Fujita T, Sato N, Ishikawa S, Kume S, Aburatani H. DNA methylation profiling of embryonic stem cell differentiation into the three germ layers. PLoS One. 2011;6:e26052.
24. Ito M, Liu Y, Yang Z, Nguyen J, Liang F, Morris RJ, Cotsarelis G. Stem cells in the hair follicle bulge contribute to wound repair but not to homeostasis of the epidermis. Nat Med. 2005;11:1351–4.
25. Kashiwagi M, Morgan BA, Georgopoulos K. The chromatin remodeler Mi-2beta is required for establishment of the basal epidermis and normal differentiation of its progeny. Development. 2007;134:1571–82.
26. Lauss M, Haq R, Cirenajwis H, Phung B, Harbst K, Staaf J, Rosengren F, Holm K, Aine M, Jirström K, et al. Genome-Wide DNA methylation analysis in melanoma reveals the importance of CpG methylation in MITF regulation. J Invest Dermatol. 2015;135:1820–8.
27. Lee HJ, Hore TA, Reik W. Reprogramming the methylome: erasing memory and creating diversity. Cell Stem Cell. 2014;14:710–9.
28. Lei M, Chuong CM. STEM CELLS. Aging, alopecia, and stem cells. Science. 2016;351:559–60.
29. Li E, Zhang Y. DNA methylation in mammals. Cold Spring Harb Perspect Biol. 2014;6:a019133.
30. Li E, Bestor TH, Jaenisch R. Targeted mutation of the DNA methyltransferase gene results in embryonic lethality. Cell. 1992;69:915–26.
31. Li J, Jiang TX, Hughes MW, Wu P, Yu J, Widelitz RB, Fan G, Chuong CM. Progressive alopecia reveals decreasing stem cell activation probability during aging of mice with epidermal deletion of DNA methyltransferase 1. J Invest Dermatol. 2012;132:2681–90.
32. Li Y, Sawalha AH, Lu Q. Aberrant DNA methylation in skin diseases. J Dermatol Sci. 2009;54:143–9.
33. Lien WH, Guo X, Polak L, Lawton LN, Young RA, Zheng D, Fuchs E. Genome-wide maps of histone modifications unwind in vivo chromatin states of the hair follicle lineage. Cell Stem Cell. 2011;9:219–32.
34. Lowdon RF, Zhang B, Bilenky M, Mauro T, Li D, Gascard P, Sigaroudinia M, Farnham PJ, Bastian BC, Tlsty TD, et al. Regulatory network decoded from epigenomes of surface ectoderm-derived cell types. Nat Commun. 2014;5:5442.
35. Matsumura H, Mohri Y, Binh NT, Morinaga H, Fukuda M, Ito M, Kurata S, Hoeijmakers J, Nishimura EK. Hair follicle aging is driven by transepidermal elimination of stem cells via COL17A1 proteolysis. Science. 2016;351:aad4395.
36. Meissner A, Mikkelsen TS, Gu H, Wernig M, Hanna J, Sivachenko A, Zhang X, Bernstein BE, Nusbaum C, Jaffe DB, et al. Genome-scale DNA methylation maps of pluripotent and differentiated cells. Nature. 2008;454:766–70.
37. Mejetta S, Morey L, Pascual G, Kuebler B, Mysliwiec MR, Lee Y, Shiekhattar R, Di Croce L, Benitah SA. Jarid2 regulates mouse epidermal stem cell activation and differentiation. EMBO J. 2011;30:3635–46.
38. Mohandas T, Sparkes RS, Shapiro LJ. Reactivation of an inactive human X chromosome: evidence for X inactivation by DNA methylation. Science. 1981;211:393–6.
39. Novershtern N, Subramanian A, Lawton LN, Mak RH, Haining WN, McConkey ME, Habib N, Yosef N, Chang CY, Shay T, et al. Densely interconnected transcriptional circuits control cell states in human hematopoiesis. Cell. 2011;144:296–309.
40. Okano M, Bell DW, Haber DA, Li E. DNA methyltransferases Dnmt3a and Dnmt3b are essential for de novo methylation and mammalian development. Cell. 1999;99:247–57.
41. Rakyan VK, Down TA, Balding DJ, Beck S. Epigenome-wide association studies for common human diseases. Nat Rev Genet. 2011;12:529–41.

42. Ratnam S, Mertineit C, Ding F, Howell CY, Clarke HJ, Bestor TH, Chaillet JR, Trasler JM. Dynamics of Dnmt1 methyltransferase expression and intracellular localization during oogenesis and preimplantation development. Dev Biol. 2002;245:304–14.
43. Rinaldi L, Avgustinova A, Martín M, Datta D, Solanas G, Prats N, Benitah SA. Loss of Dnmt3a and Dnmt3b does not affect epidermal homeostasis but promotes squamous transformation through PPAR-γ. elife. 2017;6(pii):e21697.
44. Rinaldi L, Datta D, Serrat J, Morey L, Solana G, Avgustinova A, Blanco E, Pons JI, Matallanas D, Von Kriegsheim A, Di Croce L, Benitah SA. Dnmt3a and Dnmt3b associate with enhancers to regulate human epidermal stem cell homeostasis. Cell Stem Cell. 2016;19:491–501.
45. Sen GL, Reuter JA, Webster DE, Zhu L, Khavari PA. DNMT1 maintains progenitor function in self-renewing somatic tissue. Nature. 2010;463:563–7.
46. Shipony Z, Mukamel Z, Cohen NM, Landan G, Chomsky E, Zeliger SR, Fried YC, Ainbinder E, Friedman N, Tanay A. Dynamic and static maintenance of epigenetic memory in pluripotent and somatic cells. Nature. 2014;513:115–9.
47. Smith ZD, Meissner A. DNA methylation: roles in mammalian development. Nat Rev Genet. 2013;14:204–20.
48. Zhang P, Zhao M, Liang G, Yin G, Huang D, Su F, Zhai H, Wang L, Su Y, Lu Q. Whole-genome DNA methylation in skin lesions from patients with psoriasis vulgaris. J Autoimmun. 2013;41:17–24.

Chapter 3
Polycomb Proteins and their Roles in Skin Development and Regeneration

Katherine L. Dauber-Decker, Idan Cohen, and Elena Ezhkova

Abbreviations

BCC	Basal cell carcinoma
cKO	conditional knockout
E	embryonic
EC-SC	Epidermal cancer stem cell
EDC	Epidermal Differentiation Complex
ESC	Embryonic stem cell
H2AK119ub1	histone H2A monoubiquitylation of K119
H3K27me3	trimethylation of histone H3 at lysine (K) 27
HF-SC	Hair follicle stem cells
KO	knockout
Krt	Cytokeratin
MCC	Merkel cell carcinoma
Mx	Matrix
ORS	Outer root sheath
P	Postnatal
PcG	Polycomb group
PRC	Polycomb repressive complex
RNAPII-S2P	RNAPII phosphorylated on Serine 2
SC	Stem cell
SCC	Squamous cell carcinoma

Katherine L. Dauber-Decker and Idan Cohen contributed equally to this work.

K. L. Dauber-Decker · I. Cohen · E. Ezhkova (✉)
Department of Cell, Developmental, and Regenerative Biology, Black Family Stem Cell Institute, Icahn School of Medicine at Mount Sinai, New York, NY, USA
e-mail: elena.ezhkova@mssm.edu

© Springer International Publishing AG, part of Springer Nature 2018

V. A. Botchkarev, S. E. Millar (eds.), *Epigenetic Regulation of Skin Development and Regeneration*, Stem Cell Biology and Regenerative Medicine,
https://doi.org/10.1007/978-3-319-16769-5_3

3.1 Introduction

Development and homeostasis of multicellular organisms rely on the ability to execute cell-specific gene expression programs. In recent years, the field of epigenetics has expanded and gained much recognition due to its importance in this context. In particular, many studies have focused on Polycomb complexes. These epigenetic chromatin regulators act as key developmental regulators controlling stem cell (SC) identity and gene expression during development in higher eukaryotes [95, 185].

Polycomb activity is provided by two main multi-subunit complexes called Polycomb repressive complex 1 (PRC1) and PRC2. These complexes exert a repressive function by directing histone-specific modifications and chromatin compaction (reviewed in [182]). For more than a decade now, the main dogma of Polycomb-mediated gene repression has been described as a strict hierarchical recruitment model. In this model, PRC2-mediated trimethylation of histone H3 at lysine (K) 27 (H3K27me3) leads to the recruitment of PRC1, which catalyzes histone H2A monoubiquitylation of K119 (H2AK119ub1) and compacts chromatin, resulting in transcriptional repression [28, 129, 200, 201].

Strikingly, recent findings have revolutionized our understanding of Polycomb, suggesting complex mechanisms behind its functions. These include a reversed PRC1/PRC2 chromatin recruitment model governed by variant PRC1 complexes, evidence for Polycomb binding to active genes, and dynamic complex composition and function (reviewed in [11]).

While there are seemingly infinite potential targets of Polycomb, one locus has been shown to be a target in many different cell types. The *Ink4b/Ink4a/Arf* locus is known to be a regulator of cell cycling and apoptosis (reviewed in [179]; reviewed in [180]). The locus encodes three separate proteins: p16/Ink4a [174] and p15/Ink4b [71], which are both inhibitors of the G1-S transition during the cell cycle, and ARF [153], an apoptotic regulator (reviewed in [179, 180]). Importantly, it has been shown that this locus is a target of Polycomb in various systems, including mouse and human bone marrow cells [18], murine pancreatic beta cells [31], and murine skin [5, 40, 48, 49].

The importance of Polycomb complexes is highlighted by their crucial role in early embryonic development, and accumulating evidence indicates the involvement of different Polycomb subunits in various human diseases, including cancers [64, 95, 167]. Thus, it is important to investigate the roles of Polycomb in mammalian tissue-specific SCs in both developmental and homeostatic contexts. One tissue system in which Polycomb has proven to be important in both of these contexts is the skin. Importantly, genetic studies have revealed key roles for Polycomb in epidermal progenitor fate, self-renewal, and differentiation [5, 40, 48, 49, 113, 118].

In this review, we will focus on the importance of Polycomb proteins in the skin. First, we will review the discovery of Polycomb, and describe the functions of its various subunits. Next, we will highlight the importance of Polycomb proteins to the development of the three distinct epidermal lineages. Finally, we will discuss what is known about Polycomb in skin cancers, aging, and the common skin disease psoriasis. Importantly, while many studies have been performed to elucidate Polycomb's roles in all of these biological phenomena, much work still remains to be completed if we are to fully understand how Polycomb works, and how this knowledge can eventually be translated for therapeutic uses. The field of Polycomb research remains, therefore, an exciting field of study with the potential for medical applications.

3.2 Polycomb Discovery and Mechanistic Overview

Polycomb Group (PcG) genes were originally identified in screens for homeotic mutant phenotypes in *Drosophila melanogaster*. The PRC1 component Polycomb (Pc) was first discovered in the fly, where it was named for the multiple sex combs that develop on the legs of mutants [65, 103, 165]. Similar genetic screens subsequently identified other Polycomb genes. PcG genes Enhancer of Zeste (*E(z)*) [82] and Suppressor of Zeste (*Su(z)12*) [9] were found to enhance or suppress the phenotypes of *zeste* mutant flies, respectively [65]. Flies mutant for *E(z)* display homeotic transformations that result in improper leg formation, as phenotypically anterior legs are patterned more towards the posterior [62, 82]. Similarly to the homeotic mutants that result from mutation of *E(z)*, flies with mutant *Su(z)12* show homeotic transformations, indicating that *Su(z)12* normally functions to suppress these transformations in the fly [9].

It was subsequently found that in combination with Esc, E(z) protein has histone methyltransferase activity due to its SET domain, and that this functions to methylate histone H3 on lysines 9 and 27 [38, 135]. Finally, it was also discovered that the original *Pc* protein recognizes H3K27me3, allowing recruitment of Polycomb [129, 201]. These findings combined with the similar homeotic transformations observed when knocking out PRC1 or PRC2 subunits in *Drosophila* led to the proposal of the classical hierarchical recruitment model prevalent in the field for many years.

3.2.1 Mammalian PRC1 Complexes

Mammalian PRC1 complexes are highly conserved throughout evolution, yet exhibit far greater diversity than the complexes seen in *Drosophila* [58, 112, 132]. Generally, PRC1 complexes are divided into two main groups: canonical and variant PRC1 complexes. At their central core, all PRC1 complexes contain an E3 ligase, Ring1a or Ring1b, and a Pcgf protein (Fig. 3.1). The six Pcgf proteins form distinct PRC1 families that differ in both composition and genomic distribution (Fig. 3.1; [58]). Although all PRC1 complexes are capable of catalyzing histone H2AK119ub1, the extent of the complexes' catalytic activity varies based on its additional subunits [10, 58]. Canonical PRC1 complexes contain Pcgf2 or Pcgf4 and are further defined by the presence of chromodomain-containing proteins (Cbx)

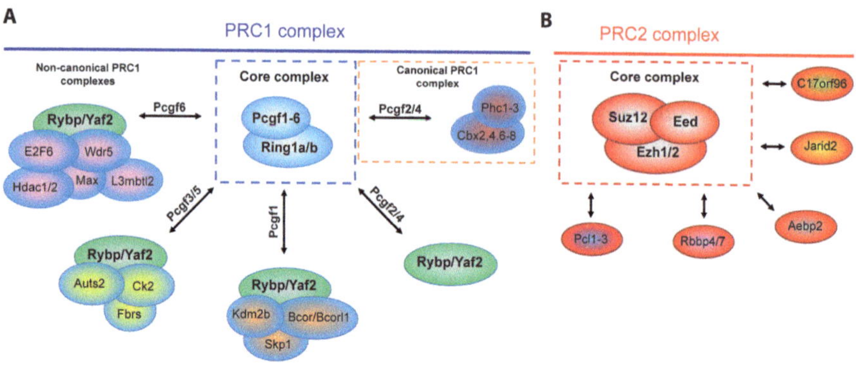

Fig. 3.1 *Schematic representation of Polycomb repressive complexes.* Polycomb complexes are divided into two principal complexes, PRC1 (A) and PRC2 (B). Both complexes have core subunits essential for complex stability and function, and accessory subunits that regulate the recruitment of the complex or modulate its activity. (**a**) Polycomb repressive complex 1 (PRC1) complexes are divided into two groups: canonical and variant (non-canonical). At their core, canonical PRC1 complexes each contain an E3 ligase – either Ring1a or Ring1b – which catalyzes the deposition of the H2AK119ub1 mark – and either Pcgf2 or Pcgf4. The additions of a chromodomain-containing protein (Cbx) and Phc protein define a canonical PRC1 complex that can recognize the PRC2-mediated H3K27me3 histone mark. In variant PRC1 complexes, Cbx and Phc proteins are replaced by Rybp/Yaf2 proteins. (**b**) PRC2 complexes contain three main core components. The activity of the complex is carried out by a histone methyltransferase – either Ezh1 or Ezh2 – which catalyzes the deposition of the H3K27me3 mark. Additional core proteins Suz12 and Eed support the function of Ezh1/2. Main accessory subunits include Rbbp4 or Rbbp7 proteins, which bind to Suz12 to enhance the activity of the complex. Other accessory subunits include Polycomb-like proteins (Phf1, Mtf2, Phf19), Jarid2, Aebp2, C17orf96, and Trim37, which allow for different functional effects

and Polyhomeotic-like (Phc) proteins. In non-canonical or variant PRC1 complexes, Cbx proteins (and Phc) are replaced by the YY1-binding proteins, Rybp or Yaf2, together with additional accessory proteins ([58]; reviewed in [182]).

3.2.2 PRC1 Recruitment and Function

In line with the canonical hierarchical recruitment model, canonical PRC1 complexes can bind to the PRC2-mediated H3K27me3 mark via the chromodomain of the Cbx subunit and facilitate PRC1 recruitment to repress target genes [129, 201]. However, while the PRC1-dependent H2AK119ub1 histone mark largely overlaps with PRC2 distribution, canonical PRC1 catalytic activity towards H2AK119ub1 deposition is very limited [10, 58]. Instead, canonical PRC1 complexes seem to exert their repressive functions by non-enzymatic chromatin compaction [11, 79]. Compelling evidence for a non-enzymatic repressive function of PRC1 is provided by several studies on Polycomb control of Hox genes. The Hox locus harbors a cluster of Polycomb-regulated homeodomain-containing transcription factors that regulate axial patterning [145]. Loss of either *Ring1b* (PRC1) or *Eed* (PRC2) leads to decompaction of the Hox locus and transcriptional derepression [46]. Interestingly, functional studies demonstrated that catalytically inactive forms of *Sce* or *Ring1b* (*Drosophila* I48A or mammalian I53A mutants, respectively) are sufficient to maintain chromatin compaction and Hox gene repression in *Drosophila Sce*-null and mouse *Ring1b*-null embryonic SC (ESC) models [46, 147]. A recent study using a mouse model of *Ring1b* I53A mutants demonstrated that the E3 ubiquitin ligase activity of Ring1b is not essential for the early stages of mouse development [77]. Surprisingly, while *Ring1b* KO mice were reported to exhibit early embryonic lethality around embryonic day (E) E10.5 [198], the *Ring1b* I53A mutant progresses through early development without marked morphological abnormalities at E12.5, yet demonstrates several fully-penetrant or partially-penetrant developmental defects at E15.5. No homozygous mutants are born alive [77]. Of note, further transcriptional profiling of *Ring1b* I53A ESCs identified minimal transcriptional alterations when compared to control cells, in contrast to *Ring1b*-null ESCs, which exhibited marked transcriptional alterations [77]. These observations support the notion of H2AK119ub1-independent transcriptional repression by canonical-PRC1 complexes. Recent evidence further suggests that H2AK119ub1-independent chromatin compaction and transcriptional repression might involve the functions of the Phc canonical PRC1 subunit. In this regard, Phc2 is able to auto-polymerize via its sterile alpha motif (SAM), and this activity is essential for PRC1 repression and chromatin compaction [79]. This is likely due to the establishment of PRC1 clustering at target genes and the stabilization of PRC1-PRC2 interactions.

The existence of variant PRC1 complexes was first evident in biochemical studies in *Drosophila* that defined the core PRC1 complex and indicated additional interacting components [175]. Later, it was shown that a distinct Kdm2-containing PRC1 complex, termed dRAF, can promote H2A ubiquitination [93]. Similarly, mammalian Ring1b-Bmi1 core PRC1 (described by [26]) was later found to be replaced by another Ring1-containing complex, termed BCOR, which can catalyze H2AK119ub1 [59]. Additional biochemical studies identified multiple paralogs for mammalian PRC1 subunits that can form six major variant PRC1 complexes [58]. In non-canonical or variant PRC1 complexes, Cbx proteins are replaced by Rybp/Yaf2 (or dKdm2 in *Drosophila*) [58, 93, 187]. This replacement abolishes the PRC1 complexes' ability to recognize the H3K27me3 mark, inconsistent with the classical hierarchical recruitment model. Although both canonical and variant complexes are capable of compacting chromatin, variant PRC1 complexes have much higher H2AK119 mono-ubiquitination activity, as Rybp stimulates Ring1b catalytic activity *in vitro* [58]. Furthermore, several recent studies have demonstrated that variant PRC1 complexes can be recruited to chromatin independent of PRC2 [187]. This is in line with the observation that H2AK119ub1 is globally maintained in PRC2-null mouse ESCs [187]. In fact, it has been shown that PRC1-mediated H2AK119ub1 can trigger the recruitment of PRC2, supporting the existence of a reversed recruitment mechanism governed by variant PRC1 complexes [10, 83].

Kdm2b, a histone H3K36 demethylase that harbors a CxxC zinc finger domain with strong affinity for CpG islands [188], provides what is perhaps the best-described mechanism of recruitment of the non-canonical PRC1 complexes. Kdm2b interacts with a Pcgf1-Ring1b-containing PRC1 complex and facilitates its recruitment to non-methylated CpG islands [50, 205]. Interestingly, depletion of Kdm2b in ESCs results in only a mild reduction in global levels of H2AK119ub1 [17, 50], supporting the existence of alternative recruitment mechanisms for PRC1. In addition to their interactions with CpG islands, PRC1 complexes may also be recruited through direct interactions with transcription factors, such as Runx1 and REST [42, 211], or through interactions with lncRNAs [210].

While all of the PRC1 complexes described above are associated with transcriptional repression, recent findings have identified a role for an Auts2-containing non-canonical PRC1 complex in gene activation [57]. Biochemical studies have demonstrated that the Pcgf5 and Ring1b proteins can interact with Auts2 to form a stable complex, which also contains Rybp/Yaf2 and casein kinase 2 (Ck2) [58]. Mechanistically, Ring1b that has been phosphorylated by Ck2 has a restricted ability to catalyze H2AK119ub1 and allows transcriptional activation through Auts2-mediated recruitment of P300 [57]. Apart from gene activation functions per se, additional variant PRC1 components seem to play essential roles in maintaining the ability to activate genes. In Polycomb-bound genes, Kdm2b is required to prevent *de novo* DNA methylation of loci co-bound by Polycomb and Kdm2b [17]. In fact, loss of Kdm2b in ESCs results in loci-specific *de novo* DNA methylation and transcriptional silencing that is restricted to Kdm2b-containing Polycomb targets. It is not clear, however, whether this property of Kdm2b is an integral Polycomb function, as Ring1b-null ESCs are protected from *de novo* methylation [17]. Of note, it is possible that Ring1a is able to compensate for the lack of Ring1b in this regard, and further studies are warranted to determine the precise involvement of PRC1 in this fascinating mechanism. Additional exciting findings of stage-specific dynamics in PRC1 composition, including accessory subunit replacement in the Pcgf2-Ring1b core complex during ESC differentiation into cardiac cells, have recently been reported to correlate with a switch from gene repression to activation [132]. However, the exact mechanism behind this observation needs to be further investigated.

Overall, the recent aforementioned studies display the complexity and challenges of elucidating the functional mechanisms of Polycomb function and the intricate interplay between Polycomb complexes, which seems to be much more intricate than what was originally proposed in the classical hierarchical recruitment model.

3.2.3 Mammalian PRC2 Complexes

Similar to the complex found in *Drosophila*, mammalian PRC2 functions to methylate H3K27. Its core includes either Ezh1 or Ezh2, which are the histone methyltransferases responsible for PRC2 enzymatic activity [119, 120]. However, their catalytic activity relies on proper complex assembly with at least two additional subunits, Suz12 and Eed [27, 86, 143]. Together, these core subunits define the canonical PRC2 complex (Fig. 3.1).

PRC2 can also contain the histone binding protein Rbbp4/7, which binds to Suz12 and further contributes to PRC2 activity [11, 138]. Additionally, PRC2 variants that include Polycomb-like (PCL) proteins (Phf1, Mtf2, Phf19), Jarid2, Aebp2, C17orf96, and Trim37 have been identified (Fig. 1; [8, 106, 107, 120]). Interestingly, each of these interactors can form a unique PRC2 complex with distinct recruitment and/or functional features.

3.2.4 Mammalian PRC2 Recruitment and Function

In the classical hierarchical recruitment model, PRC2 promotes the recruitment of PRC1 and the establishment of Polycomb domains [35]. Thus, much attention has been dedicated to elucidating the mechanisms by which PRC2 can be recruited to DNA.

Several molecules have been proposed to be involved in PRC2 targeting in a gene- and cell-specific fashion. For example, Jarid2, a member of the Jumonji protein family with an AT-rich interaction domain (ARID) that can bind DNA, acts as a genomic recruiter of PRC2 with a preference for GC-rich DNA regions [144, 146]. Jarid2 is also found in another variant PRC2 complex with Aebp2; this complex can recognize PRC1-mediated H2AK119ub1, thus providing a link between PRC1 and PRC2 [83]. However, although Jarid2 depletion leads to reduced recruitment of PRC2 in ESCs, the effect on H3K27me3 levels is subtle and lacks the marked derepression alterations seen in PRC2-null ESCs [94, 177].

The PCL proteins comprise an additional class of important PRC2 recruiters. PCL proteins are recruited to DNA via their TUDOR domain, which recognizes the histone H3K36me3 modification associated with actively transcribed gene bodies [4, 20]. This suggests that PCL-mediated recruitment of PRC2 might act as a switch to turn off transcription. Several studies have also identified a role for *Drosophila* and mammalian PCL/PHF1 in stimulating PRC2 catalytic conversion of H3K27me2 to H3K27me3, thus promoting PRC2-mediated repression [136, 166]. There are three mammalian orthologs of *Drosophila* PCL. However, the accessory proteins in this family can each form separate PRC2 complexes and have distinct biological roles in ESCs [75, 199]. Another important feature of PCL proteins is the presence of PHD finger domains, which are thought to interact with histone tails and are also required for PRC2 recruitment [30]. Further characterization of PCL PHD finger specificity might contribute to a better mechanistic understanding of their recruitment specificities.

Additional locus-specific recruitment mechanisms can include interactions with transcription factors or long non-coding RNAs (lncRNAs). For example, the transcriptional repressor factor Snail1 recruits PRC2 to repress E-cadherin gene expression [74]. Similarly, lncRNAs have emerged as potential recruiters of Polycomb to specific loci [67, 216], although the precise mechanisms of recruitment remain unclear (reviewed in [21]).

Strikingly, some work has revealed an unexpected role for non-canonical PRC1 complexes in supporting PRC2 recruitment. This is highlighted by recent findings demonstrating that *de novo* binding of PRC1 is sufficient for the recruitment of PRC2 and H3K27me3 in an H2AK119ub1-dependent manner [10]. Furthermore, loss of PRC1 and H2AK119ub1 results in a drastic reduction in PRC2 binding and H3K27me3 levels [10, 44, 147]. Recent studies of catalytically inactive PRC1 complexes (*Drosophila Sce*; mouse *Ring1b*) have highlighted the essential role of H2AK119ub1 in the efficient PRC2 binding and H3K27me3 deposition, as well as in the normal development and viability in both *Drosophila* and mice [77, 147].

PRC2-interactors may also function to restrict PRC2 recruitment and function. Indeed, recent evidence suggests that C17orf96 acts as a negative regulator of PRC2: at CpG islands enriched for PRC2, C17orf96 directly interacts with Suz12 and interferes with PRC2 catalytic activity, whereas its depletion results in increased PRC2 recruitment and H3K27me3 levels [107].

The hallmark of PRC2 activity, H3K27me3, is considered to be the main histone modification associated with Polycomb-mediated transcriptional repression (reviewed in [182]). However, PRC2 catalytic activity towards H3K27 also includes monomethylation and dimethylation [178, 181]. In line with Polycomb repressive activity, dimethylated H3K27 (H3K27me2) is widely distributed throughout the chromatin and is believed to prevent random H3K27 acetylation [53, 99]. Paradoxically, however, monomethylated H3K27 (H3K27me1) is localized within actively transcribed gene bodies [99, 195]. Importantly, PRC2 loss-of-function in ESCs significantly reduces H3K27me1 levels, although unlike its effects on H3K27me2 and H3K27me3, it does not completely abrogate H3K27me1 [53, 178]. Reduced H3K27me1 causes decreased expression of target genes in these cells [53]. Recently, Ezh1, the other histone methyltransferase of PRC2, was found to colocalize with the H3K4me3 active mark in proliferating myoblasts, and to be required for transcriptional activation of myogenic genes [134]. Moreover, Ezh1 interacts directly with RNA Polymerase in its RNAPII phosphorylated on Serine 2 (RNAPII-S2P) modified form, thus linking Ezh1 to transcriptional elongation. Ezh1 appears to be required for proper RNA polymerase II occupancy, and additional loss-of-function studies have demonstrated that the histone methyltransferase catalytic domain of Ezh1 promotes myogenic gene activation [134]. Taken together, these studies identify unexpected evidence for a role of PRC2 in gene activation. While much effort has been made to unravel the mechanisms governing PRC2 recruitment, it remains largely unknown how different PRC2 interactors facilitate or modify the activity of the complex towards different H3K27 methylated moieties, which differentially impact the transcriptional state.

3.3 The Roles of Polycomb in Skin Epithelium Differentiation and Development

The Polycomb complex serves to repress genes in a tissue-specific manner, enabling cells to modulate gene expression to either promote or prevent differentiation into a variety of cell types. Interestingly, by repressing its target genes, Polycomb helps to regulate the development of the epidermal lineages [5, 40, 48, 49].

3.3.1 Epidermis

The mammalian epidermis serves as a protective barrier against environmental threats, dehydration, infection, and poisoning [29, 47, 72, 73, 84, 189]. Murine epidermis consists of several cellular layers. The innermost layer, the basal layer, gives rise to the suprabasal layers of the epidermis ([55]; reviewed in [5, 12, 49]). The basal layer is marked by expression of cytokeratins (Krt) 5 and Krt14 ([37, 131, 137]; reviewed in [56]; reviewed in [39]; reviewed in [13, 49, 55]). Basal cells progressively differentiate to form the spinous layer, the granular layer, and the stratum corneum (reviewed in [13, 49, 170]). The different suprabasal layers express specific cellular markers. For example, the spinous layer is marked by Krt1 and Krt10 expression (reviewed in [39]; reviewed in [13]), the granular layer is marked by Loricrin and Involucrin, and the stratum corneum layer is marked by Filaggrin (reviewed in [13]).

In developing murine epidermis, Polycomb component Ezh2 binds to and represses epidermal terminal differentiation genes [49]. Epidermis-specific deletion of *Ezh2*, using a conditional *Ezh2* allele combined with a *Krt14-Cre* transgene that is specifically expressed in epidermal basal progenitor cells starting from E12 [194], showed that *Ezh2* functions to suppress premature basal cell differentiation [49]. Premature differentiation in *Ezh2* cKO embryos was indicated by accelerated expression of the suprabasal marker Filaggrin in E16 *Ezh2* cKO but not control embryos. A functional barrier assay showed that *Ezh2* cKO mice have a fully developed skin barrier that is able to exclude dye at E16, while control animals fail to exclude dye until E17 [49]. Therefore, loss of Ezh2 results in accelerated formation of the suprabasal layers and the skin barrier. Interestingly, neonatal *Ezh2 cKO* mice appear to have normal epidermis and do not display alterations in epidermal function [49]. Similar experiments have also been performed in epidermis-conditional knockout mice for PRC2 core proteins Eed and Suz12 (*Eed* cKO and *Suz12* cKO, respectively). Importantly, these studies showed that the loss of either *Eed* or *Suz12* resulted in early completion of epidermal barrier formation, but normal epidermal phenotypes by postnatal day (P)0 ([40]; reviewed in [16]).

Further investigation of the mechanism of Ezh2 function revealed that this Polycomb protein antagonizes the transcription factor AP1 to regulate expression of Epidermal Differentiation Complex (EDC) genes [49]. The EDC is a cluster of genes that are required for epidermal barrier formation [41, 43, 122, 130, 197, 215]. When Ezh2 is absent, AP1 binds to and activates EDC genes earlier, resulting in the premature formation of differentiated skin layers. However, in control animals, Ezh2 prevents AP1 from binding to and activating epidermal differentiation genes. This balance between Polycomb repression and transcriptional activation ensures the proper timing of epidermal differentiation during development [49].

Interestingly, epidermal expression of Ezh2 decreases postnatally, whereas expression of Ezh1, the other catalytic subunit of PRC2, increases. Loss of *Ezh1* in mice, however, results in no obvious phenotype, and these mutant mice display normal levels of H3K27me3, indicating that Ezh2 can compensate for the absence of Ezh1 [48]. Analysis of neonatal epidermis of mice lacking both catalytic subunits (*Ezh1/2*) in the epidermis also did not reveal alterations [48]. However, epidermal thickening was observed later in postnatal development [48]. It will be thus important to delve deeper into the roles of Ezh1/2 in adult epidermal function in the future.

Although the mechanisms controlling Polycomb's recruitment to chromatin are not yet fully understood, Jarid2 appears to be important for this process in the epidermis, as loss of *Jarid2* leads to reduced levels of H3K27me3 in neonatal epidermis [125]. Studies in murine epidermis revealed that Jarid2 is not required for the development of the epidermis. As the mouse continues to grow, Jarid2 becomes important for repressing epidermal differentiation, as *Jarid2*-null epidermis shows hyper-thickening of the Filaggrin-positive granular layer by P3. ChIP analysis in Jarid2-null neonatal keratinocytes shows reduced Suz12 and H3K27me3 at epidermis-related genes, including EDC genes like *Filaggrin*, *Loricrin*, and some *Lce1* genes. The expression of these EDC genes was also upregulated in *Jarid2*-null neonatal basal cells. Additionally, H3K27me3 and Suz12 occupancies were also reduced at *p16* (*Ink4a*). Finally, although the expression of EDC genes was not significantly altered in cells isolated from basal keratinocytes from 8-week-old mice, *p16* was upregulated [125]. Therefore, while Jarid2 seems to be dispensable for the initial development of the epidermis in mice, it is necessary postnatally for controlling differentiation through mechanisms of Polycomb repression and regulation of the *Ink4a* gene.

The H3K27 demethylase enzyme JMJD3, a member of the Jumanji C (JmjC) domain-containing family of proteins, acts in opposition to the function of PRC2 proteins [206]. As opposed to PRC2's role in depositing the H3K27me3 mark at EDC loci in the murine epidermis [48, 49], JMJD3 is important for removing the PRC2-dependent H3K27me3 mark from epidermal differentiation genes [172]. ChIP experiments in cultured human epidermal cells showed that JMJD3 binds to the promoter regions of epidermal differentiation genes *KRT1* and *S100A8* at higher levels in differentiated cells than in progenitors, whereas H3K27me3 is present at higher levels at these sites in undifferentiated progenitor cells. Knockdown of *JMJD3* in organotypic culture of epidermal cells resulted in decreased levels of KRT1 and KRT10, showing that Jmjd3 is important for the induction of these differentiation genes. Interestingly, expression analysis in keratinocytes expressing Jmjd3 with or without its demethylase activity revealed that Jmjd3 can induce expression of *KRT10* and *Filaggrin* in the absence of H3K27me3 [172]. Overall,

JMJD3 is important for epidermal differentiation, as it can derepress expression of epidermal differentiation genes through its opposition to the methylating function of Polycomb proteins.

A recent study revealed the link between mechanical force, actin polymerization, and PRC2-dependent gene repression in the epidermis [96]. Le and colleagues discovered that the application of a mechanical strain to epidermal progenitor cells *in vitro* led to an increase in H3K27me3 at lineage-specific genes, and lower levels of RNAPII-S2P [96], which is present at active genes [22]. Probing further, Le et al. found that these effects of the strain were caused by changes in actin polymerization within the cells. The addition of strain led to increased F-actin polymerization, and interfering with actin polymerization reversed the increase in H3K27me3 that they had seen in response to a strain. By analyzing epidermal lineage-specific genes, the authors found that mechanical strain results in H3K27me3-mediated repression of epidermal genes, leading to the inhibition of differentiation. Finally, to replicate these findings *in vivo*, Le et al. analyzed epidermis-conditional knockout mice for myosin heavy polypeptide 9, non-muscle ($Myh9^{EKO}$). As expected, because these mice lack the NMIIA heavy chain [36], they exhibited decreased levels of H3K27me3 at epidermal differentiation genes, resulting in increased expression of these and other Polycomb target genes [96]. Phenotypically, this resulted in epidermal thinning at P0, and observed expression of Krt10 and Loricrin, epidermal differentiation markers, in the basal layer [96]. Therefore, in epidermal progenitor cells, through mechanisms of F-actin polymerization, mechanical stress is important to maintain high levels of H3K27me3 at epidermal differentiation genes, thus preventing epidermal lineage commitment.

Our knowledge of the role of PRC1 in epidermal development is minimal and limited to the role of PRC1 subunit Cbx4. In human skin, the PRC1 protein CBX4 is important for mediating the balance between slow-cycling and active epidermal stem cells [113]. Cbx4 is an unusual PRC1 subunit that is known to have both PRC1-dependent and -independent roles ([113, 162]). Studies of chromodomain mutants have been performed to elucidate the Polycomb-related functions of CBX4 protein. Interestingly, mutation of the chromodomain of CBX4 resulted in senescence of keratinocytes. However, the chromodomain region of CBX4 was not necessary for proliferation or differentiation, as genome-wide analysis showed that most of the genes involved in these processes did not increase in expression in mutant compared to control cells. Instead, these processes are controlled by CBX4's SUMO E3 ligase activity independently of PRC1 activity. A point mutation of the ligase domain results in upregulation of proliferation and differentiation, but not quiescence genes. Human keratinocytes with a knockdown of *CBX4* fail to form properly differentiated skin, as the regenerated skin lacked formation of an involucrin-positive layer [113]. Therefore, CBX4 is essential for normal development and differentiation of human skin. It possesses both Polycomb-dependent and Polycomb-independent functions in regulating human epidermal stem cells, blocking senescence through its Polycomb functions, while preventing proliferation and differentiation through its SUMO E3 Ligase activity.

Recent studies have elucidated how Cbx4 functions *in vivo* to regulate murine skin development ([118]; reviewed in [33]). Analyses of *Cbx4*-null mice revealed precocious epidermal differentiation, decreased epidermal proliferation, and a thinner epidermis. Further experiments in which either the chromodomain or the SUMO E3 ligase-interacting domain of Cbx4 was ablated in keratinocytes showed that Cbx4's control of epidermal proliferation and differentiation occurs mainly through its SUMO E3 ligase activity, while the repression of non-epidermal genes is Polycomb-dependent, and occurs through the chromodomain. Interestingly, Mardaryev et al. also determined that Cbx4 functions downstream of the transcription factor p63, which was already known to be a crucial regulator of epidermal differentiation ([54, 117, 128, 209]; reviewed in [89]). They showed that p63 and Cbx4 repress many of the same genes in the mouse epidermis, including the non-epidermal neural gene *Nefl*. By analyzing the effects of *Cbx4* overexpression in *p63* heterozygous mice treated with *p63* shRNA, the authors found that Cbx4 was able to re-establish some of the proliferation that was lost when p63 was absent. Cbx4 was also able to upregulate Krt10 protein levels and repress expression of the *Nefl* gene, which had been expressed in the absence of p63. Therefore, Cbx4 is a p63 target that has essential chromodomain-dependent and -independent functions in maintaining epidermal proliferation and differentiation ([118]; reviewed in [33]). Interestingly, the observed effects of Cbx4 on basal cell proliferation were temporal, suggesting that other Cbx proteins could be compensating for the loss of Cbx4. Indeed, Cbx4 is one of eight mammalian Cbx proteins (reviewed in [85, 202]; reviewed in [15]) and their roles in skin control are unknown. Moreover, there are other types of PRC1 complexes that do not contain Cbx proteins ([58]; reviewed in [182]). Thus, the role of PRC1 and its sub-complexes in the control of skin development and homeostasis remains to be determined.

3.3.2 Hair Follicles

The mature mouse hair follicle consists of multiple cellular populations. The bulge contains hair follicle stem cells (HF-SCs), which are marked by their expression of both Sox9 and Lhx2 ([139, 160, 196]; reviewed in [12, 116]). HF-SCs of the bulge give rise to additional populations of the hair, including the outer root sheath (ORS), which forms the outer lining of the hair, and the matrix (Mx), which is located at the bottom of the follicle, and contains transient-amplifying (TA) cells. The TA cells form the medulla, which is part of the hair shaft, as well as other differentiated structures of the mature hair follicle (reviewed in [12]; reviewed in [48, 168]). Importantly, hair follicles undergo a cycle with three phases: anagen (growth), catagen (destruction), and telogen (rest). Following telogen, the hair cycle begins again, and a new hair follicle begins to grow (reviewed in [12]; reviewed in [48, 168]).

While initial studies of *Ezh2* cKO mouse skin provided a wealth of information about barrier formation and epidermal differentiation, these studies also suggested that Ezh1 and Ezh2 compensate for each other in the skin [48, 49].

Immunofluorescence staining for H3K27me3 at P0 showed retention of the mark in the suprabasal layers of *Ezh2* cKO mice, and in both the basal and suprabasal layers of *Ezh1* KO mice [48]. In order to completely ablate epidermal H3K27me3, it was necessary to generate mice null for both *Ezh1* and *Ezh2* (*Ezh1/2* 2KO) [48]. These *Ezh1/2* 2KO mice were used to study the development and maintenance of the hair follicle lineage. Hair follicle development is not complete when mice are initially born, and *Ezh1/2* 2KO mice do not survive past P0 due to their inability to eat and form milk spots [48]. Therefore, in order to examine the effects of combined loss of *Ezh1* and *Ezh2* on the developing hair follicles, it was necessary to perform grafting experiments, in which P0 skin was grafted onto the backs of *Nude* host mice [48]. Hair follicles were subsequently analyzed 14 days after engraftment, at which time mature hair follicles have formed [48].

While *Ezh1/2*-null hair follicles appear normal at P0, 14 days after engraftment, they are significantly shorter than control hair follicles. Additionally, 2KO hair follicles have a collapsed medulla structure, as shown by AE13 immunofluorescence staining. Further investigation of hair follicle stem cell (HF-SC) specific markers, Lhx2 and Sox9, revealed that HF-SCs are still present in 2KO hair follicles, showing that the observed defect is not due to a lack of HF-SCs [48]. Similar results were also found when PRC2 core proteins Eed or Suz12 were ablated in embryonic epidermal progenitors. Here, *Eed* cKO and *Suz12* cKO mice also had shorter, defective hair follicles, but HF-SCs were present [40].

Ezh1/2 2KO hair follicles fail to undergo hair cycling and degenerate over time. By P34 or P56, control hair follicles had already been through one cycle, whereas *Ezh1/2*-null or *Eed*-null hair follicles did not cycle [40, 48]. At P156, when control hair follicles have cycled three times, hair follicles are no longer present in the *Ezh1/2* 2KO skin [48]. Defective hair follicle development and degeneration resulted from decreased proliferation and increased apoptosis [40, 48]. BrdU analysis showed that there is a reduced percentage of BrdU+ cells in both the Mx [40, 48] and the ORS [48] of PRC2-null hair follicles compared to control follicles. This proliferation defect was recapitulated *in vitro*, as cultured hair follicle cells isolated from *Ezh1/2* 2KO mice failed to proliferate [48]. In addition, along the back skin of *Ezh1/2* cKO, *Eed* cKO, and *Suz12* cKO mice, a significantly greater percentage of hair follicles were Caspase 3-positive, demonstrating increased apoptosis in PRC2-null hair follicles compared to control. Further analysis of epidermal progenitor, ORS, and Mx cells of P14 *Ezh1/2* 2KO mice showed upregulation of genes in the *Ink4b/Ink4a/Arf* locus [40, 48]. The locus was more highly activated in the ORS and Mx than in the epidermis, which corresponds to the observed differences in apoptosis and decreased proliferation between the epidermis and hair follicles [48]. Similar upregulation of this locus was observed in *Eed* cKO and *Suz12* cKO ORS cells [40]. Importantly, knockdown of *Ink4b/Ink4a/Arf* resulted in the reversion of the proliferation defect that was observed *in vitro* in *Ezh1/2* 2KO hair follicle cells [48]. Therefore, in contrast to PRC2's role in repressing barrier formation and differentiation of the epidermal layers, PRC2 functions to promote proliferation, and to repress apoptosis in hair follicle cells by repressing the *Ink4b/Ink4a/Arf* locus.

In a 2011 study, Lien and colleagues analyzed the chromatin markers present at genes in quiescent HF-SC (qHF-SC) isolated from hair follicles in telogen, active HF-SC (aHF-SCs) isolated from hair follicles in anagen, and transient-amplifying matrix cells (HF-TACs) isolated from hair follicles in anagen [108]. In each of these populations, Lien and colleagues profiled chromatin marks H3K4me3 and H3K79me2, which are present at active genes ([81, 152, 161, 190]; reviewed in [105, 127, 169]), as well as the PRC2-mediated H3K27me3 mark. They found that only a very small number of genes are bivalent – showing both H3K27me3 and H3K4me3 occupancy [7, 108] – compared to the larger number of bivalent genes seen in ESCs [108, 127]. Importantly, the study revealed that PRC2 regulation is important for the transition between stem cell and TAC in the hair follicle, but not for the activation of qHF-SCs. Specifically, epigenetic profiling of the H3K4me3, H3K79me2, and H3K27me3 marks showed that genes had similar occupancies by these marks in qHF-SC and aHF-SC populations. In both of these HF-SC populations, HF-SCs genes are not occupied by Polycomb, while HF-TAC genes are occupied and have the H3K27me3 mark. In HF-TACs, however, the profiling changes so that HF-SC genes are Polycomb repressed, while HF-TAC genes are lacking the PRC2 mark and are expressed [108]. Overall, these findings suggest an important role for Polycomb repression in the transition between stem cell and TAC fate, which is essential for hair follicle growth. However, future functional studies of PRC2 in the control of hair follicle homeostasis and cycling are warranted.

In contrast to the findings of Lien et al., another recent study showed that compared to proliferative HF-SCs, quiescent HF-SCs had a global reduction in histone marks H3K4me3, H3K9me3, and H3K27me3 [100]. Lee and colleagues isolated HF-SCs during different stages of the hair cycle and found that the stage with the lowest levels of each of these marks was catagen. Interestingly however, changes in the levels of these marks did not correlate with changes in mRNA levels, or changes in gene expression between early anagen and catagen. Next, the authors investigated the necessity of this drop in methylation levels for normal hair follicle cycling. By treating mice with topical demethylase inhibitors during catagen, the authors prevented the loss of H3K4me3, H3K9me3, and H3K27me3 in the skin and hair follicles. This resulted in a failure of the hair follicles to continue cycling, as indicated by the increase in follicles in telogen, and the decrease in those in anagen [100]. BMP is known to inhibit the proliferation of HF-SCs (reviewed in [98]). Interestingly, by treating mice with BMP antagonist Noggin [151] during catagen, the authors were able to increase the levels of H3K4me3, H3K9me3, and H3K27me3 in the HF-SCs of the bulge [100]. Therefore, global hypomethylation of H3K4, H3K9, and H3K27 in bulge HF-SCs during catagen, and thus the transition of HF-SCs to quiescence, is regulated by BMP signaling and is essential for normal cycling of the mouse hair follicle.

As in the epidermis, Jarid2 has also been shown to affect later stages of development of the hair follicle [125]. Jarid2 is dispensable for embryonic and early neonatal hair follicle morphogenesis, but is required postnatally for the onset of anagen. In *Jarid2* cKO mice, hair follicles enter anagen later than control hair follicles. This delay may be caused by defective cell division, as there is a decrease in the number of Ki67-positive cells in the bulges of *Jarid2* cKO hairs compared to control. However, this result was not confirmed by analysis of BrdU incorporation. Finally, p16/Ink4a protein levels in the hair follicles of adult *Jarid2* cKO mice are higher than in control hair follicles [125]. The reduction in cell division is similar to what was shown in the PRC2-null postnatal hair follicles [40, 48]; however, the overall hair follicle phenotype of the *Jarid2* cKO mouse is milder than the complete degeneration seen in PRC2-null hair follicles [40, 48, 125].

3.3.3 Merkel Cells

Merkel cells are innervated, touch-sensitive cells located in the skin, whisker pads, and footpads in the mouse [5, 68, 69, 114, 121, 126]. These cells work in concert with neurites: when pressure is applied to the skin, Merkel cells enable an action potential to be fired and sustained in the accompanying neurites, leading to the perception of touch [76, 115, 204]. Importantly, Merkel cells are known to express specific signature genes, including neuronal transcription factors *Sox2* and *Isl1* [5, 69, 148], and *Atoh1*, which is essential for Merkel cell formation [121, 148].

As is true for the epidermis and hair follicles, Merkel cells originate from Krt14-positive embryonic epidermal progenitor cells during development [5, 133, 192]. Interestingly, when PRC2's *Ezh1/2*, *Eed*, or *Suz12* are conditionally ablated in the embryonic epidermal progenitors, Merkel cells are formed prematurely, and there is a significant increase in the number of Merkel cells in knockout skin compared to control skin [5, 40]. Lack of PRC2 leads to an increase in differentiation into the Merkel cell lineage and is due to derepression of genes that are critical for the specification and maturation of Merkel cells [5, 40]. Specifically, *Ezh1/2* 2KO mice show precocious activation of Sox2 in Merkel cell precursors at E15, whereas in control back skin, Sox2 will be activated only at E16 [5]. The *Ezh1/2*-null epidermal progenitors displayed decreased H3K27me3 at the *Sox2* gene, as well as increased expression of *Sox2*. Concomitant deletion of *Sox2* rescued the phenotype observed in *Ezh1/2* 2KO mice, indicating that excess Merkel cells are formed in *Ezh1/2* 2KO embryos via de-repression of *Sox2*. Merkel gene expression analysis also showed transcriptional activation of *Atoh1* in *Ezh1/2*-null epidermal progenitors *in vivo,* or upon *Sox2* overexpression *in vitro* [5]. The functional significance of one of these Merkel cell genes, *Atoh1*, was later confirmed by over-expressing this gene in the epidermis, which resulted in the formation of ectopic Merkel cells [141]. Therefore, PRC2 proteins are important for controlling the specification of the Merkel cell lineage, as they repress *Sox2*, which in turn represses *Atoh1* and other Merkel-cell-specific genes in the mouse epidermis. PRC2's ability to repress expression of these genes maintains the number of Merkel cells by exerting control over Merkel cell differentiation [5, 40, 148].

Interestingly, analysis of *Ezh1/2* 2KO and *Eed* cKO skin revealed that compared to control back skin, in which Merkel cells only form around primary, or guard hairs (reviewed in [142]), PRC2-null back skin contains Merkel cells around all hair follicle types [149]. Recent studies showed that the Shh signaling that is required for HF morphogenesis ([32, 66, 140, 184]; reviewed in [173]) is also essential for Merkel cell specification [149, 207]. In the absence of both epidermal *Eed* and *Smo*, the number of Merkel cells produced in thc mouse back skin was reduced compared to controls, suggesting that the ectopic formation of Merkel cells around other hair follicle is also dependent on Shh signaling [149]. While it is clear that epidermal Shh signaling and PRC2 interact to control Merkel cell specification, future studies will unveil the molecular mechanisms by which Merkel cell formation is restricted to primary hairs.

3.4 The Role of Polycomb in Skin Wound Healing and Regeneration

Epidermal wound healing occurs through a balance between proliferation and migration of epithelial cells to cover the wound, and differentiation to re-form the protective barrier (reviewed in [123]). Investigations into the potential roles of Polycomb proteins in wound healing have suggested a potential role for PRC2 proteins [176]. In these experiments, it was found that after the wound is introduced, the PRC2 proteins Eed, Ezh2, and Suz12 are all downregulated, whereas the demethylases Jmjd3 and Utx are upregulated. Interestingly, it has also been shown that Eed suppresses the expression of the *Myc* and *Egfr* genes that are important for wound healing. Post-injury, Eed's occupancy at the promoters of *Myc* and *Egfr* is reduced, allowing them to be upregulated for wound healing [176]. These studies suggested that PRC2 proteins normally repress the expression of genes involved in wound healing, and are removed in response to injury, allowing for a regenerative response in the epidermis. It will be important in the future to perform functional studies in mice lacking PRC2 repression in the skin epithelium to confirm the significance of these observations.

Although HF-SCs do not contribute to the interfollicular epidermis in homeostasis [101], following wounding, their progeny exit the bulge area and participate in wound repair [102, 139]. In murine epidermis, PRC2 proteins play an essential role in the regenerative process, and specifically, in the regenerative function of HF-SCs [48]. Ezhkova et al. performed split-thickness grafting to investigate the effects of loss of PRC2 proteins Ezh1/2 on bulge stem cell-dependent wound healing. In this method, instead of grafting the entire epidermis and dermis, only the dermis, which contains hair follicle appendages, is transplanted onto *Nude* recipient mice. While transplantation of control dermis/HF-SC grafts resulted in the contribution of HF-SCs to epidermal repair, transplantation of *Ezh1/2* 2KO skin did not allow for this. Here, the new epidermis that formed was only from the *Nude* host mouse, as *Ezh1/2*-null HF-SCs failed to proliferate and contribute to epidermal re-epithelialization [48]. These observations are particularly interesting, as they suggest differences in the roles of PRC2 proteins in wound repair initiated by interfollicular basal cells versus hair follicle bulge stem cells [176].

3.5 Polycomb Proteins in Cancer

Polycomb proteins appear to play important roles in the progression of cancer, as expression of these proteins is altered in many cancer types (reviewed in [167, 183]). For example, Polycomb components are upregulated in breast cancer [34, 87, 150, 154, 212] and prostate cancer [24, 193]. In patients with myeloid diseases, *EZH2* mutations have been identified that interfere with its histone methyltransferase activity [45]. Analysis of glioblastoma patients revealed that gliomas can contain a mutation in which H3K27 is replaced with a methionine residue, which decreases the global methylation status of H3K27 [104]. Additionally, Bmi1 was associated with metastasis of invasive ductal breast cancer [87]. Therefore, in a number of cancers, Polycomb is involved in the development and metastasis of the disease.

While the roles of Polycomb have begun to be studied in cancers of other organs, we still only have limited information about the roles of Polycomb in skin cancers. A small number of studies have been performed to elucidate Polycomb's role in the pathogenesis of Basal Cell Carcinoma (BCC). BCC is the most frequent malignancy and skin cancer that develops in Caucasian populations ([60, 97]; reviewed in [203]; reviewed in [164]). While it is not highly metastatic, BCC can cause extensive local damage, as BCC tumors can migrate downward into bone and preorbital tissues (reviewed in [203]). In a 2007 study, Reinisch et al. found that BCC tumors have a major upregulation in the expression of BMI1 protein ([158]; reviewed in [171]). This is in line with previous findings in other cancer types, in which BMI1 is upregulated (reviewed in [167, 183, 191]). The upregulation of BMI1 in BCC suggests that it could be important for the pathology of this disease; however, further studies are necessary to determine its functional relevance in this context.

Polycomb has also been shown to play a role in the development of skin Squamous Cell Carcinoma (SCC). Also known as cutaneous SCC, this disease is most commonly developed in Caucasian populations, and is usually caused by UV exposure from the sun ([78, 155, 163]; reviewed in [63]). Cutaneous SCC can be avoided by protecting skin from the sun and can be treated (reviewed in [63]); however, a deeper understanding of its mechanisms could lead to improved treatments and survival in the future. Immunohistochemical analyses of SCC tumor samples showed that compared to normal tissue, the tumors have higher expression of EZH2 protein [208]. It was also shown that a population of stem-cell-marker-expressing epidermal squamous cell carcinoma cells exists that has increased expression of EZH2. By *in vitro* spheroid formation and *in vivo* tumor formation analyses, it was shown that these cells have a greater ability to form SCC tumors [2]. These cells were later referred to as epidermal cancer stem cells (EC-SCs) [3]. Adhikary et al. showed that the ability of this EC-SC population to form spheroids and contribute to tumor formation and survival depends on EZH2 expression, as knockdown of *EZH2* in culture or treatment with EZH2 inhibitors *in vitro* or *in vivo* interferes with these activities [3]. Therefore, the PRC2 protein EZH2 is essential for SCC formation, and could be a possible therapeutic target for treating SCC tumors in the future.

A limited number of studies have also been performed to further our understanding of Polycomb's role in Merkel cell carcinoma (MCC). Merkel cell carcinoma is a highly metastatic form of neuroendocrine cancer [23, 61, 70, 156]. It has been shown that BMI1 is expressed in diseased MCC tissue. For example, in a study performed on human MCC tumors, 75% of the samples had BMI1 expression. While this study was largely observational and did not delve into the mechanism of BMI1 function in MCC, the authors suggested that targeting BMI1 may be therapeutically beneficial [23]. In another study, Kouzmina et al. investigated the expression of BMI1 in MCC samples, as well as the presence of Merkel cell polyoma virus (MCV), which is thought to play a causative role in many MCCs [51, 90]. Kouzmina and colleagues found that 50% of the analyzed tumor samples were positive for BMI1 staining. Interestingly, they also found a direct correlation between BMI1 expression in the tumor and metastases forming in the lymph nodes. Finally, it was shown that BMI1 expression in tumor cells is inversely correlated to MCV's presence in a tumor sample, as a greater percentage of cells with positive BMI1 staining was found in tumors without MCV. This suggests that there are different forms of MCC, based on BMI1 staining and MCV presence – BMI1 appears to play a role in MCC development when MCV is not present in the tissue [90]. Indeed, a more recent study indicated that MCC tumors positive for MCV have lower levels of H3K27me3, compared to MCV-negative tumors [25]. This provides further evidence for a potential role of Polycomb in MCC, particularly when MCV is not present.

Overall, these studies suggested the importance of BMI1 in BCC and MCC formation and its connection to metastasis. It will be important to further establish the functional relevance of BMI1 and elucidate its mechanisms of action in order to determine whether it is a potentially significant therapeutic target. Pharmacological inhibitors of Polycomb proteins are beginning to emerge as possible treatments for cancer [91, 124, 186]. One recent example is GSK126, an EZH2-inhibitor that was studied as a possible treatment for lymphoma [124]. Another potential drug that inhibits BMI1 has shown potential as a treatment for colorectal cancer, targeting cancer-initiating cells (CICs) [91].

3.6 Polycomb Proteins in Skin Aging

Epidermal aging is a phenomenon that occurs due to a combination of temporal and environmental factors and is characterized by thinning of the epidermis and dermis, wrinkling, a decrease in hair follicle growth, and the presence of solar lentigines ([19, 52]; reviewed in [14]). One of the most common environmental factors that contributes to skin aging is damage from the sun and its UV radiation [6].

Activation of expression of the *INK4a/INK4b* locus, which encodes the cell cycle inhibitors p16 [174] and p19 [153], is an important characteristic of aging in multiple tissues and organs (reviewed in [88]; reviewed in [1, 180]). This is at least in part due to loss of repression by the Polycomb protein BMI1 [80, 92, 157, 159]. In the skin, p16^{INK4A} expression is upregulated in the aging epidermis. In an *in vitro* model of the skin, immunohistochemical staining for p16 shows higher expression of the protein in skin made from aged cells than in skin reconstructed from younger cells; this is particularly evident in the basal and the lower spinous layers [1]. Another study showed that expression levels of BMI1 in aged skin are lower than those observed in younger skin. This corresponds to an increase in p16^{INK4A} expression and senescence, as BMI1 normally prevents senescence and p16^{INK4A} expression [159]. In a 2006 study, Ressler et al. analyzed BMI1 and p16^{INK4A} expression in human skin from three different groups aged 0–20 years, 21–70 years, and 71–95 years. By immunohistochemical staining, Ressler and colleagues showed that BMI1 expression was decreased and p16^{INK4A} expression was increased in the oldest compared to the youngest age group. This is consistent with other reports of p16^{INK4A} and BMI1 expression in aging and senescence ([92, 157]; reviewed in [110]). Therefore, BMI1 downregulation and upregulation of p16^{INK4A} are observed in aging skin, suggesting they might be important to the aging process in this tissue. However, the precise functions of these proteins in skin aging remain to be elucidated.

3.7 Polycomb Proteins in Psoriasis

Psoriasis is an autoimmune inflammatory ailment of the skin, characterized by red flaking and plaques. Some of the key traits of this disease include elevated epidermal proliferation, defective keratinocyte differentiation, and an increase in the presence of infiltrating immune cells ([214]; reviewed in [111]). While treatments exist that can alleviate the symptoms of this disease, these are not entirely effective in all individuals (reviewed in [111]). Thus, it remains important to elucidate the molecular mechanisms underlying this very common skin disease.

In a 2001 study, Zhang et al. found that there was no significant global change in methylation at H3K27 in psoriatic skin lesions. Interestingly, however, expression of *EZH2* was significantly increased in these patient samples. This suggests that EZH2 may play a role in psoriasis, but if so, this role might be H3K27me3-independent [214].

In addition to the upregulation of *EZH2* in psoriatic skin, there is also a decrease in expression of the transcriptional activator AP1 [213]. Interestingly, *EZH2* and AP1 have opposing functions in murine skin: when Ezh2 is expressed, AP1 cannot bind to epidermal differentiation gene promoters, and this prevents differentiation [49]. Therefore, the increase in proliferation and decrease in differentiation that characterize the psoriatic phenotype may be, at least in part, due to the elevated EZH2 and lowered AP1 levels in the skin.

A 2011 study by Liu and colleagues also showed an increase in the expression of *EZH2*, this time in a comparison between lesional psoriatic skin, uninvolved psoriatic skin, and normal skin. In lesional skin, *EZH2* expression was upregulated when compared to the other two skin types analyzed. Liu and colleagues also analyzed

EZH2 expression in lesional psoriatic skin that had been treated with Etanercept, an established therapeutic for this disease. In the presence of Etanercept, *EZH2* expression decreased [109]. Together, this evidence suggests *EZH2* as a possible target for psoriasis treatment.

3.8 Concluding Remarks

A multitude of studies have highlighted the importance of the Polycomb components in development, disease, and aging. Since their initial discovery in *Drosophila*, much progress has been made in our understanding of the basic mechanisms behind Polycomb transcriptional control. Emerging evidence shows that instead of functioning in concert in a step-wise recruitment model, Polycomb factors operate via intricate interactions with stage-specific complex dynamics, and may perform dual roles as transcriptional repressors and activators. Unfortunately, due to early embryonic lethality in Polycomb knockouts, Polycomb complexes' key roles in maintaining stem cell identity and tissue development have mostly been elucidated in ESCs and *Drosophila*. We currently have only a poor understanding of their role in tissue-specific stem cells and adult stem cells.

The skin provides an excellent, well-characterized system in which to study tissue development, homeostasis, and stem cell control. Indeed, several Polycomb proteins and Polycomb-related proteins have been shown to play key roles in epidermal lineage control, development, disease, and aging. However, most of the studies performed in the skin to date have focused on PRC2, and the role of PRC1 remains to be elucidated. Importantly, while Polycomb complexes have been studied in the context of development, the role of Polycomb complexes in adult stem cell and tissue homeostasis remains almost completely unknown. Addressing these issues using the skin as a model system could be highly beneficial to our understanding of the more general roles and underlying mechanisms governed by these key epigenetic regulators. Additionally, a complete understanding of Polycomb's function in the skin could lead to improved treatment modalities for common maladies in which Polycomb is implicated. Overall, there is still much to be discovered in terms of Polycomb's function in the skin and how this knowledge can be applied to improve medical treatments.

Acknowledgments Katherine L. Dauber-Decker was a trainee of the NIDCR-Interdisciplinary Training Program in Systems and Developmental Biology and Birth Defects T32HD075735. Elena Ezhkova is supported by the National Institute of Arthritis and Musculoskeletal and Skin Diseases of the National Institutes of Health under award numbers R01 AR063724 and R01 AR069078. The content is solely the responsibility of the authors and does not necessarily represent the official views of the National Institutes of Health.

References

1. Adamus J, Aho S, Meldrum H, Bosko C, Lee JM. p16INK4A influences the aging phenotype in the living skin equivalent. J Invest Dermatol. 2014;134(4):1131.
2. Adhikary G, Grun D, Kerr C, Balasubramanian S, Rorke EA, Vemuri M, et al. Identification of a population of epidermal squamous cell carcinoma cells with enhanced potential for tumor formation. PLoS One. 2013;8(12):e84324.
3. Adhikary G, Grun D, Balasubramanian S, Kerr C, Huang JM, Eckert RL. Survival of skin cancer stem cells requires the Ezh2 Polycomb group protein. Carcinogenesis. 2015;36(7):800–10. https://doi.org/10.1684/ejd.2013.2219.
4. Ballaré C, Lange M, Lapinaite A, Martin GM, Morey L, Pascual G, et al. Phf19 links methylated Lys36 of histone H3 to regulation of Polycomb activity. *Nature structural & molecular biology*. 2012;19(12):1257–65.
5. Bardot ES, Valdes VJ, Zhang J, Perdigoto CN, Nicolis S, Hearn SA, et al. Polycomb subunits Ezh1 and Ezh2 regulate the Merkel cell differentiation program in skin stem cells. EMBO J. 2013;32(14):1990–2000.
6. Benedetto AV. The environment and skin aging. Clin Dermatol. 1998;16(1):129–39.
7. Bernstein BE, Mikkelsen TS, Xie X, Kamal M, Huebert DJ, Cuff J, et al. A bivalent chromatin structure marks key developmental genes in embryonic stem cells. Cell. 2006;125:315–26.
8. Bhatnagar S, Gazin C, Chamberlain L, Ou J, Zhu X, Tushir JS, et al. TRIM37 is a new histone H2A ubiquitin ligase and breast cancer oncoprotein. Nature. 2014;516(7529):116–20.
9. Birve A, Sengupta AK, Beuchle D, Larsson J, Kennison JA, Rasmuson-Lestander Å, Müller J. Su (z) 12, a novel Drosophila Polycomb group gene that is conserved in vertebrates and plants. Development. 2001;128(17):3371–9.
10. Blackledge NP, Farcas AM, Kondo T, King HW, McGouran JF, Hanssen LL, et al. Variant PRC1 complex-dependent H2A ubiquitylation drives PRC2 recruitment and polycomb domain formation. Cell. 2014;157(6):1445–59.
11. Blackledge NP, Rose NR, Klose RJ. Targeting Polycomb systems to regulate gene expression: modifications to a complex story. Nat Rev Mol Cell Biol. 2015;16(11):643–9.
12. Blanpain C, Fuchs E. Epidermal homeostasis: a balancing act of stem cells in the skin. Nat Rev Mol Cell Biol. 2009;10(3):207–17.
13. Blanpain C, Fuchs E. Epidermal stem cells of the skin. Annu Rev Cell Dev Biol. 2006;22:339.
14. Bolognia JL. Aging skin. Am J Med. 1995;98(1):S99–S103.
15. Botchkarev VA, Gdula MR, Mardaryev AN, Sharov AA, Fessing MY. Epigenetic regulation of gene expression in keratinocytes. J Invest Dermatol. 2012;132(11):2505–21.
16. Botchkarev VA, Mardaryev AN. Repressing the keratinocyte genome: how the Polycomb complex subunits operate in concert to control skin and hair follicle development. J Invest Dermatol. 2016;136:1538–40. https://doi.org/10.1016/j.jid.2016.04.026.
17. Boulard M, Edwards JR, Bestor TH. FBXL10 protects Polycomb-bound genes from hypermethylation. Nat Genet. 2015;47(5):479–85.
18. Bracken AP, Kleine-Kohlbrecher D, Dietrich N, Pasini D, Gargiulo G, Beekman C, et al. The Polycomb group proteins bind throughout the INK4A-ARF locus and are disassociated in senescent cells. Genes Dev. 2007;21(5):525–30.
19. Branchet MC, Boisnic S, Frances C, Robert AM. Skin thickness changes in normal aging skin. Gerontology. 1990;36(1):28–35.
20. Brien GL, Gambero G, O'Connell DJ, Jerman E, Turner SA, Egan CM, et al. Polycomb PHF19 binds H3K36me3 and recruits PRC2 and demethylase NO66 to embryonic stem cell genes during differentiation. Nat Struct Mol Biol. 2012;19(12):1273–81.
21. Brockdorff N. Noncoding RNA and Polycomb recruitment. RNA. 2013;19(4):429–42.
22. Brookes E, de Santiago I, Hebenstreit D, Morris KJ, Carroll T, Xie SQ, et al. Polycomb associates genome-wide with a specific RNA polymerase II variant, and regulates metabolic genes in ESCs. Cell Stem Cell. 2012;10(2):157–70. https://doi.org/10.1016/j.stem.2011.12.017.

23. Brunner M, Thurnher D, Pammer J, Geleff S, Heiduschka G, Reinisch CM, et al. Expression of VEGF-A/C, VEGF-R2, PDGF-α/β, c-kit, EGFR, Her-2/Neu, Mcl-1 and BMI1 in Merkel cell carcinoma. Mod Pathol. 2008;21(7):876–84.
24. Bryant RJ, Cross NA, Eaton CL, Hamdy FC, Cunliffe VT. EZH2 promotes proliferation and invasiveness of prostate cancer cells. Prostate. 2007;67(5):547–56.
25. Busam KJ, Pulitzer MP, Coit DC, Arcila M, Leng D, Jungbluth AA, Wiesner T. Reduced H3K27me3 expression in Merkel cell polyoma virus-positive tumors. Mod Pathol. 2017;30:877–83.
26. Cao R, Tsukada YI, Zhang Y. Role of BMI1 and Ring1A in H2A ubiquitylation and Hox gene silencing. Mol Cell. 2005;20(6):845–54.
27. Cao R, Zhang Y. SUZ12 is required for both the histone methyl-transferase activity and the silencing function of the EED-EZH2 complex. Mol Cell. 2004;15:57–67.
28. Cao R, Wang L, Wang H, Xia L, Erdjument-Bromage H, Tempst P, et al. Role of histone H3 lysine 27 methylation in Polycomb-group silencing. Science. 2002;298(5595):1039–43.
29. Cartlidge P. The epidermal barrier. Semin Neonatol. 2000;5:273–80.
30. Casanova M, Preissner T, Cerase A, Poot R, Yamada D, Li X, et al. Polycomblike 2 facilitates the recruitment of PRC2 Polycomb group complexes to the inactive X chromosome and to target loci in embryonic stem cells. Development. 2011;138(8):1471–82.
31. Chen H, Gu X, Su IH, Bottino R, Contreras JL, Tarakhovsky A, Kim SK. Polycomb protein Ezh2 regulates pancreatic beta-cell Ink4a/Arf expression and regeneration in diabetes mellitus. Genes Dev. 2009;23(8):975–85. https://doi.org/10.1101/gad.1742509.
32. Chiang C, Swan RZ, Grachtchouk M, Bolinger M, Litingtung Y, Robertson EK, et al. Essential role for Sonic hedgehog during hair follicle morphogenesis. Dev Biol. 1999;205:1–9.
33. Cohen I, Ezhkova E. Cbx4: a new guardian of p63's domain of epidermal control. J Cell Biol. 2016;212(1):9–11. https://doi.org/10.1083/jcb.201512032.
34. Collett K, Eide GE, Arnes J, Stefansson IM, Eide J, Braaten A, et al. Expression of enhancer of zeste homologue 2 is significantly associated with increased tumor cell proliferation and is a marker of aggressive breast cancer. Clin Cancer Res. 2006;12(4):1168–74.
35. Comet I, Helin K. Revolution in the Polycomb hierarchy. Nat Struct Mol Biol. 2014;21(7):573–5.
36. Conti MA, Even-Ram S, Liu C, Yamada KM, Adelstein RS. Defects in cell adhesion and the visceral endoderm following ablation of non-muscle myosin heavy chain II-A in mice. J Biol Chem. 2004;279:41263–6.
37. Coulombe PA, Kopan R, Fuchs E. Expression of keratin K14 in the epidermis and hair follicle: insights into complex programs of differentiation. J Cell Biol. 1989;109(5):2295–312.
38. Czermin B, Melfi R, McCabe D, Seitz V, Imhof A, Pirrotta V. Drosophila enhancer of Zeste/ESC complexes have a histone H3 methyltransferase activity that marks chromosomal Polycomb sites. Cell. 2002;111(2):185–96.
39. Dai X, Segre JA. Transcriptional control of epidermal specification and differentiation. Curr Opin Genet Dev. 2004;14(5):485–91.
40. Dauber KL, Perdigoto CN, Valdes VJ, Santoriello FJ, Cohen I, Ezhkova E. Dissecting the roles of Polycomb repressive complex 2 subunits in the control of skin development. J Invest Dermatol. 2016;136(8):1647–55. https://doi.org/10.1016/j.jid.2016.02.809.
41. de Guzman Strong C, Conlan S, Deming CB, Cheng J, Sears KE, Segre JA. A milieu of regulatory elements in the epidermal differentiation complex syntenic block: implications for atopic dermatitis and psoriasis. Hum Mol Genet. 2010;19(8):1453–60.
42. Dietrich N, Lerdrup M, Landt E, Agrawal-Singh S, Bak M, Tommerup N, et al. REST-mediated recruitment of polycomb repressor complexes in mammalian cells. PLoS Genet. 2012;8(3):e1002494.
43. Eckert RL, Broome AM, Ruse M, Robinson N, Ryan D, Lee K. S100 proteins in the epidermis. J Invest Dermatol. 2004;123:23–33.
44. Endoh M, Endo TA, Endoh T, Isono KI, Sharif J, Ohara O, et al. Histone H2A monoubiquitination is a crucial step to mediate PRC1-dependent repression of developmental genes to maintain ES cell identity. PLoS Genet. 2012;8(7):e1002774.

45. Ernst T, Chase AJ, Score J, Hidalgo-Curtis CE, Bryant C, Jones AV, et al. Inactivating mutations of the histone methyltransferase gene EZH2 in myeloid disorders. Nat Genet. 2010;42(8):722–6.
46. Eskeland R, Leeb M, Grimes GR, Kress C, Boyle S, Sproul D, et al. Ring1B compacts chromatin structure and represses gene expression independent of histone ubiquitination. Mol Cell. 2010;38(3):452–64.
47. Evans NJ, Rutter N. Development of the epidermis in the newborn. Biol Neonate. 1986;49:74–80.
48. Ezhkova E, Lien WH, Stokes N, Pasolli HA, Silva JM, Fuchs E. EZH1 and EZH2 cogovern histone H3K27 trimethylation and are essential for hair follicle homeostasis and wound repair. Genes Dev. 2011;25(5):485–98.
49. Ezhkova E, Pasolli HA, Parker JS, Stokes N, Su IH, Hannon G, et al. Ezh2 orchestrates gene expression for the stepwise differentiation of tissue-specific stem cells. Cell. 2009;136(6):1122–35.
50. Farcas AM, Blackledge NP, Sudbery I, Long HK, McGouran JF, Rose NR, et al. KDM2B links the Polycomb repressive complex 1 (PRC1) to recognition of CpG islands. elife. 2012;1:e00205.
51. Feng H, Shuda M, Chang Y, Moore PS. Clonal integration of a polyomavirus in human Merkel cell carcinoma. Science. 2008;319(5866):1096–100.
52. Fenske NA, Lober CW. Structural and functional changes of normal aging skin. J Am Acad Dermatol. 1986;15(4):571–85.
53. Ferrari KJ, Scelfo A, Jammula S, Cuomo A, Barozzi I, Stützer A, et al. Polycomb-dependent H3K27me1 and H3K27me2 regulate active transcription and enhancer fidelity. Mol Cell. 2014;53(1):49–62.
54. Fessing MY, Mardaryev AN, Gdula MR, Sharov AA, Sharova TY, Rapisarda V, et al. p63 regulates *Satb1* to control tissue-specific chromatin remodeling during development of the epidermis. J Cell Biol. 2011;194(6):825–39.
55. Fuchs E. Scratching the surface of skin development. Nature. 2007;445(7130):834–42.
56. Fuchs E, Raghavan S. Getting under the skin of epidermal morphogenesis. Nat Rev Genet. 2002;3(3):199–209.
57. Gao Z, Lee P, Stafford JM, von Schimmelmann M, Schaefer A, Reinberg D. An AUTS2-Polycomb complex activates gene expression in the CNS. Nature. 2014;516(7531):349–54.
58. Gao Z, Zhang J, Bonasio R, Strino F, Sawai A, Parisi F, et al. PCGF homologs, CBX proteins, and RYBP define functionally distinct PRC1 family complexes. Mol Cell. 2012;45(3):344–56.
59. Gearhart MD, Corcoran CM, Wamstad JA, Bardwell VJ. Polycomb group and SCF ubiquitin ligases are found in a novel BCOR complex that is recruited to BCL6 targets. Mol Cell Biol. 2006;26(18):6880–9.
60. Goldberg LH. Basal cell carcinoma. Lancet. 1996;347(9002):663–7.
61. Gollard R, Weber R, Kosty MP, Greenway HT, Massullo V, Humberson C. Merkel cell carcinoma. Cancer. 2000;88(8):1842–51.
62. Graze RM, Barmina O, Tufts D, Naderi E, Harmon KL, Persianinova M, Nuzhdin SV. New candidate genes for sex-comb divergence between Drosophila mauritiana and Drosophila simulans. Genetics. 2007;176(4):2561–76.
63. Green AC, Olsen CM. Cutaneous squamous cell carcinoma: an epidemiological review. Br J Dermatol. 2017. https://doi.org/10.1111/bjd.15324.
64. Greer EL, Shi Y. Histone methylation: a dynamic mark in health, disease and inheritance. Nat Rev Genet. 2012;13(5):343–57.
65. Grimaud C, Nègre N, Cavalli G. From genetics to epigenetics: the tale of Polycomb group and trithorax group genes. Chromosom Res. 2006;14(4):363–75.
66. Gritli-Linde A, Hallberg K, Harfe BD, Reyahi A, Kannius-Janson M, Nilsson J, et al. Abnormal hair development and apparent follicular transformation to mammary gland in the absence of hedgehog signaling. Dev Cell. 2007;12:99–112. pmid:17199044

67. Gupta RA, Shah N, Wang KC, Kim J, Horlings HM, Wong DJ, et al. Long non-coding RNA HOTAIR reprograms chromatin state to promote cancer metastasis. Nature. 2010;464(7291):1071–6.
68. Haeberle H, Lumpkin EA. Merkel cells in somatosensation. Chemosens Percept. 2008;1(2):110–8.
69. Haeberle H, Fujiwara M, Chuang J, Medina MM, Panditrao MV, Bechstedt S, et al. Molecular profiling reveals synaptic release machinery in Merkel cells. Proc Natl Acad Sci U S A. 2004;101(40):14503–8.
70. Hanke CW, Conner AC, Temofeew RK, Lingeman RE. Merkel cell carcinoma. Arch Dermatol. 1989;125(8):1096–100.
71. Hannon GJ, Beach D. $p15^{INK4b}$ is a potential effector of cell cycle arrest mediated by TGF-beta-induced cell cycle arrest. Nature. 1994;371:257–61.
72. Hardman MJ, Moore L, Ferguson MWJ, Byrne C. Barrier formation in the human fetus is patterned. J Invest Dermatol. 1999;113:1106–13.
73. Harpin VA, Rutter N. Barrier properties of the newborn infant's skin. J Pediatr. 1983;102:419–25.
74. Herranz N, Pasini D, Díaz VM, Francí C, Gutierrez A, Dave N, et al. Polycomb complex 2 is required for E-cadherin repression by the Snail1 transcription factor. Mol Cell Biol. 2008;28(15):4772–81.
75. Hunkapiller J, Shen Y, Diaz A, Cagney G, McCleary D, Ramalho-Santos M, et al. Polycomb-like 3 promotes polycomb repressive complex 2 binding to CpG islands and embryonic stem cell self-renewal. PLoS Genet. 2012;8(3):e1002576.
76. Ikeda R, Cha M, Ling J, Jia Z, Coyle D, Gu JG. Merkel cells transduce and encode tactile stimuli to drive aβ-afferent impulses. Cell. 2014;157(3):664–75.
77. Illingworth RS, Moffat M, Mann AR, Read D, Hunter CJ, Pradeepa MM, et al. The E3 ubiquitin ligase activity of RING1B is not essential for early mouse development. Genes Dev. 2015;29(18):1897–902.
78. International Agency for Research on Cancer. Solar and UV radiation. *IARC Monographs on the Evaluation of Carcinogenic Risks to Humans: Radiation*. 2012;100D:35–101.
79. Isono K, Endo TA, Ku M, Yamada D, Suzuki R, Sharif J, et al. SAM domain polymerization links subnuclear clustering of PRC1 to gene silencing. Dev Cell. 2013;26(6):565–77.
80. Jacobs JJ, Kieboom K, Marino S, DePinho RA, van Lohuizen M. The oncogene and Polycomb-group gene BMI1 regulates cell proliferation and senescence through the ink4a locus. Nature. 1999;397(6715):164–8.
81. Jenuwein T, Allis CD. Translating the histone code. Science. 2001;293(5532):1074–80.
82. Jones RS, Gelbart WM. Genetic analysis of the enhancer of zeste locus and its role in gene regulation in Drosophila melanogaster. Genetics. 1990;126(1):185.
83. Kalb R, Latwiel S, Baymaz HI, Jansen PW, Müller CW, Vermeulen M, Müller J. Histone H2A monoubiquitination promotes histone H3 methylation in Polycomb repression. Nat Struct Mol Biol. 2014;21(6):569–71.
84. Kalia YN, Nonato LB, Lund CH, Guy RH. Development of skin barrier function in premature infants. J Invest Dermatol. 1998;111:320–6.
85. Kaustov L, Ouyang H, Amaya M, Lemak A, Nady N, Duan S, et al. Recognition and specificity determinants of the human Cbx Chromodomains. J Biol Chem. 2011;286(1):521–9.
86. Ketel CS, Andersen EF, Vargas ML, Suh J, Strome S, Simon JA. Subunit contributions to histone methyltransferase activities of fly and worm polycomb group complexes. Mol Cell Biol. 2005;25:6857–68.
87. Kim JH, Yoon SY, Jeong SH, Kim SY, Moon SK, Joo JH, et al. Overexpression of BMI1 oncoprotein correlates with axillary lymph node metastases in invasive ductal breast cancer. Breast. 2004;13(5):383–8.
88. Kim WY, Sharpless NE. The regulation of INK4/ARF in cancer and aging. Cell. 2006;127(2):265–75.
89. Kouwenhoven EN, van Bokhoven H, Zhou H. Gene regulatory mechanisms orchestrated by p63 in epithelial development and related disorders. Biochim Biophys Acta. 2015;1849:590–600. https://doi.org/10.1016/j.bbagrm.2015.03.003.

90. Kouzmina M, Häyry V, Leikola J, Haglund C, Böhling T, Koljonen V, Hagström J. BMI1 expression identifies subtypes of Merkel cell carcinoma. Virchows Arch. 2012;461(6):647–53.
91. Kreso A, van Galen P, Pedley NM, Lima-Fernandes E, Frelin C, Davis T, et al. Self-renewal as a therapeutic target in human colorectal cancer. Nat Med. 2014;20(1):29–36.
92. Krishnamurthy J, Torrice C, Ramsey MR, Kovalev GI, Al-Regaiey K, Su L, Sharpless NE. Ink4a/Arf expression is a biomarker of aging. J Clin Investig. 2004;114(9):1299.
93. Lagarou A, Mohd-Sarip A, Moshkin YM, Chalkley GE, Bezstarosti K, Demmers JA, Verrijzer CP. dKDM2 couples histone H2A ubiquitylation to histone H3 demethylation during Polycomb group silencing. Genes Dev. 2008;22(20):2799–810.
94. Landeira D, Sauer S, Poot R, Dvorkina M, Mazzarella L, Jørgensen HF, et al. Jarid2 is a PRC2 component in embryonic stem cells required for multi-lineage differentiation and recruitment of PRC1 and RNA Polymerase II to developmental regulators. Nat Cell Biol. 2010;12(6):618–24.
95. Laugesen A, Helin K. Chromatin repressive complexes in stem cells, development, and cancer. Cell Stem Cell. 2014;14(6):735–51.
96. Le HQ, Ghatak S, Yeung CY, Tellkamp F, Günschmann C, Dieterich C, et al. Mechanical regulation of transcription controls Polycomb-mediated gene silencing during lineage commitment. Nat Cell Biol. 2016;18(8):864–75.
97. Lear JT, Smith AG. Basal cell carcinoma. Postgrad Med J. 1997;73(863):538–42.
98. Lee J, Tumbar T. Hairy tale of signaling in hair follicle development and cycling. Semin Cell Dev Biol. 2012;23:906–16.
99. Lee HG, Kahn TG, Simcox A, Schwartz YB, Pirrotta V. Genome-wide activities of Polycomb complexes control pervasive transcription. Genome Res. 2015a;25(8):1170–81. https://doi.org/10.1101/gr.188920.114.
100. Lee J, Kang S, Lilja KC, Colletier JJ, Scheitz CJF, Zhang YV, Tumbar T. Signaling couples hair follicle stem cell quiescence with reduced histone H3 K4/K9/K27me3 for proper tissue homeostasis. Nat Commun. 2015b;7:11278. https://doi.org/10.1038/ncomms11278.
101. Levy V, Lindon C, Harfe BD, Morgan BA. Distinct stem cell populations regenerate the follicle and interfollicular epidermis. Dev Cell. 2005;9(6):855–61.
102. Levy V, Lindon C, Zheng Y, Harfe BD, Morgan BA. Epidermal stem cells arise from the hair follicle after wounding. FASEB J. 2007;21(7):1358–66.
103. Lewis EB. A gene complex controlling segmentation in Drosophila. Nature. 1978;276(5688):565–70.
104. Lewis PW, Müller MM, Koletsky MS, Cordero F, Lin S, Banaszynski LA, et al. Inhibition of PRC2 activity by a gain-of-function H3 mutation found in pediatric glioblastoma. Science. 2013;340(6134):857–61.
105. Li B, Carey M, Workman JL. The role of chromatin during transcription. Cell. 2007;128:707–19.
106. Li G, Margueron R, Ku M, Chambon P, Bernstein BE, Reinberg D. Jarid2 and PRC2, partners in regulating gene expression. Genes Dev. 2010;24(4):368–80.
107. Liefke R, Shi Y. The PRC2-associated factor C17orf96 is a novel CpG island regulator in mouse ES cells. Cell Disc. 2015;1:15008.
108. Lien WH, Guo X, Polak L, Lawton LN, Young RA, Zheng D, Fuchs E. Genome-wide maps of histone modifications unwind in vivo chromatin states of the hair follicle lineage. Cell Stem Cell. 2011;9(3):219–32.
109. Liu Y, Luo W, Chen S. Comparison of gene expression profiles reveals aberrant expression of FOXO1, Aurora A/B and EZH2 in lesional psoriatic skins. Mol Biol Rep. 2011;38(6):4219–24.
110. López-Otín C, Blasco MA, Partridge L, Serrano M, Kroemer G. The hallmarks of aging. Cell. 2013;153(6):1194–217.
111. Lowes MA, Bowcock AM, Krueger JG. Pathogenesis and therapy of psoriasis. Nature. 2007;445(7130):866–73.
112. Luis NM, Morey L, Di Croce L, Benitah SA. Polycomb in stem cells: PRC1 branches out. Cell Stem Cell. 2012;11(1):16–21.

113. Luis NM, Morey L, Mejetta S, Pascual G, Janich P, Kuebler B, et al. Regulation of human epidermal stem cell proliferation and senescence requires polycomb-dependent and-independent functions of Cbx4. Cell Stem Cell. 2011;9(3):233–46.
114. Lumpkin EA, Caterina MJ. Mechanisms of sensory transduction in the skin. Nature. 2007;445(7130):858–65.
115. Maksimovic S, Nakatani M, Baba Y, Nelson AM, Marshall KL, Wellnitz SA, et al. Epidermal Merkel cells are mechanosensory cells that tune mammalian touch receptors. Nature. 2014;509(7502):617–21.
116. Mardaryev AN, Meier N, Poterlowicz K, Sharov AA, Sharova TY, Ahmed MI, et al. Lhx2 differentially regulates Sox9, Tcf4 and Lgr5 in hair follicle stem cels to promote epidermal regeneration after injury. Development. 2011;138:4843–52. https://doi.org/10.1242/dev.070284.
117. Mardaryev AN, Gdula MR, Yarker JL, Emelianov VU, Poterlowicz K, Sharov AA, et al. p63 and Brg1 control developmentally regulated higher-order chromatin remodeling at the epidermal differentiation complex locus in epidermal progenitor cells. Development. 2014;141:101–11.
118. Mardaryev AN, Liu B, Rapisarda V, Poterlowicz K, Malashchuk I, Rudolf J, et al. Cbx4 maintains the epithelial lineage identity and cell proliferation in the developing stratified epithelium. J Cell Biol. 2016;212(1):77–89. https://doi.org/10.1083/jcb.201506065.
119. Margueron R, Li G, Sarma K, Blais A, Zavadil J, Woodcock CL, Dynlacht BD, Reinberg D. Ezh1 and Ezh2 maintain repressive chromatin through different mechanisms. Mol Cell. 2008;32:503–18.
120. Margueron R, Reinberg D. The Polycomb complex PRC2 and its mark in life. Nature. 2011;469(7330):343–9.
121. Maricich SM, Wellnitz SA, Nelson AM, Lesniak DR, Gerling GJ, Lumpkin EA, Zoghbi HY. Merkel cells are essential for light-touch responses. Science. 2009;324(5934):1580–2.
122. Marshall D, Hardman MJ, Nield KM, Byrne C. Differentially expressed late constituents of the epidermal cornified envelope. Proc Natl Acad Sci U S A. 2001;98:13031–6.
123. Martin P. Wound healing--aiming for perfect skin regeneration. Science. 1997;276(5309):75–81.
124. McCabe MT, Ott HM, Ganji G, Korenchuk S, Thompson C, Van Aller GS, et al. EZH2 inhibition as a therapeutic strategy for lymphoma with EZH2-activating mutations. Nature. 2012;492(7427):108–12.
125. Mejetta S, Morey L, Pascual G, Kuebler B, Mysliwiec MR, Lee Y, et al. Jarid2 regulates mouse epidermal stem cell activation and differentiation. EMBO J. 2011;30(17):3635–46.
126. Merkel F. Tastzellen und Tastkörperchen bei den Hausthieren und beim Menschen. Arch Mikrosk Anat. 1875;11:636–52.
127. Mikkelsen TS, Ku M, Jaffe DB, Issac B, Lieberman E, Giannoukos G, et al. Genome-wide maps of chromatin state in pluripotent and lineage-committed cells. Nature. 2007;448(7153):553–60. https://doi.org/10.1038/nature06008.
128. Mills AA, Zheng B, Wang XJ, Vogel H, Roop DR, Bradley A. p63 is a p53 homologue required for limb and epidermal morphogenesis. Nature. 1999;398:708–13.
129. Min J, Zhang Y, Xu RM. Structural basis for specific binding of Polycomb chromodomain to histone H3 methylated at Lys 27. Genes Dev. 2003;17(15):1823–8.
130. Mischke D, Korge BP, Marenholz I, Volz A, Ziegler A. Genes encoding structural proteins of epidermal cornification and S100 calcium-binding proteins form a gene complex (‘epidermal differentiation complex’) on human chromosome 1q21. J Invest Dermatol. 1996;106:989–92.
131. Moll R, Franke WW, Schiller DL, Geiger B, Krepler R. The catalog of human cytokeratins: patterns of expression in normal epithelia, tumors and cultured cells. Cell. 1982;31(1):11–24.
132. Morey L, Santanach A, Blanco E, Aloia L, Nora EP, Bruneau BG, Di Croce L. Polycomb regulates mesoderm cell fate-specification in embryonic stem cells through activation and repression mechanisms. Cell Stem Cell. 2015;17(3):300–15.

133. Morrison KM, Miesegaes GR, Lumpkin EA, Maricich SM. Mammalian Merkel cells are descended from the epidermal lineage. Dev Biol. 2009;336:76–83.
134. Mousavi K, Zare H, Wang AH, Sartorelli V. Polycomb protein Ezh1 promotes RNA polymerase II elongation. Mol Cell. 2012;45(2):255–62.
135. Müller J, Hart CM, Francis NJ, Vargas ML, Sengupta A, Wild B, et al. Histone methyltransferase activity of a Drosophila Polycomb group repressor complex. Cell. 2002;111(2):197–208.
136. Nekrasov M, Klymenko T, Fraterman S, Papp B, Oktaba K, Köcher T, et al. Pcl-PRC2 is needed to generate high levels of H3-K27 trimethylation at Polycomb target genes. EMBO J. 2007;26(18):4078–88.
137. Nelson WG, Sun TT. The 50-and 58-kdalton keratin classes as molecular markers for stratified squamous epithelia: cell culture studies. J Cell Biol. 1983;97(1):244–51.
138. Nowak AJ, Alfieri C, Stirnimann CU, Rybin V, Baudin F, Ly-Hartig N, et al. Chromatin-modifying complex component Nurf55/p55 associates with histones H3 and H4 and polycomb repressive complex 2 subunit Su (z) 12 through partially overlapping binding sites. J Biol Chem. 2011;286(26):23388–96.
139. Nowak JA, Polak L, Pasolli HA, Fuchs E. Hair follicle stem cells are specified and function in early skin morphogenesis. Cell Stem Cell. 2008;3(1):33–43.
140. Oro AE, Higgins K. Hair cycle regulation of Hedgehog signal reception. Dev Biol. 2003;255:238–48. pmid:12648487
141. Ostrowski SM, Wright MC, Bolock AM, Geng X, Maricich SM. Ectopic Atoh1 expression drives Merkel cell production in embryonic, postnatal and adult mouse epidermis. Development. 2015;142(14):2533–44.
142. Owens DM, Lumpkin EA. Diversification and specialization of touch receptors in skin. Cold Spring Harb Perspect Med. 2014;4:a013656.
143. Pasini D, Bracken AP, Jensen MR, Denchi EL, Helin K. Suz12 is essential for mouse development and for EZH2 histone methyltransferase activity. EMBO J. 2004;23(20):4061–71.
144. Pasini D, Cloos PA, Walfridsson J, Olsson L, Bukowski JP, Johansen JV, et al. JARID2 regulates binding of the Polycomb repressive complex 2 to target genes in ES cells. Nature. 2010;464(7286):306–10.
145. Pearson JC, Lemons D, McGinnis W. Modulating Hox gene functions during animal body patterning. Nat Rev Genet. 2005;6(12):893–904.
146. Peng JC, Valouev A, Swigut T, Zhang J, Zhao Y, Sidow A, Wysocka J. Jarid2/Jumonji coordinates control of PRC2 enzymatic activity and target gene occupancy in pluripotent cells. Cell. 2009;139(7):1290–302.
147. Pengelly AR, Kalb R, Finkl K, Müller J. Transcriptional repression by PRC1 in the absence of H2A monoubiquitylation. Genes Dev. 2015;29(14):1487–92.
148. Perdigoto CN, Bardot ES, Valdes VJ, Santoriello FJ, Ezhkova E. Embryonic maturation of epidermal Merkel cells is controlled by a redundant transcription factor network. Development. 2014;141(24):4690–6.
149. Perdigoto CN, Dauber KL, Bar C, Tsai P-C, Valdes VJ, Cohen I, et al. Polycomb-mediated repression and Sonic Hedgehog signaling interact to regulate Merkel cell specification during skin development. PLoS Genet. 2016;12(7):e1006151. https://doi.org/10.1371/journal.pgen.1006151.
150. Pietersen AM, Horlings HM, Hauptmann M, Langerod A, Ajouaou A, Cornelissen-Steijger P, et al. EZH2 and BMI1 inversely correlate with prognosis and TP53 mutation in breast cancer. Breast Cancer Res. 2008;10(6):R109.
151. Plikus MV, Mayer JA, de la Cruz D, Baker RE, Maini PK, Maxson R, Chuong CM. Cyclic dermal BMP signaling regulates stem cell activation during hair regeneration. Nature. 2008;451(7176):340–4.
152. Pokholok DK, Harbison CT, Levine S, Cole M, Hannett NM, Lee TI, et al. Genome-wide map of nucleosome acetylation and methylation in yeast. Cell. 2005;122(4):517–27.

153. Quelle DE, Zindy F, Ashmun RA, Sherr CJ. Alternative reading frames of the *INK4a* tumor suppressor gene encode two unrelated proteins capable of inducing cell cycle arrest. Cell. 1995;83:993–1000.
154. Raaphorst FM, Meijer CJ, Fieret E, Blokzijl T, Mommers E, Buerger H, et al. Poorly differentiated breast carcinoma is associated with increased expression of the human polycomb group EZH2 gene. Neoplasia. 2003;5(6):481–8.
155. Ramos J, Villa J, Ruiz A, Armstrong R, Matta J. UV dose determines key characteristics of nonmelanoma skin cancer. Cancer Epidemiol Biomark Prev. 2004;13(12):2006–11.
156. Ratner D, Nelson BR, Brown MD, Johnson TM. Merkel cell carcinoma. J Am Acad Dermatol. 1993;29(2):143–56.
157. Rayess H, Wang MB, Srivatsan ES. Cellular senescence and tumor suppressor gene p16. Int J Cancer. 2012;130(8):1715–25.
158. Reinisch CM, Uthman A, Erovic BM, Pammer J. Expression of BMI-1 in normal skin and inflammatory and neoplastic skin lesions. J Cutan Pathol. 2007;34(2):174–80.
159. Ressler S, Bartkova J, Niederegger H, Bartek J, Scharffetter-Kochanek K, Jansen-Dürr P, Wlaschek M. p16INK4A is a robust in vivo biomarker of cellular aging in human skin. Aging Cell. 2006;5(5):379–89.
160. Rhee H, Polak L, Fuchs E. Lhx2 maintains stem cell character in hair follicles. Science. 2006;312(5782):1946–9.
161. Ringrose L, Paro R. Epigenetic regulation of cellular memory by the Polycomb and Trithorax group proteins. Annu Rev Genet. 2004;38:413–43.
162. Roscic A, Möller MA, Calzado F, Renner VC, Wimmer E, Gresko KS, et al. Phosphorylation-dependent control of Pc2 SUMO E3 ligase activity by its substrate protein HIKP2. Mol Cell. 2006;24(1):77–89.
163. Rosso S, Zanetti R, Martinez C, Tormo MJ, Schraub S, Sancho-Garnier H, et al. The multicenter south European study 'Helios'. II: different sun exposure patterns in the aetiology of basal cell and squamous cell carcinomas of the skin. Br J Cancer. 1996;73(11):1447–54.
164. Rubin AI, Chen EH, Ratner D. Basal-cell carcinoma. N Engl J Med. 2005;353(21):2262–9.
165. Ruden DM, Lu X. Hsp90 affecting chromatin remodeling might explain transgenerational epigenetic inheritance in Drosophila. Curr Genomics. 2008;9(7):500.
166. Sarma K, Margueron R, Ivanov A, Pirrotta V, Reinberg D. Ezh2 requires PHF1 to efficiently catalyze H3 lysine 27 trimethylation in vivo. Mol Cell Biol. 2008;28(8):2718–31.
167. Sauvageau M, Sauvageau G. Polycomb group proteins: multi-faceted regulators of somatic stem cells and cancer. Cell Stem Cell. 2010;7(3):299–313.
168. Schneider MR, Schmidt-Ullrich R, Paus R. The hair follicle as a dynamic miniorgan. Curr Biol. 2009;19:R132–42. https://doi.org/10.1016/j.cub.2008.12.005.
169. Schuettengruber B, Chourrout D, Vervoort M, Leblanc B, Cavalli G. Genome regulation by polycomb and trithorax proteins. Cell. 2007;128:735–45.
170. Segre JA. Epidermal barrier formation and recovery in skin disorders. J Clin Investig. 2006;116(5):1150.
171. Sellheyer K. Basal cell carcinoma: cell of origin, cancer stem cell hypothesis and stem cell markers. Br J Dermatol. 2011;164(4):696–711.
172. Sen GL, Webster DE, Barragan DI, Chang HY, Khavari PA. Control of differentiation in a self-renewing mammalian tissue by the histone demethylase JMJD3. Genes Dev. 2008;22(14):1865–70.
173. Sennett R, Rendl M. Mesenchymal–epithelial interactions during hair follicle morphogenesis and cycling. Semin Cell Dev Biol. 2012;23:917–27.
174. Serrano M, Gomez-Lahoz E, DePinho RA, Beach D, Bar- Sagi D. Inhibition of ras-induced proliferation and cellular transformation by $p16^{INK4}$. Science. 1995;267:249–52.
175. Shao Z, Raible F, Mollaaghababa R, Guyon JR, Wu CT, Bender W, Kingston RE. Stabilization of chromatin structure by PRC1, a Polycomb complex. Cell. 1999;98(1):37–46.
176. Shaw T, Martin P. Epigenetic reprogramming during wound healing: loss of polycomb-mediated silencing may enable upregulation of repair genes. EMBO Rep. 2009;10(8):881–6.

177. Shen X, Kim W, Fujiwara Y, Simon MD, Liu Y, Mysliwiec MR, et al. Jumonji modulates polycomb activity and self-renewal versus differentiation of stem cells. Cell. 2009;139(7):1303–14.
178. Shen X, Liu Y, Hsu YJ, Fujiwara Y, Kim J, Mao X, et al. EZH1 mediates methylation on histone H3 lysine 27 and complements EZH2 in maintaining stem cell identity and executing pluripotency. Mol Cell. 2008;32(4):491–502.
179. Sherr CJ. The *INK4a/ARF* network in tumour suppression. Nat Rev. 2001;2:731–7.
180. Sherr CJ. Ink4-Arf locus in cancer and aging. Wiley Interdiscip Rev Dev Biol. 2012;1:731–41.
181. Simon JA, Kingston RE. Mechanisms of polycomb gene silencing: knowns and unknowns. Nat Rev Mol Cell Biol. 2009;10(10):697–708.
182. Simon JA, Kingston RE. Occupying chromatin: Polycomb mechanisms for getting to genomic targets, stopping transcriptional traffic, and staying put. Mol Cell. 2013;49(5):808–24.
183. Sparmann A, van Lohuizen M. Polycomb silencers control cell fate, development and cancer. Nat Rev Cancer. 2006;6(11):846–56.
184. St-Jacques B, Dassule HR, Karavanova I, Botchkarev VA, Li J, Danielian PS, et al. Sonic hedgehog signaling is essential for hair development. Curr Biol. 1998;8:1058–68.
185. Surface LE, Thornton SR, Boyer LA. Polycomb group proteins set the stage for early lineage commitment. Cell Stem Cell. 2010;7(3):288–98.
186. Tan J, Yang X, Zhuang L, Jiang X, Chen W, Lee PL, et al. Pharmacologic disruption of Polycomb-repressive complex 2-mediated gene repression selectively induces apoptosis in cancer cells. Genes Dev. 2007;21(9):1050–63.
187. Tavares L, Dimitrova E, Oxley D, Webster J, Poot R, Demmers J, et al. RYBP-PRC1 complexes mediate H2A ubiquitylation at polycomb target sites independently of PRC2 and H3K27me3. Cell. 2012;148(4):664–78.
188. Thomson JP, Skene PJ, Selfridge J, Clouaire T, Guy J, Webb S, et al. CpG islands influence chromatin structure via the CpG-binding protein Cfp1. Nature. 2010;464(7291):1082–6.
189. Tursken K, Troy T-C. Permeability barrier dysfunction in transgenic mice overexpressing claudin 6. Development. 2002;129:1775–84.
190. Vakoc CR, Sachdeva MM, Wang H, Blobel GA. Profile of histone lysine methylation across transcribed mammalian chromatin. Mol Cell Biol. 2006;26:9185–95.
191. Valk-Lingbeek ME, Bruggeman SW, van Lohuizen M. Stem cells and cancer: the polycomb connection. Cell. 2004;118(4):409–18.
192. Van Keymeulen A, Mascre G, Youseff KK, Harel I, Michaux C, De Geest N, et al. Epidermal progenitors give rise to Merkel cells during embryonic development and adult homeostasis. J Cell Biol. 2009;187(1):91–100.
193. van Leenders GJ, Dukers D, Hessels D, van den Kieboom SW, Hulsbergen CA, Witjes JA, et al. Polycomb-group oncogenes EZH2, BMI1, and RING1 are overexpressed in prostate cancer with adverse pathologic and clinical features. Eur Urol. 2007;52(2):455–63.
194. Vasioukhin V, Degenstein L, Wise B, Fuchs E. The magical touch: genome targeting in epidermal stem cells induced by tamoxifen application to mouse skin. Proc Natl Acad Sci U S A. 1999;96:8551–6.
195. Vavouri T, Lehner B. Human genes with CpG island promoters have a distinct transcription-associated chromatin organization. Genome Biol. 2012;13(11):R110.
196. Vidal VP, Chaboissier MC, Lützkendorf S, Cotsarelis G, Mill P, Hui CC, et al. Sox9 is essential for outer root sheath differentiation and the formation of the hair stem cell compartment. Curr Biol. 2005;15(15):1340–51.
197. Volz A, Korge BP, Compton JG, Ziegler A, Steinert PM, Mischke D. Physical mapping of a functional cluster of epidermal differentiation genes on chromosome 1q21. Genomics. 1993;18:92–9.
198. Voncken JW, Roelen BA, Roefs M, de Vries S, Verhoeven E, Marino S, et al. Rnf2 (Ring1b) deficiency causes gastrulation arrest and cell cycle inhibition. Proc Natl Acad Sci. 2003;100(5):2468–73.

199. Walker E, Chang WY, Hunkapiller J, Cagney G, Garcha K, Torchia J, et al. Polycomb-like 2 associates with PRC2 and regulates transcriptional networks during mouse embryonic stem cell self-renewal and differentiation. Cell Stem Cell. 2010;6(2):153–66.
200. Wang H, Wang L, Erdjument-Bromage H, Vidal M, Tempst P, Jones RS, Zhang Y. Role of histone H2A ubiquitination in Polycomb silencing. Nature. 2004a;431(7010):873–8.
201. Wang L, Brown JL, Cao R, Zhang Y, Kassis JA, Jones RS. Hierarchical recruitment of polycomb group silencing complexes. Mol Cell. 2004b;14(5):637–46.
202. Whitcomb SJ, Basu A, Allis CD, Bernstein E. Polycomb group proteins: an evolutionary perspective. Trends Genet. 2007;23(10):494–502.
203. Wong CSM, Strange RC, Lear JT. Basal cell carcinoma. BMJ [Br. Med. J.]. 2003;327(7418):794.
204. Woo SH, Ranade S, Weyer AD, Dubin AE, Baba Y, Qiu Z, et al. Piezo2 is required for Merkel-cell mechanotransduction. Nature. 2014;509(7502):622–6.
205. Wu X, Johansen JV, Helin K. Fbxl10/Kdm2b recruits polycomb repressive complex 1 to CpG islands and regulates H2A ubiquitylation. Mol Cell. 2013;49(6):1134–46.
206. Xiang Y, Zhu Z, Han G, Lin H, Xu L, Chen CD. JMJD3 is a histone H3K27 demethylase. Cell Res. 2007;17(10):850–7.
207. Xiao Y, Thoresen DT, Miao L, Williams JS, Wang C, Atit RP, et al. A Cascade of Wnt, Eda, and Shh signaling is essential for touch dome Merkel cell development. PLoS Genet. 2016;12(7):e1006150. https://doi.org/10.1371/journal.pgen.1006150.
208. Xie Q, Wang H, Heilman ER, Walsh MG, Haseeb MA, Gupta R. Increased expression of enhancer of Zeste homolog 2 (EZH2) differentiates squamous cell carcinoma from normal skin and actinic keratosis. Eur J Dermatol. 2014;24(1):41–5. https://doi.org/10.1684/ejd.2013.2219.
209. Yang A, Schweitzer R, Sun D, Kaghad M, Walker N, Bronson RT, et al. p63 is essential for regenerative proliferation in limb, craniofacial and epithelial development. Nature. 1999;398:714–8.
210. Yap KL, Li S, Muñoz-Cabello AM, Raguz S, Zeng L, Mujtaba S, et al. Molecular interplay of the noncoding RNA ANRIL and methylated histone H3 lysine 27 by polycomb CBX7 in transcriptional silencing of INK4a. Mol Cell. 2010;38(5):662–74.
211. Yu M, Mazor T, Huang H, Huang HT, Kathrein KL, Woo AJ, et al. Direct recruitment of polycomb repressive complex 1 to chromatin by core binding transcription factors. Mol Cell. 2012;45(3):330–43.
212. Zeidler M, Kleer CG. The Polycomb group protein Enhancer of Zeste 2: its links to DNA repair and breast cancer. J Mol Histol. 2006;37(5–7):219–23.
213. Zenz R, Eferl R, Kenner L, Florin L, Hummerich L, Mehic D, et al. Psoriasis-like skin disease and arthritis caused by inducible epidermal deletion of Jun proteins. Nature. 2005;437(7057):369–75.
214. Zhang P, Su Y, Zhao M, Huang W, Lu Q. Abnormal histone modifications in PBMCs from patients with psoriasis vulgaris. Eur J Dermatol. 2011;21(4):552–7.
215. Zhao XP, Elder JT. Positional cloning of novel skin-specific genes from the human epidermal differentiation complex. Genomics. 1997;45:250–8.
216. Zhao J, Sun BK, Erwin JA, Song JJ, Lee JT. Polycomb proteins targeted by a short repeat RNA to the mouse X chromosome. Science. 2008;322(5902):750–6.

Chapter 4
Trithorax Genes in the Control of Keratinocyte Differentiation

Rachel Herndon Klein and Bogi Andersen

4.1 The history of Trithorax: Developmental Regulation Requires a Balance Between Trithorax and Polycomb Group Regulation

Trithorax and its antagonizing protein complex Polycomb were initially characterized as regulators of *Drosophila* body segment identity during development. This complex process is regulated in part by several classes of homeotic genes, which are organized into clusters along the chromosomes, where their expression is spatiotemporally regulated [1]. In these clusters, gene activation proceeds linearly across the cluster; the first genes in the cluster are expressed earlier in development in the more anterior body segments; genes at the end of the cluster are expressed later in development in the more posterior segments. Once HOX gene expression in a body segment is established, it must be maintained throughout development. Polycomb, and later Trithorax, were identified through the study of mutations that caused disruption of segmental identity maintenance [2–5]. Mutations that caused loss Polycomb protein function resulted in upregulation of genes in HOX clusters, causing body segments to acquire a more posterior segment identity. Mutations in Trithorax complex proteins were found to cause an opposite phenotype to Polycomb mutations: loss of homeotic gene expression, with body segments acquiring more anterior segment identities.

Further studies identified these developmental regulators as two multi-protein complexes, each harboring a chromatin-modifying enzyme. Although they mainly have antagonistic functions in development, Trithorax and Polycomb are both histone methyltransferases of the same class; their enzymatic activity comes from a

R. H. Klein · B. Andersen (✉)
Departments of Biological Chemistry and Medicine, University of California, Irvine, CA, USA
e-mail: bogi@uci.edu

© Springer International Publishing AG, part of Springer Nature 2018

V. A. Botchkarev, S. E. Millar (eds.), *Epigenetic Regulation of Skin Development and Regeneration*, Stem Cell Biology and Regenerative Medicine,
https://doi.org/10.1007/978-3-319-16769-5_4

SET domain, so named because it is the catalytic domain of the *Drosophila* methyltransferases **S**u(var) 3–9, **E**nhancer of zeste (the polycomb methyltransferase), and **T**rithorax [6]. Each enzyme differs in the lysine it targets: Su(var) 3–9 methylates lysine 9, Enhancer of Zeste methylates lysine 27, and Trithorax methylates lysine 4. Whereas certain histone modifications such as phosphorylation and acetylation activate gene expression through neutralizing the positive charge on histone proteins, causing a decrease in the association between protein and DNA, histone methylation can be either activating or repressive. The effect of a given histone methylation largely depends on the specific "chromatin readers" that recognize and bind to it; these readers then recruit additional factors to bring about a change in gene regulation. For example, in humans, the activating methyl mark H3K4me3 at promoters recruits chromatin readers containing PHD domains, including ING3, which is stabilized at the promoter through interaction with the H3K4me3 modified tail. This allows ING3 to recruit histone acetyltransferases, further intensifying the active chromatin marks at gene promoters [7, 8]. Further signal amplification is provided by the Trithorax component protein WDR5, that can bind to H3K4me3 marked chromatin [9], helping to maintain and spread the H3K4me3 signal.

While the antagonistic roles of Polycomb and Trithorax are well established, recent work also points to a more complex and flexible definition of Polycomb and Trithorax gene regulatory functions, one that depends highly on context. In certain regulatory situations, and at certain gene loci, a subset of Polycomb group proteins appears to interact genetically with Trithorax proteins, increasing the severity of the Trithorax mutant phenotype [10]. Additionally, many transcriptional regulators can recruit either Polycomb or Trithorax to the genes they regulate, depending on the context. The defining role of Trithorax as an activating complex and Polycomb as a repressor is also not absolute. More recent studies have shown a role for Trithorax regulation in attenuating Hox gene expression, by activating transcription of non-coding RNAs from the HOX locus [11].

In addition, Trithorax and Polycomb complexes work together to create bivalent domains. This phenomenon has been particularly well characterized in mammalian embryonic stem cells, where H3K4me3 activating and H3K27me3 repressive marks co-occur at gene promoters. This allows for genes to be "poised" for rapid expression upon receiving a signal to differentiate. Intriguingly, these bivalent domains have not been found in *Drosophila*, but are common in Zebrafish and mammals, highlighting a difference in Polycomb and Trithorax function among different species [12, 13].

4.2 Trithorax Complex Members Regulate Methylation of Lysine 4 on Histone H3

Lysine 4 on the N-terminal tail of histone H3 can be mono-, di-, or tri-methylated. While all three modifications can be found at active promoters, trimethylation (H3K4me3), which localizes directly to the proximal promoter, has been most strongly associated with gene transcription. H3K4 dimethylation is found at active promoters, at enhancers, and within the body of genes undergoing transcription [14]. H3K4 monomethylation is present at distal regulatory regions, including active and poised enhancers; it has also been identified more recently at the 3 prime ends of actively transcribed genes [15]. Recent research points to a role of H3K4me1, which often encompasses broad domains that fully contain other marks, in restricting readers of other chromatin marks to specific domains, for example, directing readers of H3K4me3 like ING proteins to the proximal promoter regions of active genes.

Whereas many different enzymes can catalyze the transfer of a methyl group to lysine residues on a protein, all Trithorax family methyltransferases are characterized by the specific SET methyltransferase domain. The SET1 family of histone lysine methyltransferases catalyze the transfer of methyl groups to lysine 4 of histone H3. In humans, there are six such enzymes: MLL1-4 and SETd1A/B. MLL1 and MLL4 are most closely related to *Drosophila* Trithorax, while MLL2 and MLL3 are homologous to *Drosophila* Trithorax-related (Fig. 4.1) [16]. Each SET1 family enzyme interacts with the WRAD complex of proteins, so named because it comprises **W**D-40 repeat protein 5 (WDR5), **R**etinoblastoma-Binding Protein 5 (RBBP5), **A**bsent Small Homeotic-2-like (ASH2L), and **D**umpy-30 (DPY-30) (Fig. 4.2) [17, 18]. Although the exact function of each of these WRAD complex members is unclear, their association with a SET1 family member dramatically increases the enzyme's methyltransferase activity [17]. Recent work has pointed to an important role for WDR5 in interacting with non-coding RNA, which has been shown to be required for proper H3K4me3 and Trithorax complex recruitment in embryonic stem cells, as well as developmental contexts [19–21].

Drosophila Set1 and its human counterparts SETD1A and SETD1B are thought to be the primary H3K4 -trimethylating SET1 enzymes, while the Trithorax-related homologs MLL4 and MLL3 have recently been shown to be responsible for H3K4 monomethylation at enhancers [22]. Trithorax homologs MLL1 and MLL4 have both been identified as mediators of H3K4 trimethylation at select loci, including bivalently-marked gene promoters in embryonic stem cells [16].

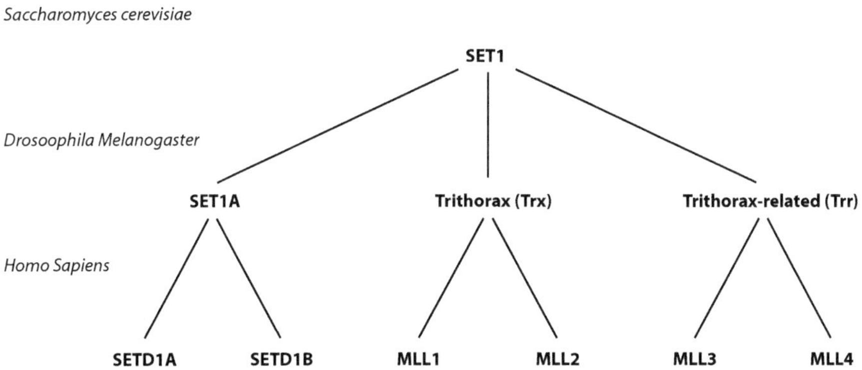

Fig. 4.1 Comparison of H3K4 methylating SET enzymes between *S. cerevisiae*, *D. melanogaster*, and *H. sapiens* [16]

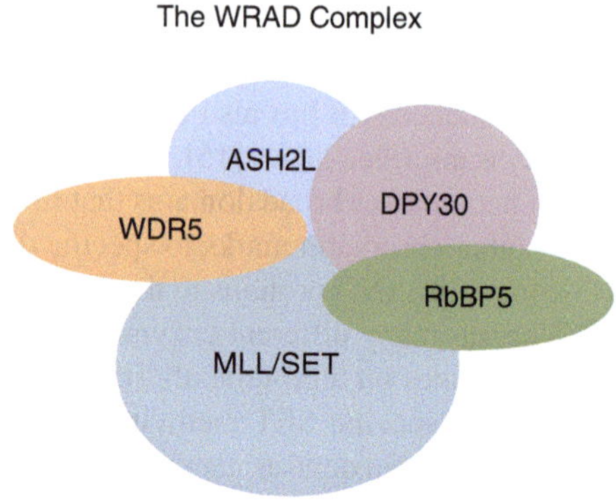

Fig. 4.2 The WRAD complex associates with Trithorax SET enzymes [17]

4.3 The Functions of Mammalian Trithorax Complexes

The importance of Trithorax complexes in development is further emphasized by the fact that knockout mice for SET1 family members are fully or partially embryonic lethal. Even heterozygotes for many Trithorax SET enzymes display overt phenotypes, indicating haploinsufficiency. *Mll1*+/− mice have a growth defect, hematological abnormalities, skeletal malformations and aberrant HOX gene expression [23]. *Mll2* (human MLL4 homolog)+/− mice have misregulated glucose homeostasis; *Mll4* (human MLL2 homolog)+/− and *Mll3*−/− mice also display growth defects and have reduced fertility [24–26]. These findings highlight the crucial role of Trithorax complexes in a wide range of developmental processes. As in *Drosophila*, Trithorax complexes target Hox genes and regulate mammalian embryonic patterning. The variable phenotypes of different SET domain gene mutants also point to the fact that, while there may be a level of redundancy between different Trithorax complexes as demonstrated by shared enzymatic activity, each has unique and vital functions in the development of the organism.

4.4 The Epidermis as a Developmental Model System

The epidermis is a continually renewing tissue that protects the organism from the external environment and prevents loss of water from the inside. As proliferative cells in the basal layer move upward in the epidermis, they activate a terminal differentiation gene expression program concurrent with a loss of proliferative capacity [27]. Cells in the rigid, uppermost layer of the epidermis, the cornified layer, are dead, continually being sloughed off from the surface only to be replaced by cells moving up from lower layers. Due to its continually replenishing nature, and well-characterized stepwise process of differentiation, the epidermis makes an ideal tissue for studying the complex regulatory events of terminal differentiation and progenitor cell maintenance. In addition, as the largest and most external organ, material is readily available for study from biopsy and surgical procedures, making the epidermis an attractive mammalian model system for the study of tissue homeostasis and disease.

To accomplish its barrier function, a differentiating keratinocyte must activate a battery of genes encoding diverse classes of proteins, including cell adhesion molecules that link cells tightly to each other, intermediate filament keratins that provide strength and structure, lipid modifying enzymes that seal the barrier and prevent moisture loss, and structural proteins that form the rigid cornified envelope [28]. Many of these genes are clustered together, possibly to facilitate their coordinated regulation during differentiation. Genes involved in cornified envelope formation are clustered together in a region of human chromosome 1 known as the epidermal differentiation complex (EDC). Keratin genes also cluster together on chromosomes 12 and 17.

The importance of chromatin regulation in epidermal differentiation dynamics has been well established through studies showing roles for a varied complement of chromatin enzymes in the epidermis. Many of these epigenetic regulators have dynamic expression patterns during epidermal differentiation. For example, the polycomb complex was found to be a repressor of the epidermal differentiation program as well as the differentiation of epidermal progenitors into Merkel cells, maintaining keratinocytes in the basal, progenitor state [29, 30]. Upon differentiation, levels of Ezh2, the catalytic component of Polycomb, decrease in keratinocytes concomitant with a release of epidermal differentiation genes from repression. Chromatin remodeling enzymes Satb1 and Brg1 are direct targets of the master epidermal regulator p63 and are required for proper remodeling of the EDC during differentiation [31, 32]. The downregulation of ACTL6A during differentiation is also required for Brg1-containing SWI/SNF complexes to remodel chromatin and activate differentiation genes [33].

4.5 Trithorax Proteins Regulate the Epidermal Differentiation Program Through Interaction with GRHL3

In contrast to other chromatin regulators whose expression levels change within a tissue in response to different physiological conditions, SET1 family members are highly expressed in the epidermis throughout development, homeostasis, and during wound healing [34]. This indicates that the presence of Trithorax complexes in the cell is not sufficient to activate the epidermal differentiation program, and that targeting of Trithorax to the proper gene loci requires additional factors. While levels of each Trithorax complex member remain relatively constant across the differentiation process, there is great variation in the levels of different Trithorax complex proteins within the epidermis. In differentiating keratinocytes, ASH2L is the most highly expressed Trithorax-associated protein, followed by WDR5 and MLL1 [34]. High expression of MLL1 and WDR5 is also observed in human whole epidermis samples; however, in the mouse, when dermis and epidermis are separated, Mll2

(MLL4 in humans) is more highly expressed in the epidermal fraction [34]. It is likely that unique combinations of Trithorax complexes are at work at the same time and in the same tissue, allowing for regulation of unique subsets of genes by these different Trithorax complexes. The composition of the Trithorax complex can also influence the enzymatic activity of the SET1 protein towards H3K4 mono-, di- or tri- methylation. Therefore, it is possible that multiple Trithorax complexes catalyze different methylations of lysine 4 at the same promoter, conferring additional layers of regulation of epidermal gene expression.

There is strong evidence that multiple Trithorax complexes are involved in epidermal differentiation, as knockdown of the SET1 enzymes MLL1, MLL2, MLL4 and the Trithorax core component WDR5 in human epidermal keratinocytes causes a reduction in expression of epidermal differentiation marker gene *TGM1* [34]. Of these Trithorax complex members, only MLL2 and WDR5 bind directly to the *TGM1* promoter, indicating that MLL2-containing Trithorax complexes are direct regulators of *TGM1*, and suggesting that the other complexes regulate *TGM1* indirectly through promotion of epidermal differentiation at other targets. It is also possible that other Trithorax complexes regulate *TGM1* through catalyzing H3K4 monomethylation at distal enhancer regions, rather than regulating H3K4 methylation in the proximal promoter.

TGM1 is also directly regulated by the epidermal differentiation-promoting transcription factor Grainyhead-like 3 (GRHL3), suggesting a potential interaction between MLL2 and GRHL3. The mammalian grainyhead family of transcription factors is homologous to a single *Drosophila* transcription factor, Grh, which plays numerous roles in development, including regulation of formation of the larval cuticle, a structure with homology to mammalian epidermis [35, 36]. *Grhl3* knockout mice die at birth with numerous abnormalities including defects in the epidermal barrier [37]. Global gene expression and ChIP-Seq experiments in mouse epidermis and human keratinocytes revealed that GRHL3 is important for coordinating the regulation of epidermal differentiation genes, as well as the activation of other transcriptional regulators. The role of GRHL3 extends beyond initial creation of the barrier. GRHL3 is also required for repair after injury, including immune-mediated insults [38], for keratinocyte migration [39], to close both embryonic and adult wounds [40], and for barrier formation in other epithelia, including bladder epithelia [41]. Depletion of GRHL3 during epidermal differentiation causes a reduction of MLL2 localization and H3K4 methylation at the promoters of genes it regulates [34]. Additional support for GRHL3 recruitment of MLL2-containing Trithorax complexes is provided by the observation that knockdown of GRHL3 or MLL2 affects a significant shared set of genes in epidermal keratinocytes; the most significant overlap is in genes that are classified as “late” differentiation genes, normally showing an increase in expression during the later stages of differentiation [34]. This is consistent with the defined role of GRHL3 in promoting terminal differentiation in the suprabasal layers of the epidermis, and indicates that GRHL3 effects gene activation through recruitment of Trithorax complexes containing MLL2 (Fig. 4.3).

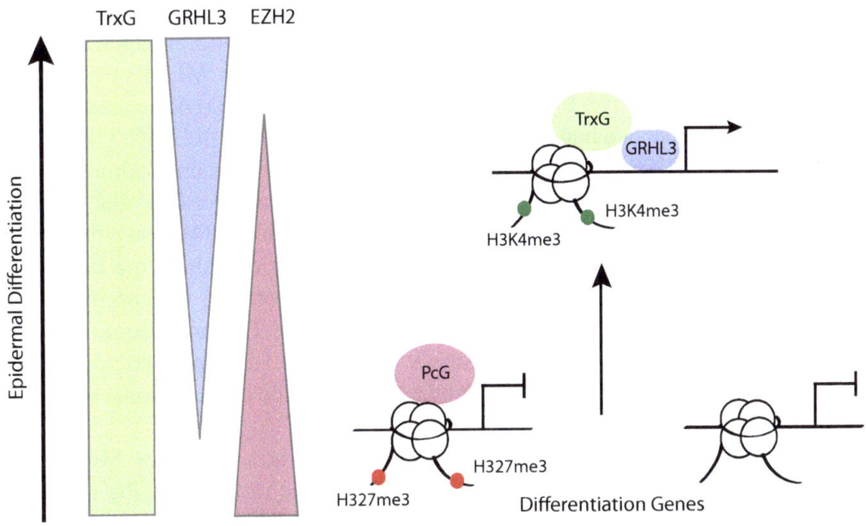

Fig. 4.3 Model for GRHL3, Polycomb, and Trithorax–mediated regulation of epidermal differentiation genes [29, 34]

In contrast to GRHL3, MLL2 regulates a large fraction of genes expressed in epidermal keratinocytes, indicating a more general role in transcriptional activation, rather than being a selective regulator of the epidermal differentiation program. Indeed, siRNA knockdown of MLL2 affects a large swath of epidermal genes regardless of their specific expression pattern, suggesting it works with a number of transcriptional regulators that may provide additional temporal and spatial specificity to gene activation [34]. The large number of genes that are affected by MLL2 in epidermal keratinocyte differentiation but that do not appear to be regulated by GRHL3 suggests that other epidermal transcription factors also may promote differentiation through recruitment of MLL2 to their target genes. Conversely, GRHL3 may recruit other Trithorax complexes to the genes it regulates independently of MLL2, or it may act through alternative mechanisms to activate these target genes.

Of note, *GRHL3* is also downregulated by MLL2 knockdown [34]; it is not known whether this represents a feed-forward positive regulatory loop, whereby GRHL3 activates its own expression through recruitment of MLL2, or whether this is an example of an alternative transcription factor working with MLL2-containing complexes to activate *GRHL3*. While GRHL3 was not detected binding to its own proximal promoter during differentiation, it does bind to enhancers within the *GRHL3* gene body, suggesting that positive auto-regulation is possible. Consistent with this idea, WDR5 localizes with GRHL3 to these enhancer regions. However, it is also found in the proximal promoter of *GRHL3* indicating that other transcriptional regulators may recruit Trithorax complexes to the *GRHL3* promoter to activate expression [34].

Further evidence that GRHL3 acts by recruiting Trithorax is provided by the observation that GRHL3 and core Trithorax member WDR5 interact directly; GRHL3 also appears to interact weakly with MLL2, but not MLL1 or SETD1A [34]. While not conclusive, this suggests that MLL1 and SETD1A are recruited by GRHL3-independent mechanisms to gene targets they share with GRHL3 in epidermal keratinocytes. The interaction between GRHL3 and certain Trithorax complexes occurs genome-wide, as ChIP-Seq experiments revealed that WDR5 co-localizes with no less than 88% of GRHL3 bound regions [34]. Forty-three percent of genes with a keratinocyte-differentiation expression pattern are bound by GRHL3 and WDR5, indicating that GRHL3 provides some of the specificity for Trithorax targeting in the epidermis. This mechanism of Trithorax recruitment by GRHL3 may also extend to other tissues where GRHL3 is expressed, and likely reflects a more general paradigm, whereby a tissue-specific transcriptional regulator recruits Trithorax to gene targets.

Intriguingly, MLL1 and SETD1A are also found at a subset of MLL2 and GRHL3 regulated genes [34]. This suggests that multiple variations of the Trithorax complex are present at active genes, and raises the question of the mechanism of recruitment for these complexes. Other transcription factors could be involved in their recruitment; the combinatorial recruitment of different Trithorax complexes by different transcription factors to the same gene targets could allow for redundancy in the system, thus amplifying transcription. The unique composition of these complexes may influence the enzymatic activity of the SETD1 enzyme for specific substrates; one Trithorax complex may catalyze trimethylation while another regulates mono- or di- methylation.

4.6 Trithorax Group Proteins Regulate Gene Clusters in Epidermal Differentiation

During development, an organism must pattern itself to allow for the proper organization and specification of cells to carry out the unique functions of each tissue. Similarly, after injury and tissue damage, patterning must be re-established to restore proper tissue function. Hox genes play important roles in many cell types during wound healing [42] and are also likely important for the re-establishment of the epidermal barrier by keratinocytes. GRHL3 is also a crucial regulator of both the embryonic and adult keratinocyte response to wounding, acting in part through regulation of the planar cell polarity pathway [40]. It is possible that GRHL3 and HOX gene regulation intersect again in this context, and that selective recruitment of Polycomb or Trithorax components by GRHL3 is required to effect the changes in HOX gene expression necessary for proper re-establishment of the epidermal barrier.

Drosophila Polycomb and Trithorax regulate Hox gene clusters during development, a role that is conserved from flies to mammals. In mammalian systems like the epidermis, other key developmental genes also exist in clusters where they are co-regulated. There is evidence that Polycomb and Trithorax complexes also regulate these regions during development and tissue differentiation. Indeed, many genes in the EDC are repressed by Polycomb in proliferating keratinocytes [29], and activated by MLL2 knockdown [34], suggesting that the classic mechanism of opposing Polycomb and Trithorax complexes at clustered gene loci is involved in regulation of the epidermal differentiation gene program.

4.7 There are Multiple Mechanisms of Epidermal Differentiation Gene Activation

Initial studies focused on the antagonistic relationship between Polycomb and Trithorax at shared gene targets. However, studies of the epidermis suggest an independent role for Trithorax complexes at a subset of genes. While the repressive H3K27me3 modification is reduced concurrent with an increase in H3K4me3 during the process of differentiation at many genes, a number of epidermal differentiation genes lack any H3K27me3, even in the undifferentiated state, when the gene is not expressed. These genes show a strong increase in H3K4me3 upon epidermal differentiation [34]. It is possible that H3K27me3 marks these genes at some point along their lineage trajectory, but that this mark is removed before acquisition of keratinocyte cell fate. The uncoupling of Polycomb repression and Trithorax activation may allow such genes to respond rapidly to signals for keratinocyte differentiation, as they do not need to recruit H3K27 demethylase enzymes to remove repressive marks before gene activation can occur. Alternatively, such genes may be primarily activated by transcription factors like GRHL3 that are only expressed in differentiating keratinocytes, and therefore do not need to be actively repressed in undifferentiated cells where they cannot be activated due to absence of the activating transcription factor. This separation of Polycomb and Trithorax regulatory functions at certain gene promoters may also be a more general mechanism active in a number of tissue types.

Recent work has identified regulatory domains, termed super-enhancers, which are characterized by the presence of clusters of typical enhancers, resulting in long stretches of active regulatory domains, and very high levels of histone modifications and transcription factor binding [43, 44]. In accordance with their high levels of H3K27ac, H3K4me3, and other active histone marks, these domains often overlap highly expressed genes. Intriguingly, these super enhancer-linked genes are not housekeeping genes, which are expressed highly in a number of cell types, but rather are tissue-specific genes that play a role in cell identity and carry out the functions of the particular cell type. Trithorax complexes have been identified at broad enhancer domains in *Drosophila* that appear comparable to the super enhancers described in mammalian systems [45]; this is perhaps not surprising as Trithorax complexes are responsible for the deposition of the H3K4me1 histone mark found at regulatory regions. Indeed, in epidermal keratinocytes, the binding of core Trithorax complex member WDR5 frequently occurs outside of proximal promoter regions [34]. GRHL3 binding in differentiating keratinocytes also frequently occurs outside of promoters and GRHL3 has been shown to bind to more than 90% of super enhancers in epidermal keratinocytes [46]. While GRHL3 alone is not responsible for the formation of the majority of super enhancers, it may work with a number of other epidermal transcriptional regulators to create super enhancer domains, possibly through its role in recruitment of chromatin regulatory factors like the Trithorax complex to sites of super enhancers. Intriguingly, GRHL3 also acts to repress the formation of spurious super enhancers in keratinocytes [46].

4.8 Evidence for GRH Interaction with PcG Proteins

In flies, Grh is highly expressed in the early embryo where it has been linked to the process of pattern specification. Fly studies have shown that Grh can interact with several Polycomb group complex members, potentially recruiting Polycomb to gene targets, including the Hox genes encoding the bithorax complex [47]. It is possible that mammalian GRHL proteins have diverged from their *Drosophila* counterpart and acquired a new mechanism for transcriptional regulation through interaction with Trithorax instead of Polycomb. Since Grh can also repress gene expression, the more likely explanation may be that Grh switches roles between an activator and repressor, binding to Trithorax or Polycomb depending on the regulatory context. This line of reasoning also opens up the possibility that mammalian GRHL proteins, including GRHL3, may interact with Polycomb in certain contexts to repress gene expression. Evidence for GRHL3 mediated repression through epigenetic mechanisms also comes from the finding that during keratinocyte migration GRHL3 binds gene targets with REST [46], a protein originally identified as a repressor or neuronal genes and cell fate [48]. This GRHL3-REST interaction promotes migration by repressing the expression of genes that inhibit keratinocyte cell movement [46] and points to an important role for GRHL3 in modulating the epigenetic environment to repress gene expression, in addition to its role in promoting gene activation through recruitment of Trithorax complexes. Many studies in numerous different cell types have demonstrated a direct interaction between REST and polycomb family members, including those of the Cbx family [49, 50]. While no direct association between GRHL3 and polycomb proteins has been established, it is certainly possible that some of the GRHL3-REST mediated gene repression occurs through interaction with polycomb complexes.

4.9 Diseases Associated with Trithorax Complexes

Aberrant epigenetic regulation has been linked to many diseases, including cancer and developmental defects. Given their roles as crucial regulators of gene expression in many cell types, it is not surprising that Trithorax and Polycomb are mutated in a variety of cancers. MLL proteins are so named because fusion proteins placing the N terminus of MLL1 together with another protein are common in mixed lineage leukemias. Intriguingly, the SET domain of MLL1 is located near the C terminus, and is not included in these fusion proteins, indicating that MLL1 promotes leukemiagenesis through mechanisms other than H3K4 methylation [51]. One common fusion is between MLL1 and the C terminus of DOT1, another methyltransferase that creates the active H3K79 methylation mark [52]. Because the MLL1 fragment can still bind DNA, this fusion results in incorrect targeting of H3K79me3 to domains that should have H3K4 methylation, an error that may promote the progression of leukemia.

In addition to roles in the progression of cancers of the hematopoietic system, MLLs are mutated in a number of solid tumors, including those of epithelial tissues. While *MLL4* is mutated in a significant proportion of hepatocellular carcinomas [53], *MLL3* is mutated in a wide range of cancers, including aggressive squamous cell carcinoma [54]. Furthermore, mutations in *MLL2*, *MLL3*, and the Polycomb enzyme EZH2 have been identified in a significant number of head and neck squamous carcinomas, a tobacco use-associated cancer that develops from epithelial mucosa in the upper aerodigestive tract [55]. In contrast to the gene fusions that occur in leukemia, MLL mutations found in epithelial cancers are inactivating, pointing to a tumor suppressive role for MLL2 and MLL3 in epithelia, possibly through promotion of terminal differentiation. Consistent with this idea, GRHL proteins suppress tumorigenesis both in the skin [56, 57] and the oral cavity [58].

Due to the role of Trithorax in regulation of HOX and other developmental clusters of genes, mutations in MLLs also cause several developmental syndromes that share similar phenotypes; syndromes caused by mutation in MLL proteins are generally characterized by growth restriction leading to short stature and low weight, intellectual disability and facial structure abnormalities [59, 60]. Patients with these syndromes also have abnormal epidermal ridges on palmoplantar epidermis and abnormal hair growth, including long eyelashes, and excessive hair on forearms and elbows [60]. These findings suggest a role for MLL proteins in the establishment of epidermal structures during development, and in the hair follicle, an epidermal appendage. These syndromes occur with the mutation of a single copy of the MLL gene in question, leading to haploinsufficiency.

4.10 Conclusion

Over the course of evolution, Trithorax group proteins have maintained many of their functions, including their role in antagonism of Polycomb group proteins and in regulation of HOX gene clusters. They have also acquired novel regulatory roles. For certain classes of epidermal genes, the processes of Polycomb-mediated repression and Trithorax-mediated activation are uncoupled, occuring in spatial or temporal isolation from each other. These loci are not marked by H3K27me3 even when their associated genes are not expressed, requiring only the addition of H3K4me3 during gene activation. In the epidermis, Trithorax proteins are broad transcriptional activators that gain specificity by interacting with tissue-specific transcription factors; GRHL3 recruitment of MLL2-containing complexes to epidermal differentiation genes provides one such example.

References

1. Mungpakdee S, Seo HC, Chourrout D. Spatio-temporal expression patterns of anterior Hox genes in Atlantic salmon (Salmo salar). Gene Expr Patterns. 2008;8(7–8):508–14. https://doi.org/10.1016/j.gep.2008.06.004. Epub 07/16. Epub 2008 Jun 21. PubMed PMID: 18620079.
2. Lewis EB. A gene complex controlling segmentation in Drosophila. Nature. 1978;276(5688):565–70. Epub 12/07. PubMed PMID: 103000.
3. Lewis EB. Control of body segment differentiation in Drosophila by the bithorax gene complex. Prog Clin Biol Res. 1982;85(Pt A):269–88. Epub 01/01. PubMed PMID: 7111279.
4. Ingham PW. Differential expression of bithorax complex genes in the absence of the extra sex combs and trithorax genes. Nature. 1983;306(5943):591–3. Epub 12/08. PubMed PMID: 24937863.
5. Ingham PW. A clonal analysis of the requirement for the trithorax gene in the diversification of segments in Drosophila. J Embryol Exp Morpholog. 1985;89:349–65. Epub 10/01. PubMed PMID: 4093752.
6. Shinsky SA, Monteith KE, Viggiano S, Cosgrove MS. Biochemical reconstitution and phylogenetic comparison of human SET1 family core complexes involved in histone methylation. J Biol Chem. 2015;290(10):6361–75. https://doi.org/10.1074/jbc.M114.627646. Epub 01/07. Epub 2015 Jan 5. PubMed PMID: 25561738.
7. Doyon Y, Selleck W, Lane WS, Tan S, Cote J. Structural and functional conservation of the NuA4 histone acetyltransferase complex from yeast to humans. Mol Cell Biol. 2004;24(5):1884–96. Epub 02/18. PubMed PMID: 14966270.
8. Pena PV, Davrazou F, Shi X, Walter KL, Verkhusha VV, Gozani O, et al. Molecular mechanism of histone H3K4me3 recognition by plant homeodomain of ING2. Nature. 2006;442(7098):100–3. https://doi.org/10.1038/nature04814. Epub 05/27. PubMed PMID: 16728977.
9. Wysocka J, Swigut T, Milne TA, Dou Y, Zhang X, Burlingame AL, et al. WDR5 associates with histone H3 methylated at K4 and is essential for H3 K4 methylation and vertebrate development. Cell. 2005;121(6):859–72. https://doi.org/10.1016/j.cell.2005.03.036. Epub 06/18. PubMed PMID: 15960974.
10. Bajusz I, Sipos L, Gyorgypal Z, Carrington EA, Jones RS, Gausz J, et al. The Trithorax-mimic allele of Enhancer of zeste renders active domains of target genes accessible to polycomb-group-dependent silencing in Drosophila melanogaster. Genetics. 2001;159(3):1135–50. Epub 12/01. PubMed PMID: 11729158.
11. Petruk S, Sedkov Y, Riley KM, Hodgson J, Schweisguth F, Hirose S, et al. Transcriptional elongation of non-coding bxd RNAs promoted by the Trithorax TAC1 complex represses Ubx by a transcriptional interference mechanism. Cell. 2006;127(6):1209–21. https://doi.org/10.1016/j.cell.2006.10.039. PubMed PMID: PMC1866366.
12. Schuettengruber B, Ganapathi M, Leblanc B, Portoso M, Jaschek R, Tolhuis B, et al. Functional anatomy of polycomb and trithorax chromatin landscapes in Drosophila embryos. PLoS Biol. 2009;7(1):e13. https://doi.org/10.1371/journal.pbio.1000013. Epub 01/16. PubMed PMID: 19143474.

13. Vastenhouw NL, Zhang Y, Woods IG, Imam F, Regev A, Liu XS, et al. Chromatin signature of embryonic pluripotency is established during genome activation. Nature. 2010;464(7290):922–6. https://doi.org/10.1038/nature08866. Epub 03/26. Epub 2010 Mar 24. PubMed PMID: 20336069.
14. Pekowska A, Benoukraf T, Zacarias-Cabeza J, Belhocine M, Koch F, Holota H, et al. H3K4 trimethylation provides an epigenetic signature of active enhancers. EMBO J. 2011;30(20):4198–210. https://doi.org/10.1038/emboj.2011.295. Epub 08/19. PubMed PMID: 21847099.
15. Heintzman ND, Stuart RK, Hon G, Fu Y, Ching CW, Hawkins RD, et al. Distinct and predictive chromatin signatures of transcriptional promoters and enhancers in the human genome. Nat Genet. 2007;39(3):311–8. https://doi.org/10.1038/ng1966. PubMed PMID: 17277777.
16. Hubert A, Henderson JM, Ross KG, Cowles MW, Torres J, Zayas RM. Epigenetic regulation of planarian stem cells by the SET1/MLL family of histone methyltransferases. Epigenetics. 2013;8(1):79–91. https://doi.org/10.4161/epi.23211. Epub 2012/12/14. PubMed PMID: 23235145; PubMed Central PMCID: PMCPmc3549883.
17. Justin N, De Marco V, Aasland R, Gamblin SJ. Reading, writing and editing methylated lysines on histone tails: new insights from recent structural studies. Curr Opin Struct Biol. 2010;20(6):730–8. https://doi.org/10.1016/j.sbi.2010.09.012. Epub 10/20. Epub 2010 Oct 16. PubMed PMID: 20956082.
18. Ernst P, Vakoc CR. WRAD: enabler of the SET1-family of H3K4 methyltransferases. Brief Funct Genomics. 2012;11(3):217–26. https://doi.org/10.1093/bfgp/els017. Epub 06/02. Epub 2012 May 30. PubMed PMID: 22652693.
19. Wang KC, Yang YW, Liu B, Sanyal A, Corces-Zimmerman R, Chen Y, et al. A long noncoding RNA maintains active chromatin to coordinate homeotic gene expression. Nature. 2011;472(7341):120–4. https://doi.org/10.1038/nature09819. Epub 2011/03/23. PubMed PMID: 21423168; PubMed Central PMCID: PMCPMC3670758.
20. Yang YW, Flynn RA, Chen Y, Qu K, Wan B, Wang KC, et al. Essential role of lncRNA binding for WDR5 maintenance of active chromatin and embryonic stem cell pluripotency. elife. 2014;3:e02046. https://doi.org/10.7554/eLife.02046. Epub 2014/02/14. PubMed PMID: 24521543; PubMed Central PMCID: PMCPMC3921674.
21. Deng C, Li Y, Zhou L, Cho J, Patel B, Terada N, et al. HoxBlinc RNA Recruits Set1/MLL Complexes to Activate Hox Gene Expression Patterns and Mesoderm Lineage Development. Cell Rep. 2016;14(1):103–14. https://doi.org/10.1016/j.celrep.2015.12.007. Epub 2016/01/05. PubMed PMID: 26725110; PubMed Central PMCID: PMCPMC4706800.
22. Hu D, Gao X, Morgan MA, Herz HM, Smith ER, Shilatifard A. The MLL3/MLL4 branches of the COMPASS family function as major histone H3K4 monomethylases at enhancers. Mol Cell Biol. 2013;33(23):4745–54. https://doi.org/10.1128/mcb.01181-13. Epub 10/02. Epub 2013 Sep 30. PubMed PMID: 24081332.
23. Schraets D, Lehmann T, Dingermann T, Marschalek R. MLL-mediated transcriptional gene regulation investigated by gene expression profiling. Oncogene. 2003;22(23):3655–68. https://doi.org/10.1038/sj.onc.1206438. Epub 06/06. PubMed PMID: 12789274.
24. Goldsworthy M, Absalom NL, Schroter D, Matthews HC, Bogani D, Moir L, et al. Mutations in Mll2, an H3K4 methyltransferase, result in insulin resistance and impaired glucose tolerance in mice. PLoS One. 2013;8(6):e61870. https://doi.org/10.1371/journal.pone.0061870. Epub 07/05. Print 2013. PubMed PMID: 23826075.
25. Lee J, Saha PK, Yang QH, Lee S, Park JY, Suh Y, et al. Targeted inactivation of MLL3 histone H3-Lys-4 methyltransferase activity in the mouse reveals vital roles for MLL3 in adipogenesis. Proc Natl Acad Sci USA. 2008;105(49):19229–34. https://doi.org/10.1073/pnas.0810100105. Epub 12/03. PubMed PMID: 19047629.
26. Glaser S, Schaft J, Lubitz S, Vintersten K, van der Hoeven F, Tufteland KR, et al. Multiple epigenetic maintenance factors implicated by the loss of Mll2 in mouse development. Development (Cambridge, UK). 2006;133(8):1423–32. https://doi.org/10.1242/dev.02302. Epub 03/17. PubMed PMID: 16540515.

27. Mascre G, Dekoninck S, Drogat B, Youssef KK, Brohee S, Sotiropoulou PA, et al. Distinct contribution of stem and progenitor cells to epidermal maintenance. Nature. 2012;489:257–62. England.
28. Fuchs E, Raghavan S. Getting under the skin of epidermal morphogenesis. Nat Rev Genet. 2002;3:199–209. England.
29. Ezhkova E, Pasolli HA, Parker JS, Stokes N, Su IH, Hannon G, et al. Ezh2 orchestrates gene expression for the stepwise differentiation of tissue-specific stem cells. Cell. 2009;136:1122–35. United States.
30. Bardot ES, Valdes VJ, Zhang J, Perdigoto CN, Nicolis S, Hearn SA, et al. Polycomb subunits Ezh1 and Ezh2 regulate the Merkel cell differentiation program in skin stem cells. EMBO J. 2013;32(14):1990–2000. https://doi.org/10.1038/emboj.2013.110. Epub 05/16. Epub 2013 May 14. PubMed PMID: 23673358.
31. Mardaryev AN, Gdula MR, Yarker JL, Emelianov VU, Poterlowicz K, Sharov AA, et al. p63 and Brg1 control developmentally regulated higher-order chromatin remodelling at the epidermal differentiation complex locus in epidermal progenitor cells. Development (Cambridge, UK). 2014;141:101–11.
32. Fessing MY, Mardaryev AN, Gdula MR, Sharov AA, Sharova TY, Rapisarda V, et al. p63 regulates Satb1 to control tissue-specific chromatin remodeling during development of the epidermis. J Cell Biol. 2011;194:825–39. United States.
33. Bao X, Tang J, Lopez-Pajares V, Tao S, Qu K, Crabtree GR, et al. ACTL6a enforces the epidermal progenitor state by suppressing SWI/SNF-dependent induction of KLF4. Cell Stem Cell. 2013;12(2):193–203. https://doi.org/10.1016/j.stem.2012.12.014. Epub 02/12. PubMed PMID: 23395444.
34. Hopkin AS, Gordon W, Klein RH, Espitia F, Daily K, Zeller M, et al. GRHL3/GET1 and Trithorax Group Members Collaborate to Activate the Epidermal Progenitor Differentiation Program. PLoS Genet. 2012;8:e1002829. United States.
35. Kim M, McGinnis W. Phosphorylation of Grainy head by ERK is essential for wound-dependent regeneration but not for development of an epidermal barrier. Proc Natl Acad Sci USA. 2011;108:650–5. United States.
36. Mace KA, Pearson JC, McGinnis W. An epidermal barrier wound repair pathway in Drosophila is mediated by grainy head. Science (New York). 2005;308(5720):381–5. https://doi.org/10.1126/science.1107573. Epub 04/16. PubMed PMID: 15831751.
37. Yu Z, Lin KK, Bhandari A, Spencer JA, Xu X, Wang N, et al. The Grainyhead-like epithelial transactivator Get-1/Grhl3 regulates epidermal terminal differentiation and interacts functionally with LMO4. Dev Biol. 2006;299:122–36. United States.
38. Gordon WM, Zeller MD, Klein RH, Swindell WR, Ho H, Espetia F, et al. A GRHL3-regulated repair pathway suppresses immune-mediated epidermal hyperplasia. J Clin Invest. 2014;124(12):5205–18. https://doi.org/10.1172/jci77138. Epub 10/28. Epub 2014 Oct 27. PubMed PMID: 25347468.
39. Yu Z, Bhandari A, Mannik J, Pham T, Xu X. Andersen B. Grainyhead-like factor Get1/Grhl3 regulates formation of the epidermal leading edge during eyelid closure. Dev Biol. 2008;319:56–67. United States.
40. Caddy J, Wilanowski T, Darido C, Dworkin S, Ting SB, Zhao Q, et al. Epidermal wound repair is regulated by the planar cell polarity signaling pathway. Dev Cell. 2010;19:138–47. United States: 2010 Elsevier Inc.
41. Yu Z, Mannik J, Soto A, Lin KK, Andersen B. The epidermal differentiation-associated Grainyhead gene Get1/Grhl3 also regulates urothelial differentiation. EMBO J. 2009;28(13):1890–903. https://doi.org/10.1038/emboj.2009.142. Epub 06/06. Epub 2009 Jun 4. PubMed PMID: 19494835.
42. Kachgal S, Mace KA, Boudreau NJ. The dual roles of homeobox genes in vascularization and wound healing. Cell Adhes Migr. 2012;6(6):457–70. https://doi.org/10.4161/cam.22164. Epub 10/19. Epub 2012 Oct 17. PubMed PMID: 23076135.

43. Hnisz D, Abraham BJ, Lee TI, Lau A, Saint-Andre V, Sigova AA, et al. Super-enhancers in the control of cell identity and disease. Cell. 2013;155(4):934–47. https://doi.org/10.1016/j.cell.2013.09.053. Epub 10/15. Epub 2013 Oct 10. PubMed PMID: 24119843.
44. Whyte WA, Orlando DA, Hnisz D, Abraham BJ, Lin CY, Kagey MH, et al. Master transcription factors and mediator establish super-enhancers at key cell identity genes. Cell. 2013;153(2):307–19. https://doi.org/10.1016/j.cell.2013.03.035. PubMed PMID: PMC3653129.
45. Tie F, Banerjee R, Saiakhova AR, Howard B, Monteith KE, Scacheri PC, et al. Trithorax monomethylates histone H3K4 and interacts directly with CBP to promote H3K27 acetylation and antagonize Polycomb silencing. Development (Cambridge, UK). 2014;141(5):1129–39. https://doi.org/10.1242/dev.102392. Epub 02/20. PubMed PMID: 24550119.
46. Klein RH, Lin Z, Hopkin AS, Gordon W, Tsoi LC, Liang Y, et al. GRHL3 binding and enhancers rearrange as epidermal keratinocytes transition between functional states. PLoS Genet. 2017;13(4):e1006745. https://doi.org/10.1371/journal.pgen.1006745. Epub 2017/04/27. PubMed PMID: 28445475; PubMed Central PMCID: PMCPMC5425218.
47. Blastyak A, Mishra RK, Karch F, Gyurkovics H. Efficient and specific targeting of Polycomb group proteins requires cooperative interaction between Grainyhead and Pleiohomeotic. Mol Cell Biol. 2006;26(4):1434–44. https://doi.org/10.1128/mcb.26.4.1434-1444.2006. Epub 02/02. PubMed PMID: 16449654.
48. Rockowitz S, Lien WH, Pedrosa E, Wei G, Lin M, Zhao K, et al. Comparison of REST cistromes across human cell types reveals common and context-specific functions. PLoS Comput Biol. 2014;10(6):e1003671. https://doi.org/10.1371/journal.pcbi.1003671. Epub 06/13. eCollection 2014 Jun. PubMed PMID: 24922058.
49. Ren X, Kerppola TK. REST interacts with Cbx proteins and regulates polycomb repressive complex 1 occupancy at RE1 elements. Mol Cell Biol. 2011;31(10):2100–10. https://doi.org/10.1128/mcb.05088-11. Epub 2011/03/16. PubMed PMID: 21402785; PubMed Central PMCID: PMCPMC3133345.
50. Mardaryev AN, Liu B, Rapisarda V, Poterlowicz K, Malashchuk I, Rudolf J, et al. Cbx4 maintains the epithelial lineage identity and cell proliferation in the developing stratified epithelium. J Cell Biol. 2016;212(1):77–89. https://doi.org/10.1083/jcb.201506065. Epub 2015/12/30. PubMed PMID: 26711500; PubMed Central PMCID: PMCPMC4700479.
51. Krivtsov AV, Armstrong SA. MLL translocations, histone modifications and leukaemia stem-cell development. Nat Rev Cancer. 2007;7(11):823–33. https://doi.org/10.1038/nrc2253. Epub 10/25. PubMed PMID: 17957188.
52. Chang MJ, Wu H, Achille NJ, Reisenauer MR, Chou CW, Zeleznik-Le NJ, et al. Histone H3 lysine 79 methyltransferase Dot1 is required for immortalization by MLL oncogenes. Cancer Res. 2010;70(24):10234–42. https://doi.org/10.1158/0008-5472.can-10-3294. Epub 12/17. PubMed PMID: 21159644.
53. Saigo K, Yoshida K, Ikeda R, Sakamoto Y, Murakami Y, Urashima T, et al. Integration of hepatitis B virus DNA into the myeloid/lymphoid or mixed-lineage leukemia (MLL4) gene and rearrangements of MLL4 in human hepatocellular carcinoma. Hum Mutat. 2008;29(5):703–8. https://doi.org/10.1002/humu.20701. Epub 03/06. PubMed PMID: 18320596.
54. Pickering CR, Zhou JH, Lee JJ, Drummond JA, Peng SA, Saade RE, et al. Mutational landscape of aggressive cutaneous squamous cell carcinoma. Clin Cancer Res. 2014;20(24):6582–92. https://doi.org/10.1158/1078-0432.ccr-14-1768. Epub 10/12. Epub 2014 Oct 10. PubMed PMID: 25303977.
55. Martin D, Abba MC, Molinolo AA, Vitale-Cross L, Wang Z, Zaida M, et al. The head and neck cancer cell oncogenome: a platform for the development of precision molecular therapies. Oncotarget. 2014;5(19):8906–23. Epub 10/03. PubMed PMID: 25275298.
56. Bhandari A, Gordon W, Dizon D, Hopkin AS, Gordon E, Yu Z, et al. The Grainyhead transcription factor Grhl3/Get1 suppresses miR-21 expression and tumorigenesis in skin: modulation of the miR-21 target MSH2 by RNA-binding protein DND1. Oncogene. 2013;32(12):1497–507. https://doi.org/10.1038/onc.2012.168. Epub 2012/05/23. PubMed PMID: 22614019; PubMed Central PMCID: PMCPmc4026359.

57. Darido C, Georgy SR, Wilanowski T, Dworkin S, Auden A, Zhao Q, et al. Targeting of the tumor suppressor GRHL3 by a miR-21-dependent proto-oncogenic network results in PTEN loss and tumorigenesis. Cancer Cell. 2011;20(5):635–48. https://doi.org/10.1016/j.ccr.2011.10.014. Epub 11/19. PubMed PMID: 22094257.
58. Mlacki M, Darido C, Jane SM, Wilanowski T. Loss of Grainy head-like 1 is associated with disruption of the epidermal barrier and squamous cell carcinoma of the skin. PLoS One. 2014;9(2):e89247. https://doi.org/10.1371/journal.pone.0089247. Epub 03/04. eCollection 2014. PubMed PMID: 24586629.
59. Miyake N, Koshimizu E, Okamoto N, Mizuno S, Ogata T, Nagai T, et al. MLL2 and KDM6A mutations in patients with Kabuki syndrome. Am J Med Genet A. 2013;161a(9):2234–43. https://doi.org/10.1002/ajmg.a.36072. Epub 08/06. Epub 2013 Aug 2. PubMed PMID: 23913813.
60. Jones WD, Dafou D, McEntagart M, Woollard WJ, Elmslie FV, Holder-Espinasse M, et al. De novo mutations in MLL cause Wiedemann-Steiner syndrome. Am J Hum Genet. 2012;91(2):358–64. https://doi.org/10.1016/j.ajhg.2012.06.008. Epub 07/17. Epub 2012 Jul 12. PubMed PMID: 22795537.

Chapter 5
Histone Deacetylase Functions in Epidermal Development, Homeostasis and Cancer

Donna M. Brennan-Crispi and Sarah E. Millar

5.1 Introduction: Epigenetics and Skin

The cellular diversity characteristic of multicellular organisms, whose cells all harbor an identical genetic blueprint, is achieved by finely orchestrating spatio-temporal patterns of gene expression. Delineating how each cell type regulates its unique and fluctuating transcriptional profile is crucial for unraveling the mechanisms underlying embryonic development and adult homeostasis. Epigenetic control of transcriptional activity has rapidly emerged as a key element in these processes.

The concept of epigenetics was first introduced by Conrad Waddington in 1942 [235], even before DNA was identified as the carrier of genetic information. Broadly, epigenetics is defined as regulation of the genome beyond the genetic code. This concept is profound in that it provides a framework for understanding how multicellular organisms with diverse cell types can develop from a single progenitor cell. Although the molecular mechanisms underpinning epigenetic regulation were not explored until several decades after Waddington's publication, his work lead to the realization that the timely regulation of genetic material is just as important as the content itself. The concept of epigenetics has evolved, and now incorporates the

D. M. Brennan-Crispi
Department of Dermatology, Perelman School of Medicine, University of Pennsylvania, Philadelphia, PA, USA

S. E. Millar (✉)
Department of Dermatology, Perelman School of Medicine, University of Pennsylvania, Philadelphia, PA, USA

Department of Cell and Developmental Biology, Perelman School of Medicine, University of Pennsylvania, Philadelphia, PA, USA

Department of Anatomy and Cell Biology, School of Dental Medicine, University of Pennsylvania, Philadelphia, PA, USA
e-mail: millars@pennmedicine.upenn.edu

© Springer International Publishing AG, part of Springer Nature 2018

V. A. Botchkarev, S. E. Millar (eds.), *Epigenetic Regulation of Skin Development and Regeneration*, Stem Cell Biology and Regenerative Medicine,
https://doi.org/10.1007/978-3-319-16769-5_5

idea that epigenetic traits can be inherited [14]. Detailed discussions of the heritability of epigenetic marks, including their mitotic inheritance from cell to cell, meiotic inheritance from generation to generation, and trans-generational inheritance, are provided in several recent reviews [16, 230]. Here, we will focus on the epigenetic functions of HDACs in modulating gene expression and determining cell fate.

The skin is easily accessible and harbors diverse cell lineages, providing an ideal system for dissecting epigenetic mechanisms. The skin epidermis in particular is an excellent model for studying cell fate decisions as it is a naturally regenerating organ that is subject to diverse disease states and contains both proliferating and differentiating cells, discrete stem cell populations, multiple epithelial-derived appendages fated to distinct regimented differentiation profiles, and hair follicles that undergo cyclical regeneration.

5.2 Chromatin Remodeling

Packaging of eukaryotic genetic material occurs through complex mechanisms and plays crucial regulatory functions in gene transcription. In order to fit into a cell nucleus, genomic DNA is highly compacted into a chromatin complex (for review, see [23]). Chromatin is comprised of nucleosomes, each containing DNA wrapped around a core histone octamer, which are separated by short strands of DNA that serve as linkers, creating a "beads on a string" structure. These are further compacted through controlled folding and higher-order loop structures. The tightly wound confirmation of DNA in the nucleosome is an impediment to active transcription; however, localized remodeling can increase its accessibility to the transcriptional machinery. Thus, control of chromatin compaction is a key element in the regulation of gene expression.

Chromatin remodeling can occur as a result of DNA methylation or hydroxymethylation (reviewed in [64]); covalent modifications to histones, including methylation, acetylation, ubiquitination, and phosphorylation (reviewed in [28, 130]); alterations in histone-DNA interactions (reviewed in [42]); and changes in higher order folding (reviewed in [46, 59]). Of these, reversible posttranslational modifications of histone tails are a readily mobilized mechanism that can influence the configuration of chromatin and alter transcriptional activity. The idea that posttranslational histone modifications could contribute to transcriptional regulation was originally proposed in the 1960s [3], but this concept remained largely unexplored until histone modifying enzymes were identified over 30 years later [26, 119]. It is now clear that histone methylation and acetylation are key gene regulatory marks, while the roles of other moieties, such as ubiquitination and phosphorylation, are less well understood. Specific histone methylation marks are associated with either repression or activation of transcription, while histone acetylation is primarily thought of as an activating mark that functions to loosen the chromatin structure and allow access for the transcriptional machinery [241]. Acetylated lysine residues on histones can also interact with bromodomain proteins and their co-activators to promote transcription [262]. Elucidating the contribution of these marks and the proteins that perpetuate them is thus vital to our understanding of transcriptional regulation during the development of normal tissues and disease states.

5.3 Histone Deacetylases (HDACs)

Histone deacetylases (HDACs) are highly conserved proteins that modify histones by catalyzing the removal of an acetyl group from lysine residues in the histone tail [73]. These enzymes function in opposition to histone acetyltransferases (HATs) which deposit acetyl groups. Histone acetylation and deacetylation plays several major roles in the cell (reviewed in [214]). Firstly, during S-phase, newly synthesized histones are acetylated by HATS, and are subsequently deacetylated and reacetylated with new combinations of acetyl marks to allow flexibility in their functions. Secondly, histone acetylation affects chromatin folding: deacetylated histones are associated with a more transcriptionally repressive environment. Thirdly, HDACs are recruited to specific sites in chromatin by transcription factors that, unlike HDACs, can bind DNA directly. HDACs and their transcriptional partners are generally found in large protein complexes that produce multi-layered control of chromatin activity. HDACs are traditionally characterized as transcriptional repressors as hypoacetylation is associated with compact chromatin and transcriptional silencing. Furthermore, HDACs can associate with polycomb-group (PcG) protein complexes to promote the formation of repressive chromatin [233]. However, ChIP-seq data reveal that HDACs as well as HATs are commonly associated with activated genes [241], suggesting that HDACs may act as a temporary brake on transcription. Interestingly, HDAC function is also important for proper initiation of transcription [214], and can promote transcription by removing acetylation marks that inhibit elongation [76].

In addition to their functions in histone deacetylation, the catalytic domains of HDACs can remove acetyl groups from lysine residues on a wide variety of protein substrates, [83]. Substrate specificity is largely determined by interactions of HDACs with a variety of binding partners. Thus, a single HDAC has many substrates; and conversely, a single substrate can be targeted by multiple HDACs. Transcription factors figure prominently among these substrates, and the effects of deacetylation on their functions and stability can vary. For instance, HDAC-mediated deacetylation of p53 inhibits its transcriptional activity [112], while HDAC1-mediated deacetylation of GLI2 potentiates its ability to activate target genes [30]. HDACs themselves can be acetylated and deacetylated, and their functions are also modulated by an array of additional post-transcriptional modifications including phosphorylation, ubiquitination and SUMOylation [210]. These mechanisms allow fine-tuning of HDAC activity. The recent recognition that some HDACs also possess enzyme-independent functions adds yet another layer of complexity to their roles [84, 87, 107, 147, 221]. The precise mechanisms underlying these non-canonical functions are under active investigation, and are likely to be isoform- and context-dependent.

In mammals, the HDAC superfamily consists of 18 proteins and can be organized into classical HDACs (HDAC1-11) and Sirtuins (Sirt1-7). The Sirtuins, also referred to as Class III HDACs, differ from the classical HDACs in that they are NAD+ dependent, can have additional enzymatic activity, and are not inhibited by Trichostatin A (TSA). This Chapter will focus on the classical HDACs; more information on Sirtuins can be found elsewhere [21, 31, 36, 94].

The 11 classical HDACs share a highly conserved deacetylase domain, and are divided into four families based on their structure, expression profiles, and enzymatic functions: Class I (HDAC1, 2, 3, and 8); Class IIa (HDAC4, 5, 7, and 9); Class IIb (HDAC6 and 10); and Class IV (HDAC11) (reviewed in [83]). Class I HDACs are broadly expressed, and have the simplest structures, consisting chiefly of the deacetylase domain. Members of this class are thought to be the primary modifiers of histone tails as they are predominantly localized to nuclei and have a high affinity for histone targets [83]. Class IIa HDACs display more restricted and cell-type specific expression patterns, and contain additional binding domains for myocyte enhancer factor 2 (MEF2) proteins and for 14-3-3, a chaperone that facilitates activation of Class IIa HDACs by promoting their nuclear localization [145, 154]. Class IIa HDACs possess minimal enzymatic activity [126], and appear to function instead by binding to and regulating the activity of transcription factors including MEF2 proteins [160, 237]. Class IIa HDACs can also complex with Class I HDACs, and may provide a scaffolding function [65, 66, 83, 153]. Class IIb HDACs are structurally the most distinct of the HDAC classes. Alone among HDACs, HDAC6 contains a zinc-finger domain and two catalytic domains; HDAC10 is characterized by a leucine rich domain [80]. HDAC6 localizes primarily to the cytoplasm, while HDAC10 can shuttle between the nucleus and the cytoplasm [80]. Class IV consists solely of HDAC11 which is structurally similar to the Class I HDACs, but its expression is limited to a few tissue types [70, 143]. To date, Class I HDACs have been the most extensively studied in the epidermis and its appendages. We therefore discuss the functions of this class in more detail below.

The importance of class I HDACs in embryonic development has been clearly established using murine global knockout models. *Hdac3* null mice die *in utero* at approximately embryonic day 9.5 (E9.5), due to gastrulation defects resulting at least in part from impaired DNA damage repair [15, 120, 169]. Loss of *Hdac8* causes severe cranial malformations and perinatal death [82]. Global ablation of *Hdac1* leads to embryonic lethality by E10.5 as a result of significant proliferation defects [125]. By contrast, *Hdac2* deletion in two different mouse models produces cardiac defects that cause either early postnatal lethality or reduced viability, depending on the model [169, 231]. The phenotypic disparities of global *Hdac2* mutants could result from differences in the efficiency of gene deletion and/or background strain differences [83].

While the phenotypes of *Hdac1* and *Hdac2* global knockout mice indicate that these genes perform specific, non-redundant roles, *Hdac1* and *Hdac2* share 83% homology and are largely co-expressed in diverse tissues [83], suggesting that they also possess overlapping functions. In line with this, ChIP-seq for HDAC1 and HDAC2 in human CD4+ T-cells revealed that these proteins bind distinct, but overlapping sets of locations in the genome [241]. Furthermore, tissue-specific knockout studies in mice have revealed that *Hdac1* and *Hdac2* function in a redundant, or partially redundant, manner in many tissues (reviewed in [117]).

5.3.1 HDACs in Repressive Complexes

To facilitate directed gene regulation, HDACs are incorporated into multi-protein co-repressor complexes that have affinity for specific targets. Incorporation into a co-repressor complex is critical for HDACs to function as histone deacetylases; furthermore, it allows for multifaceted remodeling of chromatin structure and transcriptional regulation based upon the components of the complex. Of the Class I HDACs, co-repressor complexes have been described for all but HDAC8.

HDAC1 and HDAC2 associate with a variety of transcriptional regulators as homo- or heterodimers. Their associations with the CoREST/REST (RE1-Silencing Transcription factor), NuRD (Nucleosome Remodeling Deacetylation) and SIN3 complexes have been studied most extensively to date. The CoREST/REST complex coordinates deacetlyase and demethylase activity through HDACs and Lysine Specific Demethylase (LSD1), respectively [135]. In the NuRD complex, HDACs are integrated with Mi2α/β, an ATP-dependent DNA helicase; the methyl-CpG-binding domain proteins MBD2 or MBD3; Metastasis-associated proteins MTA1, MTA2, or MTA3 that mediate binding to HDAC1/2 and DNA; and the histone-binding proteins Rbbp7 and Rbbp4 [10, 127, 250]. SIN3 is a modular, multifunctional protein containing six conserved domains that include four Paired Amphipathic Helices (PAH 1–4), a Histone Deacetylase Interaction Domain (HID) and a Highly Conserved Region (HCR). SIN3 lacks intrinsic DNA-binding capacity and functions as a molecular scaffold, coordinating a wide array of proteins to form a co-repressor complex [113].

HDAC3 is distinct from HDAC1/2 in that it is found in the N-CoR and SMRT (nuclear receptor co-repressor/silencing mediator of retinoic acid and thyroid hormone receptor) complexes [117]. These complexes are named for their ability to interact with nuclear receptors and typically mediate the transcriptional repression associated with these receptors in the absence of ligands [155]. Unlike HDAC1/2 complexes, N-CoR/SMRT assemblies can also recruit class IIa HDACs including HDAC4 [65, 66, 153].

While specific multi-protein HDAC complexes are defined by their unique core constituents, some proteins such as retinoblastoma (pRb) are found in multiple assemblies [144]. Specific interactions are dependent in part on the phosphorylation state of pRb [149]. As mentioned above, HDACs, as well as other components of the complexes, are also subject to an array of post-translational modifications, providing added levels of control. Further specificity is conferred through their association with protein paralogs such as SIN3A/B, MTA1-3, and CoREST1-3. For example, inclusion of MBD2 or MBD3 in the NuRD complex is mutually exclusive; these proteins bind methylated and non-methylated DNA respectively [89], thus targeting HDAC activity to different sites. Furthermore, in the NuRD complex MTA2 and MTA3 mediate opposite effects on epithelial-mesenchymal transition (EMT) by interacting with different transcription factors [69]. Similarly, the related factors SIN3A and SIN3B have unique gene targets, clearly demonstrated by phenotypic differences between *Sin3a* and *Sin3b* knockout mice [45, 49]. Finally, HDAC composition can vary among complexes that may contain either HDAC1/2 homo- or heterodimers, or in the case of HDAC3-containing complexes, additional Class IIa HDACs. Together these variations provide for an almost endless combination of sub-units that allow exquisite fine-tuning of HDAC targeting, and context-dependent physiological functions.

The functional significance of individual components of the various co-repressor complexes is underscored by global knockout studies in mice: for instance deletions of *Sin3A* [45], *Mbd3* [90], *Lsd1* [239], or *NCoR* [108] cause distinct phenotypes and embryonic lethality. Importantly, aberrant expression of several complex constituents is associated with human diseases ranging from neurological pathologies to cancer. It should be noted that while these complexes are traditionally referred to as co-repressors, recent studies have provided significant evidence that certain variants also have the ability to function as transcriptional activators [11, 193]. Together, these findings indicate that the precise compositions and modifications of co-repressor complex subunits permit specificity of HDAC targeting and functions.

5.3.2 Non-histone Targets

As mentioned above, in addition to their functions in chromatin remodeling, Class I HDACs can also directly modify non-histone proteins, including transcription factors and their co-factors. Transcription factor targets such as p53, STAT3, and RELA are associated with Class I HDACs in co-repressor complexes, providing easy access for their modification, and another mechanism by which HDACs can modulate gene expression (see Table 5.1 for a selected list of targets). Interestingly, HDAC3 can also associate directly with the HAT co-activator CREB-binding protein (CBP) resulting in deacetylation of HAT, and attenuation of its activity [39, 44, 77]. Thus, HDACs can alter transcriptional activity by modifying HAT function. These findings further suggest that HDACs and HATs can function within the same multi-protein complexes.

The studies summarized above reveal how HDAC enzymatic activity is dynamically regulated by a host of interacting proteins to provide a mechanism in which subtle alterations in complex constituents permit selection of discrete sets of target genes, and can have profound effects on their expression. This machinery is therefore well adapted to regulate cell fate during tissue development, homeostasis, and disease.

Table 5.1 List of selected non-histone substrates of class I HDACs

Enzyme	Substrate	Description	References
HDAC1	E2F1	Transcription factor	[263]
	GLI1	Transcription factor	[30]
	GLI2	Transcription factor	[30]
	p53	Transcription factor	[93, 102, 224]
	STAT3	Transcription factor	[261]
	YY1	Transcription factor	[256]
HDAC2	Glucocorticoid Receptor	Nuclear receptor	[103]
	STAT3	Transcription factor	[261]
	YY1	Transcription factor	[256]
HDAC3	MEF2	Transcription factor	[77]
	p300/CBP	Histone acetyl transferase	[77]
	PCAF	Histone acetyl transferase	[77]
	PPARγ	Nuclear receptor	[111]
	RELA (NFκB p65)	Transcription factor	[35]
	STAT3	Transcription factor	[261]

5.4 Skin Development and Disease

The embryonic or early postnatal lethality observed in most global Class I HDAC knockout mice precludes careful assessment of the roles of these enzymes in the development, homeostasis and diseases of specific tissues. Thus, the generation of tissue-specific mutant mice has been a priority. The epidermis of the skin is a particularly attractive model system for delineating HDAC functions and their underlying mechanisms because of its accessibility, the availability of multiple effective tools for genetic manipulation, its natural regenerative capacity, and the clinical importance of its diseases.

The interfollicular epidermis (IFE) is derived from a single layered surface ectoderm that stratifies during embryogenesis, giving rise to a functional barrier that prevents dehydration and bars the entry of pathogens and noxious substances. Embryonic surface ectodermal cells also interact with the underlying mesenchyme to give rise to complex ectodermal appendage structures such as hair follicles, sebaceous glands, and sweat glands. In adult life, the outer, cornified layer of the epidermis is constantly shed, and is renewed through the proliferative activity of basal stem cells. Hair follicles undergo repeated cycles of growth (anagen), quiescence (telogen) and regression (catagen) throughout life, relying on bulge stem cells that are specifically dedicated to the hair follicle, and only contribute to the epidermis during wound healing [12, 200]. Skin cancers are the most common form of cancer in Caucasian populations, and the skin is the site of a host of other common diseases such as psoriasis and atopic dermatitis, that involve dysregulation of genes that control cellular proliferation and differentiation.

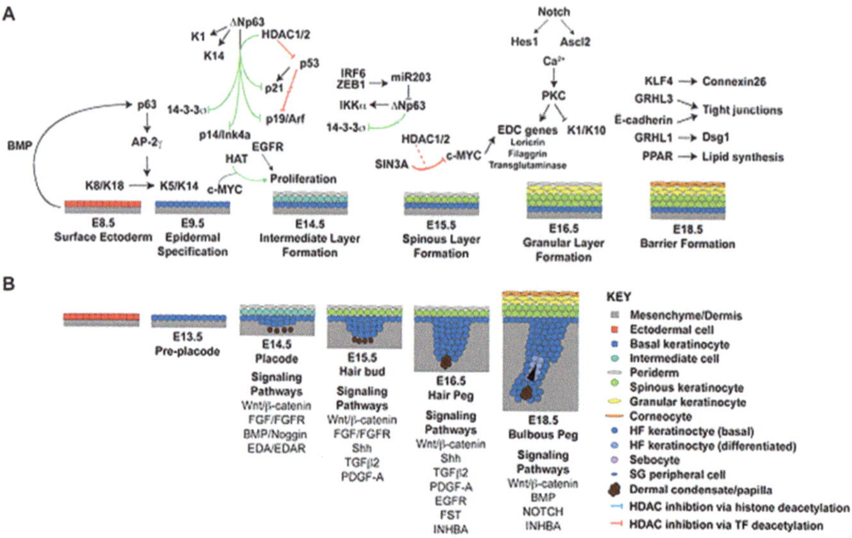

Fig. 5.1 Key events and signals in the development of embryonic epidermis (a) and hair follicles (b). (For reviews, please see [22, 162, 164, 213])

5.4.1 IFE: Development and Homeostasis

Epidermal development is a well-characterized process (Fig. 5.1a; [68, 124]). The developing embryo is initially covered by the ectoderm, a single cell layer that will give rise to the mature epidermis and its appendages. Around embryonic day 8.5 (E8.5) of mouse embryogenesis, a mesenchymal signal initiates epidermal specification. By E 9.5, cells lose expression of markers of the primitive ectoderm [167] and start to express the basal keratinocyte proteins keratin 5 and keratin 14 (K5 and K14) [27]. At E14.5 the epidermis stratifies to form the intermediate layer. This transitional layer expresses keratin 1 (K1), a marker of differentiated keratinocytes, but retains the capacity to proliferate. The intermediate layer is short-lived and differentiates into the spinous layer at around E15.5, a switch that is accompanied by a loss of proliferative capacity. At this stage, keratinocytes enter into a terminal differentiation program, a process that ultimately establishes the mature epidermis. The spinous to granular transition (E16.5), is characterized by the expression of cornified envelope (CE) proteins, which are critical to barrier formation, the last stage in epidermal development. A functional barrier that prevents trans-epidermal water loss and protects against pathogens is established by approximately E18.5. This process involves coordination of desquamation, crosslinking of CE proteins, and lipid synthesis [201, 211]. Severe disruption of barrier function is incompatible with life; thus deregulation during any stage of epidermal development that significantly impairs proper barrier formation can result in embryonic/perinatal lethality. Once established, the IFE is maintained through the proliferation of basal keratinocytes which give rise to daughter cells that can either self-renew or enter the terminal differentiation pathway (reviewed in [18]).

Epidermal morphogenesis has been well-studied, and many key players have been identified. The transcription factor p63 is expressed as epidermal lineages are specified, and loss of its expression results in failure of epidermal morphogenesis [123]. P63 acts upstream of other early regulators of ectodermal appendage specification, epidermal proliferation and asymmetric cell division, and epidermal development, and is thus considered a master regulator of these processes [122, 129, 132, 165, 232, 254]. Two different isoforms, TAp63 which contains a p53-like transactivation domain and ΔNp63 which lacks this region [253], have unique and complementary roles in development [29]. ΔNp63, the predominant isoform in embryonic epidermis, can function either as an activator or a repressor; for example it activates expression of the basal keratin genes *K5* and *K14* [198, 199], and promotes epidermal progenitor cell proliferation by directly repressing the anti-proliferative target genes *p21* and *14-3-3σ* ([232, 267]). The EGFR pathway also promotes proliferation of basal cells, but is not required for epidermal specification [159, 217].

Intermediate layer formation requires ΔNp63 [122] and Notch signaling [19, 191]. Differentiation of the spinous layer is facilitated by a number of players including 14-3-3σ [92, 139], IRF6 [101, 194], miR-203 [175, 258], and ΔNp63α-mediated activation of IKKα [122]. The spinous to granular transition is associated with expression of the CE proteins loricin, fillagrin and transglutaminase [51, 54, 255]. The Notch signaling pathway is also implicated in regulating the spinous-granular transition [170]. The terminal differentiation program of keratinocytes is regulated in part by extracellular calcium (Ca^{2+}) concentrations [57, 157, 158]. Maturation of the epidermis concludes with barrier formation. Transcriptional regulators such as KLF4 [106, 212], GRHL1 [166], GRHL3 [229, 260], PPAR proteins [151, 203, 236], and GATA-3 [50] each mediate essential aspects of barrier function including cellular adhesion and lipid metabolism. Many of these embryonic regulators are re-utilized in adult life to ensure constant renewal of the epidermis and maintenance of barrier function [18].

5.4.2 *Hair Follicle Morphogenesis and Cycling*

Hair follicle development is initiated by signals from the mesenchyme that result in the formation of epithelial thickenings, known as placodes. p63 is a master regulator of skin appendage formation (reviewed in [123, 161]) and loss of p63 leads to failure of hair follicle specification and placode formation [165, 253]. During normal development, placodes are initially irregular in their sizes and spacing, and gradually become refined to form an orderly pattern. This process involves a competition between placode-promoting signals such as Wnt/β-catenin and EDA/EDAR and inhibitory signals including Dickkopf (DKK) family Wnt inhibitors, and Bone Morphogenic Proteins (BMPs) (reviewed in [134, 162, 213]). In response to epithelial FGF signaling, dermal cells condense under each placode [100]. The placode gradually invaginates into the dermis, eventually almost engulfing the dermal condensate to form a dermal papilla. Reciprocal signaling from the dermal papilla stimulates proliferation and differentiation of the hair follicle epithelium to generate the hair shaft and its surrounding root sheaths (Fig. 5.1a; reviewed in [134, 162, 164]). Disruption of Sonic hedgehog (SHH) signaling perturbs proliferation leading to blunted hair follicle development after the germ stage [48, 86, 163], while perturbation of EGFR activity results in defects in follicular shape and differentiation [85, 146]. Multiple transcription factors, including MSX2, HOXC13, and FOXN1 are required for hair shaft differentiation [43, 148, 207].

Postnatally, hair follicles enter undergo repeated cycles of growth (anagen), regression (catagen) and rest (telogen), that are governed by many of the same signaling pathways that are active during morphogenesis (Fig. 5.2; for reviews see [4, 43, 134, 215]). The hair follicle growth cycle is a tightly regulated process involving stem cell activation; progenitor cell proliferation, migration and differentiation; and controlled apoptosis and cellular quiescence. BMP signaling maintains hair follicle stem cells in a quiescent state during telogen [121, 187]. At anagen onset, inhibition

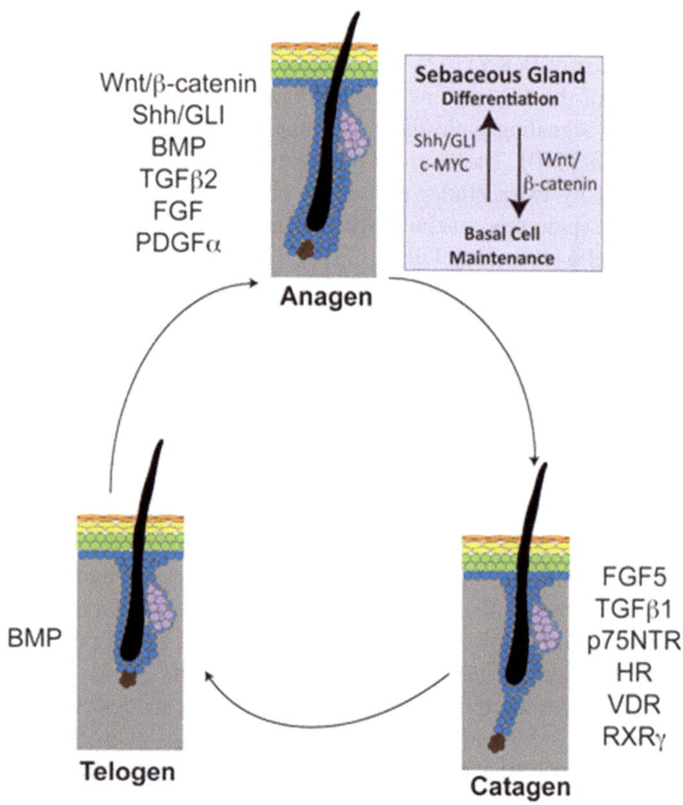

Fig. 5.2 **The adult hair follicle growth cycle**. Factors playing key roles at each stage are listed. (For reviews, please see [43, 213])

of BMP signaling and activation of Wnt/β-catenin signaling initiate proliferation of progenitor cells in the secondary hair germ of the follicle, followed by transient proliferation of bulge stem cells [37, 121, 141, 187]. Sonic hedgehog signaling is required for continued proliferation of transient amplifying cells, which multiply rapidly in the hair bulb, and subsequently differentiate along specific pathways to form the hair shaft and inner root sheath [180]. Normal timing of catagen requires activation of FGF5 and TGFβ1 signaling, among other pathways (reviewed in [43]). Once the regression phase is complete, the hair follicle enters telogen and BMP signaling functions once again to maintain follicular quiescence.

5.4.3 Sebaceous Gland Development and Maintenance

The sebaceous gland develops from follicular keratinocytes during late embryogenesis. A peripheral cell layer of basal keratinocytes surrounds the sebaceous gland and undergoes a distinctive differentiation program to produce mature sebocytes, which release their sebum through a holocrine mechanism. Sebum is essential for proper emergence of the hair shaft, and also contributes to the barrier function of the skin [266]. The transcription factor SOX9 is required for sebaceous gland morphogenesis [177]. Low levels of Wnt/β-catenin signaling, and high levels of c-MYC, are thought to direct follicular progenitors towards a sebaceous, rather than hair follicle fate, (reviewed in [174]).

5.4.4 Cancer Development

Cancers are characterized by excessive proliferation, often resulting from dysregulation of developmental pathways. Similarly, epithelial-mesenchymal transition (EMT) is a process widely employed by tissues during development, and its improper induction can promote malignant transformation and tumor metastasis. Cutaneous carcinomas can derive from multiple epithelial compartments. Basal cell carcinoma (BCC) is the most commonly diagnosed form of skin cancer. Although BCC typically does not metastasize, it is locally invasive and can cause extensive tissue damage if left untreated. These lesions can arise from hair follicle epithelium or interfollicular epidermis [17], and are caused by inherited or sporadic mutations that cause inappropriate activation of the Hedgehog (Hh) signaling pathway [58, 185]. Advanced cases may be treated with the Hh pathway inhibitor Vismodegib [53]. Squamous cell carcinomas (SCC), which can arise from any stratified epithelium, are highly heterogeneous and possess metastatic potential. Similar to BCC, papillomas and SCCs of the skin appear to arise from either the IFE or the hair follicle bulge [17]. Not surprisingly, EMT is a common feature of these tumors. While the mutational landscape of cutaneous SCC is varied, mutations in *p53*, *HRAS*, *KNSTRN*, *CDKN2A* and Notch pathway genes appear to be important drivers of tumorigenesis [133, 186]. As increased EGFR activity is common among many SCCs, EGFR inhibition by small molecules or humanized antibodies has been proposed for treatment of unresectable cutaneous SCC [13]. Tumors of the sebaceous gland are relatively rare, but in some cases are associated with inactivating mutations in the Wnt signaling effector LEF1 [222].

5.5 HDACs in Epidermal Development, Homeostasis, and Disease

Early work using cultured keratinocytes showed that depletion of aryl hydrocarbon receptor nuclear translocator (ARNT) enhanced expression of Class I HDAC proteins, and induced the expression of several key differentiation markers. These effects were abolished by treatment with the HDAC inhibitor TSA [196]. In line with this, topical treatment of murine skin with TSA repressed expression of the CE protein profilaggrin [152]. Analysis of histone H4 acetylation in quiescent stem cells of the hair follicle revealed that chromatin is hypoacetylated in these cells. Treatment with TSA reversed the hypoacetylation, and induced proliferation of stem cells and their exit from the bulge niche [67]. These data using a broadly acting inhibitor suggested that HDACs can influence the proliferation and differentiation of diverse epithelial cell types in the skin, likely through multiple different, context-dependent mechanisms. These observations underscore the importance of genetic studies to delineate the specific functions of individual HDACs in the skin.

5.5.1 HDAC1/2 in Epidermal Development

During early development, HDAC1 and HDAC2 are expressed throughout the epidermis. Interestingly, by E16.5, despite ubiquitous expression, nuclear localization is more prominent in differentiated keratinocytes and in cells of the hair follicle bud [131]. However, it is not yet clear whether increased HDAC localization in certain cell types within the embryonic skin has functional consequences. As early lethality of global *Hdac1* and *Hdac2* knockout mice [125, 169, 231] complicated analysis of skin phenotypes, investigators employed tissue-specific deletion strategies to elucidate the functions of these genes in the epidermis. These studies demonstrated that, similar to the situation in other tissues [33, 105, 169, 252], HDAC1 and HDAC2 play both redundant and divergent roles in the skin.

LeBoeuf et al. showed that *K14-Cre*-mediated epidermal-specific deletion of either *Hdac1* or *Hdac2* did not impair embryonic development of the epidermis or hair follicles. However, concomitant deletion of both genes resulted in perinatal lethality due to a complete failure of epidermal stratification [131]. In double conditional knockout (DcKO) embryos, a single layered epidermis was formed that expressed K14 and K5 basal keratins, but suparbasal layers were absent and hair follicles failed to be specified. These defects were associated with a gradual decrease in basal cell proliferation, suggesting failure of progenitor cell self-renewal, as well as progressively increased apoptosis. The defects observed in DcKO mice were reminiscent of the effects of p63 deletion [129, 165, 254]. However, levels of p63 protein were unchanged in DcKO embryonic epidermis, suggesting that loss of HDAC1/2 might interfere with p63's functions as a transcriptional regulator. In line with this, while expression of positively regulated targets of ΔNp63 remained unchanged in DcKO embryos, repressed targets such as *p21*, *14-3-3σ*, and *p16/INK4a* showed elevated expression [131]. HDAC1 and HDAC2 were found to associate specifically with the promoters of these ΔNp63-repressed gene targets. Furthermore, these promoter regions exhibited increased levels of acetylated histone H3 in TSA-treated keratinocytes, indicating that they are normally deacetylated by HDACs. These data supported a model in which the repressive functions of ΔNp63 are mediated via complex formation with HDAC1/2. In support of this, ΔNp63, HDAC1 and HDAC2 form a trimeric complex in SCC cells [190].

HDAC1 represses p53 activity through direct deacetylation of p53 protein [93, 102, 224]. In DcKO mice, levels of acetylated-p53 were increased and *in vitro* experiments in cultured keratinocytes confirmed that loss of HDAC activity resulted in increased acetylation of p53 and elevated expression of its target gene *p21* [131]. Thus, hyperacetylated p53 likely contributes to the elevated p21 levels observed in DcKO epidermis. Interestingly, p53 can upregulate several p63-repressive targets [56, 91]. The ability of HDAC1/2 to differentially modify distinct subsets of genes is illustrated by p19/ARF. Like p16/INK4a, p19/ARF is a repressive target of ΔNp63 [220], but unlike p16/INK4a, its levels are not increased in DcKO mice. A likely mechanism underlying this observation involves hyperacetylation of p53, which has been shown to indirectly suppress p19/ARF expression [114]. Thus this work demonstrates how the effects of HDAC1/2 on p63 and p53 functions can result in nuanced regulation of overlapping target genes.

The co-repressor complexes responsible for directing HDAC1/2 to specific target genes in embryonic epidermis have not yet been identified. Deletion of Mi-2β, a component of the NuRD complex, in embryonic epidermis causes proliferation defects in basal keratinocytes and failure of hair follicle specification [116], similar to the effects of *Hdac1/2* deletion, suggesting possible association of HDAC1/2 with Mi-2β in this tissue. The severe effects of *Hdac1/2* deletion on embryonic epidermal basal cell proliferation and survival may mask additional functions of *Hdac1/2*. Concomitant deletion of *p16* and/or *p53* might be expected to partially rescue proliferation and cell survival, permitting analysis of such roles.

5.5.2 HDAC1/2 in Epidermal Homeostasis and Stem Cell Regulation

While deletion of either *Hdac1* or *Hdac2* alone had no observable consequences in the embryonic epidermis, two groups independently reported that epidermal *Hdac1* mutants develop skin phenotypes after birth [99, 246]. By contrast, epidermal

Table 5.2 Epidermal phenotypes of *Hdac* mutant mice

Genotype	Phenotype	Reference
K5-Cre Hdac1$^{fl/fl}$	Hyperproliferative scarring on tails in 24% of mice	[246]
K14-Cre Hdac1$^{fl/fl}$	Small body size, alopecia, progressive follicular dystrophy and cyst formation, loss of hair-type subsets, abnormal pigmentation due to melanocyte expansion, supernumerary claws, abnormal eyelid formation, deregulated facial pelage, epidermal hyperplasia and hyperkeratosis	[99]
K14/5-Cre Hdac2$^{fl/fl}$	None	[99, 246]
K14/5-Cre Hdac1$^{fl/+}$ *Hdac2*$^{fl/+}$	None	[99, 246]
K14-Cre Hdac1$^{fl/fl}$ *Hdac2*$^{fl/+}$	Generally more severe versions of phenotype described for *K14-Cre Hdac1*$^{fl/fl}$ mice	[99]
K5-Cre Hdac1$^{fl/fl}$ *Hdac2*$^{fl/+}$	Smaller/reduced weight gain, progressive alopecia, shorter whiskers, scaly tail regions, hyperkeratosis, hyperproliferative epidermis, enlarged sebaceous glands, spontaneous papilloma-like lesions	[246]
K5-Cre Hdac1$^{fl/+}$ *Hdac2*$^{fl/fl}$	None	[246]
K14-Cre Hdac1$^{fl/fl}$ *Hdac2*$^{fl/fl}$	Not viable due to severe epidermal developmental defects.	[99, 131, 246]
	Lack of embryonic epidermal stratification, lack of hair follicle specification, lack of filiform papillae in the tongue, abnormal dental structures, failure of eyelid fusion, failure of limb-digit separation	[131]

Hdac2 mutants remained phenotypicially normal in adult life, even when combined with heterozygous deletion of *Hdac1*, indicating that one copy of HDAC1 can compensate for HDAC2's functions in postnatal epidermis. Supporting this hypothesis, expression levels of HDAC1 protein increase upon deletion of HDAC2 and vice versa [246].

Interestingly, the epidermal *Hdac1* mutants generated in these two studies exhibited slightly different phenotypes (Table 5.2). In one report, the only observed anomaly was hyperproliferative scaring on the tails of a subset of mice [246]. However, in the second study significant pathologies were observed, including alopecia, progressive follicular dystrophy and cyst formation, deregulated facial pelage, enlarged Meibomian glands, and epidermal hyperplasia and hyperkeratosis [99]. These discrepancies may result from the use of different conditional alleles of *Hdac1* [169, 252], and Cre drivers (*K5-Cre* versus *K14-Cre*), possibly resulting in subtle differences in spatio-temporal expression of Cre-induced recombination. As each group used different mixed strain backgrounds, strain-specific phenotypic variation [217, 225] might also have contributed to the distinct observed phenotypes.

In both cases, however, additional loss of a single copy of *Hdac2* in *Hdac1* mutant epidermis increased the severity of skin phenotypes including progressive alopecia and epidermal hyperplasia and hyperkeratosis [99, 246]. Winter et al. also reported the development of spontaneous papilloma-like lesions in these combined mutants [246] (Table 5.2). Hyperkeratosis appeared to result from elevated expression of a specific subset of EDC genes including members of the *Sprr* and *CE* families [246]. In contrast, hair follicle abnormalities resulted from failure to enter a normal hair growth cycle, likely due to increased p53 expression and apoptosis that lead to hair follicle degeneration. Winter and colleagues further observed down-regulation of the hair follicle stem cell markers CD34, K15, LGR5, and SOX9, and an increase in expression of LGR6 which marks stem cells that contribute to the sebaceous gland and IFE [209, 219, 246]. In line with this, label-retaining bulge stem cell numbers were also decreased in the *Hdac1/2* compound mutants [246]. Thus, loss of *Hdac1/2* appears to affect lineage determination in the skin.

Interestingly, the SIN3A, MTA2/NuRD, and CoREST complexes were disturbed in compound epidermal *Hdac1/2* mutants. In addition to remodeling histones, the SIN3A complex deacetylates c-MYC, which destabilizes the protein and keeps its expression in check [173]. In line with this, compound epidermal *Hdac1/2* mutants displayed increased c-MYC levels, which could at least partially account for the epidermal hyperplasia, papilloma formation, and sebaceous gland enlargement observed in these mice [246]. These phenotypes contrast with the gradual failure of proliferation seen in embryonic epidermis following complete removal of both *Hdac1* and *Hdac2*. These distinct outcomes likely reflect the competing functions of diverse HDAC1/2 targets: in postnatal cells lacking both copies of *Hdac1* and one copy of *Hdac2* the effects of hyperacetylated c-MYC predominate, whereas in the complete absence of both proteins in embryonic epidermis the overiding phenotypes may be a consequence of elevated p16 and p21 levels. We hypothesize that complete removal of HDAC1 and HDAC2 in adult skin would similarly lead to failure of epidermal stem cell proliferation and self-renewal. This hypothesis may be tested in future experiments using inducible homozygous epidermal-specific deletion of both *Hdac1* and *Hdac2*.

5.5.3 Evidence for Epidermal Functions of HDAC3

As outlined above, HDAC3, another member of the Class I HDAC family, associates with different multi-protein complexes than those containing HDAC1/2, suggesting that it performs distinct functions. While the requirements for HDAC3 in embryonic epidermal development have not yet been reported, recent data suggest that this protein plays key roles in inflammatory responses in adult skin. The glucocorticoid receptor (GR), a common target for anti-inflammatory therapy, is a nuclear hormone receptor that regulates target gene transcription via several mechanisms [192]. Firstly, following glucocorticoid binding, GR regulates transcription through direct interaction with GRE binding sites in the DNA; secondly, GR can mediate transrepression by associating with conserved inverted repeated negative response elements (IR nGRE DNA binding sequences); and thirdly GR can effect tethered indirect transrepression via interaction with transactivators such as AP1/STAT3 bound to their cognate DNA binding sequences. Pierre Chambon's group showed recently that glucocorticoid-induced tethered repression requires SUMOylation of GR and formation of a SUMO-SMRT/NCoR1-HDAC3 repressive complex, and this is essential for IR nGRE-mediated transrepression. In line with this, deletion of *Hdac3* in epidermal keratinocytes prevents glucocorticoid-induced direct repression [95, 96].

HDAC3 has also been recently shown to suppress expression of Aquaporin-3, a water and glycerol channel involved in epidermal homeostasis, wound healing and barrier repair [38]. A review of the literature reveals additional potential roles for HDAC3 in the epidermis. In humans and mice, mutations in the *Hairless (hr)* gene result in alopecia [2, 184]. HR is functionally similar to the nuclear co-repressors N-CoR and SMRT [188, 228], and can associate with HDAC3 [52, 188]. In *hr* mutant mice, hair follicles prematurely enter a severely deregulated catagen phase resulting in their destruction [184]. As HDAC3 can mediate at least some of the repressive effects of HR [188], it may be important in catagen regulation.

Another potential role for HDAC3 in the epidermis is in mediating skin barrier function. HDAC3 regulates lipid metabolism via-PPARγ in adipocytes and other tissues [20, 61, 128, 138], and can deacetylate PPARγ to inhibit target gene expression [111]. In the skin PPARs have definitive roles in epidermal barrier formation and in sebocyte differentiation; therefore, it will be interesting to determine whether HDAC3 regulates PPARs in the epidermis.

5.5.4 HDAC8 Plays Roles in Epidermal Innate Immunity

Epidermal keratinocytes produce antimicrobial peptides and cytokines that elicit inflammatory responses to danger- or pathogen-associated molecular patterns (PAMPs). However, these responses must be dampened in order for the skin to tolerate contact with the diverse array of microorganisms that form the normal microbiome. A recent study showed that exposure to the short-chain fatty acid sodium butyrate, a potent HDAC inhibitor, increases pro-inflammatory gene expression in keratinocytes exposed to the TLR2/6 ligand macrophage-activating lipopeptide 2 (MALP-2), thus breaking normal tolerance [205]. Using systematic depletion of each of the HDACs expressed in keratinocytes, the investigators found that loss of HDAC8 or HDAC9 mimicked the effects of butyrate, identifying these HDACs as responsible for suppressing excessive inflammation in response to MALP-2. Further investigation revealed that the commensal skin bacterium *Propionibacterium acnes* synthesizes short-chain fatty acids when provided with a lipid source and grown under hypoxic conditions. These molecules inhibit HDAC activity in a similar manner to sodium butyrate, enhancing host defense mechanisms in mouse skin *in vivo* [205]. These interesting findings suggest an HDAC8-mediated epigenetic mechanism by which *P. acnes* can elicit inflammation when present in the hypoxic environment of the hair follicle.

5.5.5 Additional Roles for Class I HDACs

The severe phenotypes evident in epidermal-deletion mutants likely mask other, subtler HDAC functions. Table 5.3 presents a selected list of players critical to epidermal development and homeostasis that have been shown to interact with HDACs in other tissues. Because HDACs regulate gene expression in a highly context-dependent manner it is impossible to know which of these mechanisms may be relevant to the epidermis without direct study. Nevertheless, this list demonstrates how pervasive HDAC activity is across signaling networks and organ systems and may identify promising targets for future epidermal research.

5.5.6 HDACs in Skin Cancer

As HDACs can drive cellular proliferation and EMT they are attractive targets for anticancer therapies. Indeed, HDAC inhibitors (HDACi) are effective in treating hematopoietic malignancies, and Pabinostat (pan-HDACi) and Vorinostat (class I/II-HDACi) are FDA-approved chemotherapeutics for acute myelocytic leukemia (AML) and cutaneous T-cell lymphoma (CTCL), respectively. The efficacy of HDACi in epithelial malignancies is currently being assessed in multiple clinical trials.

Recent work provides a mechanism by which HDACs may be relevant to BCC. BCCs are the result of aberrant activation of the Hh pathway. Loss of the inhibitor Ptch1 leads to constitutive activation of the signal transducer Smoothened (SMO), which promotes activation of GLI transcription factors [195]. The SMO inhibitor Vismodegib is approved for treatment of advanced BCC, although adverse reactions and drug-resistance can mitigate its effectiveness [5, 223]. GLI1 and GLI2 must be deacetylated in order to function effectively, and HDACi thus attenuates Hh pathway activity [30]. Recently, a chimeric Vismodegib-Vorinostat compound designed to inhibit both SMO and HDAC was found to be more effective than individual drugs at down-regulating Hh pathway activity [265]. This may be due to the ability of HDACi to directly suppress the transcriptional activity of GLI2. Dual targeting of HDAC and Hh activity as a potentially effective anti-cancer treatment has also been demonstrated in aerodigestive and pancreatic cancer cell lines [40, 41]. Despite the promising results of these *in vitro* studies, more work is needed before this approach can be applied clinically.

HDAC1/2 expression has been linked to poor prognosis in cohorts of SCC [32, 227], and HDACi are currently being tested in clinical trials for SCC treatment both alone and in combination with anti-EGFR therapies. One rationale for compound therapy centers around the EMT observed in SCC. EMT is characterized by cadherin switching, a process by which E-cadherin expression is replaced by N-cadherin. E-cadherin can function to restrain EGFR activity [8, 238]; thus its loss is thought to partially account for enhanced EGFR signaling in cancer. Additionally, EGFR

Table 5.3 List of selected epidermal regulators that interact with HDACs in other tissue types

Epidermal Regulators	Function in Epidermis	HDAC	System	Findings	Reference
14-3-3σ	Required for keratinocyte commitment to terminal differentiation (for review see [124])	HDAC4/5 (Class IIa)	Jurkat cells (T-cell line)	14-3-3σ binds to phosphorylated HDAC4/5 and sequesters them in the cytoplasm	[79]
		HDAC4 (Class IIa)	MCF7 (breast cancer) & COS7 (monkey kidney fibroblasts) cells	14-3-3σ binds to phosphorylated HDAC4, inhibiting its nuclear localization	[176]
BMP	BMP signaling regulates proliferation and differentiation in the IFE and hair follicle, and also has important roles in mediating hair follicle cycling (for review see [24])	HDAC1	Mouse osteoblast precursor cell line	BMP signaling is inhibited by an HDAC/TWIST1/SMAD4 repressor complex.	[88]
BMP2, BMP4		Class I (HDAC2)	Murine prostate	HDACs regulate the expression of *Bmp* genes to regulate prostate morphogenesis	[208]
E2F/pRb	Regulates keratinocyte differentiation, skin regeneration, and malignant transformation (for review see [104])	HDAC1	Assorted cell lines	pRB recruits HDAC1 to mediate target gene repression.	[25, 150]
GRHL1	Regulates *Dsg1* expression; its deletion results in barrier defects and enhances SCC tumor progression [166]	HDAC3	Neuroblastoma	HDAC3 and MYCN cooperate to repress *Grhl1* gene expression.	[60]

Hedgehog Pathway	Regulates hair follicle development and cycling; its aberrant activation leads to BCC [7]				
SMO	GPCR signal transducer of the Hh pathway, regulates degradation and activation of GLI transcription factors [195]; constitutively active mutants develop basaloid hamratomas [75]	Classical HDACs	Brain	Constitutively active Smo mutant mice exhibit deregulated expression of various HDACs with changes at both the RNA and protein level. HDAC1 is expressed in normal cerebellum, and HDAC1/2/3 all demonstrated increased protein expression in medulloblastomas. HDAC activity correlated with proliferative capacity of cells.	[136]
GLI1, GLI2	GLI2 is primarily a transcriptional activator; GLI1 is induced upon Hh pathway activation and is exclusively an activator of target gene expression [195]; GLI2 over expression leads to tumor development [74]	HDAC1	NIH-3 T3, HEK293T, RKE3E, Daoy cell lines, mouse cerebellum	GLI1 and GLI2 are HDAC1 targets; deacetylation enhances activity of these transcription factors.	[30]
GLI3	Primarily a transcriptional repressor inhibiting Hh target gene expression.	HDAC1	293 T & MNS-70 cells;	HDAC1 associates with SKI and GLI3 in a repressor complex; SKI is required for GLI3-mediated gene repression	[47]
IKKα	Regulates epidermal and appendage development [218]	HDAC3	Fibroblasts	IKKα mediates the removal of HDAC3/SMRT complexes from select NFκB promoter targets	[72]

(continued)

Table 5.3 (continued)

Epidermal Regulators	Function in Epidermis	HDAC	System	Findings	Reference
IRF6	Required for keratinocyte commitment to terminal differentiation (reviewed in [124])	HDAC6, HDAC10	porcine monocyte-derived dendritic cells	Inhibition of HDACs with Sulforaphane suppresses IRF6 and TGF-β1 production.	[189]
KLF4	Mediates repression of junctional proteins and promotes barrier formation [212]	Classical HDACs	Lung cancer	HDACi increase KLF4 expression, and decrease cell proliferation	[259]
		HDAC1	Acute myeloid leukemia (AML)	HDAC1 negatively correlates with KLF4 in AML, and elevated HDAC1 levels are associated with better prognosis; HDAC1 and KILF4 competitively bind *Klf4* promoter and oppositely regulate *Klf4* expression	[98]
		Class I	Prostate cancer	HDACs mediate *Klf4* repression by altering H3K4 methylation status	[97]
Notch	Critical player in development of IFE and hair follicles; is deregulated in subsets of SCC (for review see [179, 182])	HDAC1	T-Cell lymphoma: cell line and mouse model	Acetylated Notch3 is targeted for Ubiquitin-mediated degradation; HDAC1 stabilizes Notch3 to promote proliferation	[181]
Notch		HDAC1	CV1 cells, Smooth muscle cells, Retina	HDAC1 mediates the repression of subsets of Notch target genes, that is alleviated upon Notch activation	[115, 223, 251]
PKC	Select PKC isoforms promote the expression of *Lce* gene (for review see [124])	HDAC5	Heart	PKC/PKCα mediates nuclear export of HDAC5	[234, 264]
		Classical HDACs	Non-small cell lung cancer and head and neck SCC	HDACi treatment decreases PKC activation	[178]
		Classical HDACs	Assorted epithelial cancers	HDACi treatment enhances PKC activation and promotes tumor progression	[142]

Wnt/β--catenin –TCF/LEF	Critical pathway involved in epidermal and hair follicle morphogenesis and homeostasis ([78, 242])				
LEF1	LEF1 is a transcriptional enhancer required for Wnt/β-cateinn-mediated transcriptional activation.	Classical HDACs	Cos-1 cells, colon cancer cell lines	Groucho repressor complexes compete with β-catenin to bind LEF1; repression is dependent on HDAC activity	[6]
TCF4/ OVOL2	TCF4 mediates activation of Wnt/β-catenin-target gene expression. OVOL2 inhibits c-MYC and Notch1 to suppress cell cycling and terminal differentiation [244]	HDAC1 Colorectal cancer	Colorectal cancer	OVOL2 recruits HDAC1 to TCF4/β-catenin complexes to repress target gene activity.	[257]
OVOL1	Downstream of β-catenin/LEF, it is required for keratinocyte commitment to terminal differentiation; negatively regulates c-MYC [172]	HDAC1/2/3	NIH-3 T3, 293 T cells (fibroblast and kidney cell lines)	OVOL1 inhibits its own transcription in part by recruiting HDAC1/2/3 to its promoter.	[171]
SP1	Regulates expression of differentiation genes including loricrin and transglutaminase [109, 197]	HDAC1	MCF7 Cells	Leutinizing hormone receptor is derepressed through PKCα/ERK-mediated SP1 phosphorylation, causing the release of an HDAC1/SIN3A complex from the promoter	[140]
STAT3	Plays important roles in wound healing, keratinocyte migration, and hair follicle growth; however, it can also contribute to skin tumor development (for review see [206])	Classical HDACs	Colorectal cancer cells	JAK2/STAT3 signaling is initiated by TSA via inducing the promoter-associated histone acetylation of suppressor of cytokine signaling genes	[249]

(continued)

Table 5.3 (continued)

Epidermal Regulators	Function in Epidermis	HDAC	System	Findings	Reference
		Class I	PC3 (prostate cancer) cell line	STAT3 acetylation is required for dimerization and activation, and HDACs mediate the removal of this acetyl moiety and inhibit STAT3 activation	[261]
		HDAC1/2	Rat renal interstitial fibroblasts, mouse embryonic fibroblasts	HDAC1/2 can mediate cellular proliferation by deacetylating STAT3 to inhibit its signaling capacity	[183]
		HDAC3	Rat renal interstitial fibroblasts, mouse embryonic fibroblasts	HDAC3 associates with STAT3; HDAC3 knockdown enhances STAT3 acetylation but preventes STAT3 phosphorylation, inhibiting cell survival	[81]
TGFβ		Class I & II	SRA01/01 & HLEB3 (Lens epithelium cells)	TGFβ2 increases HDAC activity; TSA inhibits proliferation and EMT	[34]
ZEB1	Represses key players involved in epidermal development including E-cadherin [245] and miR-203 [243]	Class 1	Pancreatic and lung cancer cell lines	The class I HDAC inhibitor mocetinostat acts epigenetically to interfere with ZEB1 function	[156]
		HDAC1/2	Pancreatic Cancer	ZEB1 recruits HDAC1/2 to repress target gene expression	[1]

inhibitors are more effective when E-cadherin is expressed [248]. E-cadherin expression is modulated by a number of factors including ZEB1 [245], which can repress target genes in an HDAC-dependent manner [1]. Furthermore, HDACi has been shown to reverse EMT in a number of cell types [137, 202, 204, 240].

Despite the findings that HDAC activity can promote EMT in the contexts outlined above, in other cases including colon [63, 110], breast [247] and HNSCC [71] cancers the opposite appears to be the case, as HDACi enhance rather than attenuate EMT in these tumors. The use of pan inhibitors makes it challenging to identify the specific HDACs that are responsible for particular physiological responses. Isozyme-specific functions may be further obscured by the fact that some broad-spectrum HDACi have varying specificity for different HDACs and even for specific HDAC complexes [9, 55, 118]. Another counterintuitive finding is that loss of HDAC1 in mouse skin enhances tumor development in a K5-SOS tumor model, which relies on constitutively active Ras signaling and is dependent on functional EGFR [216, 246]. Given that repressor complexes are responsible for targeting of HDAC activity, context-dependent co-repressor complements may contribute to these apparently paradoxical data. These findings suggest that caution should be employed when targeting HDACs in clinical treatments.

5.6 Future Directions

An expanding body of evidence has unequivocally identified HDACs as key regulators of epidermal development, homeostasis, and disease. HDACs are at the crux of numerous cell signaling pathways during epidermal morphogenesis, and can integrate the responses between signaling cascades. Additionally, as HDACs can target multiple proteins within a pathway, they can facilitate both positive and negative feedback loops. Therefore, these proteins are uniquely situated to precisely modulate diverse signaling networks, giving them the capacity to both maintain stem cell quiescence and mobilize global changes in a spatio-temporal manner to regulate proliferation, differentiation, and even disease development.

Direct and indirect evidence allows us to infer that HDACs may function to modulate an extensive list of epidermal regulators (Fig. 5.1a, Table 5.3). Despite progress in delineating the roles of HDAC1/2 in embryonic and adult epidermis, we have just begun to scratch the surface of HDAC functions in the skin. The putative contributions of the remaining class I HDACs, HDAC3 and HDAC8, in epidermal development have yet to be elucidated. The members of other classes of HDACs may also prove to play critical roles in the skin, either as chromatin modulators or though their actions on transcription factors and other regulatory proteins.

Another consideration is that most functional studies have specifically focused on HDAC activity in the keratinocytes. However, given the dynamic interplay between the dermis and epidermis, it is reasonable to hypothesize that HDAC activity in fibroblasts also plays critical roles, including in modulating signaling from the dermis to the epidermis. Such mechanisms may have major impacts on epidermal development, hair follicle morphogenesis and cycling, and cancer. It is also possible that a specific HDAC paralog may oppositely regulate a process depending upon the compartment in which it is expressed. Should this prove true, targeting with HDACi may need to be cell type-specific.

A limitation to many of the studies performed to date is the use of pan-HDACi. While convenient for identifying some processes regulated by HDACs, these are blunt tools that cannot effectively distinguish contributions of individual HDACs, and subtle changes may be masked by stronger phenotypes. Novel isozyme-specific inhibitors are being developed, but only a few reports have been published using these compounds [62, 168, 226]. Together with genetic studies, the development of specific HDAC inhibitors will profoundly contribute to our understanding of the roles of individual isozymes to cellular functions, and has the potential to improve clinical treatments while mitigating off-target effects. However, even this approach comes with caveats, for to fully grasp the multi-faceted and sometimes contradictory effects of HDACs, we must also consider the context in which these proteins are expressed.

Elaborate spatio-temporal differences in HDAC functions have been well-documented, EMT being a prime example. These divergent effects are likely mediated by variation in the compositions of HDAC repressor complexes, as is the case with the N-CoR complex which can promote or inhibit EMT depending upon which MTA variant resides within the assembly [69]. The ubiquitous nature of HDACs expression coupled with their ability to act upon multiple substrates may limit the therapeutic effects of pan- or even paralog-specific HDACi. To provide the most tailored response on HDAC-mediated cellular processes highly specific, targeted disruptions of particular HDAC-repressor complex interactions may be more desirable. However, employment of such a strategy relies on elucidating the constituents of individual HDAC complexes in specific disease contexts, along with their functional effects.

Acknowledgements The authors thank Fang Liu for critically reading this manuscript.

References

1. Aghdassi A, Sendler M, Guenther A, Mayerle J, Behn CO, Heidecke CD, Friess H, Buchler M, Evert M, Lerch MM, et al. Recruitment of histone deacetylases HDAC1 and HDAC2 by the transcriptional repressor ZEB1 downregulates E-cadherin expression in pancreatic cancer. Gut. 2012;61:439–48.
2. Ahmad W, Faiyaz ul Haque M, Brancolini V, Tsou HC, ul Haque S, Lam H, Aita VM, Owen J, deBlaquiere M, Frank J, et al. Alopecia universalis associated with a mutation in the human hairless gene. Science (New York). 1998;279:720–4.
3. Allfrey VG, Faulkner R, Mirsky AE. Acetylation And Methylation Of Histones And Their Possible Role In The Regulation Of Rna Synthesis. Proc Natl Acad Sci USA. 1964;51:786–94.
4. Alonso L, Fuchs E. The hair cycle. J Cell Sci. 2006;119:391–3.

5. Amin SH, Motamedi KK, Ochsner MC, Song TE, Hybarger CP. Mechanisms and efficacy of vismodegib in the treatment of basal cell carcinoma. Discov Med. 2013;16:229–32.
6. Arce L, Pate KT, Waterman ML. Groucho binds two conserved regions of LEF-1 for HDAC-dependent repression. BMC Cancer. 2009;9:159.
7. Athar M, Tang X, Lee JL, Kopelovich L, Kim AL. Hedgehog signalling in skin development and cancer. Exp Dermatol. 2006;15:667–77.
8. Bae GY, Choi SJ, Lee JS, Jo J, Lee J, Kim J, Cha HJ. Loss of E-cadherin activates EGFR-MEK/ERK signaling, which promotes invasion via the ZEB1/MMP2 axis in non-small cell lung cancer. Oncotarget. 2013;4:2512–22.
9. Bantscheff M, Hopf C, Savitski MM, Dittmann A, Grandi P, Michon AM, Schlegl J, Abraham Y, Becher I, Bergamini G, et al. Chemoproteomics profiling of HDAC inhibitors reveals selective targeting of HDAC complexes. Nat Biotechnol. 2011;29:255–65.
10. Basta J, Rauchman M. The nucleosome remodeling and deacetylase complex in development and disease. Transl Res. 2015;165:36–47.
11. Baymaz HI, Karemaker ID, Vermeulen M. Perspective on unraveling the versatility of 'co-repressor' complexes. Biochim Biophys Acta. 2015;1849:1051–6.
12. Beck B, Blanpain C. Mechanisms regulating epidermal stem cells. EMBO J. 2012;31:2067–75.
13. Bejar C, Maubec E. Therapy of advanced squamous cell carcinoma of the skin. Curr Treat Options in Oncol. 2014;15:302–20.
14. Berger SL, Kouzarides T, Shiekhattar R, Shilatifard A. An operational definition of epigenetics. Genes Dev. 2009;23:781–3.
15. Bhaskara S, Chyla BJ, Amann JM, Knutson SK, Cortez D, Sun ZW, Hiebert SW. Deletion of histone deacetylase 3 reveals critical roles in S phase progression and DNA damage control. Mol Cell. 2008;30:61–72.
16. Bheda P, Schneider R. Epigenetics reloaded: the single-cell revolution. Trends Cell Biol. 2014;24:712–23.
17. Blanpain C. Tracing the cellular origin of cancer. Nat Cell Biol. 2013;15:126–34.
18. Blanpain C, Fuchs E. Epidermal homeostasis: a balancing act of stem cells in the skin. Nat Rev Mol Cell Biol. 2009;10:207–17.
19. Blanpain C, Lowry WE, Pasolli HA, Fuchs E. Canonical notch signaling functions as a commitment switch in the epidermal lineage. Genes Dev. 2006;20:3022–35.
20. Bochkis IM, Przybylski D, Chen J, Regev A. Changes in nucleosome occupancy associated with metabolic alterations in aged mammalian liver. Cell Rep. 2014;9:996–1006.
21. Bosch-Presegue L, Vaquero A. Sirtuin-dependent epigenetic regulation in the maintenance of genome integrity. FEBS J. 2015;282:1745–67.
22. Botchkarev VA. Integration of the Transcription Factor-Regulated and Epigenetic Mechanisms in the Control of Keratinocyte Differentiation. J Investig Dermatol Symp Proc. 2015;17:30–2.
23. Botchkarev VA, Gdula MR, Mardaryev AN, Sharov AA, Fessing MY. Epigenetic regulation of gene expression in keratinocytes. J Invest Dermatol. 2012;132:2505–21.
24. Botchkarev VA, Sharov AA. BMP signaling in the control of skin development and hair follicle growth. Differentiation. 2004;72:512–26.
25. Brehm A, Miska EA, McCance DJ, Reid JL, Bannister AJ, Kouzarides T. Retinoblastoma protein recruits histone deacetylase to repress transcription. Nature. 1998;391:597–601.
26. Brownell JE, Zhou J, Ranalli T, Kobayashi R, Edmondson DG, Roth SY, Allis CD. Tetrahymena histone acetyltransferase A: a homolog to yeast Gcn5p linking histone acetylation to gene activation. Cell. 1996;84:843–51.
27. Byrne C, Tainsky M, Fuchs E. Programming gene expression in developing epidermis. Development. 1994;120:2369–83.
28. Campos EI, Reinberg D. Histones: annotating chromatin. Annu Rev Genet. 2009;43:559–99.
29. Candi E, Rufini A, Terrinoni A, Dinsdale D, Ranalli M, Paradisi A, De Laurenzi V, Spagnoli LG, Catani MV, Ramadan S, et al. Differential roles of p63 isoforms in epidermal development: selective genetic complementation in p63 null mice. Cell Death Differ. 2006;13:1037–47.

30. Canettieri G, Di Marcotullio L, Greco A, Coni S, Antonucci L, Infante P, Pietrosanti L, De Smaele E, Ferretti E, Miele E, et al. Histone deacetylase and Cullin3-REN(KCTD11) ubiquitin ligase interplay regulates Hedgehog signalling through Gli acetylation. Nat Cell Biol. 2010;12:132–42.
31. Chalkiadaki A, Guarente L. The multifaceted functions of sirtuins in cancer. Nat Rev Cancer. 2015;15:608–24.
32. Chang HH, Chiang CP, Hung HC, Lin CY, Deng YT, Kuo MY. Histone deacetylase 2 expression predicts poorer prognosis in oral cancer patients. Oral Oncol. 2009;45:610–4.
33. Chen S, Bellew C, Yao X, Stefkova J, Dipp S, Saifudeen Z, Bachvarov D, El-Dahr SS. Histone deacetylase (HDAC) activity is critical for embryonic kidney gene expression, growth, and differentiation. J Biol Chem. 2011a;286:32775–89.
34. Chen X, Xiao W, Chen W, Luo L, Ye S, Liu Y. The epigenetic modifier trichostatin A, a histone deacetylase inhibitor, suppresses proliferation and epithelial-mesenchymal transition of lens epithelial cells. Cell Death Dis. 2013;4:e884.
35. Chen Y, Wang H, Yoon SO, Xu X, Hottiger MO, Svaren J, Nave KA, Kim HA, Olson EN, Lu QR. HDAC-mediated deacetylation of NF-kappaB is critical for Schwann cell myelination. Nat Neurosci. 2011b;14:437–41.
36. Choi JE, Mostoslavsky R. Sirtuins, metabolism, and DNA repair. Curr Opin Genet Dev. 2014;26:24–32.
37. Choi YS, Zhang Y, Xu M, Yang Y, Ito M, Peng T, Cui Z, Nagy A, Hadjantonakis AK, Lang RA, et al. Distinct functions for Wnt/beta-catenin in hair follicle stem cell proliferation and survival and interfollicular epidermal homeostasis. Cell Stem Cell. 2013;13:720–33.
38. Choudhary V, Olala LO, Kagha K, Pan ZQ, Chen X, Yang R, Cline A, Helwa I, Marshall L, Kaddour-Djebbar I, et al. Regulation of the glycerol transporter, aquaporin-3, by histone deacetylase-3 and p53 in keratinocytes. J Investig Dermatol. 2017;137:1935.
39. Chuang HC, Chang CW, Chang GD, Yao TP, Chen H. Histone deacetylase 3 binds to and regulates the GCMa transcription factor. Nucleic Acids Res. 2006;34:1459–69.
40. Chun SG, Park H, Pandita RK, Horikoshi N, Pandita TK, Schwartz DL, Yordy JS. Targeted inhibition of histone deacetylases and hedgehog signaling suppress tumor growth and homologous recombination in aerodigestive cancers. Am J Cancer Res. 2015;5:1337–52.
41. Chun SG, Zhou W, Yee NS. Combined targeting of histone deacetylases and hedgehog signaling enhances cytoxicity in pancreatic cancer. Cancer Biol Ther. 2009;8:1328–39.
42. Clapier CR, Cairns BR. The biology of chromatin remodeling complexes. Annu Rev Biochem. 2009;78:273–304.
43. Cotsarelis G, Millar SE. Towards a molecular understanding of hair loss and its treatment. Trends Mol Med. 2001;7:293–301.
44. Cowger JJ, Torchia J. Direct association between the CREB-binding protein (CBP) and nuclear receptor corepressor (N-CoR). Biochemistry. 2006;45:13150–62.
45. Cowley SM, Iritani BM, Mendrysa SM, Xu T, Cheng PF, Yada J, Liggitt HD, Eisenman RN. The mSin3A chromatin-modifying complex is essential for embryogenesis and T-cell development. Mol Cell Biol. 2005;25:6990–7004.
46. Cremer T, Cremer C. Chromosome territories, nuclear architecture and gene regulation in mammalian cells. Nat Rev Genet. 2001;2:292–301.
47. Dai P, Shinagawa T, Nomura T, Harada J, Kaul SC, Wadhwa R, Khan MM, Akimaru H, Sasaki H, Colmenares C, et al. Ski is involved in transcriptional regulation by the repressor and full-length forms of Gli3. Genes Dev. 2002;16:2843–8.
48. Dassule HR, Lewis P, Bei M, Maas R, McMahon AP. Sonic hedgehog regulates growth and morphogenesis of the tooth. Development. 2000;127:4775–85.
49. David G, Grandinetti KB, Finnerty PM, Simpson N, Chu GC, Depinho RA. Specific requirement of the chromatin modifier mSin3B in cell cycle exit and cellular differentiation. Proc Natl Acad Sci USA. 2008;105:4168–72.

50. de Guzman Strong C, Wertz PW, Wang C, Yang F, Meltzer PS, Andl T, Millar SE, Ho IC, Pai SY, Segre JA. Lipid defect underlies selective skin barrier impairment of an epidermal-specific deletion of Gata-3. J Cell Biol. 2006;175:661–70.
51. Denning MF, Dlugosz AA, Williams EK, Szallasi Z, Blumberg PM, Yuspa SH. Specific protein kinase C isozymes mediate the induction of keratinocyte differentiation markers by calcium. Cell Growth Differ. 1995;6:149–57.
52. Djabali K, Christiano AM. Hairless contains a novel nuclear matrix targeting signal and associates with histone deacetylase 3 in nuclear speckles. Differentiation. 2004;72:410–8.
53. Dlugosz A, Agrawal S, Kirkpatrick P. Vismodegib. Nat Rev Drug Discov. 2012;11:437–8.
54. Dlugosz AA, Yuspa SH. Protein kinase C regulates keratinocyte transglutaminase (TGK) gene expression in cultured primary mouse epidermal keratinocytes induced to terminally differentiate by calcium. J Invest Dermatol. 1994;102:409–14.
55. Dokmanovic M, Perez G, Xu W, Ngo L, Clarke C, Parmigiani RB, Marks PA. Histone deacetylase inhibitors selectively suppress expression of HDAC7. Mol Cancer Ther. 2007;6:2525–34.
56. el-Deiry WS, Tokino T, Velculescu VE, Levy DB, Parsons R, Trent JM, Lin D, Mercer WE, Kinzler KW, Vogelstein B. WAF1, a potential mediator of p53 tumor suppression. Cell. 1993;75:817–25.
57. Elias PM, Nau P, Hanley K, Cullander C, Crumrine D, Bench G, Sideras-Haddad E, Mauro T, Williams ML, Feingold KR. Formation of the epidermal calcium gradient coincides with key milestones of barrier ontogenesis in the rodent. J Invest Dermatol. 1998;110:399–404.
58. Epstein EH. Basal cell carcinomas: attack of the hedgehog. Nat Rev Cancer. 2008;8:743–54.
59. Eskiw CH, Cope NF, Clay I, Schoenfelder S, Nagano T, Fraser P. Transcription factories and nuclear organization of the genome. Cold Spring Harb Symp Quant Biol. 2010;75:501–6.
60. Fabian J, Lodrini M, Oehme I, Schier MC, Thole TM, Hielscher T, Kopp-Schneider A, Opitz L, Capper D, von Deimling A, et al. GRHL1 acts as tumor suppressor in neuroblastoma and is negatively regulated by MYCN and HDAC3. Cancer Res. 2014;74:2604–16.
61. Fajas L, Egler V, Reiter R, Hansen J, Kristiansen K, Debril MB, Miard S, Auwerx J. The retinoblastoma-histone deacetylase 3 complex inhibits PPARgamma and adipocyte differentiation. Dev Cell. 2002;3:903–10.
62. Falkenberg KJ, Johnstone RW. Histone deacetylases and their inhibitors in cancer, neurological diseases and immune disorders. Nat Rev Drug Discov. 2014;13:673–91.
63. Feng J, Cen J, Li J, Zhao R, Zhu C, Wang Z, Xie J, Tang W. Histone deacetylase inhibitor valproic acid (VPA) promotes the epithelial mesenchymal transition of colorectal cancer cells via up regulation of Snail. Cell Adhes Migr. 2015;9:495–501.
64. Feng S, Jacobsen SE, Reik W. Epigenetic reprogramming in plant and animal development. Science (New York). 2010;330:622–7.
65. Fischle W, Dequiedt F, Hendzel MJ, Guenther MG, Lazar MA, Voelter W, Verdin E. Enzymatic activity associated with class II HDACs is dependent on a multiprotein complex containing HDAC3 and SMRT/N-CoR. Mol Cell. 2002;9:45–57.
66. Fischle W, Kiermer V, Dequiedt F, Verdin E. The emerging role of class II histone deacetylases. Biochem Cell Biol. 2001;79:337–48.
67. Frye M, Fisher AG, Watt FM. Epidermal stem cells are defined by global histone modifications that are altered by Myc-induced differentiation. PLoS One. 2007;2:e763.
68. Fuchs E. Scratching the surface of skin development. Nature. 2007;445:834–42.
69. Fujita N, Jaye DL, Kajita M, Geigerman C, Moreno CS, Wade PA. MTA3, a Mi-2/NuRD complex subunit, regulates an invasive growth pathway in breast cancer. Cell. 2003;113:207–19.
70. Gao L, Cueto MA, Asselbergs F, Atadja P. Cloning and functional characterization of HDAC11, a novel member of the human histone deacetylase family. J Biol Chem. 2002;277:25748–55.
71. Giudice FS, Pinto DS Jr, Nor JE, Squarize CH, Castilho RM. Inhibition of histone deacetylase impacts cancer stem cells and induces epithelial-mesenchyme transition of head and neck cancer. PLoS One. 2013;8:e58672.
72. Gloire G, Horion J, El Mjiyad N, Bex F, Chariot A, Dejardin E, Piette J. Promoter-dependent effect of IKKalpha on NF-kappaB/p65 DNA binding. J Biol Chem. 2007;282:21308–18.

73. Glozak MA, Seto E. Histone deacetylases and cancer. Oncogene. 2007;26:5420–32.
74. Grachtchouk M, Mo R, Yu S, Zhang X, Sasaki H, Hui CC, Dlugosz AA. Basal cell carcinomas in mice overexpressing Gli2 in skin. Nat Genet. 2000;24:216–7.
75. Grachtchouk V, Grachtchouk M, Lowe L, Johnson T, Wei L, Wang A, de Sauvage F, Dlugosz AA. The magnitude of hedgehog signaling activity defines skin tumor phenotype. EMBO J. 2003;22:2741–51.
76. Greer CB, Tanaka Y, Kim YJ, Xie P, Zhang MQ, Park IH, Kim TH. Histone Deacetylases Positively Regulate Transcription through the Elongation Machinery. Cell Rep. 2015;13:1444–55.
77. Gregoire S, Xiao L, Nie J, Zhang X, Xu M, Li J, Wong J, Seto E, Yang XJ. Histone deacetylase 3 interacts with and deacetylates myocyte enhancer factor 2. Mol Cell Biol. 2007;27:1280–95.
78. Grigoryan T, Wend P, Klaus A, Birchmeier W. Deciphering the function of canonical Wnt signals in development and disease: conditional loss- and gain-of-function mutations of beta-catenin in mice. Genes Dev. 2008;22:2308–41.
79. Grozinger CM, Schreiber SL. Regulation of histone deacetylase 4 and 5 and transcriptional activity by 14-3-3-dependent cellular localization. Proc Natl Acad Sci USA. 2000;97:7835–40.
80. Guardiola AR, Yao TP. Molecular cloning and characterization of a novel histone deacetylase HDAC10. J Biol Chem. 2002;277:3350–6.
81. Gupta M, Han JJ, Stenson M, Wellik L, Witzig TE. Regulation of STAT3 by histone deacetylase-3 in diffuse large B-cell lymphoma: implications for therapy. Leukemia. 2012;26:1356–64.
82. Haberland M, Mokalled MH, Montgomery RL, Olson EN. Epigenetic control of skull morphogenesis by histone deacetylase 8. Genes Dev. 2009a;23:1625–30.
83. Haberland M, Montgomery RL, Olson EN. The many roles of histone deacetylases in development and physiology: implications for disease and therapy. Nat Rev Genet. 2009b;10:32–42.
84. Hagelkruys A, Mattes K, Moos V, Rennmayr M, Ringbauer M, Sawicka A, Seiser C. Essential Nonredundant Function of the Catalytic Activity of Histone Deacetylase 2 in Mouse Development. Mol Cell Biol. 2016;36:462–74.
85. Hansen LA, Alexander N, Hogan ME, Sundberg JP, Dlugosz A, Threadgill DW, Magnuson T, Yuspa SH. Genetically null mice reveal a central role for epidermal growth factor receptor in the differentiation of the hair follicle and normal hair development. Am J Pathol. 1997;150:1959–75.
86. Hardcastle Z, Mo R, Hui CC, Sharpe PT. The Shh signalling pathway in tooth development: defects in Gli2 and Gli3 mutants. Development. 1998;125:2803–11.
87. Harris LG, Wang SH, Mani SK, Kasiganesan H, Chou CJ, Menick DR. Evidence for a non-canonical role of HDAC5 in regulation of the cardiac Ncx1 and Bnp genes. Nucleic Acids Res. 2016;44:3610–7.
88. Hayashi M, Nimura K, Kashiwagi K, Harada T, Takaoka K, Kato H, Tamai K, Kaneda Y. Comparative roles of Twist-1 and Id1 in transcriptional regulation by BMP signaling. J Cell Sci. 2007;120:1350–7.
89. Hendrich B, Bird A. Identification and characterization of a family of mammalian methyl-CpG binding proteins. Mol Cell Biol. 1998;18:6538–47.
90. Hendrich B, Guy J, Ramsahoye B, Wilson VA, Bird A. Closely related proteins MBD2 and MBD3 play distinctive but interacting roles in mouse development. Genes Dev. 2001;15:710–23.
91. Hermeking H, Lengauer C, Polyak K, He TC, Zhang L, Thiagalingam S, Kinzler KW, Vogelstein B. 14-3-3sigma is a p53-regulated inhibitor of G2/M progression. Mol Cell. 1997;1:3–11.
92. Herron BJ, Liddell RA, Parker A, Grant S, Kinne J, Fisher JK, Siracusa LD. A mutation in stratifin is responsible for the repeated epilation (Er) phenotype in mice. Nat Genet. 2005;37:1210–2.
93. Higashitsuji H, Higashitsuji H, Masuda T, Liu Y, Itoh K, Fujita J. Enhanced deacetylation of p53 by the anti-apoptotic protein HSCO in association with histone deacetylase 1. J Biol Chem. 2007;282:13716–25.

94. Houtkooper RH, Pirinen E, Auwerx J. Sirtuins as regulators of metabolism and healthspan. Nat Rev Mol Cell Biol. 2012;13:225–38.
95. Hua G, Ganti KP, Chambon P. Glucocorticoid-induced tethered transrepression requires SUMOylation of GR and formation of a SUMO-SMRT/NCoR1-HDAC3 repressing complex. Proc Natl Acad Sci USA. 2016a;113:E635–43.
96. Hua G, Paulen L, Chambon P. GR SUMOylation and formation of an SUMO-SMRT/NCoR1-HDAC3 repressing complex is mandatory for GC-induced IR nGRE-mediated transrepression. Proc Natl Acad Sci USA. 2016b;113:E626–34.
97. Huang PH, Chen CH, Chou CC, Sargeant AM, Kulp SK, Teng CM, Byrd JC, Chen CS. Histone deacetylase inhibitors stimulate histone H3 lysine 4 methylation in part via transcriptional repression of histone H3 lysine 4 demethylases. Mol Pharmacol. 2011;79:197–206.
98. Huang Y, Chen J, Lu C, Han J, Wang G, Song C, Zhu S, Wang C, Li G, Kang J, et al. HDAC1 and Klf4 interplay critically regulates human myeloid leukemia cell proliferation. Cell Death Dis. 2014;5:e1491.
99. Hughes MW, Jiang TX, Lin SJ, Leung Y, Kobielak K, Widelitz RB, Chuong CM. Disrupted ectodermal organ morphogenesis in mice with a conditional histone deacetylase 1, 2 deletion in the epidermis. J Invest Dermatol. 2014;134:24–32.
100. Huh SH, Narhi K, Lindfors PH, Haara O, Yang L, Ornitz DM, Mikkola ML. Fgf20 governs formation of primary and secondary dermal condensations in developing hair follicles. Genes Dev. 2013;27:450–8.
101. Ingraham CR, Kinoshita A, Kondo S, Yang B, Sajan S, Trout KJ, Malik MI, Dunnwald M, Goudy SL, Lovett M, et al. Abnormal skin, limb and craniofacial morphogenesis in mice deficient for interferon regulatory factor 6 (Irf6). Nat Genet. 2006;38:1335–40.
102. Ito A, Kawaguchi Y, Lai CH, Kovacs JJ, Higashimoto Y, Appella E, Yao TP. MDM2-HDAC1-mediated deacetylation of p53 is required for its degradation. EMBO J. 2002;21:6236–45.
103. Ito K, Yamamura S, Essilfie-Quaye S, Cosio B, Ito M, Barnes PJ, Adcock IM. Histone deacetylase 2-mediated deacetylation of the glucocorticoid receptor enables NF-kappaB suppression. J Exp Med. 2006;203:7–13.
104. Ivanova IA, D'Souza SJ, Dagnino L. Signalling in the epidermis: the E2F cell cycle regulatory pathway in epidermal morphogenesis, regeneration and transformation. Int J Biol Sci. 2005;1:87–95.
105. Jacob C, Christen CN, Pereira JA, Somandin C, Baggiolini A, Lotscher P, Ozcelik M, Tricaud N, Meijer D, Yamaguchi T, et al. HDAC1 and HDAC2 control the transcriptional program of myelination and the survival of Schwann cells. Nat Neurosci. 2011;14:429–36.
106. Jaubert J, Cheng J, Segre JA. Ectopic expression of kruppel like factor 4 (Klf4) accelerates formation of the epidermal permeability barrier. Development. 2003;130:2767–77.
107. Jensen ED, Schroeder TM, Bailey J, Gopalakrishnan R, Westendorf JJ. Histone deacetylase 7 associates with Runx2 and represses its activity during osteoblast maturation in a deacetylation-independent manner. J Bone Miner Res. 2008;23:361–72.
108. Jepsen K, Solum D, Zhou T, McEvilly RJ, Kim HJ, Glass CK, Hermanson O, Rosenfeld MG. SMRT-mediated repression of an H3K27 demethylase in progression from neural stem cell to neuron. Nature. 2007;450:415–9.
109. Jessen BA, Phillips MA, Hovnanian A, Rice RH. Role of Sp1 response element in transcription of the human transglutaminase 1 gene. J Invest Dermatol. 2000;115:113–7.
110. Ji M, Lee EJ, Kim KB, Kim Y, Sung R, Lee SJ, Kim DS, Park SM. HDAC inhibitors induce epithelial-mesenchymal transition in colon carcinoma cells. Oncol Rep. 2015;33:2299–308.
111. Jiang X, Ye X, Guo W, Lu H, Gao Z. Inhibition of HDAC3 promotes ligand-independent PPARgamma activation by protein acetylation. J Mol Endocrinol. 2014;53:191–200.
112. Juan LJ, Shia WJ, Chen MH, Yang WM, Seto E, Lin YS, Wu CW. Histone deacetylases specifically down-regulate p53-dependent gene activation. J Biol Chem. 2000;275:20436–43.
113. Kadamb R, Mittal S, Bansal N, Batra H, Saluja D. Sin3: insight into its transcription regulatory functions. Eur J Cell Biol. 2013;92:237–46.

114. Kamijo T, Weber JD, Zambetti G, Zindy F, Roussel MF, Sherr CJ. Functional and physical interactions of the ARF tumor suppressor with p53 and Mdm2. Proc Natl Acad Sci USA. 1998;95:8292–7.
115. Kao HY, Ordentlich P, Koyano-Nakagawa N, Tang Z, Downes M, Kintner CR, Evans RM, Kadesch T. A histone deacetylase corepressor complex regulates the Notch signal transduction pathway. Genes Dev. 1998;12:2269–77.
116. Kashiwagi M, Morgan BA, Georgopoulos K. The chromatin remodeler Mi-2beta is required for establishment of the basal epidermis and normal differentiation of its progeny. Development. 2007;134:1571–82.
117. Kelly RD, Cowley SM. The physiological roles of histone deacetylase (HDAC) 1 and 2: complex co-stars with multiple leading parts. Biochem Soc Trans. 2013;41:741–9.
118. Khan N, Jeffers M, Kumar S, Hackett C, Boldog F, Khramtsov N, Qian X, Mills E, Berghs SC, Carey N, et al. Determination of the class and isoform selectivity of small-molecule histone deacetylase inhibitors. Biochem J. 2008;409:581–9.
119. Kleff S, Andrulis ED, Anderson CW, Sternglanz R. Identification of a gene encoding a yeast histone H4 acetyltransferase. J Biol Chem. 1995;270:24674–7.
120. Knutson SK, Chyla BJ, Amann JM, Bhaskara S, Huppert SS, Hiebert SW. Liver-specific deletion of histone deacetylase 3 disrupts metabolic transcriptional networks. EMBO J. 2008;27:1017–28.
121. Kobielak K, Pasolli HA, Alonso L, Polak L, Fuchs E. Defining BMP functions in the hair follicle by conditional ablation of BMP receptor IA. J Cell Biol. 2003;163:609–23.
122. Koster MI, Dai D, Marinari B, Sano Y, Costanzo A, Karin M, Roop DR. p63 induces key target genes required for epidermal morphogenesis. Proc Natl Acad Sci USA. 2007;104:3255–60.
123. Koster MI, Roop DR. p63 and epithelial appendage development. Differentiation. 2004;72:364–70.
124. Koster MI, Roop DR. Mechanisms regulating epithelial stratification. Annu Rev Cell Dev Biol. 2007;23:93–113.
125. Lagger G, O'Carroll D, Rembold M, Khier H, Tischler J, Weitzer G, Schuettengruber B, Hauser C, Brunmeir R, Jenuwein T, et al. Essential function of histone deacetylase 1 in proliferation control and CDK inhibitor repression. EMBO J. 2002;21:2672–81.
126. Lahm A, Paolini C, Pallaoro M, Nardi MC, Jones P, Neddermann P, Sambucini S, Bottomley MJ, Lo Surdo P, Carfi A, et al. Unraveling the hidden catalytic activity of vertebrate class IIa histone deacetylases. Proc Natl Acad Sci USA. 2007;104:17335–40.
127. Lai AY, Wade PA. Cancer biology and NuRD: a multifaceted chromatin remodelling complex. Nat Rev Cancer. 2011;11:588–96.
128. Lai PH, Wang WL, Ko CY, Lee YC, Yang WM, Shen TW, Chang WC, Wang JM. HDAC1/HDAC3 modulates PPARG2 transcription through the sumoylated CEBPD in hepatic lipogenesis. Biochim Biophys Acta. 2008;1783:1803–14.
129. Laurikkala J, Mikkola ML, James M, Tummers M, Mills AA, Thesleff I. p63 regulates multiple signalling pathways required for ectodermal organogenesis and differentiation. Development. 2006;133:1553–63.
130. Lawrence M, Daujat S, Schneider R. Lateral thinking: how histone modifications regulate gene expression. Trends Genet. 2016;32:42–56.
131. LeBoeuf M, Terrell A, Trivedi S, Sinha S, Epstein JA, Olson EN, Morrisey EE, Millar SE. Hdac1 and Hdac2 act redundantly to control p63 and p53 functions in epidermal progenitor cells. Dev Cell. 2010;19:807–18.
132. Lechler T, Fuchs E. Asymmetric cell divisions promote stratification and differentiation of mammalian skin. Nature. 2005;437:275–80.
133. Lee CS, Bhaduri A, Mah A, Johnson WL, Ungewickell A, Aros CJ, Nguyen CB, Rios EJ, Siprashvili Z, Straight A, et al. Recurrent point mutations in the kinetochore gene KNSTRN in cutaneous squamous cell carcinoma. Nat Genet. 2014;46:1060–2.
134. Lee J, Tumbar T. Hairy tale of signaling in hair follicle development and cycling. Semin Cell Dev Biol. 2012;23:906–16.

135. Lee MG, Wynder C, Bochar DA, Hakimi MA, Cooch N, Shiekhattar R. Functional interplay between histone demethylase and deacetylase enzymes. Mol Cell Biol. 2006;26:6395–402.
136. Lee SJ, Lindsey S, Graves B, Yoo S, Olson JM, Langhans SA. Sonic hedgehog-induced histone deacetylase activation is required for cerebellar granule precursor hyperplasia in medulloblastoma. PLoS One. 2013;8:e71455.
137. Lei W, Zhang K, Pan X, Hu Y, Wang D, Yuan X, Shu G, Song J. Histone deacetylase 1 is required for transforming growth factor-beta1-induced epithelial-mesenchymal transition. Int J Biochem Cell Biol. 2010;42:1489–97.
138. Li D, Wang X, Ren W, Ren J, Lan X, Wang F, Li H, Zhang F, Han Y, Song T, et al. High expression of liver histone deacetylase 3 contributes to high-fat-diet-induced metabolic syndrome by suppressing the PPAR-gamma and LXR-alpha-pathways in E3 rats. Mol Cell Endocrinol. 2011;344:69–80.
139. Li Q, Lu Q, Estepa G, Verma IM. Identification of 14-3-3sigma mutation causing cutaneous abnormality in repeated-epilation mutant mouse. Proc Natl Acad Sci USA. 2005;102:15977–82.
140. Liao M, Zhang Y, Dufau ML. Protein kinase Calpha-induced derepression of the human luteinizing hormone receptor gene transcription through ERK-mediated release of HDAC1/Sin3A repressor complex from Sp1 sites. Mol Endocrinol (Baltimore). 2008;22:1449–63.
141. Lien WH, Fuchs E. Wnt some lose some: transcriptional governance of stem cells by Wnt/beta-catenin signaling. Genes Dev. 2014;28:1517–32.
142. Lin KT, Wang YW, Chen CT, Ho CM, Su WH, Jou YS. HDAC inhibitors augmented cell migration and metastasis through induction of PKCs leading to identification of low toxicity modalities for combination cancer therapy. Clin Cancer Res. 2012;18:4691–701.
143. Liu H, Hu Q, Kaufman A, D'Ercole AJ, Ye P. Developmental expression of histone deacetylase 11 in the murine brain. J Neurosci Res. 2008;86:537–43.
144. Loyola A, Almouzni G. Histone chaperones, a supporting role in the limelight. Biochim Biophys Acta. 2004;1677:3–11.
145. Lu J, McKinsey TA, Zhang CL, Olson EN. Regulation of skeletal myogenesis by association of the MEF2 transcription factor with class II histone deacetylases. Mol Cell. 2000;6:233–44.
146. Luetteke NC, Phillips HK, Qiu TH, Copeland NG, Earp HS, Jenkins NA, Lee DC. The mouse waved-2 phenotype results from a point mutation in the EGF receptor tyrosine kinase. Genes Dev. 1994;8:399–413.
147. Ma C, D'Mello SR. Neuroprotection by histone deacetylase-7 (HDAC7) occurs by inhibition of c-jun expression through a deacetylase-independent mechanism. J Biol Chem. 2011;286:4819–28.
148. Ma L, Liu J, Wu T, Plikus M, Jiang TX, Bi Q, Liu YH, Muller-Rover S, Peters H, Sundberg JP, et al. 'Cyclic alopecia' in Msx2 mutants: defects in hair cycling and hair shaft differentiation. Development (Cambridge, UK). 2003;130:379–89.
149. Macdonald JI, Dick FA. Posttranslational modifications of the retinoblastoma tumor suppressor protein as determinants of function. Genes Cancer. 2012;3:619–33.
150. Magnaghi-Jaulin L, Groisman R, Naguibneva I, Robin P, Lorain S, Le Villain JP, Troalen F, Trouche D, Harel-Bellan A. Retinoblastoma protein represses transcription by recruiting a histone deacetylase. Nature. 1998;391:601–5.
151. Man MQ, Barish GD, Schmuth M, Crumrine D, Barak Y, Chang S, Jiang Y, Evans RM, Elias PM, Feingold KR. Deficiency of PPARbeta/delta in the epidermis results in defective cutaneous permeability barrier homeostasis and increased inflammation. J Invest Dermatol. 2008;128:370–7.
152. Markova NG, Karaman-Jurukovska N, Pinkas-Sarafova A, Marekov LN, Simon M. Inhibition of histone deacetylation promotes abnormal epidermal differentiation and specifically suppresses the expression of the late differentiation marker profilaggrin. J Invest Dermatol. 2007;127:1126–39.
153. Martin M, Kettmann R, Dequiedt F. Class IIa histone deacetylases: conducting development and differentiation. Int J Dev Biol. 2009;53:291–301.

154. McKinsey TA, Zhang CL, Lu J, Olson EN. Signal-dependent nuclear export of a histone deacetylase regulates muscle differentiation. Nature. 2000;408:106–11.
155. McQuown SC, Wood MA. HDAC3 and the molecular brake pad hypothesis. Neurobiol Learn Mem. 2011;96:27–34.
156. Meidhof S, Brabletz S, Lehmann W, Preca BT, Mock K, Ruh M, Schuler J, Berthold M, Weber A, Burk U, et al. ZEB1-associated drug resistance in cancer cells is reversed by the class I HDAC inhibitor mocetinostat. EMBO Mol Med. 2015;7:831–47.
157. Menon GK, Elias PM, Lee SH, Feingold KR. Localization of calcium in murine epidermis following disruption and repair of the permeability barrier. Cell Tissue Res. 1992;270:503–12.
158. Menon GK, Grayson S, Elias PM. Ionic calcium reservoirs in mammalian epidermis: ultrastructural localization by ion-capture cytochemistry. J Invest Dermatol. 1985;84:508–12.
159. Miettinen PJ, Berger JE, Meneses J, Phung Y, Pedersen RA, Werb Z, Derynck R. Epithelial immaturity and multiorgan failure in mice lacking epidermal growth factor receptor. Nature. 1995;376:337–41.
160. Mihaylova MM, Vasquez DS, Ravnskjaer K, Denechaud PD, Yu RT, Alvarez JG, Downes M, Evans RM, Montminy M, Shaw RJ. Class IIa histone deacetylases are hormone-activated regulators of FOXO and mammalian glucose homeostasis. Cell. 2011;145:607–21.
161. Mikkola ML. p63 in skin appendage development. Cell Cycle (Georgetown). 2007;6:285–90.
162. Mikkola ML, Millar SE. The mammary bud as a skin appendage: unique and shared aspects of development. J Mammary Gland Biol Neoplasia. 2006;11:187–203.
163. Mill P, Mo R, Fu H, Grachtchouk M, Kim PC, Dlugosz AA, Hui CC. Sonic hedgehog-dependent activation of Gli2 is essential for embryonic hair follicle development. Genes Dev. 2003;17:282–94.
164. Millar SE. Molecular mechanisms regulating hair follicle development. J Invest Dermatol. 2002;118:216–25.
165. Mills AA, Zheng B, Wang XJ, Vogel H, Roop DR, Bradley A. p63 is a p53 homologue required for limb and epidermal morphogenesis. Nature. 1999;398:708–13.
166. Mlacki M, Darido C, Jane SM, Wilanowski T. Loss of Grainy head-like 1 is associated with disruption of the epidermal barrier and squamous cell carcinoma of the skin. PLoS One. 2014;9:e89247.
167. Moll R, Franke WW, Schiller DL, Geiger B, Krepler R. The catalog of human cytokeratins: patterns of expression in normal epithelia, tumors and cultured cells. Cell. 1982;31:11–24.
168. Moll UM, Slade N. p63 and p73: roles in development and tumor formation. Mol Cancer Res. 2004;2:371–86.
169. Montgomery RL, Potthoff MJ, Haberland M, Qi X, Matsuzaki S, Humphries KM, Richardson JA, Bassel-Duby R, Olson EN. Maintenance of cardiac energy metabolism by histone deacetylase 3 in mice. J Clin Invest. 2008;118:3588–97.
170. Moriyama M, Durham AD, Moriyama H, Hasegawa K, Nishikawa S, Radtke F, Osawa M. Multiple roles of Notch signaling in the regulation of epidermal development. Dev Cell. 2008;14:594–604.
171. Nair M, Bilanchone V, Ortt K, Sinha S, Dai X. Ovol1 represses its own transcription by competing with transcription activator c-Myb and by recruiting histone deacetylase activity. Nucleic Acids Res. 2007;35:1687–97.
172. Nair M, Teng A, Bilanchone V, Agrawal A, Li B, Dai X. Ovol1 regulates the growth arrest of embryonic epidermal progenitor cells and represses c-myc transcription. J Cell Biol. 2006;173:253–64.
173. Nascimento EM, Cox CL, MacArthur S, Hussain S, Trotter M, Blanco S, Suraj M, Nichols J, Kubler B, Benitah SA, et al. The opposing transcriptional functions of Sin3a and c-Myc are required to maintain tissue homeostasis. Nat Cell Biol. 2011;13:1395–405.
174. Niemann C, Horsley V. Development and homeostasis of the sebaceous gland. Semin Cell Dev Biol. 2012;23:928–36.
175. Ning MS, Andl T. Concise review: custodians of the transcriptome: how microRNAs guard stemness in squamous epithelia. Stem Cells (Dayton). 2015;33:1047–54.

176. Nishino TG, Miyazaki M, Hoshino H, Miwa Y, Horinouchi S, Yoshida M. 14-3-3 regulates the nuclear import of class IIa histone deacetylases. Biochem Biophys Res Commun. 2008;377:852–6.
177. Nowak JA, Polak L, Pasolli HA, Fuchs E. Hair follicle stem cells are specified and function in early skin morphogenesis. Cell Stem Cell. 2008;3:33–43.
178. Oh SH, Whang YM, Min HY, Han SH, Kang JH, Song KH, Glisson BS, Kim YH, Lee HY. Histone deacetylase inhibitors enhance the apoptotic activity of insulin-like growth factor binding protein-3 by blocking PKC-induced IGFBP-3 degradation. Int J Cancer. 2012;131:2253–63.
179. Okuyama R, Tagami H, Aiba S. Notch signaling: its role in epidermal homeostasis and in the pathogenesis of skin diseases. J Dermatol Sci. 2008;49:187–94.
180. Ouspenskaia T, Matos I, Mertz AF, Fiore VF, Fuchs E. WNT-SHH antagonism specifies and expands stem cells prior to niche formation. Cell. 2016;164:156–69.
181. Palermo R, Checquolo S, Giovenco A, Grazioli P, Kumar V, Campese AF, Giorgi A, Napolitano M, Canettieri G, Ferrara G, et al. Acetylation controls Notch3 stability and function in T-cell leukemia. Oncogene. 2012;31:3807–17.
182. Panelos J, Massi D. Emerging role of Notch signaling in epidermal differentiation and skin cancer. Cancer Biol Ther. 2009;8:1986–93.
183. Pang M, Ma L, Liu N, Ponnusamy M, Zhao TC, Yan H, Zhuang S. Histone deacetylase 1/2 mediates proliferation of renal interstitial fibroblasts and expression of cell cycle proteins. J Cell Biochem. 2011;112:2138–48.
184. Panteleyev AA, Paus R, Ahmad W, Sundberg JP, Christiano AM. Molecular and functional aspects of the hairless (hr) gene in laboratory rodents and humans. Exp Dermatol. 1998;7:249–67.
185. Peterson SC, Eberl M, Vagnozzi AN, Belkadi A, Veniaminova NA, Verhaegen ME, Bichakjian CK, Ward NL, Dlugosz AA, Wong SY. Basal cell carcinoma preferentially arises from stem cells within hair follicle and mechanosensory niches. Cell Stem Cell. 2015;16:400–12.
186. Pickering CR, Zhou JH, Lee JJ, Drummond JA, Peng SA, Saade RE, Tsai KY, Curry JL, Tetzlaff MT, Lai SY, et al. Mutational landscape of aggressive cutaneous squamous cell carcinoma. Clin Cancer Res. 2014;20:6582–92.
187. Plikus MV, Mayer JA, de la Cruz D, Baker RE, Maini PK, Maxson R, Chuong CM. Cyclic dermal BMP signalling regulates stem cell activation during hair regeneration. Nature. 2008;451:340–4.
188. Potter GB, Beaudoin GM 3rd, DeRenzo CL, Zarach JM, Chen SH, Thompson CC. The hairless gene mutated in congenital hair loss disorders encodes a novel nuclear receptor corepressor. Genes Dev. 2001;15:2687–701.
189. Qu X, Proll M, Neuhoff C, Zhang R, Cinar MU, Hossain MM, Tesfaye D, Grosse-Brinkhaus C, Salilew-Wondim D, Tholen E, et al. Sulforaphane epigenetically regulates innate immune responses of porcine monocyte-derived dendritic cells induced with lipopolysaccharide. PLoS One. 2015;10:e0121574.
190. Ramsey MR, He L, Forster N, Ory B, Ellisen LW. Physical association of HDAC1 and HDAC2 with p63 mediates transcriptional repression and tumor maintenance in squamous cell carcinoma. Cancer Res. 2011;71:4373–9.
191. Rangarajan A, Talora C, Okuyama R, Nicolas M, Mammucari C, Oh H, Aster JC, Krishna S, Metzger D, Chambon P, et al. Notch signaling is a direct determinant of keratinocyte growth arrest and entry into differentiation. EMBO J. 2001;20:3427–36.
192. Ratman D, Vanden Berghe W, Dejager L, Libert C, Tavernier J, Beck IM, De Bosscher K. How glucocorticoid receptors modulate the activity of other transcription factors: a scope beyond tethering. Mol Cell Endocrinol. 2013;380:41–54.
193. Reynolds N, O'Shaughnessy A, Hendrich B. Transcriptional repressors: multifaceted regulators of gene expression. Development. 2013;140:505–12.

194. Richardson RJ, Dixon J, Malhotra S, Hardman MJ, Knowles L, Boot-Handford RP, Shore P, Whitmarsh A, Dixon MJ. Irf6 is a key determinant of the keratinocyte proliferation-differentiation switch. Nat Genet. 2006;38:1329–34.
195. Robbins DJ, Fei DL, Riobo NA. The Hedgehog signal transduction network. Sci Signal. 2012;5:re6.
196. Robertson ED, Weir L, Romanowska M, Leigh IM, Panteleyev AA. ARNT controls the expression of epidermal differentiation genes through HDAC- and EGFR-dependent pathways. J Cell Sci. 2012;125:3320–32.
197. Romano RA, Birkaya B, Sinha S. Defining the regulatory elements in the proximal promoter of DeltaNp63 in keratinocytes: potential roles for Sp1/Sp3, NF-Y, and p63. J Invest Dermatol. 2006;126:1469–79.
198. Romano RA, Ortt K, Birkaya B, Smalley K, Sinha S. An active role of the DeltaN isoform of p63 in regulating basal keratin genes K5 and K14 and directing epidermal cell fate. PLoS One. 2009;4:e5623.
199. Romano RA, Solomon LW, Sinha S. Tp63 in oral development, neoplasia, and autoimmunity. J Dent Res. 2012;91:125–32.
200. Rompolas P, Greco V. Stem cell dynamics in the hair follicle niche. Semin Cell Dev Biol. 2014;25-26:34–42.
201. Roop D. Defects in the barrier. Science (New York). 1995;267:474–5.
202. Ruscetti M, Dadashian EL, Guo W, Quach B, Mulholland DJ, Park JW, Tran LM, Kobayashi N, Bianchi-Frias D, Xing Y, Nelson PS, Wu H. HDAC inhibition impedes epithelial-mesenchymal plasticity and suppresses metastatic, castration-resistant prostate cancer. Oncogene. 2016;35:3781–95.
203. Sahu RP, DaSilva SC, Rashid B, Martel KC, Jernigan D, Mehta SR, Mohamed DR, Rezania S, Bradish JR, Armstrong AB, et al. Mice lacking epidermal PPARgamma exhibit a marked augmentation in photocarcinogenesis associated with increased UVB-induced apoptosis, inflammation and barrier dysfunction. Int J Cancer. 2012;131:E1055–66.
204. Sakamoto T, Kobayashi S, Yamada D, Nagano H, Tomokuni A, Tomimaru Y, Noda T, Gotoh K, Asaoka T, Wada H, et al. A histone deacetylase inhibitor suppresses epithelial-mesenchymal transition and attenuates chemoresistance in biliary tract cancer. PLoS One. 2016;11:e0145985.
205. Sanford JA, Zhang L-J, Williams MR, Gangoiti JA, Huang C-M, Gallo RL. Inhibition of HDAC8 and HDAC9 by microbial short-chain fatty acids breaks immune tolerance of the epidermis to TLR ligands. Sci Immunol. 2016;1:1–10.
206. Sano S, Chan KS, DiGiovanni J. Impact of Stat3 activation upon skin biology: a dichotomy of its role between homeostasis and diseases. J Dermatol Sci. 2008;50:1–14.
207. Satokata I, Ma L, Ohshima H, Bei M, Woo I, Nishizawa K, Maeda T, Takano Y, Uchiyama M, Heaney S, et al. Msx2 deficiency in mice causes pleiotropic defects in bone growth and ectodermal organ formation. Nat Genet. 2000;24:391–5.
208. Scholl C, Weibetamuller K, Holenya P, Shaked-Rabi M, Tucker KL, Wolfl S. Distinct and overlapping gene regulatory networks in BMP- and HDAC-controlled cell fate determination in the embryonic forebrain. BMC Genomics. 2012;13:298.
209. Schuijers J, Clevers H. Adult mammalian stem cells: the role of Wnt, Lgr5 and R-spondins. EMBO J. 2012;31:2685–96.
210. Segre CV, Chiocca S. Regulating the regulators: the post-translational code of class I HDAC1 and HDAC2. J Biomed Biotechnol. 2011;2011:690848.
211. Segre J. Complex redundancy to build a simple epidermal permeability barrier. Curr Opin Cell Biol. 2003;15:776–82.
212. Segre JA, Bauer C, Fuchs E. Klf4 is a transcription factor required for establishing the barrier function of the skin. Nat Genet. 1999;22:356–60.
213. Sennett R, Rendl M. Mesenchymal-epithelial interactions during hair follicle morphogenesis and cycling. Semin Cell Dev Biol. 2012;23:917–27.

214. Shahbazian MD, Grunstein M. Functions of site-specific histone acetylation and deacetylation. Annu Rev Biochem. 2007;76:75–100.
215. Shimomura Y, Christiano AM. Biology and genetics of hair. Annu Rev Genomics Hum Genet. 2010;11:109–32.
216. Sibilia M, Fleischmann A, Behrens A, Stingl L, Carroll J, Watt FM, Schlessinger J, Wagner EF. The EGF receptor provides an essential survival signal for SOS-dependent skin tumor development. Cell. 2000;102:211–20.
217. Sibilia M, Wagner EF. Strain-dependent epithelial defects in mice lacking the EGF receptor. Science (New York). 1995;269:234–8.
218. Sil AK, Maeda S, Sano Y, Roop DR, Karin M. IkappaB kinase-alpha acts in the epidermis to control skeletal and craniofacial morphogenesis. Nature. 2004;428:660–4.
219. Snippert HJ, Haegebarth A, Kasper M, Jaks V, van Es JH, Barker N, van de Wetering M, van den Born M, Begthel H, Vries RG, et al. Lgr6 marks stem cells in the hair follicle that generate all cell lineages of the skin. Science (New York). 2010;327:1385–9.
220. Su X, Cho MS, Gi YJ, Ayanga BA, Sherr CJ, Flores ER. Rescue of key features of the p63-null epithelial phenotype by inactivation of Ink4a and Arf. EMBO J. 2009;28:1904–15.
221. Sun Z, Feng D, Fang B, Mullican SE, You SH, Lim HW, Everett LJ, Nabel CS, Li Y, Selvakumaran V, et al. Deacetylase-independent function of HDAC3 in transcription and metabolism requires nuclear receptor corepressor. Mol Cell. 2013;52:769–82.
222. Takeda H, Lyle S, Lazar AJ, Zouboulis CC, Smyth I, Watt FM. Human sebaceous tumors harbor inactivating mutations in LEF1. Nat Med. 2006;12:395–7.
223. Tang JY, Mackay-Wiggan JM, Aszterbaum M, Yauch RL, Lindgren J, Chang K, Coppola C, Chanana AM, Marji J, Bickers DR, et al. Inhibiting the hedgehog pathway in patients with the basal-cell nevus syndrome. N Engl J Med. 2012;366:2180–8.
224. Tang Y, Zhao W, Chen Y, Zhao Y, Gu W. Acetylation is indispensable for p53 activation. Cell. 2008;133:612–26.
225. Teng A, Nair M, Wells J, Segre JA, Dai X. Strain-dependent perinatal lethality of Ovol1-deficient mice and identification of Ovol2 as a downstream target of Ovol1 in skin epidermis. Biochim Biophys Acta. 2007;1772:89–95.
226. Thaler F, Mercurio C. Towards selective inhibition of histone deacetylase isoforms: what has been achieved, where we are and what will be next. ChemMedChem. 2014;9:523–6.
227. Theocharis S, Klijanienko J, Giaginis C, Rodriguez J, Jouffroy T, Girod A, Alexandrou P, Sastre-Garau X. Histone deacetylase-1 and -2 expression in mobile tongue squamous cell carcinoma: associations with clinicopathological parameters and patients survival. J Oral Pathol Med. 2011;40:706–14.
228. Thompson CC, Beaudoin GMJ III. Hairless: a nuclear receptor corepressor essential for skin function. In: Advances in developmental biology. Amsterdam: Elsevier; 2006. p. 357–87.
229. Ting SB, Caddy J, Wilanowski T, Auden A, Cunningham JM, Elias PM, Holleran WM, Jane SM. The epidermis of grhl3-null mice displays altered lipid processing and cellular hyperproliferation. Organogenesis. 2005;2:33–5.
230. Trerotola M, Relli V, Simeone P, Alberti S. Epigenetic inheritance and the missing heritability. Hum Genomics. 2015;9:17.
231. Trivedi CM, Luo Y, Yin Z, Zhang M, Zhu W, Wang T, Floss T, Goettlicher M, Noppinger PR, Wurst W, et al. Hdac2 regulates the cardiac hypertrophic response by modulating Gsk3 beta activity. Nat Med. 2007;13:324–31.
232. Truong AB, Khavari PA. Control of keratinocyte proliferation and differentiation by p63. Cell Cycle (Georgetown). 2007;6:295–9.
233. van der Vlag J, Otte AP. Transcriptional repression mediated by the human polycomb-group protein EED involves histone deacetylation. Nat Genet. 1999;23:474–8.
234. Vega RB, Harrison BC, Meadows E, Roberts CR, Papst PJ, Olson EN, McKinsey TA. Protein kinases C and D mediate agonist-dependent cardiac hypertrophy through nuclear export of histone deacetylase 5. Mol Cell Biol. 2004;24:8374–85.
235. Waddington CH. The Epigenotpye. Endeavour. 1942;1:18–20.

236. Wallmeyer L, Lehnen D, Eger N, Sochorova M, Opalka L, Kovacik A, Vavrova K, Hedtrich S. Stimulation of PPARalpha normalizes the skin lipid ratio and improves the skin barrier of normal and filaggrin deficient reconstructed skin. J Dermatol Sci. 2015;80:102–10.
237. Wang B, Moya N, Niessen S, Hoover H, Mihaylova MM, Shaw RJ, Yates JR 3rd, Fischer WH, Thomas JB, Montminy M. A hormone-dependent module regulating energy balance. Cell. 2011a;145:596–606.
238. Wang D, Su L, Huang D, Zhang H, Shin DM, Chen ZG. Downregulation of E-Cadherin enhances proliferation of head and neck cancer through transcriptional regulation of EGFR. Mol Cancer. 2011b;10:116.
239. Wang J, Scully K, Zhu X, Cai L, Zhang J, Prefontaine GG, Krones A, Ohgi KA, Zhu P, Garcia-Bassets I, et al. Opposing LSD1 complexes function in developmental gene activation and repression programmes. Nature. 2007;446:882–7.
240. Wang X, Xu J, Wang H, Wu L, Yuan W, Du J, Cai S. Trichostatin A, a histone deacetylase inhibitor, reverses epithelial-mesenchymal transition in colorectal cancer SW480 and prostate cancer PC3 cells. Biochem Biophys Res Commun. 2015;456:320–6.
241. Wang Z, Zang C, Cui K, Schones DE, Barski A, Peng W, Zhao K. Genome-wide mapping of HATs and HDACs reveals distinct functions in active and inactive genes. Cell. 2009;138:1019–31.
242. Watt FM, Collins CA. Role of beta-catenin in epidermal stem cell expansion, lineage selection, and cancer. Cold Spring Harb Symp Quant Biol. 2008;73:503–12.
243. Wellner U, Schubert J, Burk UC, Schmalhofer O, Zhu F, Sonntag A, Waldvogel B, Vannier C, Darling D, zur Hausen A, et al. The EMT-activator ZEB1 promotes tumorigenicity by repressing stemness-inhibiting microRNAs. Nat Cell Biol. 2009;11:1487–95.
244. Wells J, Lee B, Cai AQ, Karapetyan A, Lee WJ, Rugg E, Sinha S, Nie Q, Dai X. Ovol2 suppresses cell cycling and terminal differentiation of keratinocytes by directly repressing c-Myc and Notch1. J Biol Chem. 2009;284:29125–35.
245. Wels C, Joshi S, Koefinger P, Bergler H, Schaider H. Transcriptional activation of ZEB1 by Slug leads to cooperative regulation of the epithelial-mesenchymal transition-like phenotype in melanoma. J Invest Dermatol. 2011;131:1877–85.
246. Winter M, Moser MA, Meunier D, Fischer C, Machat G, Mattes K, Lichtenberger BM, Brunmeir R, Weissmann S, Murko C, et al. Divergent roles of HDAC1 and HDAC2 in the regulation of epidermal development and tumorigenesis. EMBO J. 2013;32:3176–91.
247. Wu S, Luo Z, Yu PJ, Xie H, He YW. Suberoylanilide hydroxamic acid (SAHA) promotes the epithelial mesenchymal transition of triple negative breast cancer cells via HDAC8/FOXA1 signals. Biol Chem. 2016;397:75–83.
248. Xin HW, Yang JH, Nguyen DM. Sensitivity to epidermal growth factor receptor tyrosine kinase inhibitor requires E-cadherin in esophageal cancer and malignant pleural mesothelioma. Anticancer Res. 2013;33:2401–8.
249. Xiong H, Du W, Zhang YJ, Hong J, Su WY, Tang JT, Wang YC, Lu R, Fang JY. Trichostatin A, a histone deacetylase inhibitor, suppresses JAK2/STAT3 signaling via inducing the promoter-associated histone acetylation of SOCS1 and SOCS3 in human colorectal cancer cells. Mol Carcinog. 2012;51:174–84.
250. Xue Y, Wong J, Moreno GT, Young MK, Cote J, Wang W. NURD, a novel complex with both ATP-dependent chromatin-remodeling and histone deacetylase activities. Mol Cell. 1998;2:851–61.
251. Yamaguchi M, Tonou-Fujimori N, Komori A, Maeda R, Nojima Y, Li H, Okamoto H, Masai I. Histone deacetylase 1 regulates retinal neurogenesis in zebrafish by suppressing Wnt and Notch signaling pathways. Development. 2005;132:3027–43.
252. Yamaguchi T, Cubizolles F, Zhang Y, Reichert N, Kohler H, Seiser C, Matthias P. Histone deacetylases 1 and 2 act in concert to promote the G1-to-S progression. Genes Dev. 2010;24:455–69.

253. Yang A, Kaghad M, Wang Y, Gillett E, Fleming MD, Dotsch V, Andrews NC, Caput D, McKeon F. p63, a p53 homolog at 3q27-29, encodes multiple products with transactivating, death-inducing, and dominant-negative activities. Mol Cell. 1998;2:305–16.
254. Yang A, Schweitzer R, Sun D, Kaghad M, Walker N, Bronson RT, Tabin C, Sharpe A, Caput D, Crum C, et al. p63 is essential for regenerative proliferation in limb, craniofacial and epithelial development. Nature. 1999;398:714–8.
255. Yang LC, Ng DC, Bikle DD. Role of protein kinase C alpha in calcium induced keratinocyte differentiation: defective regulation in squamous cell carcinoma. J Cell Physiol. 2003;195:249–59.
256. Yao YL, Yang WM, Seto E. Regulation of transcription factor YY1 by acetylation and deacetylation. Mol Cell Biol. 2001;21:5979–91.
257. Ye GD, Sun GB, Jiao P, Chen C, Liu QF, Huang XL, Zhang R, Cai WY, Li SN, Wu JF, et al. OVOL2, an inhibitor of WNT signaling, reduces invasive activities of human and mouse cancer cells and is down-regulated in human colorectal tumors. Gastroenterology. 2016;150:659–671.e616.
258. Yi R, Poy MN, Stoffel M, Fuchs E. A skin microRNA promotes differentiation by repressing 'stemness. Nature. 2008;452:225–9.
259. Yu T, Chen X, Zhang W, Liu J, Avdiushko R, Napier DL, Liu AX, Neltner JM, Wang C, Cohen D, et al. KLF4 regulates adult lung tumor-initiating cells and represses K-Ras-mediated lung cancer. Cell Death Differ. 2016;23:207–15.
260. Yu Z, Lin KK, Bhandari A, Spencer JA, Xu X, Wang N, Lu Z, Gill GN, Roop DR, Wertz P, et al. The Grainyhead-like epithelial transactivator Get-1/Grhl3 regulates epidermal terminal differentiation and interacts functionally with LMO4. Dev Biol. 2006;299:122–36.
261. Yuan ZL, Guan YJ, Chatterjee D, Chin YE. Stat3 dimerization regulated by reversible acetylation of a single lysine residue. Science (New York). 2005;307:269–73.
262. Zeng L, Zhou MM. Bromodomain: an acetyl-lysine binding domain. FEBS Lett. 2002;513:124–8.
263. Zhang W, Ji W, Liu X, Ouyang G, Xiao W. ELL inhibits E2F1 transcriptional activity by enhancing E2F1 deacetylation via recruitment of histone deacetylase 1. Mol Cell Biol. 2014;34:765–75.
264. Zhang Y, Matkovich SJ, Duan X, Diwan A, Kang MY, Dorn GW 2nd. Receptor-independent protein kinase C alpha (PKCalpha) signaling by calpain-generated free catalytic domains induces HDAC5 nuclear export and regulates cardiac transcription. J Biol Chem. 2011;286:26943–51.
265. Zhao J, Quan H, Xie C, Lou L. NL-103, a novel dual-targeted inhibitor of histone deacetylases and hedgehog pathway, effectively overcomes vismodegib resistance conferred by Smo mutations. Pharmacol Res Perspect. 2014;2:e00043.
266. Zheng Y, Eilertsen KJ, Ge L, Zhang L, Sundberg JP, Prouty SM, Stenn KS, Parimoo S. Scd1 is expressed in sebaceous glands and is disrupted in the asebia mouse. Nat Genet. 1999;23:268–70.
267. Westfall MD, Mays DJ, Sniezek JC, Pietenpol JA. The Delta Np63 alpha phosphoprotein binds the p21 and 14-3-3 sigma promoters in vivo and has transcriptional repressor activity that is reduced by Hay-Wells syndrome-derived mutations. Mol Cell Biol. 2003;23:2264–76.

Chapter 6
The Role of ATP-dependent Chromatin Remodeling in the Control of Epidermal Differentiation and Skin Stem Cell Activity

Gitali Ganguli-Indra and Arup K. Indra

Abbreviations

ACTL6a	Actin-like 6a
AER	Apical ectodermal ridge
ATP	Adenosine tri-phosphate
ATP-DCR	ATP-dependent chromatin remodeling
BAF	BRG1/BRM associated factors
C elegans	*Caenorhabditis.elegans*
EDC	Epidermal differentiation complex
EPB	Epidermal permeability barrier
HF	Hair follicle;

G. Ganguli-Indra
Department of Pharmaceutical Sciences, College of Pharmacy, Oregon State University, Corvallis, OR 97331, USA

Molecular Cell Biology Program, Oregon State University, Corvallis, OR 97331, USA

Knight Cancer Institute, Oregon Health & Science University (OHSU), Portland, OR 97239, USA

A. K. Indra (✉)
Department of Pharmaceutical Sciences, College of Pharmacy, Oregon State University, Corvallis, OR 97331, USA

Molecular Cell Biology Program, Oregon State University, Corvallis, OR 97331, USA

Knight Cancer Institute, Oregon Health & Science University (OHSU), Portland, OR 97239, USA

Department of Biochemistry and Biophysics, Oregon State University, Corvallis, OR 97331, USA

Linus Pauling Institute, Oregon State University, Corvallis, OR 97331, USA

Departments of Dermatology, OHSU, Portland, OR 97239, USA
e-mail: arup.indra@oregonstate.edu

© Springer International Publishing AG, part of Springer Nature 2018

V. A. Botchkarev, S. E. Millar (eds.), *Epigenetic Regulation of Skin Development and Regeneration*, Stem Cell Biology and Regenerative Medicine,
https://doi.org/10.1007/978-3-319-16769-5_6

HFSCs	Hair follicle stem cells
ISWI	Imitation SWI2
KLF4	Kruppel-like factor 4
SNF2	Sucrose Non Fermentation
SWI2	Mating type Switching 2

6.1 Introduction

Regulation of gene expression requires interplay between the transcription machinery and the multi-subunit protein complexes that are involved in remodeling of chromatin, thereby controlling the spatial and temporal accessibility of sequence specific transcription factors to their corresponding target genes. Such complexes either covalently alter histones or DNA, or dismantle histone-DNA contacts through ATP-dependent nucleosome remodeling. ATP-dependent chromatin remodeling (ATP-DCR) complexes contain multiple subunits, including an ATPase of the SNF2 family that hydrolyzes ATP in order to modify or reshape the histone-DNA interaction within the nucleosomes, leading to nucleosome sliding, removal of histones, and/or exchange of histone variants [1]. The first chromatin remodeling complex was discovered during screens in yeast for molecules involved in the signal transduction responsible for mating type switching, and is also known as Switch or SWI [2].

In different species, several subfamilies of the SNF2 family, including SWI2/SNF2, ISWI and CHD/Mi-2, have been identified based on the status of their catalytic ATPase subunit. SWI2/SNF2, also known as the BRG1/BRM-associated factor (BAF) complex, is one of the best characterized nucleosome remodeling complex subfamilies, and is highly conserved among eukaryotes. Animal SWI2/SNF2 complexes contain one of the two catalytic ATPase subunits, BRG1 (SNF2α) or BRM (SNF2β), and a variable, eight to fourteen regulatory subunit configuration of BAFs [3]. These subunits can be encoded by at least 20 different genes, resulting in over 200 amalgamations of complexes. These target different DNA sequences and may have different functions [1, 4]. All of these alterations to nucleosome architecture result in an open chromatin structure and transcriptional activation, leading to gene expression.

The following Chapter focuses on the *in vivo* role of ATP-dependent chromatin remodeling in the control of epidermal differentiation and skin stem cell activity during hair follicle morphogenesis, hair cycling and wound healing in mammals such as mice and humans.

6.2 Control of Epidermal Differentiation

Skin is the largest organ of the body and protects us from toxic and arid external hazards by establishing and maintaining the epidermal permeability barrier (EPB). Proliferating cells in the epidermal basal layer undergo stepwise terminal differentiation, giving rise to a stratified epithelium. Cells in the outermost layer of the epidermis become enucleated and form the stratum corneum, whose cells are sloughed off periodically and are replenished from the proliferative basal layer [5].

Prior *in vitro* studies suggested that BRM and BRG1, the two ATPases of the SWI2/SNF2 family, regulate the expression of independent sets of genes [6]. However, viability of *Brm-null* mice indicated that these two factors are functionally redundant *in vivo* and that BRG1 can compensate for loss of BRM [7]. In contrast, as fibroblasts lacking *Brg1* are viable but *Brg1-null* embryos die very early during development (during the peri-implantation stage), BRG1 might exert cell-specific functions in early development [8].

In order to elucidate the *in vivo* role of these factors in the later stages of development, including in skin development and maintenance of skin homeostasis, mice bearing LoxP-flanked (floxed) *Brg1*-alleles were established. *Brg1/SNF2α* was ablated in the forming epidermis using K14-Cre [9] or K14-Cre-ER^{T2} [10] transgenic mice that express either the bacteriophage P1 Cre-recombinase or the ligand-dependent Cre-ER^{T2} recombinase driven by the human K14 promoter, which is active in the surface ectoderm and the basal layer of the epidermis [11]. BRG1 is expressed in the surface ectoderm including that of the outgrowing limbs as early as embryonic day 10 (E10) of development [12]. At E18.5, BRG1 is strongly expressed in most, if not all, basal cells, as well as in about 70% of the spinous and 30% of the granular cells of the developing epidermis. Using constitutively active Cre recombinase, *Brg1* was efficiently ablated in epidermal keratinocytes and in the ectodermal layer of the limbs before E12 [12, 13]. Ablation of *Brg1* in the surface ectoderm induced severe hindlimb defects due to lack of maintenance of the apical ectodermal ridge (AER). The absence of forelimb defects in constitutive Cre mutants most likely reflects the earlier development of the forelimb, which occurs before efficient expression of the Cre recombinase [14], rather than differential participation of BRG1 in fore- and hindlimb development. Hindlimb defects can be avoided by inducing *Brg1* ablation in postnatal skin. Indra and co-workers demonstrated that BRG1 is dispensable for formation of embryonic epidermis, but is essential for establishment of the epidermal permeability barrier (EPB). Interestingly, temporal ablation of *Brg1* in the epidermis of mice lacking BRM showed that the BRM/BRG1 ATP-dependent chromatin remodeling complex is not required for epidermal proliferation and "early" differentiation, and revealed partial redundancy between BRM and BRG1 in regulating "late" terminal differentiation of the epidermal keratinocytes E12 [13].

Taken together, these results suggested that BRG1 selectively controls the expression of genes involved in the epithelial mesenchymal interactions required for limb patterning [15] and in terminal differentiation of keratinocytes during development. Similar to the situation in undifferentiated F9 embryonal carcinoma cells and in peri-implantation embryos [8, 16], BRM, which is dispensable for epidermis and limb formation, cannot functionally replace BRG1 in these processes. However, BRM can partially substitute for BRG1 in keratinocytes undergoing terminal differentiation.

It was recently shown that ACTL6a (actin-like 6a), a protein also known as BAF53a/INO80K/Arp4, modulates the SWI/SNF complex to suppress differentiation in the epidermis [17]. ACTL6 expression is downregulated during epidermal differentiation and is most strongly expressed in the less differentiated cells close to the epidermal basement membrane [17]. Spatio-temporal ablation of the epidermal ACTL6a resulted in reduced progenitor functions, premature terminal differentiation and epidermal thinning (hypoplasia) during epidermal development and also in adult tissue homeostasis [17]. Significant derepression of specific well-characterized differentiation related genes at the mRNA and protein levels was observed upon loss of ACTL6a [17]. ACTL6a target gene characterization identified KLF4 (Kruppel-like factor 4), a known activator of epidermal differentiation, as a key target of ACTL6a repression. A large number of genes that are regulated by ACTL6a were also identified as targets of KLF4. In line with this, KLF4 loss together with deletion of ACTL6a significantly compensated for the defects caused by ACTL6a depletion in progenitors [17].

Recent studies also suggest that ACTL6a can associate with different epigenetic regulators, including the Tip60 HAT complexes, the KAT2a HAT complexes, and the SWI/SNF chromatin-remodeling complex [18–20]. Depletion of either *KAT2a* or *Tip60* failed to significantly alter expression of differentiation related genes. However, ablation of the largest component of the SWI/SNF complex, *BAF250a/ARID1A*, but not *BAF250b/ARID1B*, produced impacts similar to ACTL6a loss: decreased clonogenic growth and premature induction of differentiation. These results indicated that ACTL6a supports the maintenance of the epidermal progenitor state by sequestering BRM1/BRG1 to prevent activation of differentiation programs.

Chromatin immuno-precipitation assays on keratinocytes with functional ACTL6a indicated that compared to undifferentiated cells, differentiated keratinocytes displayed enhanced binding by BRM/BRG1 and RNA polymerase II at the promoters of differentiation genes, including KLF4 as well as KRT10, S100A9, SPRR3, and BMP6. Loss of ACTL6a in undifferentiated progenitor populations enhanced the binding of both BRM and BRG1 as well as RNA polymerase II to differentiation-gene promoters but failed to alter binding to other gene promoters (Fig. 6.1). Thus, ACTL6a helps to maintain the undifferentiated state by inhibiting SWI/SNF chromatin remodeling complex attachment to and activation of KLF4 and other differentiation related gene promoters [17].

Similar to the SWI2/SNF2 chromatin remodeling complex, the ISWI-containing NURF complex and the CHD-containing NuRD complex are important for maintenance of human keratinocyte stem cells in an undifferentiated state [21]. Although there are no other described functions for these complexes in epidermal differentiation and barrier formation, they are required, along with SWI/SNF, for DNA repair. Specifically, the NURF and NuRD complexes are essential for the exposure of damaged DNA bases to the repair machinery and for the repair of double-stranded breaks through recombination [22]. These events are critical for reducing the risk of DNA damage in skin by environmental insults such as solar UV-irradiation, and for diminishing the risk of developing melanoma and non-melanoma skin cancer.

The ATP dependent chromatin remodelers Mi-2α and Mi-2β, alternatively known as CHD3 and CHD4, are expressed in a wide range of developing tissues including skin and mucosal epithelia (Fig. 6.1) [23]. Mi-2β is expressed in epidermis and hair follicle placodes during embryonic development, and in the matrix of mature hair follicles (Fig. 6.1) [24]. Extensive studies of Mi-2β expression at the level of RNA were performed during the epidermal differentiation process. At E10.5, Mi-2β mRNA is uniformly expressed in the single layer ectoderm, and at E14.5 expression is observed in basal and suprabasal layers as well as in the hair peg and matrix of differentiating hair follicles (HF) (Fig. 6.2). By contrast, Mi-2β mRNA expression is very low in mature inter-follicular epithelium (IFE).

In order to study the role of Mi-2β/CHD4 in skin development and homeostasis in detail, Georgopoulos and coworkers generated mice with specific deletion of Mi-2β in keratinocytes using *K14-Cre* mediated cell-specific recombination [24].

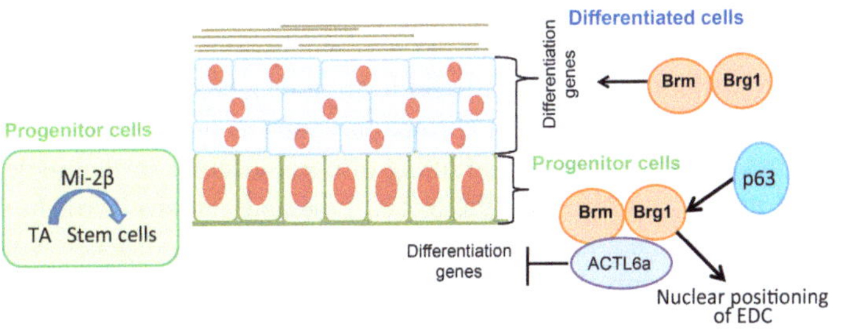

Fig. 6.1 ATP-dependent chromatin remodeling proteins in the control of epidermal homeostasis during development

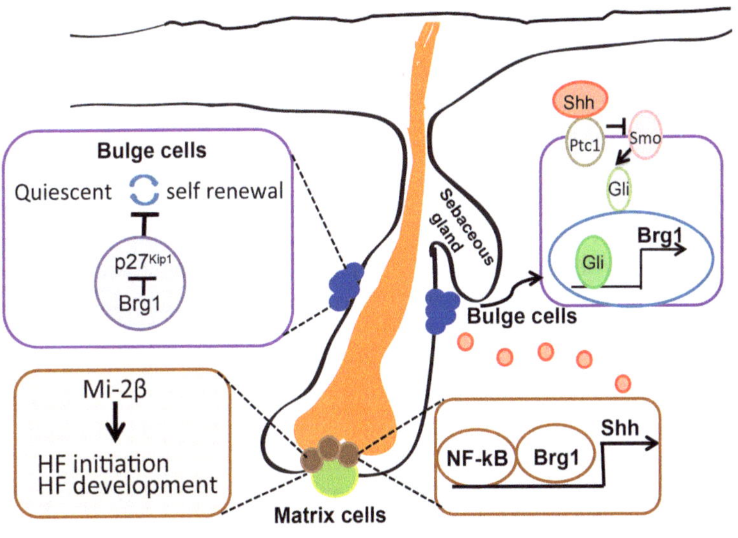

Fig. 6.2 ATP-dependent chromatin remodelers in the control of hair follicle morphogenesis, hair cycling, and stem cell activity

Homozygous mice died within 24 h. of birth with shiny and flaky skin suggesting a possible role for Mi-2β in barrier formation. In addition, the mutants exhibited reduced numbers of HFs, abnormal whisker hairs, and curly tails (to be discussed later). These mice displayed striking difference in the phenotypes between the dorsal and ventral skin in terms of epidermal structure and presence of hair follicles. Early (E10.5) depletion of Mi-2β/CHD4 in the ventral epidermis resulted in reduction of epidermal suprabasal layers and depletion of the basal layers later in embryogenesis. In contrast, later (E13.5) loss of Mi-2β/CHD4 did not affect epidermal differentiation or maintenance of the basal layer, but induction of HFs was blocked. The ventral skin phenotype was caused by deletion of Mi-2β at early stages of epidermal development, whereas the distinct dorsal skin phenotype was due to removal of Mi-2β after initiation of epidermal morphogenesis [24].

During the process of epidermal differentiation and permeability barrier formation, it has been suggested that the nucleolus transitions from a state of active transcription in proliferating basal cells to a fully inactive state in the stratum corneum [25]. Current advances in analyses of 3D genomic organization using (a) confocal microscopy after fluorescent *in situ* hybridization or labeling with transgenic chimeric fluorescent proteins, and (b) chromatin conformation capture (3C, 4C and Hi-C) revealed that the relative positioning of chromosomes in the nucleus is not random, and instead is rather specific to cell type and cell size [26]. In the interphase nucleus, chromosomes reside in defined domain(s) and genes within these regions are non-randomly positioned relative to each other and to nuclear suborganelles [26, 27]. As an example, in mouse epidermis the position of chromosome 3 in keratinocytes of the basal and suprabasal layers is more peripheral than that of chromosome 11 [25]. Massive remodeling of the higher-order chromatin structure of the epidermal differentiation complex (EDC) on mouse chromosome 3 occurs during epidermal morphogenesis. The locus moves away from the nuclear periphery and towards the nuclear interior into a location that is rich in SC35 (Srsf2)-positive nuclear speckles. This realignment of the EDC locus occurs before transcriptional activation of EDC genes that drive terminal differentiation of keratinocytes. The transcription factor p63, a master regulator of epidermal development [28, 29], orchestrates this lineage-specific, developmentally controlled event in the epidermis. In *p63*-null mouse skin, significant changes in expression of EDC genes are associated with alteration of the developmentally controlled relocation of the EDC within the nucleus [30]. In epidermal basal cells, *p63* directly regulates expression of the ATP-dependent chromatin remodeler Brg1, which binds to distinct domains within the EDC and is required for repositioning of the EDC and *Loricrin* loci towards the nuclear interior (Fig. 6.1) [30]. This combinatorial effect of p63 and BRG1 drives higher order chromatin remodeling, 3D-genomic organization and efficient gene expression of the EDC genes in epidermal precursor cells during epidermal morphogenesis (Fig. 6.1) [30].

Studies have also shown that transcription factors MAF:MAFB are regulated by lncRNAs TINCR and ANCR, besides p63, to give rise to a complex regulatory gene network for epidermal differentiation [31]. Using DeepCAGE, genome-wide profiling of histone modifications and retroviral integration analysis, Cavazza et al., showed that most of the active promoters are differentially controlled in progenitor and differentiated keratinocytes, while nearly 50% of the enhancers and super-enhancers mediate, in a stage specific manner, the epigenetic changes in differentiated keratinocytes [32]. They also observed that p63 binds to and controls cell specific super-enhancers in both proliferating and differentiating keratinocytes.

6.3 Stem Cell Activity During Hair Follicle Morphogenesis and Cycling

Hair follicle (HF) morphogenesis is initiated during embryonic development and involves epithelial-mesenchymal interactions that require the activity of the Wnt/β--catenin, Shh, Notch, BMP and Edar signaling pathways [33]. In adult life, HFs cycle periodically through anagen (growth), catagen (regression) and telogen (resting) phases [25, 34–36]. HF stem cells (HFSCs) in the secondary hair germ and bulge regions are stimulated to proliferate at anagen onset. The secondary hair germ generates the proliferative HF matrix, which produces the hair shaft and its surrounding inner root sheath. Bulge stem cells give rise to the HF outer root sheath, and these contribute to the matrix in the subsequent HF growth cycle [33, 37, 38]. In homeostasis, bulge stem cells contribute only to the HF, but following skin wounding their progeny exit the hair follicle and contribute transiently to epidermal repair [34–36]. The ability of HFSCs to self-renew is critical to ensure that HFs can continue to cycle throughout life [33, 39].

HFSCs in the bulge and secondary hair germ are characterized by expression of distinct sets of markers ($Krt15^+$, $Lgr5^+$, $CD34^+$, $Sox9^+$, $Lhx2^+$, $Tcf3^+$, $Nfatc1^+$ for the bulge, and $Krt15^+$, $Gli1^+$, $Lgr5^+$ for the secondary hair germ). Additional stem cell populations reside in the junctional zone ($Lrig1^+$), sebaceous glands ($Blimp1^+$), and isthmus ($Lgr6^+$, $Plet1^+$, $Gli1^+$) [34–36, 40–42]. Lrig1, Lhx2 and Nfatc1 are thought to play roles in lineage maintenance of these regions [40, 41, 43].

Chromatin remodeling involving the SWI/SNF complex is important in controlling the activities of genes that regulate stem cell functions. Brahma related gene 1 (BRG1), an ATPase component of the BAF chromatin remodeling structure, plays key roles in normal hair regeneration [44] and is dynamically expressed in HF at different stages of the hair cycle [44]. Low levels of BRG1 expression are observed in late telogen, while early anagen is marked with increased expression mainly in the lower bulge, and at later stages expression declines [44]. To elucidate the physiological role of BRG1 in hair regeneration, *Nfatc-Cre* mice, which express Cre recombinase specifically in the bulge [43], were combined with a conditional allele of *Brg1* [44]. Deletion of *Brg1* in the bulge caused decreased matrix cell proliferation, retarded hair growth, and progressive hair loss. BRG1 functions by suppressing p27kip1 and recruiting NF-kB, which in turn activates Shh in matrix cells promoting their proliferation [45, 46] (Fig. 6.2). Shh signaling through Gli activates BRG1 in bulge cells, creating a positive feedback loop. Thus, gene regulation by chromatin restructuring plays a key role in HFSC activation.

The ATP dependent chromatin remodeler Mi-2β and its role in epidermal differentiation have been described in the above section. In addition to its expression in the developing epidermis, Mi-2β is also expressed in the hair placode and matrix of developing HF [24]. Loss of Mi-2β in embryonic dorsal epidermis prevents induction of hair follicle placodes. After initiation of the follicle, markers of follicular morphogenesis such as Edar, Shh, Bmp2 and β-catenin [24, 33, 38] are expressed, and some subsequent morphogenesis of the hair peg proceeds in the absence of Mi-2β; however production of the progenitors that give rise to the inner layers of the hair follicle and hair shaft is impaired [24].

6.4 Stem Cell Activity During Wound Healing

Wound healing is a complex process that requires coordination of inflammatory, proliferative and remodeling mechanisms. HFSCs contribute to tissue repair and regeneration processes, particularly in re-epithelialization and re-establishment of skin homeostasis [36]. For instance, in response to full-thickness wounding, several different HFSC populations are activated, and their progeny migrate out from the HFs to participate in wound re-epithelialization [35, 40, 42, 47–49]. Gene regulation mechanisms are controlled by histone modifications, and are essential for normal physiological processes. A recent article by Na et al. (2016) demonstrates that histone modifications are crucial during the wound healing process [50]. Expression of Histone H3K27 demethylase [known as Jumonji domain containing protein D3 (JMJD3)], which has an important role in keratinocyte differentiation, is upregulated in the wound edges and it's inactivation leads to aberrant wound healing [50]. Specifically, JMJD3 induces keratinocyte activation during the re-epithelialization process by interacting with NF-kB targets on inflammatory, MMP, and growth factor gene promoters [50, 51]. Similarly, expression of the chromatin remodeling protein BRG1 in HF bulge and bulge-derived cells increases following depilation or full thickness wounding in mouse skin [44]. Brg1 functions to facilitate emergence of bulge stem cells from the hair follicle to contribute to wound repair [34, 36, 44, 52].

6.5 Skin Pathophysiology

Exposure to solar UV irradiation can cause sunburn in the short term, and skin photoaging and skin cancer due to DNA damage in the longer term. Despite these negative effects, controlled UV irradiation can be beneficial in the treatment of skin diseases such as eczema and psoriasis. Brahma (BRM) is a component of the SWI/SNF chromatin remodeling complex. Human non-melanoma skin cancers are reported to have hot spot mutations in the *Brm* gene, caused by ultraviolet (UV) exposure [53], and recent studies have demonstrated the protective functions of BRM as a tumor

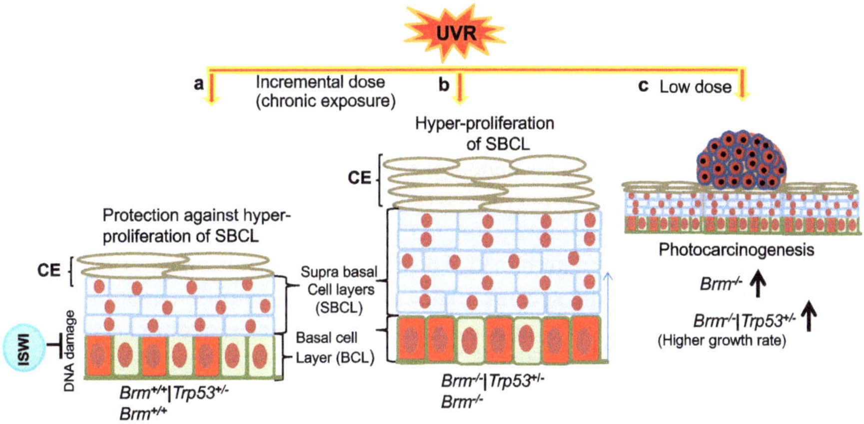

Fig. 6.3 ATP-dependent chromatin remodelers in the control of UV irradiation induced epidermal proliferation and photo-carcinogenesis

suppressor [54]. In line with these findings, irradiation of *Brm*-null mice with low dose ultraviolet (UV) light revealed their increased susceptibility to skin photocarcinognesis compared with controls, but did not alter the protective apoptotic response to UV-induced sunburn [54]. In the absence of one allele of the tumor suppressor p53, loss of BRM did not further increase tumor incidence, but did result in a higher growth rate of the tumors [54, 55]. Importantly, increased cell division occurred predominantly in the differentiated suprabasal layer of the epidermis, rather than in the basal layer in *Brm-/-* mice, revealing that BRM protects suprabasal cells from UV induced proliferation (Fig. 6.3) [55]. Suprabasal cells are more exposed to UV than basal cells and are therefore highly susceptible to mutation [56]. Moreover, UV-induced mutation occurs when cells divide without repairing damaged DNA [54, 56]. These observations may help to explain why the authors' initial studies with *Brm-null* mice showed increased photo-carcinogenesis, while in later studies where they used a UV dose that simulated chronic sunlight exposure with mild sunburn damage, similar to the effects of natural sun exposure during normal activities in humans, they observed increased UV-induced cellular hyper-proliferation.

By contrast, UVB irradiation of *C. elegans* nematodes lacking the *Brm* analog *psa-4* caused increased UV-sensitivity and cell death. This finding may reflect differences between the *C. elegans* and mammalian genomes and/or differences in sensitivity to UVB between *C. elegans* and mammalian keratinocytes [57].

6.6 Conclusions

The studies summarized above suggest that all subfamilies of the SNF2 family, including SWI2/SNF2 (BRG1/BRM), ISWI and CHD/Mi-2β, play critical roles in the maintenance of epidermal homeostasis by controlling the balance between proliferation and differentiation, and also in controlling epidermal permeability barrier formation. Future studies will further explore the roles of epigenetic mechanisms and their cross-talk with other regulatory pathways in controlling keratinocyte proliferation and the switch to differentiation in healthy and diseased skin.

In particular, the mechanisms underlying targeting of BRG1 to specific domains of the EDC, and its potential interaction with key epidermal transcription factors, such as p63, AP-1, and Klf4, that regulate EDC gene expression in epidermal progenitor cells remain to be determined. The cooperative involvement of BRG1 and SATB1 for establishing the specific EDC configuration in differentiating epidermal keratinocytes needs to be further investigated. Similarly, additional studies to explore the mechanisms by which Mi-2β exerts selective effects on development of the epidermis and related skin appendages will further elucidate the role that Mi-2 β-like chromatin remodelers play in establishing lineage specific stem cell identity. Future studies on the functions of other actin-like proteins, similar to ACTL6a, in multiple tissues will provide insights into developmental regulation by these modulators and the epigenetic regulatory complexes with which they associate.

The BRG1 chromatin remodeler plays crucial roles in maintaining the bulge stem cell pool, controlling hair cycling, permitting normal skin homeostasis, and facilitating repair and regeneration processes (Figs. 6.1 and 6.2). Other factors involved in the process of Gli-mediated activation of BRG1 during hair cycling are currently unknown, and are an important subject for further study (Fig. 6.2). It would be interesting to examine the regulation of BRG1 by other bulge stem cell factors such as Lhx2 and Tcf3 in this context. Improved understanding of BRG1 mediated regulatory controls of other cutaneous stem cell markers such as CD34, CD133 and Lrig1 during hair cycling and wound healing may help identify novel approaches to improve wound repair and tissue regeneration. Regulation of JMJD3 activity may provide a therapeutic approach for treatment of chronic wounds. Similarly, future mechanistic studies will provide more information regarding the role of Mi-2β in mobilization of stem cells during hair follicle morphogenesis and hair cycling, as well as in wound repair.

In the future, a better understanding of the epigenetic control of other key regulators by BRM and/or BRG1 in photo-carcinogenesis, UV-induced DNA damage and UV-induced skin inflammation will be necessary to gain more insights into these processes. Elucidation of detailed molecular mechanisms, together with gene expression profiling of coding and non-coding small and long-RNAs in UVB induced skin, may lead to discovery of potential biomarkers in photocarcinogenesis.

Last but not least, many of the studies discussed here highlight the power of Cre-ER^{T2}-mediated recombination technology for creating spatio-temporally controlled targeted somatic mutations that allow dissection of gene functions throughout skin morphogenesis and in adult life [12, 58, 59].

References

1. Clapier CR, Cairns BR. The biology of chromatin remodeling complexes. Annu Rev Biochem. 2009;78:273–304. https://doi.org/10.1146/annurev.biochem.77.062706.153223. Epub 2009/04/10. PubMed PMID: 19355820.
2. Stern M, Jensen R, Herskowitz I. Five SWI genes are required for expression of the HO gene in yeast. J Mol Biol. 1984;178(4):853–68. Epub 1984/10/05. PubMed PMID: 6436497.
3. Eberharter A, Becker PB. ATP-dependent nucleosome remodelling: factors and functions. J Cell Sci. 2004;117(Pt 17):3707–11. https://doi.org/10.1242/jcs.01175. Epub 2004/08/03. PubMed PMID: 15286171.
4. Wu JI, Lessard J, Crabtree GR. Understanding the words of chromatin regulation. Cell. 2009;136(2):200–6. https://doi.org/10.1016/j.cell.2009.01.009. Epub 2009/01/27. PubMed PMID: 19167321; PubMed Central PMCID: PMC2770578.
5. Kalinin AE, Kajava AV, Steinert PM. Epithelial barrier function: assembly and structural features of the cornified cell envelope. BioEssays. 2002;24(9):789–800. https://doi.org/10.1002/bies.10144. Epub 2002/09/05. PubMed PMID: 12210515.
6. Kadam S, Emerson BM. Transcriptional specificity of human SWI/SNF BRG1 and BRM chromatin remodeling complexes. Mol Cell. 2003;11(2):377–89. Epub 2003/03/07. PubMed PMID: 12620226.
7. Reyes JC, Barra J, Muchardt C, Camus A, Babinet C, Yaniv M. Altered control of cellular proliferation in the absence of mammalian brahma (SNF2alpha). EMBO J. 1998;17(23):6979–91. https://doi.org/10.1093/emboj/17.23.6979. Epub 1998/12/08. PubMed PMID: 9843504; PubMed Central PMCID: PMC1171046.
8. Bultman S, Gebuhr T, Yee D, La Mantia C, Nicholson J, Gilliam A, Randazzo F, Metzger D, Chambon P, Crabtree G, Magnuson T. A Brg1 null mutation in the mouse reveals functional differences among mammalian SWI/SNF complexes. Mol Cell. 2000;6(6):1287–95. Epub 2001/02/13. PubMed PMID: 11163203.
9. Li M, Chiba H, Warot X, Messaddeq N, Gerard C, Chambon P, Metzger D. RXR-alpha ablation in skin keratinocytes results in alopecia and epidermal alterations. Development. 2001;128(5):675–88. Epub 2001/02/15. PubMed PMID: 11171393.
10. Li M, Indra AK, Warot X, Brocard J, Messaddeq N, Kato S, Metzger D, Chambon P. Skin abnormalities generated by temporally controlled RXRalpha mutations in mouse epidermis. Nature. 2000;407(6804):633–6. https://doi.org/10.1038/35036595. Epub 2000/10/18. PubMed PMID: 11034212.
11. Vassar R, Rosenberg M, Ross S, Tyner A, Fuchs E. Tissue-specific and differentiation-specific expression of a human K14 keratin gene in transgenic mice. Proc Natl Acad Sci U S A. 1989;86(5):1563–7. Epub 1989/03/01. PubMed PMID: 2466292; PubMed Central PMCID: PMC286738.
12. Indra AK, Li M, Brocard J, Warot X, Bornert JM, Gerard C, Messaddeq N, Chambon P, Metzger D. Targeted somatic mutagenesis in mouse epidermis. Horm Res. 2000;54(5–6):296–300. Epub 2001/10/12. doi: 53275. PubMed PMID: 11595821.
13. Indra AK, Dupe V, Bornert JM, Messaddeq N, Yaniv M, Mark M, Chambon P, Metzger D. Temporally controlled targeted somatic mutagenesis in embryonic surface ectoderm and fetal epidermal keratinocytes unveils two distinct developmental functions of BRG1 in limb morphogenesis and skin barrier formation. Development. 2005;132(20):4533–44. https://doi.org/10.1242/dev.02019. Epub 2005/09/30. PubMed PMID: 16192310.
14. Kaufman MH, BLB J. The anatomical basis of mouse development. San Diego: Academic; 1999. 291 p.
15. Byrne C, Hardman M, Nield K. Covering the limb – formation of the integument. J Anat. 2003;202(1):113–23. Epub 2003/02/18. PubMed PMID: 12587926; PubMed Central PMCID: PMC1571060.
16. Sumi-Ichinose C, Ichinose H, Metzger D, Chambon P. SNF2beta-BRG1 is essential for the viability of F9 murine embryonal carcinoma cells. Mol Cell Biol. 1997;17(10):5976–86. Epub 1997/10/07. PubMed PMID: 9315656; PubMed Central PMCID: PMC232446.

17. Bao X, Tang J, Lopez-Pajares V, Tao S, Qu K, Crabtree GR, Khavari PA. ACTL6a enforces the epidermal progenitor state by suppressing SWI/SNF-dependent induction of KLF4. Cell Stem Cell. 2013;12(2):193–203. https://doi.org/10.1016/j.stem.2012.12.014. Epub 2013/02/12. PubMed PMID: 23395444; PubMed Central PMCID: PMC3661004.
18. Park J, Wood MA, Cole MD. BAF53 forms distinct nuclear complexes and functions as a critical c-Myc-interacting nuclear cofactor for oncogenic transformation. Mol Cell Biol. 2002;22(5):1307–16. Epub 2002/02/13. PubMed PMID: 11839798; PubMed Central PMCID: PMC134713.
19. Tea JS, Luo L. The chromatin remodeling factor Bap55 functions through the TIP60 complex to regulate olfactory projection neuron dendrite targeting. Neural Dev. 2011;6:5. https://doi.org/10.1186/1749-8104-6-5. Epub 2011/02/03. PubMed PMID: 21284845; PubMed Central PMCID: PMC3038883.
20. Zhao K, Wang W, Rando OJ, Xue Y, Swiderek K, Kuo A, Crabtree GR. Rapid and phosphoinositol-dependent binding of the SWI/SNF-like BAF complex to chromatin after T lymphocyte receptor signaling. Cell. 1998;95(5):625–36. Epub 1998/12/09. PubMed PMID: 9845365.
21. Mulder KW, Wang X, Escriu C, Ito Y, Schwarz RF, Gillis J, Sirokmany G, Donati G, Uribe-Lewis S, Pavlidis P, Murrell A, Markowetz F, Watt FM. Diverse epigenetic strategies interact to control epidermal differentiation. Nat Cell Biol. 2012;14(7):753–63. https://doi.org/10.1038/ncb2520. Epub 2012/06/26. PubMed PMID: 22729083.
22. Bao Y. Chromatin response to DNA double-strand break damage. Epigenomics. 2011;3(3):307–21. https://doi.org/10.2217/epi.11.14. Epub 2011/11/30. PubMed PMID: 22122340.
23. Kim J, Sif S, Jones B, Jackson A, Koipally J, Heller E, Winandy S, Viel A, Sawyer A, Ikeda T, Kingston R, Georgopoulos K. Ikaros DNA-binding proteins direct formation of chromatin remodeling complexes in lymphocytes. Immunity. 1999;10(3):345–55. Epub 1999/04/16. PubMed PMID: 10204490.
24. Kashiwagi M, Morgan BA, Georgopoulos K. The chromatin remodeler Mi-2beta is required for establishment of the basal epidermis and normal differentiation of its progeny. Development. 2007;134(8):1571–82. https://doi.org/10.1242/dev.001750. Epub 2007/03/16. PubMed PMID: 17360773.
25. Botchkarev VA, Gdula MR, Mardaryev AN, Sharov AA, Fessing MY. Epigenetic regulation of gene expression in keratinocytes. J Invest Dermatol. 2012;132(11):2505–21. https://doi.org/10.1038/jid.2012.182. Epub 2012/07/06. PubMed PMID: 22763788; PubMed Central PMCID: PMC3650472.
26. Cremer T, Cremer M. Chromosome territories. In: Misteli T, Spector D, editors. The Nucleus. New York: Cold Spring Harbor Laboratory Press; 2011. p. 93–114.
27. Sanyal A, Bau D, Marti-Renom MA, Dekker J. Chromatin globules: a common motif of higher order chromosome structure? Curr Opin Cell Biol. 2011;23(3):325–31. https://doi.org/10.1016/j.ceb.2011.03.009. Epub 2011/04/15. PubMed PMID: 21489772; PubMed Central PMCID: PMC3109114.
28. Koster MI, Roop DR. Mechanisms regulating epithelial stratification. Annu Rev Cell Dev Biol. 2007;23:93–113. https://doi.org/10.1146/annurev.cellbio.23.090506.123357. Epub 2007/05/11. PubMed PMID: 17489688.
29. Truong AB, Khavari PA. Control of keratinocyte proliferation and differentiation by p63. Cell Cycle. 2007;6(3):295–9. Epub 2007/02/01. PubMed PMID: 17264679.
30. Mardaryev AN, Gdula MR, Yarker JL, Emelianov VU, Poterlowicz K, Sharov AA, Sharova TY, Scarpa JA, Joffe B, Solovei I, Chambon P, Botchkarev VA, Fessing MY. p63 and Brg1 control developmentally regulated higher-order chromatin remodelling at the epidermal differentiation complex locus in epidermal progenitor cells. Development. 2014;141(1):101–11. https://doi.org/10.1242/dev.103200. Epub 2013/12/19. PubMed PMID: 24346698; PubMed Central PMCID: PMC3865752.
31. Lopez-Pajares V, Qu K, Zhang J, Webster DE, Barajas BC, Siprashvili Z, Zarnegar BJ, Boxer LD, Rios EJ, Tao S, Kretz M, Khavari PA. A LncRNA-MAF:MAFB transcription fac-

tor network regulates epidermal differentiation. Dev Cell. 2015;32(6):693–706. https://doi.org/10.1016/j.devcel.2015.01.028. PubMed PMID: 25805135; PubMed Central PMCID: PMCPMC4456036.
32. Cavazza A, Miccio A, Romano O, Petiti L, Malagoli Tagliazucchi G, Peano C, Severgnini M, Rizzi E, De Bellis G, Bicciato S, Mavilio F. Dynamic transcriptional and epigenetic regulation of human epidermal keratinocyte differentiation. Stem Cell Reports. 2016;6(4):618–32. https://doi.org/10.1016/j.stemcr.2016.03.003. PubMed PMID: 27050947; PubMed Central PMCID: PMC4834057.
33. Rishikaysh P, Dev K, Diaz D, Qureshi WM, Filip S, Mokry J. Signaling involved in hair follicle morphogenesis and development. Int J Mol Sci. 2014;15(1):1647–70. https://doi.org/10.3390/ijms15011647. Epub 2014/01/24. PubMed PMID: 24451143; PubMed Central PMCID: PMC3907891.
34. Bhattacharya S, Wheeler H, Leid M, Ganguli-Indra G, Indra AK. Transcription factor CTIP2 maintains hair follicle stem cell pool and contributes to altered expression of LHX2 and NFATC1. J Invest Dermatol. 2015. https://doi.org/10.1038/jid.2015.281. Epub 2015/07/16. PubMed PMID: 26176759.
35. Ito M, Liu Y, Yang Z, Nguyen J, Liang F, Morris RJ, Cotsarelis G. Stem cells in the hair follicle bulge contribute to wound repair but not to homeostasis of the epidermis. Nat Med. 2005;11(12):1351–4. https://doi.org/10.1038/nm1328. Epub 2005/11/17. PubMed PMID: 16288281.
36. Liang X, Bhattacharya S, Bajaj G, Guha G, Wang Z, Jang HS, Leid M, Indra AK, Ganguli-Indra G. Delayed cutaneous wound healing and aberrant expression of hair follicle stem cell markers in mice selectively lacking Ctip2 in epidermis. PLoS One. 2012;7(2):e29999. https://doi.org/10.1371/journal.pone.0029999. Epub 2012/03/03. PubMed PMID: 22383956; PubMed Central PMCID: PMC3283611.
37. Botchkarev VA, Paus R. Molecular biology of hair morphogenesis: development and cycling. J Exp Zool B Mol Dev Evol. 2003;298(1):164–80. https://doi.org/10.1002/jez.b.33. Epub 2003/09/02. PubMed PMID: 12949776.
38. Millar SE. Molecular mechanisms regulating hair follicle development. J Invest Dermatol. 2002;118(2):216–25. https://doi.org/10.1046/j.0022-202x.2001.01670.x. Epub 2002/02/14. PubMed PMID: 11841536.
39. Suda T, Arai F. Wnt signaling in the niche. Cell. 2008;132(5):729–30. https://doi.org/10.1016/j.cell.2008.02.017. Epub 2008/03/11. PubMed PMID: 18329358.
40. Jaks V, Kasper M, Toftgard R. The hair follicle-a stem cell zoo. Exp Cell Res. 2010;316(8):1422–8. https://doi.org/10.1016/j.yexcr.2010.03.014. Epub 2010/03/27. PubMed PMID: 20338163.
41. Mardaryev AN, Meier N, Poterlowicz K, Sharov AA, Sharova TY, Ahmed MI, Rapisarda V, Lewis C, Fessing MY, Ruenger TM, Bhawan J, Werner S, Paus R, Botchkarev VA. Lhx2 differentially regulates Sox9, Tcf4 and Lgr5 in hair follicle stem cells to promote epidermal regeneration after injury. Development. 2011;138(22):4843–52. https://doi.org/10.1242/dev.070284. Epub 2011/10/27. PubMed PMID: 22028024; PubMed Central PMCID: PMC4067271.
42. Plikus MV, Gay DL, Treffeisen E, Wang A, Supapannachart RJ, Cotsarelis G. Epithelial stem cells and implications for wound repair. Semin Cell Dev Biol. 2012;23(9):946–53. https://doi.org/10.1016/j.semcdb.2012.10.001. Epub 2012/10/23. PubMed PMID: 23085626; PubMed Central PMCID: PMC3518754.
43. Horsley V, Aliprantis AO, Polak L, Glimcher LH, Fuchs E. NFATc1 balances quiescence and proliferation of skin stem cells. Cell. 2008;132(2):299–310. https://doi.org/10.1016/j.cell.2007.11.047. Epub 2008/02/05. PubMed PMID: 18243104; PubMed Central PMCID: PMC2546702.
44. Xiong Y, Li W, Shang C, Chen RM, Han P, Yang J, Stankunas K, Wu B, Pan M, Zhou B, Longaker MT, Chang CP. Brg1 governs a positive feedback circuit in the hair follicle for tissue regeneration and repair. Dev Cell. 2013;25(2):169–81. https://doi.org/10.1016/j.devcel.2013.03.015. Epub 2013/04/23. PubMed PMID: 23602386.

45. Alonso L, Fuchs E. The hair cycle. J Cell Sci. 2006;119(Pt 3):391–3. https://doi.org/10.1242/jcs02793. Epub 2006/01/31. PubMed PMID: 16443746.
46. Blanpain C, Lowry WE, Geoghegan A, Polak L, Fuchs E. Self-renewal, multipotency, and the existence of two cell populations within an epithelial stem cell niche. Cell. 2004;118(5):635–48. https://doi.org/10.1016/j.cell.2004.08.012. Epub 2004/09/02. PubMed PMID: 15339667.
47. Levy V, Lindon C, Zheng Y, Harfe BD, Morgan BA. Epidermal stem cells arise from the hair follicle after wounding. FASEB J. 2007;21(7):1358–66. https://doi.org/10.1096/fj.06-6926com. Epub 2007/01/27. PubMed PMID: 17255473.
48. Nowak JA, Polak L, Pasolli HA, Fuchs E. Hair follicle stem cells are specified and function in early skin morphogenesis. Cell Stem Cell. 2008;3(1):33–43. https://doi.org/10.1016/j.stem.2008.05.009. Epub 2008/07/03. PubMed PMID: 18593557; PubMed Central PMCID: PMC2877596.
49. Snippert HJ, Haegebarth A, Kasper M, Jaks V, van Es JH, Barker N, van de Wetering M, van den Born M, Begthel H, Vries RG, Stange DE, Toftgard R, Clevers H. Lgr6 marks stem cells in the hair follicle that generate all cell lineages of the skin. Science. 2010;327(5971):1385–9. https://doi.org/10.1126/science.1184733. Epub 2010/03/13. PubMed PMID: 20223988.
50. Na J, Lee K, Na W, Shin JY, Lee MJ, Yune TY, Lee HK, Jung HS, Kim WS, Ju BG. Histone H3K27 Demethylase JMJD3 in cooperation with NF-kappaB regulates keratinocyte wound healing. J Invest Dermatol. 2016;136(4):847–58. https://doi.org/10.1016/j.jid.2015.11.029. PubMed PMID: 26802933.
51. Odorisio T. Epigenetic control of skin Re-Epithelialization: the NF-kB/JMJD3 connection. J Invest Dermatol. 2016;136(4):738–40. https://doi.org/10.1016/j.jid.2016.01.010. PubMed PMID: 27012558.
52. Menchon C, Edel MJ, Izpisua Belmonte JC. The cell cycle inhibitor p27Kip(1) controls self-renewal and pluripotency of human embryonic stem cells by regulating the cell cycle, Brachyury and Twist. Cell Cycle. 2011;10(9):1435–47. Epub 2011/04/12. PubMed PMID: 21478681; PubMed Central PMCID: PMC3685623.
53. Moloney FJ, Lyons JG, Bock VL, Huang XX, Bugeja MJ, Halliday GM. Hotspot mutation of Brahma in non-melanoma skin cancer. J Invest Dermatol. 2009;129(4):1012–5. https://doi.org/10.1038/jid.2008.319. Epub 2008/10/17. PubMed PMID: 18923443.
54. Halliday GM, Zhou Y, Sou PW, Huang XX, Rana S, Bugeja MJ, Painter N, Scolyer RA, Muchardt C, Di Girolamo N, Lyons JG. The absence of Brm exacerbates photocarcinogenesis. Exp Dermatol. 2012;21(8):599–604. https://doi.org/10.1111/j.1600-0625.2012.01522.x. Epub 2012/07/11. PubMed PMID: 22775994.
55. Hassan NM, Painter N, Howlett CR, Farrell AW, Di Girolamo N, Lyons JG, Halliday GM. Brm inhibits the proliferative response of keratinocytes and corneal epithelial cells to ultraviolet radiation-induced damage. PLoS One. 2014;9(9):e107931. https://doi.org/10.1371/journal.pone.0107931. Epub 2014/09/26. PubMed PMID: 25254962; PubMed Central PMCID: PMC4177874.
56. Halliday GM, Cadet J. It's all about position: the basal layer of human epidermis is particularly susceptible to different types of sunlight-induced DNA damage. J Invest Dermatol. 2012;132(2):265–7. https://doi.org/10.1038/jid.2011.281. Epub 2012/01/14. PubMed PMID: 22241442.
57. Lans H, Marteijn JA, Schumacher B, Hoeijmakers JH, Jansen G, Vermeulen W. Involvement of global genome repair, transcription coupled repair, and chromatin remodeling in UV DNA damage response changes during development. PLoS Genet. 2010;6(5):e1000941. https://doi.org/10.1371/journal.pgen.1000941. Epub 2010/05/14. PubMed PMID: 20463888; PubMed Central PMCID: PMC2865526.
58. Indra AK, Warot X, Brocard J, Bornert JM, Xiao JH, Chambon P, Metzger D. Temporally-controlled site-specific mutagenesis in the basal layer of the epidermis: comparison of the recombinase activity of the tamoxifen-inducible Cre-ER(T) and Cre-ER(T2) recombinases.

Nucleic Acids Res. 1999;27(22):4324–7. Epub 1999/10/28. PubMed PMID: 10536138; PubMed Central PMCID: PMC148712.
59. Metzger D, Indra AK, Li M, Chapellier B, Calleja C, Ghyselinck NB, Chambon P. Targeted conditional somatic mutagenesis in the mouse: temporally-controlled knock out of retinoid receptors in epidermal keratinocytes. Methods Enzymol. 2003;364:379–408. Epub 2003/11/25. PubMed PMID: 14631857.

Chapter 7
Orchestrated Role of microRNAs in Skin Development and Regeneration

Natalia V. Botchkareva and Rui Yi

7.1 Regulation of miRNA Biogenesis and Functions

The biogenesis pathway of miRNAs has been extensively investigated. These studies have delineated a canonical pathway for miRNA biogenesis that is mediated sequentially by nuclear cleavage of primary miRNAs by the Drosha/Dgcr8 microprocessor, nuclear export of pre-miRNAs by Xpo5, cytoplasmic cleavage of pre-miRNAs by Dicer together with TRBP and/or PACT, and the incorporation of single-stranded mature miRNA into the Argonaute (Ago) proteins to form the RNA-induced silencing complex (RISC). In addition, several alternative pathways have been identified. We refer our readers to a recent review for detailed description of these studies [1].

In this Chapter, we will focus on recent research and aim to provide new insights to stimulate future studies of miRNAs' roles in the skin. In particular, it is clear that miRNA biogenesis is under extensive regulation and integrating these regulatory mechanisms into studies of miRNA functions will be critical for our understanding of these fascinating small molecules.

Although it is widely accepted that pri-miRNAs are transcribed by RNA polymerase II, the transcription start site, relative location of mature miRNAs on the pri-miRNA transcripts, and the transcription termination site are generally ill-defined. Global annotation of miRNA primary transcripts in mouse and human [2] has revealed that expression of individual miRNAs within a cluster is modulated

N. V. Botchkareva (✉)
Centre for Skin Sciences, University of Bradford, Richmond Road, Bradford, BD7 1DP, UK
e-mail: N.Botchkareva@bradford.ac.uk

R. Yi
Department of Molecular, Cellular and Developmental Biology, University of Colorado, Boulder, Denver, CO, USA
e-mail: Rui.Yi@Colorado.edu

© Springer International Publishing AG, part of Springer Nature 2018

V. A. Botchkarev, S. E. Millar (eds.), *Epigenetic Regulation of Skin Development and Regeneration*, Stem Cell Biology and Regenerative Medicine,
https://doi.org/10.1007/978-3-319-16769-5_7

through the use of alternative transcription start sites. This finding implicates involvement of alternative promoter sequences and transcription factors, providing a mechanism by which individual miRNA expression from a genomic cluster can be regulated in a cell type-specific manner. In addition, expression of some miRNAs, such as human miR-205, a miRNA that is highly expressed in the skin, can be adjusted via alternative splicing [3, 4]. Mature miRNAs can be expressed from either intronic or exonic sequences. However, in the case of human miR-205, alternative splicing can slice the pre-miR-205 at the loop region, likely abolishing mature miR-205 production from the spliced transcript. Although it is not clear how this alternative splicing mechanism regulates miR-205 expression and function in human cells, this observation establishes a basis to link alternative splicing to miRNA expression. In our characterization of miR-205 primary transcripts in mouse, however, we do not detect any potential interference from splicing patterns [4]. This implies that such regulatory mechanisms may be human-specific even though mature miR-205 is fully conserved across vertebrates.

It is also unclear how the transcriptional output of primary miRNA transcripts is related to accumulation of mature miRNAs, which can be readily measured by Northern blotting, qRT-PCR, and deep sequencing. In our recent efforts to deeply sequence the skin transcriptome by RNA-seq, we have observed accumulation of primary miRNA transcripts 3′ to the pre-miRNA hairpin for some highly abundant miRNAs; however we were not able to detect the entire transcript even for these highly expressed miRNAs. This suggests that primary transcripts are efficiently processed by the microprocessor complex to release the miRNA hairpin. Of note, these miRNAs are typically highly expressed in the skin.

miR-205 has recently been shown to be expressed at high levels in hair follicle stem cells (HF-SCs), with an estimated copy number of ~50,000 per cell, and plays an important role in promoting HF-SC expansion [4]. Interestingly, the miR-205 locus harbors super-enhancers, suggesting combinatorial control of this locus by multiple transcription factors [5]. Similarly, miRNAs that are specific to embryonic stem cells, such as miR-290~295, harbor super-enhancers encompassing numerous binding sites for Nanog, OCT4, SOX2 and TCF transcription factors [6]. These observations suggest that highly expressed and cell type-specific miRNAs are under strong transcriptional regulation.

Upon transcription, the recruitment of the microprocessor to primary miRNA transcripts is also highly regulated, at least for a subset of miRNAs, and such regulation can affect the abundance of mature miRNAs. Interestingly, super-enhancers can facilitate recruitment of the Drosha/Dgcr8 microprocessor to primary miRNA precursors for nuclear cleavage, providing an intimate link between miRNA transcription and processing [7]. Earlier studies suggested that nuclear factors such as the transcription factors p53 and SMAD and the tumor suppressor BRCA1 can act to enhance recruitment of the microprocessor complex to primary miRNA transcripts [7–10]. A recent study revealed that YAP1, another transcription factor, can inhibit recognition and processing of primary transcripts by sequestering p72/DDX10, a co-factor for the microprocessor [11]. Interestingly, Yap1 overexpression in the skin produces a hair follicle evagination phenotype, similar to that observed in epidermis conditionally deleted for Dicer1, Dgcr8 or Ago1/2. Although miRNA levels were not assessed in Yap1 transgenic mice, it is possible that Yap1 overexpression interferes with the miRNA biogenesis pathway. Furthermore, N^6-methyladenosine is reported to mark some primary miRNAs to enhance their recognition by Dgcr8 and subsequent processing by the microprocessor complex [12]. However, this effect appears to be subtle. It will be interesting to define the number of miRNAs that are subject to such regulation and to determine the extent to which levels of mature miRNAs can be altered independent of transcriptional control.

Although Xpo5 has been identified as a key factor mediating nuclear export of pre-miRNAs, it remains unclear whether this process is subject to regulation. Earlier studies suggested that Xpo5 is required for miRNA maturation [13, 14] and demonstrated that overexpression of Xpo5 can enhance miRNA functions, presumably through the enhanced miRNA transportation [15]. However, more recent investigations have shown that some less abundant miRNAs can completely bypass Xpo5-mediated nuclear export and instead utilize an Xpo1-dependent mechanism [16]. The functional significance of this alternate pathway has not yet been tested in animal models. Genetic deletion of key factors for miRNA biogenesis in the skin epidermis causes major defects in epidermal development [17–21]. Future studies deleting Xpo5 and Xpo1 in the skin and comparing the resulting phenotypes to those described in Dicer1, Dgcr8 and Ago conditional deletion mutants will yield important insights into the functional importance of Xpo5-dependent and -independent pathways.

When pre-miRNA hairpins reach the cytoplasm, Dicer1 further processes the hairpin and generates a dsRNA duplex, one strand of which is incorporated into the RISC by binding to its constituent Ago proteins [22, 23]. Dicer1 can also bind to numerous RNA species including mRNAs and lncRNAs, and such binding events seem to stabilize these transcripts [24].

The requirement of Ago proteins for miRNA accumulation has been demonstrated by genetic studies. Deletion of Ago1 and Ago2 leads to the loss of ~80% of total miRNAs in the epidermis, correlating with the higher abundance of Ago1 and Ago2 compared with other Ago proteins expressed in the skin [19]. Interestingly, when mature miRNAs are abolished in Dicer1 KO fibroblasts, Ago proteins are also significantly depleted [25]. Furthermore, absolute copy numbers of Ago proteins and miRNAs are comparable [19, 26]. These observations suggest that mature miRNAs and Ago proteins are required to stabilize each other and form the RISC. The loading preferences of individual Ago proteins were recently examined in the skin by Ago-IP followed by deep sequencing. Based on these results, it seems that miRNAs are randomly loaded onto available Ago proteins without substantial preference for the different isoforms. However, additional studies have identified a mechanism mediated by AUF1, an RNA binding protein that can facilitate loading of the let-7b miRNA onto Ago2 [27]. It is yet to be determined whether this mechanism also helps to load let-7b onto other Ago proteins. Another open question is whether different Ago proteins associate with distinct effectors such as GW proteins, also known as TNRC6a/b/c in mammals, as well as PAN2-PAN3 and the CCR4-NOT complex. Different Ago proteins are thought to be subject to distinct post-translational modifications, which may influence their ability to interact with specific downstream effectors. Recently, Ago2 was found to undergo a phosphorylation cycle regulating target engagement [28]. This exciting discovery suggests a dynamic model of miRNA-mediated regulation and provokes further studies of how and when miRNAs regulate their targets, perhaps in an Ago-specific manner. In addition, Ago-bound miRNAs are found to exist in low molecular weight complexes that are not associated with mRNAs in many adult tissues [29]. During T-cell activation, signaling pathways stimulate Ago-bound miRNAs to form high molecular weight complexes, which bind mRNAs and regulate their expression. This finding provides an intriguing example in which miRNA-mediated function is regulated by signaling pathways in a cell context-specific manner.

Finally, the mechanisms controlling miRNA decay are incompletely understood. In general, miRNAs seem to enjoy a relatively long half-life. However, some miRNAs, especially those involved in regulation of the cell cycle, are actively degraded [30]. Biochemical and genetic studies have shown that uridylation and, to a lesser extent, adenylation of mature miRNAs impacts their accumulation and/or function [31–33]. Furthermore, it is unclear how individual miRNAs are distinguished when associated with Ago proteins. We speculate that Ago modifications together with motifs contained within individual miRNAs could specifically mark miRNAs for modification and subsequent degradation.

When oncogenic stimuli are activated in quiescent HF-SCs, their global miRNA expression profile changes from a quiescent to an activated signature [34]. Similarly, when an oncogenic form of Hras ($Hras^{G12D}$) is introduced into keratinocytes, it leads to a profound shift in global miRNA expression, including significant alterations in the expression levels of several highly expressed miRNAs [35]. Collectively, these results suggest that global miRNA profiles can be dramatically shifted upon cellular transformation. To accomplish these changes, it is likely that many miRNAs and/or the RISC are subjected to selective degradation. We speculate that the miRNA degradation machinery is activated during these transitions to facilitate reprogramming of the RISC.

Skin is one of the few systems in which the functions of most of the key components of the miRNA biogenesis pathway, including Drosha, Dgcr8, Dicer1 and Argonaute, have been examined using loss of function genetic mouse models [17–21]. Epidermal-specific targeting of *Dicer1* or *Dgcr8* causes dramatic skin phenotypes [17, 20, 21]. Upon *Dicer1* or *Dgcr8* deletion the epithelial cells of the developing hair follicle become more susceptible to apoptosis and, strikingly, the dermal papillae abnormally migrate into the epidermis, eventually leading to the formation of hair germ-like cysts. Similar abnormalities in skin morphogenesis are seen following ablation of the essential RISC components Ago1 and Ago2 [19]. Pigment producing melanocytes are also highly dependent on miRNAs: *Dicer1* deletion in melanocytes induces their apoptosis, resulting in formation of unpigmented hair. Interestingly, the master regulator of melanocyte development, activity and survival, microphthalmia-associated transcription factor (MITF), controls Dicer1 transcription in melanocytes [36].

These early studies illuminated miRNAs as essential actors controlling the execution of gene expression programs in a cell lineage specific manner, and established a solid foundation for an explosion of research in the area of miRNAs in normal skin and hair follicle development, and their implications in a variety of skin disorders.

7.2 MicroRNA Functions in Epithelial Stem Cells

With well-defined lineages during development and in adult tissues, skin is an ideal model system to investigate miRNA functions in specification, maintenance and differentiation of tissue stem cells. Multiple populations of stem and progenitor cells exist to fuel tissue homeostasis in the skin. In the basal epidermis, unipotent, interfollicular progenitors maintain the epidermal lineage, while in the hair follicle, bulge stem cells contribute to all epithelial lineages of the hair follicle. During wound healing, bulge stem cells can also migrate towards the epidermis and contribute to the epidermal lineage [37]. Several additional stem cell populations have been identified in hair follicles, and other skin appendages [38] (Fig. 7.1).

The miRNA pathway is dispensable for interfollicular progenitors *in vivo*. In grafted Dicer1 cKO skin and Ago1/2 double cKO skin, the epidermal lineage remains relatively intact and only shows signs of mild hyperproliferation [19, 39].

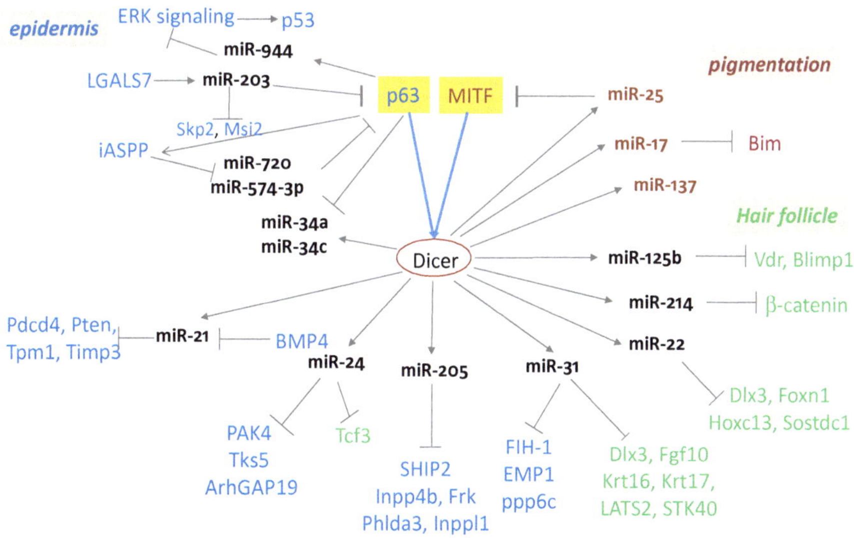

Fig. 7.1 Functions of miRNAs in the formation and maintenance of functional skin and its appendages. miRNA/mRNA interactions play key roles in the control of epidermal homeostasis, hair follicle cycling, and pigmentation. miRNA target genes involved in the control of epidermal homeostasis and the hair cycle are shown by blue and green colours, respectively. miRNAs regulating melanocyte activity and pigmentation are shown in brown

In sharp contrast, however, bulge HF-SCs are not maintained in Dicer1 and Dgcr8 cKO skin [17, 20, 21]. Although they are specified, judged by their initial expression of SOX9, NFATC1 and TCF3/4, all of which are key regulators of HF-SCs, these cells quickly lose expression of these transcription factors and the hair follicle degenerates [4]. Quiescent HF-SCs appear to be more resistant to the loss of Dgcr8 or Drosha, whereas activated HF-SCs perish rapidly [18]. These observations suggest that actively dividing HF-SCs are particularly sensitive to the loss of miRNAs. These findings suggest differential requirements for miRNAs in interfollicular progenitors and bulge HF-SCs, and should guide future efforts to identify individual miRNA functions in each lineage.

For the past several years, much effort has been dedicated to studying the functions of individual miRNAs in skin stem cells. Several miRNAs including miR-203, miR-205, miR-125b, miR-24, miR-21, miR-22, and miR-103/107 have been examined in detail in the skin in vivo or in cultured keratinocytes. All of these miRNAs are highly expressed in their cellular contexts. For example, miR-205 and miR-125b are the most and second most abundantly expressed miRNAs in neonatal HF-SCs [4, 40]; miR-203 is the most abundantly expressed miRNA in suprabasal epidermal cells; and miR-214 and miR-31 are highly expressed in the transient amplifying cells in the hair matrix [41, 42]. Among these miRNAs, the functions of miR-203 and miR-205 have been examined in vivo using genetic loss of function mouse models [4, 39].

While initial studies reported strong effects of miR-203 in repressing progenitor proliferation using gain-of-function approaches in cultured keratinocytes and in vivo [39, 43], genetic deletion of mR-203 causes only a mild increase in epidermal proliferation during embryonic development when the proliferation rate is relatively high [35]. This result is not surprising, as the interfollicular epidermis is relatively unaffected by deletion of Dicer1, which should abolish most miRNAs. Interestingly, however, expression of a mutant, oncogenic form of Hras (*Hras*^{G12D}) is found to suppress miR-203. Furthermore, identification of miR-203 target mRNAs using expression analyses and direct capture of RISC-associated mRNAs, yielded a number of genes involved in Ras signaling and cell proliferation. Although the mechanism of Hras^{G12D} mediated miR-203 repression is not clear, Hras^{G12D} and miR-203 appear to antagonize each other and this interaction plays an important role in Hras^{G12D} initiated tumorigenesis [35].

The importance of highly expressed miRNAs in HF-SCs has been demonstrated by studies on miR-205 and miR-125b using genetically engineered mouse models. In neonatal HF-SCs, miR-205 and miR-125b account for ~30% and ~15% of the entire miRNA pool, respectively. Genetic deletion of miR-205 compromises PI(3)K signaling and significantly reduces pAkt levels in both epidermal progenitors and hair follicle stem cells. Of note, the reduction in pAkt levels observed in miR-205 KO mice is similar to that in *Dicer1* cKO cells, identifying miR-205 as a major contributor to miRNA-mediated regulation of pAkt levels in skin stem cells. Consistent with this notion, hair follicle growth is similarly stunted in miR-205 KO and Dicer1 cKO mice. However, the phenotype of hair germ evagination, a hallmark of *Dicer1* cKO skin, is not observed in the miR-205 KO despite high levels of miR-205 expression in wild-type hair germs. This suggests that loss of other miRNAs may be responsible for this defect. Interestingly, a super-enhancer for miR-205 has recently been identified in HF-SCs, underscoring the extensive transcription regulation of miR-205 in these cells [5]. However, the precise role of miR-205 in adult HF-SCs remains to be elucidated.

The functions of miR-125b, the mammalian lin-4 homologue, have been examined in an inducible overexpression mouse model. Enhanced expression of miR-125b leads to hair loss and enlarged sebaceous gland formation [40]. Intriguingly, hair loss is not caused by depletion of HF-SCs but instead results from their adoption of sebaceous gland rather than hair follicle fates. These gain of function data suggest that miR-125b may function to balance the fate decision of HF-SCs between sebaceous gland and hair follicle; however, definitive loss of function genetic studies will be required to confirm this. It will also be interesting to determine the requirements for the related miRNA, for miR-125a, in maintenance and differentiation of HF-SCs using loss of function mouse models. In addition, current evidence suggests that miR-103/107 miRNAs may promote epithelial stem cell fate by regulating multiple targets that are involved in cell cycle control, signaling, cell adhesion and cell-cell communication, respectively [44].

Collectively, these studies have begun to unravel important functions for individual miRNAs in the maintenance and differentiation of skin stem cells *in vivo* and *in vitro* (Table 7.1). Several insights can be drawn from these studies. Firstly, highly expressed miRNAs tend to regulate a larger number of targets and/or enforce stronger regulation to individual target genes than less abundant miRNAs. In the future, it will be interesting to determine whether miRNAs that are expressed at low or intermediate levels can also regulate functionally important targets. Secondly, individual miRNAs frequently have mild effects, often producing a less than two-fold decrease in the abundance of any given target mRNA. For example, in our studies of miR-203 using both gain- and loss-of-function approaches, we found that the majority of miR-203 targets only changed mildly when miR-203 levels were modulated. Of note, overexpression appears to have a stronger effect on mRNA levels than loss of function [35]. Thirdly, although most miRNA-mediated repression (66–90%) can be explained by mRNA destabilization [45], individual target genes can be regulated through alternate mechanisms. For example, *Trp63* is a *bona fide* miR-203 target [35, 39, 43], but its mRNA levels barely change when miR-203 is either overexpressed or depleted [35, 39, 43]. We suggest that combinatorial approaches involving the detection of miRNA:mRNA physical interactions using HITS-CLIP or PAR-CLIP [46, 47] in parallel with quantitation of mRNA levels by RNA-seq or microarray will be important for identifying bona fide miRNA targets.

7.3 MicroRNAs in Epidermal Homeostasis

The p63 transcription factor is a major regulator of epidermal development and differentiation [48–50]. p63 knockout mice fail to form a stratified epidermis and lack expression of several epidermal-specific genes [48, 50]. In human epidermis, p63 is required for both keratinocyte proliferation and differentiation, as p63 loss leads to severe epidermal hypoproliferation and suppressed differentiation [49].

Table 7.1 Highly expressed miRNA in the skin

Neonatal HF-SCs	Reads	%	Neonatal epidermis	Reads	%
mmu-miR-205-5p	4,031,674	31.89	mmu-miR-205-5p	4,667,769	56.67
mmu-miR-125b-5p	1,249,019	9.88	mmu-miR-24-3p	324,667	3.94
mmu-miR-23b-3p	725,453	5.74	mmu-miR-23a-3p	311,354	3.78
mmu-miR-16-5p	609,242	4.82	mmu-miR-125b-5p	244,631	2.97
mmu-miR-24-3p	577,532	4.57	mmu-miR-23b-3p	219,017	2.66
mmu-miR-23a-3p	457,381	3.62	mmu-miR-125a-5p	174,940	2.12
mmu-miR-200b-3p	413,024	3.27	mmu-miR-92a-3p	155,639	1.89
mmu-let-7c-5p	365,932	2.89	mmu-miR-16-5p	154,776	1.88
mmu-miR-125a-5p	316,604	2.50	mmu-miR-200b-3p	132,136	1.60
mmu-miR-15b-5p	246,980	1.95	mmu-miR-26a-5p	124,997	1.52
mmu-miR-26a-5p	214,916	1.70	mmu-miR-15b-5p	122,425	1.49
mmu-miR-92a-3p	188,567	1.49	mmu-let-7c-5p	115,616	1.40
mmu-let-7b-5p	187,297	1.48	mmu-miR-365-3p	109,799	1.33
mmu-miR-10a-5p	178,175	1.41	mmu-miR-200c-3p	100,050	1.21
mmu-miR-30c-5p	165,916	1.31	mmu-miR-193b-3p	89,412	1.09
mmu-miR-27b-3p	164,910	1.30	mmu-let-7b-5p	80,145	0.97
mmu-miR-99a-5p	164,350	1.30	mmu-miR-30c-5p	70,153	0.85
mmu-let-7a-5p	162,480	1.29	mmu-let-7a-5p	70,105	0.85
mmu-miR-200c-3p	144,591	1.14	mmu-miR-484	52,723	0.64
mmu-miR-484	112,283	0.89	mmu-miR-30d-5p	46,317	0.56
mmu-miR-203-3p	107,218	0.85	mmu-miR-191-5p	44,268	0.54
mmu-let-7d-5p	91,780	0.73	mmu-miR-27b-3p	40,119	0.49
mmu-miR-191-5p	82,213	0.65	mmu-miR-10a-5p	37,719	0.46
mmu-miR-30d-5p	81,974	0.65	mmu-let-7d-5p	37,705	0.46
mmu-miR-27a-3p	71,561	0.57	mmu-miR-203-3p	37,240	0.45
mmu-let-7f-5p	70,739	0.56	mmu-miR-99a-5p	31,848	0.39
mmu-miR-365-3p	59,946	0.47	mmu-let-7f-5p	31,593	0.38
mmu-miR-342-3p	58,978	0.47	mmu-miR-25-3p	29,977	0.36
mmu-miR-25-3p	58,859	0.47	mmu-miR-27a-3p	29,926	0.36
mmu-miR-195a-5p	53,110	0.42	mmu-miR-181a-5p	21,770	0.26
mmu-miR-181a-5p	52,592	0.42	mmu-miR-200a-3p	17,613	0.21
mmu-miR-19b-3p	45,866	0.36	mmu-miR-19b-3p	16,686	0.20
mmu-miR-99b-5p	41,926	0.33	mmu-miR-429-3p	15,486	0.19
mmu-miR-378a-3p	38,430	0.30	mmu-miR-222-3p	14,841	0.18
mmu-miR-200a-3p	38,100	0.30	mmu-miR-652-3p	14,537	0.18
mmu-miR-30b-5p	37,974	0.30	mmu-miR-320-3p	14,298	0.17
mmu-miR-193b-3p	37,973	0.30	mmu-miR-744-5p	13,577	0.16
mmu-let-7e-5p	35,881	0.28	mmu-let-7d-3p	13,546	0.16
mmu-miR-15a-5p	33,207	0.26	mmu-miR-30b-5p	13,343	0.16
mmu-let-7g-5p	30,712	0.24	mmu-miR-182-5p	12,715	0.15
mmu-miR-429-3p	30,247	0.24	mmu-miR-22-3p	12,499	0.15
mmu-miR-182-5p	26,061	0.21	mmu-let-7i-5p	11,750	0.14

(continued)

Table 7.1 (continued)

Neonatal HF-SCs	Reads	%	Neonatal epidermis	Reads	%
mmu-miR-148a-3p	25,718	0.20	mmu-miR-99b-5p	11,627	0.14
mmu-miR-652-3p	23,893	0.19	mmu-let-7g-5p	11,099	0.13
mmu-miR-744-5p	23,699	0.19	mmu-let-7e-5p	11,008	0.13
mmu-miR-320-3p	21,925	0.17	mmu-miR-26b-5p	10,116	0.12
mmu-let-7i-5p	21,911	0.17	mmu-miR-425-5p	10,067	0.12
mmu-miR-130a-3p	18,658	0.15	mmu-miR-141-3p	9654	0.12
mmu-miR-17-5p	18,601	0.15	mmu-miR-342-3p	9321	0.11
mmu-miR-425-5p	18,245	0.14	mmu-miR-15a-5p	9125	0.11

Recent studies demonstrated that expression of p63 is post-transcriptionally regulated by several miRNAs; in turn, p63 controls the expression of specific miRNAs. As discussed above, miR-203 contributes to the control of epidermal differentiation, at least in part, by targeting p63. miR-203 expression is restricted to differentiating cells of the epidermis in both mouse and human skin [39, 43, 51], and its expression is induced in keratinocytes during calcium-induced differentiation in vitro [43]. Galectin-7 positively regulates miR-203 expression by binding to and stabilizing MAP kinase JNK1 [52]. Thus, miR-203 operates as a molecular switch between proliferative basal and terminally-differentiating suprabasal cells by suppressing their proliferative potential and inducing cell-cycle exit.

p63 expression is also controlled by miR-720 and miR-574-3p, which, like miR203, are expressed in suprabasal layers of the epidermis [53]. Expression of miR-720 and miR-574-3p is repressed by iASPP, an inhibitory member of the ASPP (apoptosis stimulating protein of p53) family that, in turn, is modulated by p63. Therefore, miR-720 and miR-574-3p participate in an auto-regulatory feedback loop between iASPP and p63 to control epidermal differentiation [53].

Notably, p63 directly regulates certain miRNAs. Intriguingly, the promoter and sequence for miR-944 are located in an intron of Tp63. The ΔNp63 transcription factor positively regulates miR-944 transcription by directly binding and stimulating the activity of its promoter. miR-944 targets genes in the ERK signaling pathway, resulting in upregulation of p53 and activation of early stages of epidermal differentiation, thus contributing to ΔNp63-mediated induction of epidermal differentiation [54].

By contrast, p63 represses the expression of a subset of miR-34 family members by binding to their promoters [55]. The miR-34 miRNAs are differentially expressed in the epidermis: miR-34a is predominantly expressed in suprabasal layers; miR-34c is mainly present in the basal layer; and miR-34b expression is not detectable. Inhibition of both miR-34a and miR-34b restores expression of cyclin D1 and Cdk4, and proliferative activity, in p63 deficient keratinocytes [55].

These findings indicate that cell cycle exit in basal epidermal keratinocytes is tightly orchestrated by an interplay between p63 and miRNAs: miR-203, miR-574-3p and miR-720 act as upstream regulators of p63, while miR-944, miR-34a and miR-34c are repressed by p63.

In addition to miRNA interactions with p63, several miRNAs have been discovered to regulate the proliferation to differentiation switch in epidermal keratinocytes in a p63-independent manner. For example, miR-24 expression is restricted to the suprabasal layers in both mouse and human epidermis [56]. Forced overexpression of miR-24 in the basal layer *in vitro* and *in vivo* reduces keratinocyte proliferation and induces the differentiation program. As a result, transgenic mice overexpressing miR-24 under a K5 promoter developed a thinner epidermis and exhibited premature expression of K10 in the basal keratinocytes, compared with controls. miR-24 is thought to operate in part by directly targeting the cytoskeletal modulators PAK4, Tks5, and ArhGAP19, thus altering actin cable formation and cell motility to promote precocious differentiation [56].

In addition to *in vivo* studies carried out using genetically engineered mouse models, *in vitro* studies with cultured human keratinocytes have also shed light on the versatile functions of miRNAs in stratified epithelial cells. For example, miR-31 directly represses factor-inhibiting hypoxia-inducible factor 1 (FIH-1) leading to activation of the Notch signaling pathway and keratinocyte differentiation [57].

Taken together, these studies indicate that miRNAs provide multi-leveled control of gene expression in epidermal keratinocytes tosustain epidermal proliferation, differentiation and stratification programs.

7.4 MicroRNA Functions in Hair Follicle Development and Cycling

Hair follicle morphogenesis and cycling are controlled by the coordinated activities of many signalling pathways, including Wnt, Hedgehog, EDAR, BMP, FGF, and Notch [58–60]. Wnt signaling acts as one of the most powerful regulators of skin development: it controls cell proliferation in both the epithelium (epidermis, hair follicle matrix) and mesenchyme, as well as regulating hair shaft differentiation and the activity of hair follicle dermal papilla cells [61–65].

miR-214 has been shown to impact multiple signaling pathways including Wnt, BMP, EDAR, and SHH as well as cell cycle-associated genes during hair follicle morphogenesis [41]. In particular, miR-214 directly targets β-catenin and compromises activation of the Wnt pathway. These findings have begun to link miRNA-mediated gene expression to major developmental signaling pathways. miR-214 and β-catenin exhibit reciprocal expression patterns in the epidermis, but are co-localized in undifferentiated epithelial cells of the hair peg and in hair matrix keratinocytes [41]. Transgenic mice over-expressing miR-214 display a skin phenotype similar to that of mice with conditional epithelial ablation of β-catenin [65–67]. Inducible overexpression of miR-214 promotes activation of an anti-proliferative program in epidermal and hair follicle keratinocytes, and leads to formation of fewer hair follicles with reduced hair bulb sizes, and production of thinner hair shafts [41]. In addition, miR-214 over-expression results in reduced numbers of hair follicle outer root sheath cells positive for the transcription factor SOX9, which is required for guiding stem cell progeny to the hair matrix [41, 68]. These phenotypes are also associated with decreased β-catenin expression, consistent with β-catenin's role as an upstream regulator of Sox9 expression in intestinal epithelium and neural crest cells [69, 70]. These studies suggest that, miR-214 may control activity of the Wnt signaling pathway and β-catenin expression in developing and postnatal skin and hair follicles, and thus contribute to the control of keratinocyte proliferation and stem cell activity. Future studies in which miR-214 is depleted in vivo will be necessary to confirm these functions. Interestingly, miR-214 functions are implicated in multiple cancer types [71–73]. Thus, further elucidation of miR-214 and other miRNA's regulatory capacity in normal skin and hair follicle development may also provide insight into their roles in pathological conditions.

The differentiation program in developing hair follicles may also be regulated at least in part by miR-24, which is predominantly expressed in differentiated keratinocytes of the inner root sheath and is absent in proliferating hair follicle cells [74]. When miR-24 is experimentally overexpressed in the outer root sheath, it causes reduced proliferation and premature differentiation in the hair follicle via altered expression of Tcf3, a transcription factor expressed in follicular stem cells and their progeny in the outer root sheath that acts to maintain an undifferentiated state [75, 76].

During postnatal development the hair follicle undergoes cyclic transformation, transiting through successive phases of rest (telogen), active growth (anagen), and regression (catagen). Hair follicle structural remodeling is driven by timely activation of stem cells, balanced proliferation and differentiation of their progeny, and coordinated apoptosis [77–79]. miRNA profiling studies identified significant variations in the expression levels of individual miRNAs during the hair follicle growth cycle. These studies identified candidate miRNAs that may be important in the regulation of proliferation and apoptosis [42].

The most striking hair cycle-associated variation was seen in the expression of miR-31. miR-31 is abundantly expressed in the proliferating hair matrix keratinocytes and the outer root sheath of anagen hair follicles, and is undetectable in telogen skin [42]. Pharmacological inhibition of miR-31 in mouse back skin promotes anagen and results in the appearance of hair shaft defects [42]. Constitutive deletion of miR-31 causes significantly increased hair follicle length and noticeable anagen prolongation [80]. By contrast, keratinocyte specific overexpression of miR-31 impairs keratinocyte functions resulting in changes in proliferation, apoptosis, and differentiation that result in loss of external hair [80]. miR-31 exerts its inhibitory effects on hair growth by modulating activity of the BMP and FGF signalling pathways [42], and by regulating Hippo and STK40 signaling [80].

In contrast to miR-31, miR-22 is involved in regulation of hair cycle termination. Expression of miR-22 predominates in the hair follicle outer and inner root sheaths, with only low levels cin the hair matrix, and levels of miR-22 expression are elevated during the hair follicle regression phase (catagen) [50]. Transgenic overexpression of miR-22 in mouse skin promotes the anagen-catagen transition and causes hair loss by repressing cell differentiation programs and expansion of keratinocyte progenitors, and by activating apoptosis. Conversely, deletion of miR-22 results in the formation of elongated anagen follicles and delayed entry into catagen, and causes acceleration of the transition from telogen to anagen. miR-22 executes these effects by targeting multiple transcriptional factors that are critical for hair follicle differentiation and hair shaft formation, including Dlx3, Foxn1, Hoxc13, and the Wnt/Bmp antagonist Sostdc1 [50]. While miR-22 induces apoptosis in follicular keratinocytes, a major feature of the anagen-catagen transition, the molecular mechanisms underlying this effect remain to be elucidated [50].

The transition of undifferentiated melanocyte progenitors to fully differentiated cells actively producing and transporting melanin to the hair shaft keratinocytes is closely coordinated with the hair follicle growth cycle [81, 82]. Microphthalmia-associated transcription factor (MITF) acts as a master regulator of melanocyte development, survival, and differentiation [83–85]. Interestingly, MITF directly regulates expression of Dicer, whose expression is upregulated during melanocyte differentiation [36]. Targeted Dicer deletion in melanocytes causes the formation of depigmented hair shafts due to melanocyte apoptosis that is mediated at least in part by miR-17. Consistent with this, forced expression of miR-17 partially rescues Dicer ablation-induced melanocyte apoptosis by targeting a critical pro-apoptotic gene Bim (BCL2L11) [36]. miR-137 has recently been identified as a possible determinant of hair pigmentation [86]: transgenic mice with the highest levels of miR-137 over-expression produce grey hair, whereas moderate elevation of miR-137 results in the appearance of a brown coat [86]. However, the mechanisms by which miR-137 contributes to hair pigmentation, and any effects of its removal on this process, remain unknown.

Collectively, these studies provide compelling evidence that individual miRNAs are essential regulators in the complex mechanisms that control hair cycle-associated changes in gene expression. By fine tuning the activity of a variety of key signalling pathways miRNAs appear to play crucial roles in maintaining a balance between proliferation, differentiation, and apoptosis during specific hair cycle stages. However, it still remains unclear whether individual miRNAs are involved in the pathogenesis of human hair follicle disorders, which include a variety of hair loss diseases as well as excessive hair growth.

7.5 miRNA Control of Wound Induced Skin Repair

Wound healing is a complex process involved in repairing skin integrity and functions, and can be divided into four overlapping phases of haemostasis, inflammation, proliferation and remodeling, each of which is essential for successful tissue repair [87–90]. These stages are characterized by coordinated intrinsic cellular responses in keratinocytes, fibroblasts, neutrophils, macrophages, and endothelial cells, that are regulated by a variety of growth factors, including platelet-derived growth factor, tumour necrosis factor-α, insulin-like growth factor-1, epidermal growth factor, TGF-α, BMPs, vascular endothelial growth factor and platelet factor-IV [87–90].

Skin injury initiates an immediate stress response in epidermal keratinocytes, which begin to proliferate and migrate towards the wound, forming a layer of hyperproliferative epithelium [88, 89, 91]. Hair follicles and sweat glands also contribute to skin repair by supplying stem cell progeny to the site of the wound [37, 92–94]. During the proliferative and remodeling phases, the dermis is re-established by fibroblast recruitment from nearby unwounded dermis, circulating fibrocytes, and bone marrow progenitor cells [95–98]. At early stages following injury, activated fibroblasts produce a variety of matrix components, such as fibronectin, hyaluronan, type III collagen, proteoglycans and vimentin, which contribute to dermal-mediated wound contraction [99–102]. At advanced wound healing stages, the composition of the extracellular matrix is altered: type III collagen is gradually replaced by type I collagen in a process controlled by a balance between matrix metalloproteinase (MMP) activity and the actions of MMP inhibitors, known as TIMPs [88].

Multiple factors and conditions can lead to impaired wound healing. Chronic non-healing wounds, such as arterial and venous ulcers and diabetic ulcers, represent a significant social and healthcare burden worldwide. Chronic wound healing disorders are a common problem among the elderly [103, 104], and acute wound healing is temporally delayed even in healthy older individuals. Age-related impairment of wound repair is associated with alterations in all major components of the healing process [104, 105], including changes in the activity of multiple growth factor pathways, such as PDGF, TGF-β, VEGF, stromal cell-derived factor 1 (SDF-1) and FGF signaling [103, 104, 106].

Several studies have demonstrated that expression of a large number of miRNAs fluctuates during acute and chronic wound healing in mouse and human skin, suggesting that miRNAs may function to alter the activity of a variety of signaling pathways during wound repair. Interestingly, miRNA biogenesis is activated in response to wounding, indicated by increased expression of Drosha, Dicer, Exportin 5 and Argounate 2 in mouse skin at the latest stages of the re-epithelialisation [107]. In line with this, perturbation of miRNA biogenesis by Dicer ablation in epidermal keratinocytes results in delayed wound closure, which is associated with compromised restoration of barrier function in the skin post-wounding [107]. Disrupted barrier function in these mice appears to be due to suppressed epidermal differentiation, as the epidermis is abnormally thin with decreased expression of major protein components of the cornified cell envelope, such as loricrin, involucrin, and filaggrin [107]. This phenotype is also associated with increased expression of cyclin-dependent kinase inhibitor 1 (p21Waf1/Cip1) [107], which is known to suppress terminal differentiation in keratinocytes [108].

Multiple individual miRNAs have been identified to play important roles at distinct phases of wound healing. In a very elegant study, Li et al. characterized dynamic changes in miRNA expression profiles during the inflammatory phase of human skin wound healing. These authors identified two TGF-beta inducible miRNAs, miR-132 and miR-31 [109, 110], whose expression is highly upregulated in keratinocytes at the wound edge during the inflammatory phase. These high expression levels are sustained during the subsequent proliferative phase. Deletion of miR-132 in mouse epidermis results in delayed wound healing due to impaired keratinocyte proliferation and severe inflammation. miR-132 exerts an anti-inflammatory effect via suppression of NF-κB signaling, by targeting multiple components of this pathway including EGF-like growth factor (HB-EGF), interleukin 8 (IL-8), chemokine (C-C motif) ligand 20 (CCL20), C-X-C motif chemokine 5 (CXCL5), IL-1A, IL-1B and TNF [110].

miR-31 is a well-established marker of activated keratinocytes: levels of miR-31 are elevated at the early stages of acute wound healing as well as under other hyperproliferative conditions, such as the anagen phase of the hair cycle, psoriasis, and cutaneous squamous cell carcinoma [42, 57, 109, 111–113]. Levels of mature miR-31 rise during the inflammatory phase, despite unchanged abundance of the pri-miR-31 primary transcript. However, during the proliferative phase, sustained miR-31 levels are achieved via an increase in miR-31 gene transcription. Increased miR-31 expression promotes keratinocyte proliferation and migration, which are essential components of the re-epithelialisation process [109]. One of miR-31's targets that could mediate its positive effects on wound closure is epithelial membrane protein 1 (EMP1) [109]. EPM1 and miR-31 exhibit mutually exclusive expression patterns during wound healing and, silencing of EMP-1 in keratinocytes causes similar effects to those produced by overexpression of miR-31 [109]. In other biological systems, EMP1 plays a key role in tight junction formation, and functions as a negative regulator of carcinogenesis [114–116].

In psoriasis, NF-κB signaling is activated by inflammatory cytokines, and induces miR-31 transcription. In turn, miR-31 targets protein phosphatase 6 (ppp6c), a negative regulator of proliferation that restricts G1 to S phase progression [113]. Conditional miR-31 deletion in epidermal keratinocytes leads not only to markedly reduced proliferation, but also to decreased inflammation in an imiquimod-induced psoriasis mouse model [113]. miR-31's functions in this regulatory circuit in psoriatic skin are consistent with previous observations regarding its stimulatory effects in wound healing, suggesting that miR-31 could facilitate skin repair not only by stimulating keratinocyte proliferation and migration, but possibly also by controlling wound healing-associated inflammation.

Several other miRNAs have been identified as actors in the dermal component of cutaneous wounds. For instance, miR-378a negatively impacts wound healing by suppressing fibroblast activity. Accelerated wound closure is observed in transgenic mice expressing an antimiR-378a sponge (miR-Pirate378a), and this is accompanied by increased alpha-smooth muscle actin expression in the transgenic wounds, suggesting increased differentiation of myofibroblasts [117]. Silencing of miR-378a in fibroblasts also leads to their increased migration and differentiation, associated with decreased expression of the miR-378a direct targets vimentin and β3 integrin. Interestingly, accelerated skin repair in miR-Pirate378a transgenic mice is additionally accompanied by activation of angiogenesis [117].

It has also been suggested that increased expression of miR-145 and miR-181b in human hypertrophic scar tissue contributes to aberrant skin repair [118, 119]. Expression of miR-145 is induced by the profibrotic protein TGF-beta 1, and regulates the differentiation and activity of myofibroblasts [118]. miR-145 induces contractile force generation and migration of myofibroblasts via elevated expression of α-1 type I collagen and TGF-β1 secretion [118]. By contrast, miR-181b inhibits myofibroblast differentiation and suppresses expression of the proteoglycan decorin [119], which is essential for collagen fibrillogenesis and successful wound healing [120, 121].

miR-29 family members are differentially regulated by TGF-beta signaling and are important modulators of extracellular matrix (ECM) components such as collagen 1 in various mouse and human organs including skin [122–124]. It has been suggested that miR29s may facilitate ECM matrix remodeling following skin injury, while their deregulation could possibly lead to scar formation [122]. miRagen has initiated a phase 1 clinical study to test the effects of a miR-29b mimic, MRG-201, on extracellular matrix remodelers and collagen synthesis in skin (http://miragen-therapeutics.com/clinical-trials/ accessed 6 December 2015). If successful, this initial study could potentially lead to the development of novel therapies for fibrotic conditions, including scar formation as a result of aberrant wound healing.

Dysregulated expression of miRNAs is thought to severely compromise skin repair. As mentioned above, several studies have suggested the involvement of miRNAs in chronic non-healing wounds, such as arterial, venous and diabetic ulcers. For instance, enhanced levels of miR-16, -20a, -21, -106a -130a, and -203 have been detected in venous ulcers [125], and forced expression of miR-130a and miR-21 is found to delay re-epithelialization in *ex vivo* acute human skin wound models. Similarly, intradermal delivery of a miR-21 mimic into rat skin prior to wounding causes a severe reduction in granulation tissue formation, increased infiltration of immune cells, and suppressed re-epithelialization. Interestingly, miR-130a and miR-21 share a common target gene, Leptin receptor [125]. Leptin signalling is known to be important for successful cutaneous repair by exerting pleiotropic effects [126–128]. Expression of miR-21 is negatively regulated by BMP4 signaling [129], which is known to suppress wound-induced skin repair [130, 131]. Together, these lines of evidence suggest that miR-21 negatively impacts wound healing and might be involved in the pathology of chronic wounds.

Chronic wounds are a severe and common complication of diabetes. miRNA profiling in skin tissue from normal and diabetic mice revealed 14 miRNAs that are differentially expressed in diabetic skin; of these, miR-146b and miR-21 show the most dramatic differences compared with controls. The expression pattern of these miRNAs is also altered during healing of diabetic wounds. Another subset of miRNAs (miR-20b, miR-10a, miR-10b, miR-96, miR-128, miR-452 and miR-541) exhibits similar basal levels in normal and diabetic skin, but is dysregulated during healing of diabetic wounds. Amongst the miRNAs studied, miR-21 showed a distinct signature with increased expression in diabetic skin but decreased expression during diabetic wound healing [132].

Interestingly, the involvement of miRNAs in wound healing is not confined to those expressed in the skin. For example, circulating fibrocytes, a subset of bone marrow-derived progenitor cells, accelerate wound healing via secretion of exosomes containing a variety of bioactive molecules including miRNAs. Specifically, fibrocyte-derived exosomes are enriched for miR-126, miR-124a, miR-125b, miR-130a, miR-132, and miR-21 [133]. Notably, expression of miR-21 and miR-130a is also elevated in the keratinocytes of venous ulcers [125]. Fibrocytic exosomes exert multiple effects; these include activation of dermal fibroblasts; stimulation of diabetic keratinocyte migration and proliferation *in vitro*; proangiogenic properties; and promotion of wound healing *in vivo* in genetically diabetic mice [133]. These findings suggest possible applications of fibrocyte-derived exosomes in the management of diabetic ulcers.

7.6 Summary

Numerous studies and extensive efforts in the last two decades have enhanced our knowledge and understanding of the complex mechanisms underlying post-transcriptional gene regulation mediated by miRNAs. Many discoveries have illuminated the significance of miRNAs in skin development and homeostasis and their roles in regulating a variety of skin pathological conditions. However, further investigation will be needed to uncover functionally important miRNA-mRNA interactions in skin and hair follicle biology. In turn, these efforts will advance our understanding of the biological roles of miRNAs in healthy organisms, but also allow for the development of novel therapeutic approaches targeting miRNAs for human diseases including skin pathologies.

References

1. Ha M, Kim VN. Regulation of microRNA biogenesis. Nat Rev Mol Cell Biol. 2014;15(8):509–24.
2. Chang TC, Pertea M, Lee S, Salzberg SL, Mendell JT. Genome-wide annotation of microRNA primary transcript structures reveals novel regulatory mechanisms. Genome Res. 2015;25(9):1401–9.
3. Landgraf P, Rusu M, Sheridan R, Sewer A, Iovino N, Aravin A, et al. A mammalian microRNA expression atlas based on small RNA library sequencing. Cell. 2007;129(7):1401–14.
4. Wang D, Zhang Z, O'Loughlin E, Wang L, Fan X, Lai EC, et al. MicroRNA-205 controls neonatal expansion of skin stem cells by modulating the PI(3)K pathway. Nat Cell Biol. 2013;15(10):1153–63.
5. Adam RC, Yang H, Rockowitz S, Larsen SB, Nikolova M, Oristian DS, et al. Pioneer factors govern super-enhancer dynamics in stem cell plasticity and lineage choice. Nature. 2015;521(7552):366–70.
6. Marson A, Levine SS, Cole MF, Frampton GM, Brambrink T, Johnstone S, et al. Connecting microRNA genes to the core transcriptional regulatory circuitry of embryonic stem cells. Cell. 2008;134(3):521–33.
7. Suzuki HI, Yamagata K, Sugimoto K, Iwamoto T, Kato S, Miyazono K. Modulation of microRNA processing by p53. Nature. 2009;460(7254):529–33.
8. Davis BN, Hilyard AC, Lagna G, Hata A. SMAD proteins control DROSHA-mediated microRNA maturation. Nature. 2008;454(7200):56–61.
9. Kawai S, Amano A. BRCA1 regulates microRNA biogenesis via the DROSHA microprocessor complex. J Cell Biol. 2012;197(2):201–8.

10. Krell J, Stebbing J, Frampton AE, Carissimi C, Harding V, De Giorgio A, et al. The role of TP53 in miRNA loading onto AGO2 and in remodelling the miRNA-mRNA interaction network. Lancet. 2015;385 Suppl 1:S15.
11. Mori M, Triboulet R, Mohseni M, Schlegelmilch K, Shrestha K, Camargo FD, et al. Hippo signaling regulates microprocessor and links cell-density-dependent miRNA biogenesis to cancer. Cell. 2014;156(5):893–906.
12. Alarcon CR, Lee H, Goodarzi H, Halberg N, Tavazoie SF. N6-methyladenosine marks primary microRNAs for processing. Nature. 2015;519(7544):482–5.
13. Lund E, Guttinger S, Calado A, Dahlberg JE, Kutay U. Nuclear export of microRNA precursors. Science. 2004;303(5654):95–8.
14. Yi R, Qin Y, Macara IG, Cullen BR. Exportin-5 mediates the nuclear export of pre-microRNAs and short hairpin RNAs. Genes Dev. 2003;17(24):3011–6.
15. Yi R, Doehle BP, Qin Y, Macara IG, Cullen BR. Overexpression of exportin 5 enhances RNA interference mediated by short hairpin RNAs and microRNAs. RNA. 2005;11(2):220–6.
16. Xie M, Li M, Vilborg A, Lee N, Shu MD, Yartseva V, et al. Mammalian 5′-capped microRNA precursors that generate a single microRNA. Cell. 2013;155(7):1568–80.
17. Andl T, Murchison EP, Liu F, Zhang Y, Yunta-Gonzalez M, Tobias JW, et al. The miRNA-processing enzyme dicer is essential for the morphogenesis and maintenance of hair follicles. Current Biol CB. 2006;16(10):1041–9.
18. Teta M, Choi YS, Okegbe T, Wong G, Tam OH, Chong MM, et al. Inducible deletion of epidermal Dicer and Drosha reveals multiple functions for miRNAs in postnatal skin. Development. 2012;139(8):1405–16.
19. Wang D, Zhang Z, O'Loughlin E, Lee T, Houel S, O'Carroll D, et al. Quantitative functions of Argonaute proteins in mammalian development. Genes Dev. 2012;26(7):693–704.
20. Yi R, O'Carroll D, Pasolli HA, Zhang Z, Dietrich FS, Tarakhovsky A, et al. Morphogenesis in skin is governed by discrete sets of differentially expressed microRNAs. Nat Genet. 2006;38(3):356–62.
21. Yi R, Pasolli HA, Landthaler M, Hafner M, Ojo T, Sheridan R, et al. DGCR8-dependent microRNA biogenesis is essential for skin development. Proc Natl Acad Sci U S A. 2009;106(2):498–502.
22. Czech B, Hannon GJ. Small RNA sorting: matchmaking for Argonautes. Nat Rev Genet. 2011;12(1):19–31.
23. Jonas S, Izaurralde E. Towards a molecular understanding of microRNA-mediated gene silencing. Nat Rev Genet. 2015;16(7):421–33.
24. Rybak-Wolf A, Jens M, Murakawa Y, Herzog M, Landthaler M, Rajewsky N. A variety of dicer substrates in human and C. elegans. Cell. 2014;159(5):1153–67.
25. Smibert P, Yang JS, Azzam G, Liu JL, Lai EC. Homeostatic control of Argonaute stability by microRNA availability. Nat Struct Mol Biol. 2013;20(7):789–95.
26. Calabrese JM, Seila AC, Yeo GW, Sharp PA. RNA sequence analysis defines Dicer's role in mouse embryonic stem cells. Proc Natl Acad Sci U S A. 2007;104(46):18097–102.
27. Yoon JH, Jo MH, White EJ, De S, Hafner M, Zucconi BE, et al. AUF1 promotes let-7b loading on Argonaute 2. Genes Dev. 2015;29(15):1599–604.
28. Golden RJ, Chen B, Li T, Braun J, Manjunath H, Chen X, et al. An Argonaute phosphorylation cycle promotes microRNA-mediated silencing. Nature. 2017;542(7640):197–202.
29. La Rocca G, Olejniczak SH, Gonzalez AJ, Briskin D, Vidigal JA, Spraggon L, et al. In vivo, Argonaute-bound microRNAs exist predominantly in a reservoir of low molecular weight complexes not associated with mRNA. Proc Natl Acad Sci U S A. 2015;112(3):767–72.
30. Rissland OS, Hong SJ, Bartel DP. MicroRNA destabilization enables dynamic regulation of the miR-16 family in response to cell-cycle changes. Mol Cell. 2011;43(6):993–1004.
31. Burroughs AM, Ando Y, de Hoon MJ, Tomaru Y, Nishibu T, Ukekawa R, et al. A comprehensive survey of 3′ animal miRNA modification events and a possible role for 3′ adenylation in modulating miRNA targeting effectiveness. Genome Res. 2010;20(10):1398–410.
32. Jones MR, Blahna MT, Kozlowski E, Matsuura KY, Ferrari JD, Morris SA, et al. Zcchc11 uridylates mature miRNAs to enhance neonatal IGF-1 expression, growth, and survival. PLoS Genet. 2012;8(11):e1003105.

33. Jones MR, Quinton LJ, Blahna MT, Neilson JR, Fu S, Ivanov AR, et al. Zcchc11-dependent uridylation of microRNA directs cytokine expression. Nat Cell Biol. 2009;11(9):1157–63.
34. White AC, Khuu JK, Dang CY, Hu J, Tran KV, Liu A, et al. Stem cell quiescence acts as a tumour suppressor in squamous tumours. Nat Cell Biol. 2014;16(1):99–107.
35. Riemondy K, Wang XJ, Torchia EC, Roop DR, Yi R. MicroRNA-203 represses selection and expansion of oncogenic Hras transformed tumor initiating cells. eLife 2015;4:e07004. https://doi.org/10.7554/eLife.07004
36. Levy C, Khaled M, Robinson KC, Veguilla RA, Chen PH, Yokoyama S, et al. Lineage-specific transcriptional regulation of DICER by MITF in melanocytes. Cell. 2010;141(6):994–1005.
37. Ito M, Liu Y, Yang Z, Nguyen J, Liang F, Morris RJ, et al. Stem cells in the hair follicle bulge contribute to wound repair but not to homeostasis of the epidermis. Nat Med. 2005;11(12):1351–4.
38. Beck B, Blanpain C. Mechanisms regulating epidermal stem cells. EMBO J. 2012;31(9):2067–75.
39. Yi R, Poy MN, Stoffel M, Fuchs E. A skin microRNA promotes differentiation by repressing 'stemness'. Nature. 2008;452(7184):225–9.
40. Zhang L, Stokes N, Polak L, Fuchs E. Specific microRNAs are preferentially expressed by skin stem cells to balance self-renewal and early lineage commitment. Cell Stem Cell. 2011;8(3):294–308.
41. Ahmed MI, Alam M, Emelianov VU, Poterlowicz K, Patel A, Sharov AA, et al. MicroRNA-214 controls skin and hair follicle development by modulating the activity of the Wnt pathway. J Cell Biol. 2014;207(4):549–67.
42. Mardaryev AN, Ahmed MI, Vlahov NV, Fessing MY, Gill JH, Sharov AA, et al. MicroRNA-31 controls hair cycle-associated changes in gene expression programs of the skin and hair follicle. FASEB J Off Publ Fed Am Soc Exp Biol. 2010;24(10):3869–81.
43. Lena AM, Shalom-Feuerstein R, Rivetti di Val Cervo P, Aberdam D, Knight RA, Melino G, et al. miR-203 represses 'stemness' by repressing DeltaNp63. Cell Death Differ. 2008;15(7):1187–95.
44. Peng H, Park JK, Katsnelson J, Kaplan N, Yang W, Getsios S, et al. microRNA-103/107 family regulates multiple epithelial stem cell characteristics. Stem Cells. 2015;33(5):1642–56.
45. Eichhorn SW, Guo H, McGeary SE, Rodriguez-Mias RA, Shin C, Baek D, et al. mRNA destabilization is the dominant effect of mammalian microRNAs by the time substantial repression ensues. Mol Cell. 2014;56(1):104–15.
46. Chi SW, Zang JB, Mele A, Darnell RB. Argonaute HITS-CLIP decodes microRNA-mRNA interaction maps. Nature. 2009;460(7254):479–86.
47. Hafner M, Landthaler M, Burger L, Khorshid M, Hausser J, Berninger P, et al. Transcriptome-wide identification of RNA-binding protein and microRNA target sites by PAR-CLIP. Cell 2010;141(1):129–41.
48. Mills AA, Zheng B, Wang XJ, Vogel H, Roop DR, Bradley A. p63 is a p53 homologue required for limb and epidermal morphogenesis. Nature. 1999;398(6729):708–13.
49. Truong AB, Kretz M, Ridky TW, Kimmel R, Khavari PA. p63 regulates proliferation and differentiation of developmentally mature keratinocytes. Genes Dev. 2006;20(22):3185–97.
50. Yuan S, Li F, Meng Q, Zhao Y, Chen L, Zhang H, et al. Post-transcriptional regulation of keratinocyte progenitor cell expansion, differentiation and hair follicle regression by miR-22. PLoS Genet. 2015;11(5):e1005253.
51. Jackson SJ, Zhang Z, Feng D, Flagg M, O'Loughlin E, Wang D, et al. Rapid and widespread suppression of self-renewal by microRNA-203 during epidermal differentiation. Development. 2013;140(9):1882–91.
52. Chen HL, Chiang PC, Lo CH, Lo YH, Hsu DK, Chen HY, et al. Galectin-7 regulates keratinocyte proliferation and differentiation through JNK-miR-203-p63 signaling. J Invest Dermatol. 2016;136(1):182–91.
53. Chikh A, Matin RN, Senatore V, Hufbauer M, Lavery D, Raimondi C, et al. iASPP/p63 autoregulatory feedback loop is required for the homeostasis of stratified epithelia. EMBO J. 2011;30(20):4261–73.

54. Kim KH, Cho EG, Yu SJ, Kang H, Kim YJ, Kim SH, et al. DeltaNp63 intronic miR-944 is implicated in the DeltaNp63-mediated induction of epidermal differentiation. Nucleic Acids Res. 2015;43(15):7462–79.
55. Antonini D, Russo MT, De Rosa L, Gorrese M, Del Vecchio L, Missero C. Transcriptional repression of miR-34 family contributes to p63-mediated cell cycle progression in epidermal cells. J Invest Dermatol. 2010;130(5):1249–57.
56. Amelio I, Lena AM, Viticchie G, Shalom-Feuerstein R, Terrinoni A, Dinsdale D, et al. miR-24 triggers epidermal differentiation by controlling actin adhesion and cell migration. J Cell Biol. 2012;199(2):347–63.
57. Peng H, Kaplan N, Hamanaka RB, Katsnelson J, Blatt H. Yang W, et al. microRNA-31/factor-inhibiting hypoxia-inducible factor 1 nexus regulates keratinocyte differentiation. Proc Natl Acad Sci U S A. 2012;109(35):14030–4.
58. Blanpain C, Fuchs E. Epidermal homeostasis: a balancing act of stem cells in the skin. Nature reviews Molecular cell biology. 2009;10(3):207–17.
59. Millar SE. Molecular mechanisms regulating hair follicle development. The Journal of investigative dermatology. 2002;118(2):216–25.
60. Schmidt-Ullrich R, Paus R. Molecular principles of hair follicle induction and morphogenesis. Bioessays. 2005;27(3):247–61.
61. Andl T, Reddy ST, Gaddapara T, Millar SE. WNT signals are required for the initiation of hair follicle development. Dev Cell. 2002;2(5):643–53.
62. Choi YS, Zhang Y, Xu M, Yang Y, Ito M, Peng T, et al. Distinct functions for Wnt/beta-catenin in hair follicle stem cell proliferation and survival and interfollicular epidermal homeostasis. Cell Stem Cell. 2013;13(6):720–33.
63. Fu J, Hsu W. Epidermal Wnt controls hair follicle induction by orchestrating dynamic signaling crosstalk between the epidermis and dermis. J Invest Dermatol. 2013;133(4):890–8.
64. Sick S, Reinker S, Timmer J, Schlake T. WNT and DKK determine hair follicle spacing through a reaction-diffusion mechanism. Science. 2006;314(5804):1447–50.
65. Tsai SY, Sennett R, Rezza A, Clavel C, Grisanti L, Zemla R, et al. Wnt/beta-catenin signaling in dermal condensates is required for hair follicle formation. Dev Biol. 2014;385(2):179–88.
66. Enshell-Seijffers D, Lindon C, Kashiwagi M, Morgan BA. Beta-catenin activity in the dermal papilla regulates morphogenesis and regeneration of hair. Dev Cell. 2010;18(4):633–42.
67. Huelsken J, Vogel R, Erdmann B, Cotsarelis G, Birchmeier W. beta-Catenin controls hair follicle morphogenesis and stem cell differentiation in the skin. Cell. 2001;105(4):533–45.
68. Vidal VP, Chaboissier MC, Lutzkendorf S, Cotsarelis G, Mill P, Hui CC, et al. Sox9 is essential for outer root sheath differentiation and the formation of the hair stem cell compartment. Curr Biol CB. 2005;15(15):1340–51.
69. Blache P, van de Wetering M, Duluc I, Domon C, Berta P, Freund JN, et al. SOX9 is an intestine crypt transcription factor, is regulated by the Wnt pathway, and represses the CDX2 and MUC2 genes. J Cell Biol. 2004;166(1):37–47.
70. Liu JA, Wu MH, Yan CH, Chau BK, So H, Ng A, et al. Phosphorylation of Sox9 is required for neural crest delamination and is regulated downstream of BMP and canonical Wnt signaling. Proc Natl Acad Sci U S A. 2013;110(8):2882–7.
71. Mitra AK, Zillhardt M, Hua Y, Tiwari P, Murmann AE, Peter ME, et al. MicroRNAs reprogram normal fibroblasts into cancer-associated fibroblasts in ovarian cancer. Cancer Discov. 2012;2(12):1100–8.
72. Penna E, Orso F, Cimino D, Tenaglia E, Lembo A, Quaglino E, et al. microRNA-214 contributes to melanoma tumour progression through suppression of TFAP2C. EMBO J. 2011;30(10):1990–2007.
73. Penna E, Orso F, Taverna D. miR-214 as a key hub that controls cancer networks: small player, multiple functions. J Invest Dermatol. 2015;135(4):960–9.
74. Amelio I, Lena AM, Bonanno E, Melino G, Candi E. miR-24 affects hair follicle morphogenesis targeting Tcf-3. Cell Death Dis. 2013;4:e922.

75. DasGupta R, Fuchs E. Multiple roles for activated LEF/TCF transcription complexes during hair follicle development and differentiation. Development. 1999;126(20):4557–68.
76. Merrill BJ, Gat U, DasGupta R, Fuchs E. Tcf3 and Lef1 regulate lineage differentiation of multipotent stem cells in skin. Genes Dev. 2001;15(13):1688–705.
77. Botchkarev VA, Kishimoto J. Molecular control of epithelial-mesenchymal interactions during hair follicle cycling. J Invest Dermatol Symp Proc Soc Invest Dermatol Inc [and] Eur Soc Dermatol Res. 2003;8(1):46–55.
78. Botchkareva NV, Ahluwalia G, Shander D. Apoptosis in the hair follicle. J Invest Dermatol. 2006;126(2):258–64.
79. Lee J, Tumbar T. Hairy tale of signaling in hair follicle development and cycling. Semin Cell Dev Biol. 2012;23(8):906–16.
80. Luan L, Shi J, Yu Z, Andl T. The major miR-31 target genes STK40 and LATS2 and their implications in the regulation of keratinocyte growth and hair differentiation. Exp Dermatol. 2017;26(6):497–504.
81. Botchkareva NV, Botchkarev VA, Gilchrest BA. Fate of melanocytes during development of the hair follicle pigmentary unit. J Invest Dermatol Symp Proc Soc Invest Dermatol Inc [and] Eur Soc Dermatol Res. 2003;8(1):76–9.
82. Botchkareva NV, Khlgatian M, Longley BJ, Botchkarev VA, Gilchrest BA. SCF/c-kit signaling is required for cyclic regeneration of the hair pigmentation unit. FASEB J Off Publ Fed Am Soc Exp Biol. 2001;15(3):645–58.
83. Hemesath TJ, Steingrimsson E, McGill G, Hansen MJ, Vaught J, Hodgkinson CA, et al. Microphthalmia, a critical factor in melanocyte development, defines a discrete transcription factor family. Genes Dev. 1994;8(22):2770–80.
84. Levy C, Khaled M, Fisher DE. MITF: master regulator of melanocyte development and melanoma oncogene. Trends Mol Med. 2006;12(9):406–14.
85. Yavuzer U, Keenan E, Lowings P, Vachtenheim J, Currie G, Goding CR. The Microphthalmia gene product interacts with the retinoblastoma protein in vitro and is a target for deregulation of melanocyte-specific transcription. Oncogene. 1995;10(1):123–34.
86. Dong C, Wang H, Xue L, Dong Y, Yang L, Fan R, et al. Coat color determination by miR-137 mediated down-regulation of microphthalmia-associated transcription factor in a mouse model. RNA. 2012;18(9):1679–86.
87. Barrientos S, Stojadinovic O, Golinko MS, Brem H, Tomic-Canic M. Growth factors and cytokines in wound healing. Wound Repair Reg Off Publ Wound Healing Soc [and] European Tissue Repair Soc. 2008;16(5):585–601.
88. Li J, Chen J, Kirsner R. Pathophysiology of acute wound healing. Clin Dermatol. 2007;25(1):9–18.
89. Schafer M, Werner S. Transcriptional control of wound repair. Annu Rev Cell Dev Biol. 2007;23:69–92.
90. Shaw TJ, Martin P. Wound repair at a glance. J Cell Sci. 2009;122(Pt 18):3209–13.
91. Taylor G, Lehrer MS, Jensen PJ, Sun TT, Lavker RM. Involvement of follicular stem cells in forming not only the follicle but also the epidermis. Cell. 2000;102(4):451–61.
92. Cotsarelis G. Epithelial stem cells: a folliculocentric view. J Invest Dermatol. 2006;126(7):1459–68.
93. Langton AK, Herrick SE, Headon DJ. An extended epidermal response heals cutaneous wounds in the absence of a hair follicle stem cell contribution. J Invest Dermatol. 2008;128(5):1311–8.
94. Levy V, Lindon C, Zheng Y, Harfe BD, Morgan BA. Epidermal stem cells arise from the hair follicle after wounding. FASEB J Off Publ Fed Am Soc Exp Biol. 2007;21(7):1358–66.
95. Abe R, Donnelly SC, Peng T, Bucala R, Metz CN. Peripheral blood fibrocytes: differentiation pathway and migration to wound sites. J Immunol. 2001;166(12):7556–62.
96. Fathke C, Wilson L, Hutter J, Kapoor V, Smith A, Hocking A, et al. Contribution of bone marrow-derived cells to skin: collagen deposition and wound repair. Stem Cells. 2004;22(5):812–22.
97. Hinz B. Formation and function of the myofibroblast during tissue repair. J Invest Dermatol. 2007;127(3):526–37.

98. Wu Y, Zhao RC, Tredget EE. Concise review: bone marrow-derived stem/progenitor cells in cutaneous repair and regeneration. Stem Cells. 2010;28(5):905–15.
99. Bitterman PB, Rennard SI, Adelberg S, Crystal RG. Role of fibronectin as a growth factor for fibroblasts. J Cell Biol. 1983;97(6):1925–32.
100. Eckes B, Colucci-Guyon E, Smola H, Nodder S, Babinet C, Krieg T, et al. Impaired wound healing in embryonic and adult mice lacking vimentin. J Cell Sci. 2000;113(Pt 13):2455–62.
101. Min LJ, Cui TX, Yahata Y, Yamasaki K, Shiuchi T, Liu HW, et al. Regulation of collagen synthesis in mouse skin fibroblasts by distinct angiotensin II receptor subtypes. Endocrinology. 2004;145(1):253–60.
102. Tiedemann K, Malmstrom A, Westergren-Thorsson G. Cytokine regulation of proteoglycan production in fibroblasts: separate and synergistic effects. Matrix Biol J Int Soc Matrix Biol. 1997;15(7):469–78.
103. Jaul E. Non-healing wounds: the geriatric approach. Arch Gerontol Geriatr. 2009;49(2):224–6.
104. Sgonc R, Gruber J. Age-related aspects of cutaneous wound healing: a mini-review. Gerontology. 2013;59(2):159–64.
105. Bentov I, Damodarasamy M, Plymate S, Reed MJ. Decreased proliferative capacity of aged dermal fibroblasts in a three dimensional matrix is associated with reduced IGF1R expression and activation. Biogerontology. 2014;15(4):329–37.
106. Gosain A, DiPietro LA. Aging and wound healing. World J Surg. 2004;28(3):321–6.
107. Ghatak S, Chan YC, Khanna S, Banerjee J, Weist J, Roy S, et al. Barrier function of the repaired skin is disrupted following arrest of dicer in keratinocytes. Mol Ther. 2015;23(7):1201–10.
108. Devgan V, Nguyen BC, Oh H, Dotto GP. p21WAF1/Cip1 suppresses keratinocyte differentiation independently of the cell cycle through transcriptional up-regulation of the IGF-I gene. J Biol Chem. 2006;281(41):30463–70.
109. Li D, Li X, Wang A, Meisgen F, Pivarcsi A, Sonkoly E, et al. MicroRNA-31 promotes skin wound healing by enhancing keratinocyte proliferation and migration. J Invest Dermatol. 2015;135(6):1676–85.
110. Li D, Wang A, Liu X, Meisgen F, Grunler J, Botusan IR, et al. MicroRNA-132 enhances transition from inflammation to proliferation during wound healing. J Clin Invest. 2015;125(8):3008–26.
111. Joyce CE, Zhou X, Xia J, Ryan C, Thrash B, Menter A, et al. Deep sequencing of small RNAs from human skin reveals major alterations in the psoriasis miRNAome. Hum Mol Genet. 2011;20(20):4025–40.
112. Wang A, Landen NX, Meisgen F, Lohcharoenkal W, Stahle M, Sonkoly E, et al. MicroRNA-31 is overexpressed in cutaneous squamous cell carcinoma and regulates cell motility and colony formation ability of tumor cells. PLoS One. 2014;9(7):e103206.
113. Yan S, Xu Z, Lou F, Zhang L, Ke F, Bai J, et al. NF-kappaB-induced microRNA-31 promotes epidermal hyperplasia by repressing protein phosphatase 6 in psoriasis. Nat Commun. 2015;6:7652.
114. Durgan J, Tao G, Walters MS, Florey O, Schmidt A, Arbelaez V, et al. SOS1 and Ras regulate epithelial tight junction formation in the human airway through EMP1. EMBO Rep. 2015;16(1):87–96.
115. Sun GG, Wang YD, Cui DW, Cheng YJ, Hu WN. Epithelial membrane protein 1 negatively regulates cell growth and metastasis in colorectal carcinoma. World J Gastroenterol. 2014;20(14):4001–10.
116. Sun GG, Wang YD, Lu YF, Hu WN. EMP1, a member of a new family of antiproliferative genes in breast carcinoma. Tumour Biol J Int Soc Oncodevelopmental Biol Med. 2014;35(4):3347–54.
117. Li H, Chang L, Du WW, Gupta S, Khorshidi A, Sefton M, et al. Anti-microRNA-378a enhances wound healing process by upregulating integrin beta-3 and vimentin. Mol Ther. 2014;22(10):1839–50.
118. Gras C, Ratuszny D, Hadamitzky C, Zhang H, Blasczyk R, Figueiredo C. miR-145 contributes to hypertrophic scarring of the skin by inducing myofibroblast activity. Mol Med. 2015;21:296–304.

119. Kwan P, Ding J, Tredget EE. MicroRNA 181b regulates decorin production by dermal fibroblasts and may be a potential therapy for hypertrophic scar. PLoS One. 2015;10(4):e0123054.
120. Danielson KG, Baribault H, Holmes DF, Graham H, Kadler KE, Iozzo RV. Targeted disruption of decorin leads to abnormal collagen fibril morphology and skin fragility. J Cell Biol. 1997;136(3):729–43.
121. Jarvelainen H, Puolakkainen P, Pakkanen S, Brown EL, Hook M, Iozzo RV, et al. A role for decorin in cutaneous wound healing and angiogenesis. Wound Repair Reg Off Publ Wound Healing Soc [and] European Tissue Repair Soc. 2006;14(4):443–52.
122. Cheng J, Wang Y, Wang D, Wu Y. Identification of collagen 1 as a post-transcriptional target of miR-29b in skin fibroblasts: therapeutic implication for scar reduction. Am J Med Sci. 2013;346(2):98–103.
123. Ciechomska M, O'Reilly S, Suwara M, Bogunia-Kubik K, van Laar JM. MiR-29a reduces TIMP-1 production by dermal fibroblasts via targeting TGF-beta activated kinase 1 binding protein 1, implications for systemic sclerosis. PLoS One. 2014;9(12):e115596.
124. Maurer B, Stanczyk J, Jungel A, Akhmetshina A, Trenkmann M, Brock M, et al. MicroRNA-29, a key regulator of collagen expression in systemic sclerosis. Arthritis Rheum. 2010;62(6):1733–43.
125. Pastar I, Khan AA, Stojadinovic O, Lebrun EA, Medina MC, Brem H, et al. Induction of specific microRNAs inhibits cutaneous wound healing. J Biol Chem. 2012;287(35):29324–35.
126. Frank S, Stallmeyer B, Kampfer H, Kolb N, Pfeilschifter J. Leptin enhances wound re-epithelialization and constitutes a direct function of leptin in skin repair. J Clin Invest. 2000;106(4):501–9.
127. Murad A, Nath AK, Cha ST, Demir E, Flores-Riveros J, Sierra-Honigmann MR. Leptin is an autocrine/paracrine regulator of wound healing. FASEB J Off Publ Fed Am Soc Exp Biol. 2003;17(13):1895–7.
128. Tadokoro S, Ide S, Tokuyama R, Umeki H, Tatehara S, Kataoka S, et al. Leptin promotes wound healing in the skin. PLoS One. 2015;10(3):e0121242.
129. Ahmed MI, Mardaryev AN, Lewis CJ, Sharov AA, Botchkareva NV. MicroRNA-21 is an important downstream component of BMP signalling in epidermal keratinocytes. J Cell Sci. 2011;124(Pt 20):3399–404.
130. Lewis CJ, Mardaryev AN, Poterlowicz K, Sharova TY, Aziz A, Sharpe DT, et al. Bone morphogenetic protein signaling suppresses wound-induced skin repair by inhibiting keratinocyte proliferation and migration. J Invest Dermatol. 2014;134(3):827–37.
131. Lewis CJ, Mardaryev AN, Sharpe DT, Botchkareva NV. Inhibition of bone morphogenetic protein signalling promotes wound healing in a human ex vivo model. Eur J Plast Surg. 2015;38(1):1–12.
133. Madhyastha R, Madhyastha H, Nakajima Y, Omura S, Maruyama M. MicroRNA signature in diabetic wound healing: promotive role of miR-21 in fibroblast migration. Int Wound J. 2012;9(4):355–61.
133. Geiger A, Walker A, Nissen E. Human fibrocyte-derived exosomes accelerate wound healing in genetically diabetic mice. Biochem Biophys Res Commun. 2015;467(2):303–9.

Chapter 8
Long Noncoding RNA and Its Role in the Control of Gene Expression in the Skin

Kevin C. Wang and Howard Y. Chang

8.1 Introduction

Epithelial tissues in animals function as a life-sustaining interface between the host and the environment. Together, the epithelial layers covering the skin, the gastrointestinal tract, the lungs, and the urogenital tract make up a surface of hundreds of square meters that constitutes the boundary between the body and the environment. One of the best-recognized functions of epithelia is to provide structural barriers that physically separate host cells from the outside world. These physical barriers prevent loss of essential body fluids, support the exchange of gases and nutrients, and protect the body from environmental, mechanical, chemical, and microbial insults.

Epithelial layers are formed by specialized sheets of cells that undergo tightly regulated proliferation and differentiation programs to ensure the renewal of epithelial tissues without compromising their functional properties [1]. These cells possess the remarkable capacity to both maintain and restore basal homeostasis as well as giving rise to differentiating cells that form mature tissues. Recent findings indicate that in addition to the usual protein players, nucleic acids such as RNA have essential roles in orchestrating the formation of epithelial tissues, often at the interface between stemness and differentiation.

In this Chapter, we discuss our current understanding of how members of a specific class of RNAs, long noncoding RNAs (lncRNAs), function in epithelial

K. C. Wang
Program in Epithelial Biology, Stanford University School of Medicine, Stanford, CA, USA

Veterans Affairs Palo Alto Healthcare System, Palo Alto, CA, USA

H. Y. Chang (✉)
Program in Epithelial Biology, Stanford University School of Medicine, Stanford, CA, USA
e-mail: howchang@stanford.edu

© Springer International Publishing AG, part of Springer Nature 2018

V. A. Botchkarev, S. E. Millar (eds.), *Epigenetic Regulation of Skin Development and Regeneration*, Stem Cell Biology and Regenerative Medicine,
https://doi.org/10.1007/978-3-319-16769-5_8

development and regulate adult stem-cell maintenance in stratified epithelial tissues such as the epidermis. During the last few years, work from multiple groups has shown that epigenetic mechanisms are involved in the control of epidermal development, keratinocyte differentiation, and melanocyte functions [2]. Furthermore, a number of recent studies have defined the expression of lncRNAs in normal and diseased skin cells and tissue, and provided plausible molecular mechanisms to explain how lncRNAs participate in maintaining homeostasis in the skin, and how their roles are altered in its transition to a diseased state.

8.2 A Functional Link Between RNA and Chromatin

The recent revolution in next generation sequencing technologies has made it possible to survey the transcriptomes of organisms to an unprecedented degree and has dramatically altered our understanding of the abundance and relevance or noncoding RNAs. Results from numerous studies utilizing these technologies have provided convincing evidence that the genomes of many organisms, including mammals, produce numerous long intergenic transcripts that have no significant protein-coding potential, collectively referred to as long noncoding RNAs (lncRNAs, defined as transcribed RNA molecules greater than 200 nucleotides in length [3–8]. LncRNAs resemble mRNAs biochemically: they are transcribed by RNA polymerase II, but do not function as templates for protein synthesis [9]. Long thought to be mostly "transcriptional noise," multiple lines of evidence from both *in vitro* experiments and animal models have demonstrated that lncRNAs affect many biological processes including regulation of gene expression, genomic imprinting, dosage compensation, and nuclear organization and compartmentalization [10–13]. In contrast to small noncoding RNAs such as siRNAs, miRNAs, and piRNAs, which are highly conserved and are involved in transcriptional and posttranscriptional gene silencing through specific base pairing with their targets, lncRNAs are poorly conserved and regulate gene expression by diverse mechanisms that are not yet fully understood [14–19].

Indeed, it now appears that lncRNAs function as RNA genes to regulate a myriad of biological processes [13, 20]. Furthermore, a number of studies have shown many lncRNAs to be dysregulated in various human diseases and disorders [21, 22]. While their precise mechanisms of action remain to be determined at the molecular level, lncRNAs have emerged as key regulators of the epigenome, influencing transcriptional networks and multicellular development through changes in epigenetic modifications such as histone methylation, acetylation, and DNA methylation [13, 20]. Interestingly, this layer of lncRNA regulation may couple chromatin, the mass of genetic material contained within DNA-protein complexes, to the factors that control its structure and function.

This intricate relationship between RNA and chromatin is not a new one—in fact, RNA had first been hypothesized to be an important influence on chromatin structure and gene regulation over 35 years ago [23]. More recently, an increasingly compelling body of work suggests that RNA plays significant roles in the maintenance of chromatin function, from providing an integral component of chromatin structure [24–26], to coating entire chromosomes while conferring epigenetic marks [27], and targeting proteins to the heterochromatin [28].

8.3 Role of Chromatin Modifiers in Development

To ensure proper development, organisms need to have in place regulatory mechanisms that allow for rapid changes in transcriptional modules to initiate, control, and maintain major developmental programs. Cells in differentiated states may, in theory, be stabilized by the establishment of epigenetic changes, such as DNA and/or histone modifications, that do not require continuous instructive signaling for their maintenance.

Several chromatin-modifying complexes have been shown to regulate renewal and differentiation of a range of stem cell types, including embryonic stem (ES) cells and hematopoietic stem cells [29–31]. The polycomb (PcG) group of proteins [32] is of particular interest in this context. PcG genes were initially characterized in *Drosophila* as repressors of homeotic Hox gene expression, and function to maintain but not initiate the repressive state of chromatin. The products of PcG genes form evolutionarily conserved multiprotein complexes referred to as Polycomb-repressive complexes (PRCs), which covalently modify histone tails and repress transcription of their target genes [33]. *In vivo* loss of PcG function in the developing skin alters epithelial stem cell proliferation and results in acceleration of the timing of skin development [34]. Studies in ES cells have found that loss of function of PcG components compromises pluripotency and the ability of ES cells to generate differentiated progeny [35, 36].

8.3.1 lncRNAs with Important Biological Functions in Skin

Epithelial tissues such as the epidermis demonstrate remarkable diversity in their structure and function, depending on their anatomic location. For example, palmoplantar skin possesses a thickened epidermal barrier that provides resistance to mechanical stress, while long terminally differentiated hairs populate the epidermis covering the scalp. Due to the constant turnover of the epidermis, the observed stability of these site-specific features raises the fundamental question of how positional identity is acquired and maintained in the skin.

Classic transplantation experiments in chick, along with more recent genomic approaches [37–39] have identified dermal fibroblasts as major players in the determination of specific epithelial cell fates. Moreover, primary adult fibroblasts retain many features of the embryonic pattern of expression of *HOX* genes [37, 39], a family of transcription factors containing DNA-binding homeodomains that act to specify positional identity during development. *HOX* genes are critical for determining body segmentation patterns and organ formation [40], are highly conserved through evolution, and are often clustered in the genome [41]. Their arrangement along the chromosome reflects their order of activation during development, as well as their positional expression along the anterior-posterior axis, a fascinating feature known as colinearity [41]. Remarkably, the patterns of *Hox* gene expression in fibroblasts from different anatomical locations faithfully recapitulate the principles of colinearity established during embryogenesis [39, 42]. In addition, transcriptional analysis has revealed that the human HOX cluster is punctuated by numerous spatiotemporally regulated lncRNAs [43], some of which overlap with the *Hox* genes themselves and are colinear with the overall anatomic expression pattern of the HOX loci. For example, HOTAIR (Hox antisense intergenic RNA), a lncRNA embedded within the HOXC locus, is expressed in cells with distal and posterior positional identities [43], while FRIGIDAIR [5], another HOXC lncRNA, has an anterior pattern of expression. In contrast, HOTTIP (*HOXA* transcript at the distal tip [44], a lncRNA found at the distal end of the human HOXA cluster, is also expressed in distal cells. The implication is that these lncRNAs and the *HOX* genes probably utilize the same enhancers, and hints at possible roles in co-regulation.

This phenomenon of co-localization is reminiscent of the existence of lncRNAs with important developmental functions near the *SOX2* loci during ES cell development [45]. In fact, approximately one third of the known conserved lncRNAs are expressed in ES cells [46, 47], and as is the case for protein-coding genes, the transcriptional landscape of long noncoding transcripts dramatically changes when ES cells differentiate, suggesting a more direct role for lncRNAs in differentiation [47]. A role of lncRNAs in differentiation and development has also been inferred from the numerous lncRNAs identified in the developing brain and differentiating T lymphocytes [48, 49].

8.3.1.1 HOTAIR

HOTAIR is expressed from the HOXC cluster in a position-dependent manner, binds to the Polycomb-group protein Polycomb Repressive Complex 2(PRC2), and causes widespread repression within the HOXD locus through histone 3 lysine 27 (H3K27) trimethylation [43]. This is a remarkable interaction in that the entire HOXD locus is on a separate chromosome to that of HOXC, indicating that HOTAIR acts to recruit the PRC2 complex to exert transcriptional repression *in trans* at specific genomic locations. In this way, lncRNAs such as HOTTIP may serve as the initial trigger of H3K27 trimethylation, with additional H3K27 methylation deposited over neighboring histones by PRC2 [50]. Such a mechanism establishes a connection between lncRNAs such as HOTAIR with the maintenance of epigenetic memory, via functional integration of external stimuli with resultant chromatin modifications [40]. In addition, targeted deletion of mouse HOTAIR leads to de-repression of hundreds of genes, resulting in homeotic transformation of the spine and malformation of metacarpal-carpal bones, revealing the function and mechanisms of HOTAIR in enforcing a silent chromatin state at *Hox* and additional genes [51].

Interestingly, expression of HOTAIR has also been associated with breast cancer metastasis [52]. Elevated HOTAIR levels are observed in human primary and metastatic breast cancer tissue. Furthermore, depletion of HOTAIR *in vitro* reduces invasiveness of cancer cells expressing high levels of PRC2 [52]. Taken together, these findings suggest that targeting of polycomb complexes by lncRNA is a critical event in breast tumorigenesis. Specifically, lncRNAs such as HOTAIR may be able to alter and regulate epigenetic states in cells by targeting chromatin-modifying complex occupancy and activity *in trans*. Indeed, a number of siRNA-mediated lncRNA depletion experiments from various cell types demonstrate not only the presence of several other PRC2-binding lncRNAs, but also significant de-repression of genes normally repressed by PRC2 [6, 53].

8.3.1.2 HOTTIP

The identification of another novel lncRNA, HOTTIP, from the human HOXA cluster [44], brings to light a potential new class of players in gene regulation of positional identity in epithelial development. Through application of high-throughput chromosome conformation capture (3C) technology, which has allowed interrogation of the spatial organization of the genome and higher order chromatin interactions at unprecedented resolution [54], HOTTIP was found to serve as a critical intermediate that activates contiguous genes on the distal HOXA loci *in cis* via its interaction with chromosomal structural configuration [44]. In other words, a lncRNA can also transmit information from higher-order chromosomal looping into modifications of chromatin. HOTTIP recruits the mammalian Trithorax homolog, Mixed-Lineage Leukemia-1 (MLL-1) complex, to bring about histone 3 lysine 4 (H3K4) trimethylation and *Hox* gene activation. The implication is that lncRNAs can function as important organizers of chromatin domains to coordinate long-range changes in gene expression, with the intriguing potential involvement in the transcriptional regulation of multiple contiguous loci such as HOX, a phenomenon termed locus control. It is possible that HOTTIP takes advantage of the existing 3-dimensional chromosomal structure at the distal HOXA loci to affect the local transcription landscape. This would be an attractive mechanism given the now-appreciated highly conserved hierarchical organization of the genome [55], in which fundamental structural units serve to guide regulatory elements—or, in this case, specific RNA transcripts—to their cognate promoters. These interactions could provide insight into the way in which the 3-dimensional organization of the genome reflects alterations in lineage and stage-specific transcriptional programs that govern cell fate, with lncRNAs such as HOTTIP acting as well-timed molecular switches. This may represent a fundamental property of mammalian gene regulatory networks; whether or not lncRNAs play a role in establishing the topological domain structure of the genome remains to be determined.

8.3.2 lncRNAs with Critical Roles in Skin Progenitor Renewal, Differentiation, Aging, and Cancer

Of special interest to skin biologists is the recent explosion of progress uncovering the widespread involvement of noncoding RNAs in seemingly every important process in cutaneous biology. Short and long noncoding RNAs have been implicated in the regulation of progenitor cell development, maintenance, remodeling, and differentiation [56, 57]. A few of the more prominent lncRNAs that play important functional roles in each of these areas are discussed below.

8.3.2.1 ANCR

Even though lncRNAs have been shown to regulate diverse processes, their potential roles in maintaining the undifferentiated state in somatic tissue progenitor cells remain largely uncharacterized. Kretz and colleagues performed transcriptome sequencing and tiling arrays to compare lncRNA expression in epidermal progenitor populations versus differentiating cells [58]. They characterized a lncRNA called ANCR (anti-differentiation ncRNA) that was down-regulated during epidermal differentiation. Depleting ANCR in progenitor-containing populations, in the absence of other stimuli, led to a rapid and robust induction of differentiation genes. Furthermore, in the epidermis, loss of ANCR abolished the normal exclusion of differentiation from the progenitor-containing compartment [58]. The ANCR lncRNA thus appears to be of critical importance in the maintenance of the undifferentiated cell state within the epidermis. This is consistent with recent data from high-throughput experiments using shRNAs in mouse ES cells [59] implicating several lncRNAs as major players in the maintenance of pluripotency, in some cases through their interactions with chromatin regulatory complexes. It is also possible that lncRNAs serve as "environmental sensors" to alert progenitor and stem cells to maintain pluripotency or to differentiate depending on changes in the environment.

8.3.2.2 TINCR

In addition to ANCR, Kretz and colleagues also recently uncovered a novel lncRNA with significant functional impact on homeostasis and differentiation of mature epidermal tissue [60]. Using cultured keratinocytes and high-throughput full transcriptome sequencing, they found a lncRNA, TINCR (terminal differentiation-induced ncRNA), that was required for maintaining high mRNA abundance of key differentiation genes such as filaggrin (FLG), loricrin(LOR), and members of the arachidonate lipoxygenases family ALOXE3 and ALOX12B, many of which are mutated in human skin diseases. TINCR-deficient epidermis lacked terminal differentiation ultrastructures, including keratohyalin granules and intact lamellar bodies (Fig. 8.1b).

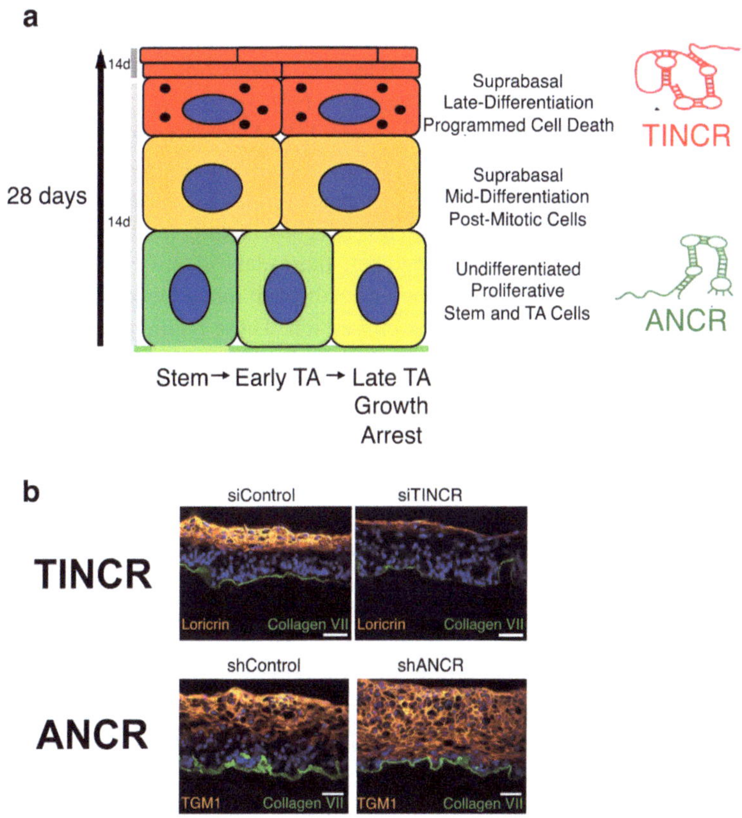

Fig. 8.1 (**a**) Schematic of mammalian epidermal differentiation. Upon exit from the cell cycle, a subset of keratinocytes undergoes a process of terminal differentiation and stratification, consisting of the upward migration of keratinocytes from the basal layer containing progenitor cells (dark green, Stem) transitioning into transient amplifying cells (light green, TA), into the spinous (orange) and granular (red) layers, to establish the outer barrier of the skin. Multiple long non-coding RNAs have been shown to control epidermal homeostasis. ANCR is highly expressed in the progenitor-containing basal layer where it reinforces the undifferentiated state, while TINCR is found most abundantly in the differentiated compartment and plays a major role in regulating the differentiation process. (Schematic courtesy of Paul A. Khavari). (**b**) Ectopic expression of epidermal differentiation proteins in the epidermal basal and lower spinous layers in TINCR and ANCR-depleted regenerating organotypic human epidermis (siTINCR and shANCR, respectively) using ANCR-specific versus scrambled control siRNA and shRNA constructs (siControl and shControl, respectively). TINCR-deficient epidermis stratified normally; however, expression of key differentiation proteins was markedly reduced. Depletion of ANCR resulted in expression of differentiation proteins in the epidermal basal layer, a compartment in which differentiation proteins are never normally found. Differentiation proteins (Loricrin and Transglutaminase1) detected by immunofluorescence (orange), basement membrane collagen VII (green), and nuclear Hoechst 33,342 (blue). Bar, 50 mm. (Photos courtesy of Markus Kretz)

Surprisingly, it turns out that TINCR controls human epidermal differentiation by a post-transcriptional mechanism. Genome-scale RNA interactome analysis revealed that TINCR interacts with a suite of differentiation mRNAs, and a high-throughput screen to analyze TINCR-protein binding revealed direct binding of TINCR RNA to the Staufen1 (STAU1) protein [60]. Interestingly, STAU1-deficient tissue recapitulated the impaired epidermal differentiation seen with TINCR depletion. Furthermore, the TINCR/STAU1 complex appears to mediate stabilization of differentiation mRNAs, such as keratin 80 (KRT80), an important structural protein in keratinocyte epithelium. Taken together, these data identify TINCR as a key lncRNA required for epithelial differentiation through inducible lncRNA binding to differentiation mRNAs to ensure their expression. These findings support a potentially important role for lncRNAs, and the proteins and mRNAs that interact with them, in the control of somatic tissue differentiation.

Lopez-Pajares and colleagues took an integrative approach by performing kinetic transcriptome analysis during regeneration of differentiated epidermis, and identified the transcription factors MAF and MAFB as downstream effectors of ANCR and TINCR lncRNAs in a genetic circuitry underlying epidermal differentiation [61]. In addition to revealing an unexpectedly prominent role for lncRNAs in this process, a first for any stratified epithelial tissue, they show that ANCR and TINCR converge on MAF and MAFB by either repressing them in progenitors or by stimulating them in differentiation. It appears that ANCR repression of MAF:MAFB expression occurs in part through binding of EZH2-repressive complexes to differentiation genes in progenitor keratinocytes, while TINCR-dependent upregulation of MAF:MAFB is mediated through its mRNA binding and stabilization [61]. A lncRNA-TF network is thus essential for epidermal differentiation, and may serve as a model for lncRNA regulatory circuits that serve similar functions in other tissues.

8.3.3 lncRNAs in Specific Skin Disorders

One of the strongest arguments in favor of lncRNA function is that many lncRNAs are associated with disease processes. For example, the lncRNA ANRIL is highly expressed in prostate cancer, and may be responsible for cancer initiation [62]. ANRIL is also associated with increased risk of heart diseases [3, 63]. As described above, HOTAIR is expressed in metastatic breast cancer and may be a key driver of cancer invasiveness [52]. Recent exciting evidence for lncRNA involvement in melanoma, psoriasis, and human aging has implicated these noncoding genes as critical players in cutaneous disorders.

8.3.4 SPRY4-IT1 and BANCR

SPRY4-IT1, derived from an intron of the SPRY4 gene, was identified as a differentially expressed lncRNA in melanoma cell lines [64]. RNA-FISH analysis showed that SPRY4-IT1 is predominantly localized in the cytoplasm of melanoma cells, and knockdown of SPRY4-IT1 transcripts results in defects in cell growth, differentiation, and higher rates of apoptosis. Increased SPRY4-IT1 expression was also detected *in vivo* in samples from patients with melanoma, suggesting that dysregulated expression of SPRY4-IT1 may have an important role in melanoma development, and could serve as an early biomarker and a key regulator for human melanoma pathogenesis. The exact molecular mechanisms through which SPRY4-IT1 regulates the gene expression programs affecting melanoma progression and metastasis remain to be determined.

Taking a similar approach, Flockhard and colleagues sought to define the impacts of oncogenic BRAF on the melanocyte transcriptome by performing massively parallel cDNA sequencing (RNA-seq) on genetically matched normal human melanocytes with and without the BRAF (V600E) mutation [65], and identified several annotated and unannotated intergenic transcripts. The transcripts display tissue-specific expression profiles with distinctive regulatory chromatin marks and transcription factor binding sites indicative of active transcription. One novel lncRNA in particular, named BANCR (for BRAF-activated non-coding RNA), was shown to be involved in the migration of melanoma cells *in vitro*. This is an example of the utility of combining RNA-seq of oncogene-expressing normal cells with RNA-seq of their corresponding human cancers as a useful approach to discover new oncogene-regulated RNA transcripts of potential clinical relevance in cancer.

Similarly, Lee and colleagues, utilizing whole transcriptome sequencing in patients with cutaneous T-cell lymphoma (CTCL), identified multiple differentially expressed unannotated transcripts termed Sézary cell-associated transcripts (SeCATs), including lncRNAs [66]. High-throughput sequencing in archived tumors from additional CTCL patients confirmed the differential expression of the lncRNAs in CTCL, lending support to the importance of lncRNA dysregulation in human malignancies.

8.3.5 lncRNAs in Psoriasis

Despite the increasingly widespread identification of lncRNAs, inference of their functional relevance and underlying molecular mechanisms of action remains a challenge. For example, one of the first attempts to identify novel genetic factors contributing to psoriasis susceptibility centered around differential gene expression profiles of epidermis from patients with psoriasis compared to those from healthy controls [67]. A lncRNA, termed the psoriasis susceptibility-related RNA gene induced by stress (PRINS), was found to be present at higher levels in the skin of psoriasis patients compared to that of normal controls. Real time reverse transcription-PCR analysis showed that stress signals such as ultraviolet-B irradiation, viral infection, and translational inhibition increased the expression levels of PRINS [67], which also appeared to be involved in the cellular stress response. This dynamic regulation of PRINS expression suggests that it may function as a noncoding regulatory RNA, modifying the expression of other genes involved in the proliferation and survival of cells exposed to stress.

More recently, Elder and colleagues used psoriasis as a disease platform and identified thousands of skin-specific expressed lncRNAs, a significant percentage of which are differentially expressed in psoriatic lesions compared with uninvolved or normal skin [68]. Furthermore, many of these differentially expressed lncRNAs are co-expressed with genes involved in immune related functions, and novel lncRNAs are enriched for localization in the epidermal differentiation complex. Some of these lncRNAs are regulated by cytokine treatment in cultured human keratinocytes, implicating them in the immunopathogenesis of psoriasis as well as potentially in other autoimmune disorders.

8.3.6 lncRNAs as Indicators of Solar Exposure

Prolonged exposure to ultraviolet B (UVB) radiation from the sun is well known to result in sunburn and premature aging, as well as in carcinogenesis; however, the underlying mechanisms for the resultant acute inflammatory reactions in the skin are not well understood. In an elegant study, Bernard et al. demonstrated that RNA released from keratinocytes after UVB exposure stimulates production of the inflammatory cytokines tumor necrosis factor alpha (TNF-α) and interleukin-6 (IL-6) from non-irradiated keratinocytes and peripheral blood mononuclear cells (PBMCs; [69]). Whole-transcriptome sequencing revealed that UVB irradiation of keratinocytes induced alterations in the double-stranded domains of some noncoding RNAs, and that the damaged RNAs serve as indicators for sun induced skin injury and participate in activating skin inflammation and repair of the skin barrier through interactions with toll-like receptor 3 (TLR3), a protein involved in pro-inflammatory responses [69]. These findings suggest the potential for noncoding RNAs as biomarkers of solar injury.

8.3.7 lncRNAs in Cell Cycle Control

Taking a more directed approach, Hung and colleagues hypothesized that the genomic loci of cell-cycle genes may harbor functional lncRNAs that are differentially regulated in response to DNA damage [70]. Using human fibroblasts as a model, they showed that DNA damage induces several lncRNAs from the cyclin-dependent kinase inhibitor 1A (*CDKN1A*) promoter, and that one particular lncRNA, named PANDA (P21 associated ncRNA DNA damage activated), is induced in a p53-dependent manner. Mechanistically, *PANDA* interacts with the transcription factor NF-YA to limit expression of pro-apoptotic genes; *PANDA* depletion markedly sensitized human fibroblasts to apoptosis by doxorubicin [70]. LncRNAs-mediated non-canonical DNA damage response pathways may ensure that the response to DNA damage is diverse and reliable depending on the cellular context. More recently, another lncRNA—damaged induced noncoding (DINO)—was identified from human fibroblasts as a component of the DNA damage response [71]. Induction of DINO, by DNA damage or through overexpression, enhanced p53 protein stability and transactivation of p53 targets, revealing a feed forward mechanism that amplifies p53 activity in response to DNA damage [71]. This is a nice example of an inducible lncRNA that can create a feedback loop with its cognate transcription factor to amplify cellular signaling networks. Thus, control of lncRNA expression is a previously unknown mechanism for cell cycle regulation, opening new avenues for understanding how stress adaptation, a major function of the skin, is accomplished in response to changing environments.

8.3.8 lncRNAs in Skin Aging

An interesting study by Chang and colleagues elucidated the gene expression program associated with human photoaging and intrinsic skin aging and the impact of broadband light (BBL) treatment by comparing transcriptomes between young and aged human subjects [72]. The authors found skin aging to be associated with significantly altered expression levels of thousands of coding and noncoding RNAs, including many lncRNAs whose expression was "rejuvenated" after BBL treatment. These lncRNAs were proximally located near several known key regulators of organismal longevity [72], providing potential insights into the role these noncoding genes play as molecular targets of the skin aging process.

8.4 Conclusions and Unanswered Questions

It is now clear that lncRNAs play critical roles in the developmental regulation of the epithelium, providing a means to integrate differentiation cues with dynamic nuclear responses, through control of epigenetic processes. It is very likely that additional functions for lncRNAs will be discovered, as only a small percentage of lncRNAs have been studied in detail to date. For instance, just over the past few months, dysregulation or mutation affecting several lncRNAs have been reported in a variety of cutaneous pathologies such as basal cell carcinoma [73], melanoma [74], squamous cell carcinoma [75, 76], scleroderma [77], keloids [78], and psoriasis [79]. Better characterization of the lncRNA-protein "interactome" will undoubtedly uncover novel mechanisms of gene regulation, both locally and at the genomic level, and will allow researchers to affect cellular developmental programs by modulating transcriptional gene expression. Identification of the functions of the specific lncRNAs discussed in this Chapter represents tremendous progress, and serves as a foundation for future in-depth functional and mechanistic studies aimed at a more complete understanding of gene expression programs in cutaneous biology.

Crucial developmental processes such as X-chromosome inactivation and genomic imprinting have been shown to be largely regulated by lncRNAs, but direct evidence for a role of the majority of lncRNAs in organismal development is scant. However, increasingly available genomic data together with accumulating experimental evidence strongly point towards key roles for lncRNAs in the determination of crucial decisions during certain stages in development. The time has come to functionally address the modes of action of these lncRNAs, utilizing state-of-the-art genomics approaches with experimental validation. Further studies are required to systematically dissect the various epigenetic activating and silencing targeting mechanisms, as well as their interdependencies. Global gene expression profiles from different model organisms and tissues, as well as from multiple disease and cell-differentiation states will undoubtedly continue to reveal well-interconnected and highly dynamic transcriptional networks underlying individual cellular functions. Recent advances in high-throughput technologies to interrogate the DNA, RNA, and protein interactomes of lncRNAs [60, 80, 81] will no doubt accelerate the elucidation of their molecular mechanisms of action. Understanding how lncRNAs interact with chromatin as global regulators of transcription, as well as the three-dimensional state of chromosomal structures at large, will be critical to deciphering these networks. The answers are unlikely to be straightforward, but the efforts towards unraveling the mysteries and molecular mechanisms will undoubtedly provide invaluable new insights into the complex nature of gene regulation and the development of multicellular organisms.

References

1. Blanpain C, Horsley V, Fuchs E. Epithelial stem cells: turning over new leaves. Cell. 2007;128:445–58. https://doi.org/10.1016/j.cell.2007.01.014.
2. Millington GWM. Epigenetics and dermatological disease. Pharmacogenomics. 2008;9:1835–50. https://doi.org/10.2217/14622416.9.12.1835.
3. Cabili MN, Trapnell C, Goff L, Koziol M, Tazon-Vega B, Regev A, Rinn JL. Integrative annotation of human large intergenic noncoding RNAs reveals global properties and specific subclasses. Genes Dev. 2011;25:1915–27. https://doi.org/10.1101/gad.17446611.
4. Carninci P, Kasukawa T, Katayama S, Gough J, Frith MC, Maeda N, Oyama R, Ravasi T, Lenhard B, Wells C, Kodzius R, Shimokawa K, Bajic VB, Brenner SE, Batalov S, Forrest ARR, Zavolan M, Davis MJ, Wilming LG, Aidinis V, Allen JE, Ambesi-Impiombato A, Apweiler R, Aturaliya RN, Bailey TL, Bansal M, Baxter L, Beisel KW, Bersano T, Bono H, Chalk AM, Chiu KP, Choudhary V, Christoffels A, Clutterbuck DR, Crowe ML, Dalla E, Dalrymple BP, de Bono B, Gatta Della G, di Bernardo D, Down T, Engstrom P, Fagiolini M, Faulkner G, Fletcher CF, Fukushima T, Furuno M, Futaki S, Gariboldi M, Georgii-Hemming P, Gingeras TR, Gojobori T, Green RE, Gustincich S, Harbers M, Hayashi Y, Hensch TK,

Hirokawa N, Hill D, Huminiecki L, Iacono M, Ikeo K, Iwama A, Ishikawa T, Jakt M, Kanapin A, Katoh M, Kawasawa Y, Kelso J, Kitamura H, Kitano H, Kollias G, Krishnan SPT, Kruger A, Kummerfeld SK, Kurochkin IV, Lareau LF, Lazarevic D, Lipovich L, Liu J, Liuni S, McWilliam S, Madan Babu M, Madera M, Marchionni L, Matsuda H, Matsuzawa S, Miki H, Mignone F, Miyake S, Morris K, Mottagui-Tabar S, Mulder N, Nakano N, Nakauchi H, Ng P, Nilsson R, Nishiguchi S, Nishikawa S, Nori F, Ohara O, Okazaki Y, Orlando V, Pang KC, Pavan WJ, Pavesi G, Pesole G, Petrovsky N, Piazza S, Reed J, Reid JF, Ring BZ, Ringwald M, Rost B, Ruan Y, Salzberg SL, Sandelin A, Schneider C, Schönbach C, Sekiguchi K, Semple CAM, Seno S, Sessa L, Sheng Y, Shibata Y, Shimada H, Shimada K, Silva D, Sinclair B, Sperling S, Stupka E, Sugiura K, Sultana R, Takenaka Y, Taki K, Tammoja K, Tan SL, Tang S, Taylor MS, Tegner J, Teichmann SA, Ueda HR, van Nimwegen E, Verardo R, Wei CL, Yagi K, Yamanishi H, Zabarovsky E, Zhu S, Zimmer A, Hide W, Bult C, Grimmond SM, Teasdale RD, Liu ET, Brusic V, Quackenbush J, Wahlestedt C, Mattick JS, Hume DA, Kai C, Sasaki D, Tomaru Y, Fukuda S, Kanamori-Katayama M, Suzuki M, Aoki J, Arakawa T, Iida J, Imamura K, Itoh M, Kato T, Kawaji H, Kawagashira N, Kawashima T, Kojima M, Kondo S, Konno H, Nakano K, Ninomiya N, Nishio T, Okada M, Plessy C, Shibata K, Shiraki T, Suzuki S, Tagami M, Waki K, Watahiki A, Okamura-Oho Y, Suzuki H, Kawai J, Hayashizaki Y, FANTOM Consortium, RIKEN Genome Exploration Research Group and Genome Science Group (Genome Network Project Core Group). The transcriptional landscape of the mammalian genome. Science. 2005;309:1559–63. https://doi.org/10.1126/science.1112014.

5. Guttman M, Amit I, Garber M, French C, Lin MF, Feldser D, Huarte M, Zuk O, Carey BW, Cassady JP, Cabili MN, Jaenisch R, Mikkelsen TS, Jacks T, Hacohen N, Bernstein BE, Kellis M, Regev A, Rinn JL, Lander ES. Chromatin signature reveals over a thousand highly conserved large non-coding RNAs in mammals. Nature. 2009;458:223–7. https://doi.org/10.1038/nature07672.
6. Khalil AM, Guttman M, Huarte M, Garber M, Raj A, Rivea Morales D, Thomas K, Presser A, Bernstein BE, van Oudenaarden A, Regev A, Lander ES, Rinn JL. Many human large intergenic noncoding RNAs associate with chromatin-modifying complexes and affect gene expression. Proc Natl Acad Sci U S A. 2009;106:11667–72. https://doi.org/10.1073/pnas.0904715106.
7. Ponting CP, Oliver PL, Reik W. Evolution and functions of long noncoding RNAs. Cell. 2009;136:629–41. https://doi.org/10.1016/j.cell.2009.02.006.
8. Djebali S, Davis CA, Merkel A, Dobin A, Lassmann T, Mortazavi A, Tanzer A, Lagarde J, Lin W, Schlesinger F, Xue C, Marinov GK, Khatun J, Williams BA, Zaleski C, Rozowsky J, Röder M, Kokocinski F, Abdelhamid RF, Alioto T, Antoshechkin I, Baer MT, Bar NS, Batut P, Bell K, Bell I, Chakrabortty S, Chen X, Chrast J, Curado J, Derrien T, Drenkow J, Dumais E, Dumais J, Duttagupta R, Falconnet E, Fastuca M, Fejes-Toth K, Ferreira P, Foissac S, Fullwood MJ, Gao H, Gonzalez D, Gordon A, Gunawardena H, Howald C, Jha S, Johnson R, Kapranov P, King B, Kingswood C, Luo OJ, Park E, Persaud K, Preall JB, Ribeca P, Risk B, Robyr D, Sammeth M, Schaffer L, See L-H, Shahab A, Skancke J, Suzuki AM, Takahashi H, Tilgner H, Trout D, Walters N, Wang H, Wrobel J, Yu Y, Ruan X, Hayashizaki Y, Harrow J, Gerstein M, Hubbard T, Reymond A, Antonarakis SE, Hannon G, Giddings MC, Ruan Y, Wold B, Carninci P, Guigo R, Gingeras TR. Landscape of transcription in human cells. Nature. 2012;489:101–8. https://doi.org/10.1038/nature11233.
9. Mercer TR, Dinger ME, Mattick JS. Long non-coding RNAs: insights into functions. Nat Rev Genet. 2009;10:155–9. https://doi.org/10.1038/nrg2521.
10. Clark MB, Mattick JS. Long noncoding RNAs in cell biology. Semin Cell Dev Biol. 2011;22:366–76. https://doi.org/10.1016/j.semcdb.2011.01.001.
11. Mattick JS, Amaral PP, Dinger ME, Mercer TR, Mehler MF. RNA regulation of epigenetic processes. BioEssays. 2009;31:51–9. https://doi.org/10.1002/bies.080099.
12. Nagano T, Fraser P. No-nonsense functions for long noncoding RNAs. Cell. 2011;145:178–81. https://doi.org/10.1016/j.cell.2011.03.014.
13. Wang KCK, Chang HYH. Molecular mechanisms of long noncoding RNAs. Mol Cell. 2011;43:904–14. https://doi.org/10.1016/j.molcel.2011.08.018.

14. Bernstein E, Allis CD. RNA meets chromatin. Genes Dev. 2005;19:1635–55. https://doi.org/10.1101/gad.1324305.
15. Bracken AP, Helin K. Polycomb group proteins: navigators of lineage pathways led astray in cancer. Nat Rev Cancer. 2009;9:773–84. https://doi.org/10.1038/nrc2736.
16. Faghihi MA, Wahlestedt C. Regulatory roles of natural antisense transcripts. Nat Rev Mol Cell Biol. 2009;10:637–43. https://doi.org/10.1038/nrm2738.
17. Whitehead J, Pandey GK, Kanduri C. Regulation of the mammalian epigenome by long noncoding RNAs. Biochim Biophys Acta. 2009;1790:936–47. https://doi.org/10.1016/j.bbagen.2008.10.007.
18. Wilusz JE, Sunwoo H, Spector DL. Long noncoding RNAs: functional surprises from the RNA world. Genes Dev. 2009;23:1494–504. https://doi.org/10.1101/gad.1800909.
19. Taft RJ, Pang KC, Mercer TR, Dinger M, Mattick JS. Non-coding RNAs: regulators of disease. J Pathol. 2010;220:126–39. https://doi.org/10.1002/path.2638.
20. Rinn JL, Chang HY. Genome regulation by long noncoding RNAs. Annu Rev Biochem. 2012;81:145–66. https://doi.org/10.1146/annurev-biochem-051410-092902.
21. Qureshi IA, Mattick JS, Mehler MF. Long non-coding RNAs in nervous system function and disease. Brain Res. 2010;1338:20–35. https://doi.org/10.1016/j.brainres.2010.03.110.
22. Wapinski O, Chang HY. Long noncoding RNAs and human disease. Trends Cell Biol. 2011;21:354–61. https://doi.org/10.1016/j.tcb.2011.04.001.
23. Paul J, Duerksen JD. Chromatin-associated RNA content of heterochromatin and euchromatin. Mol Cell Biochem. 1975;9:9–16.
24. Bernstein BE, Mikkelsen TS, Xie X, Kamal M, Huebert DJ, Cuff J, Fry B, Meissner A, Wernig M, Plath K, Jaenisch R, Wagschal A, Feil R, Schreiber SL, Lander ES. A bivalent chromatin structure marks key developmental genes in embryonic stem cells. Cell. 2006;125:315–26. https://doi.org/10.1016/j.cell.2006.02.041.
25. Pandey RR, Mondal T, Mohammad F, Enroth S, Redrup L, Komorowski J, Nagano T, Mancini-Dinardo D, Kanduri C. Kcnq1ot1 antisense noncoding RNA mediates lineage-specific transcriptional silencing through chromatin-level regulation. Mol Cell. 2008;32:232–46. https://doi.org/10.1016/j.molcel.2008.08.022.
26. Rodríguez-Campos A, Azorín F. RNA is an integral component of chromatin that contributes to its structural organization. PLoS One. 2007;2:e1182. https://doi.org/10.1371/journal.pone.0001182.
27. Clemson CM, McNeil JA, Willard HF, Lawrence JB. XIST RNA paints the inactive X chromosome at interphase: evidence for a novel RNA involved in nuclear/chromosome structure. J Cell Biol. 1996;132:259–75.
28. Muchardt C, Guilleme M, Seeler J-S, Trouche D, Dejean A, Yaniv M. Coordinated methyl and RNA binding is required for heterochromatin localization of mammalian HP1alpha. EMBO Rep. 2002;3:975–81. https://doi.org/10.1093/embo-reports/kvf194.
29. Majewski IJ, Blewitt ME, de Graaf CA, McManus EJ, Bahlo M, Hilton AA, Hyland CD, Smyth GK, Corbin JE, Metcalf D, Alexander WS, Hilton DJ. Polycomb repressive complex 2 (PRC2) restricts hematopoietic stem cell activity. PLoS Biol. 2008;6:e93. https://doi.org/10.1371/journal.pbio.0060093.
30. Spivakov M, Fisher AG. Epigenetic signatures of stem-cell identity. Nat Rev Genet. 2007;8:263–71. https://doi.org/10.1038/nrg2046.
31. Surani MA, Hayashi K, Hajkova P. Genetic and epigenetic regulators of pluripotency. Cell. 2007;128:747–62. https://doi.org/10.1016/j.cell.2007.02.010.
32. Schuettengruber B, Chourrout D, Vervoort M, Leblanc B, Cavalli G. Genome regulation by polycomb and trithorax proteins. Cell. 2007;128:735–45. https://doi.org/10.1016/j.cell.2007.02.009.
33. Schwartz YB, Pirrotta V. Polycomb silencing mechanisms and the management of genomic programmes. Nat Rev Genet. 2007;8:9–22. https://doi.org/10.1038/nrg1981.
34. Ezhkova E, Pasolli HA, Parker JS, Stokes N, Su I-H, Hannon G, Tarakhovsky A, Fuchs E. Ezh2 orchestrates gene expression for the stepwise differentiation of tissue-specific stem cells. Cell. 2009;136:1122–35. https://doi.org/10.1016/j.cell.2008.12.043.

35. Boyer LA, Plath K, Zeitlinger J, Brambrink T, Medeiros LA, Lee TI, Levine SS, Wernig M, Tajonar A, Ray MK, Bell GW, Otte AP, Vidal M, Gifford DK, Young RA, Jaenisch R. Polycomb complexes repress developmental regulators in murine embryonic stem cells. Nature. 2006;441:349–53. https://doi.org/10.1038/nature04733.
36. Pasini D, Bracken AP, Hansen JB, Capillo M, Helin K. The polycomb group protein Suz12 is required for embryonic stem cell differentiation. Mol Cell Biol. 2007;27:3769–79. https://doi.org/10.1128/MCB.01432-06.
37. Chang HY, Chi J-T, Dudoit S, Bondre C, van de Rijn M, Botstein D, Brown PO. Diversity, topographic differentiation, and positional memory in human fibroblasts. Proc Natl Acad Sci U S A. 2002;99:12877–82. https://doi.org/10.1073/pnas.162488599.
38. Chi J-T, Chang HY, Wang NN, Chang DS, Dunphy N, Brown PO. Genomewide view of gene silencing by small interfering RNAs. Proc Natl Acad Sci U S A. 2003;100:6343–6. https://doi.org/10.1073/pnas.1037853100.
39. Rinn JL, Bondre C, Gladstone HB, Brown PO, Chang HY. Anatomic demarcation by positional variation in fibroblast gene expression programs. PLoS Genet. 2006;2:e119. https://doi.org/10.1371/journal.pgen.0020119.
40. Wang KC, Helms JA, Chang HY. Regeneration, repair and remembering identity: the three Rs of Hox gene expression. Trends Cell Biol. 2009;19:268–75. https://doi.org/10.1016/j.tcb.2009.03.007.
41. Lemons D, McGinnis W. Genomic evolution of Hox gene clusters. Science. 2006;313:1918–22. https://doi.org/10.1126/science.1132040.
42. Rinn JL, Wang JK, Allen N, Brugmann SA, Mikels AJ, Liu H, Ridky TW, Stadler HS, Nusse R, Helms JA, Chang HY. A dermal HOX transcriptional program regulates site-specific epidermal fate. Genes Dev. 2008;22:303–7. https://doi.org/10.1101/gad.1610508.
43. Rinn JL, Kertesz M, Wang JK, Squazzo SL, Xu X, Brugmann SA, Goodnough LH, Helms JA, Farnham PJ, Segal E, Chang HY. Functional demarcation of active and silent chromatin domains in human HOX loci by noncoding RNAs. Cell. 2007;129:1311–23. https://doi.org/10.1016/j.cell.2007.05.022.
44. Wang KC, Yang YW, Liu B, Sanyal A, Corces-Zimmerman R, Chen Y, Lajoie BR, Protacio A, Flynn RA, Gupta RA, Wysocka J, Lei M, Dekker J, Helms JA, Chang HY. A long noncoding RNA maintains active chromatin to coordinate homeotic gene expression. Nature. 2011;472:120–4. https://doi.org/10.1038/nature09819.
45. Amaral PP, Neyt C, Wilkins SJ, Askarian-Amiri ME, Sunkin SM, Perkins AC, Mattick JS. Complex architecture and regulated expression of the Sox2ot locus during vertebrate development. RNA. 2009;15:2013–27. https://doi.org/10.1261/rna.1705309.
46. Araki R, Fukumura R, Sasaki N, Kasama Y, Suzuki N, Takahashi H, Tabata Y, Saito T, Abe M. More than 40,000 transcripts, including novel and noncoding transcripts, in mouse embryonic stem cells. Stem Cells. 2006;24:2522–8. https://doi.org/10.1634/stemcells.2006-0005.
47. Dinger ME, Amaral PP, Mercer TR, Pang KC, Bruce SJ, Gardiner BB, Askarian-Amiri ME, Ru K, Solda G, Simons C, Sunkin SM, Crowe ML, Grimmond SM, Perkins AC, Mattick JS. Long noncoding RNAs in mouse embryonic stem cell pluripotency and differentiation. Genome Res. 2008;18:1433–45. https://doi.org/10.1101/gr.078378.108.
48. Mercer TR, Dinger ME, Sunkin SM, Mehler MF, Mattick JS. Specific expression of long noncoding RNAs in the mouse brain. Proc Natl Acad Sci U S A. 2008;105:716–21. https://doi.org/10.1073/pnas.0706729105.
49. Pang KC, Dinger ME, Mercer TR, Malquori L, Grimmond SM, Chen W, Mattick JS. Genome-wide identification of long noncoding RNAs in CD8+ T cells. J Immunol. 2009;182:7738–48. https://doi.org/10.4049/jimmunol.0900603.
50. Margueron R, Justin N, Ohno K, Sharpe ML, Son J, Drury WJ, Voigt P, Martin SR, Taylor WR, De Marco V, Pirrotta V, Reinberg D, Gamblin SJ. Role of the polycomb protein EED in the propagation of repressive histone marks. Nature. 2009;461:762–7. https://doi.org/10.1038/nature08398.
51. Li L, Liu B, Wapinski OL, Tsai M-C, Qu K, Zhang J, Carlson JC, Lin M, Fang F, Gupta RA, Helms JA, Chang HY. Targeted disruption of hotair leads to homeotic transformation and gene derepression. Cell Rep. 2013;5:3–12. https://doi.org/10.1016/j.celrep.2013.09.003.

52. Gupta RA, Shah N, Wang KC, Kim J, Horlings HM, Wong DJ, Tsai M-C, Hung T, Argani P, Rinn JL, Wang Y, Brzoska P, Kong B, Li R, West RB, van de Vijver MJ, Sukumar S, Chang HY. Long non-coding RNA HOTAIR reprograms chromatin state to promote cancer metastasis. Nature. 2010;464:1071–6. https://doi.org/10.1038/nature08975.
53. Zhao J, Ohsumi TK, Kung JT, Ogawa Y, Grau DJ, Sarma K, Song JJ, Kingston RE, Borowsky M, Lee JT. Genome-wide identification of polycomb-associated RNAs by RIP-seq. Mol Cell. 2010;40:939–53. https://doi.org/10.1016/j.molcel.2010.12.011.
54. Sanyal A, Baù D, Martí-Renom MA, Dekker J. Chromatin globules: a common motif of higher order chromosome structure? Curr Opin Cell Biol. 2011;23:325–31. https://doi.org/10.1016/j.ceb.2011.03.009.
55. Lupiáñez DG, Spielmann M, Mundlos S. Breaking TADs: how alterations of chromatin domains result in disease. Trends Genet. 2016;32:225–37. https://doi.org/10.1016/j.tig.2016.01.003.
56. Hu W, Dominguez JRA, Lodish HF. Regulation of mammalian cell differentiation by long non-coding RNAs. EMBO Rep. 2012;13:971–83. https://doi.org/10.1038/embor.2012.145.
57. Yi R, Fuchs E. A miR image of stem cells and their lineages. Curr Top Dev Biol. 2012;99:175–99. https://doi.org/10.1016/B978-0-12-387038-4.00007-0.
58. Kretz M, Webster DE, Flockhart RJ, Lee CS, Zehnder A, Lopez-Pajares V, Qu K, Zheng GXY, Chow J, Kim GE, Rinn JL, Chang HY, Siprashvili Z, Khavari PA. Suppression of progenitor differentiation requires the long noncoding RNA ANCR. Genes Dev. 2012;26:338–43. https://doi.org/10.1101/gad.182121.111.
59. Guttman M, Donaghey J, Carey BW, Garber M, Grenier JK, Munson G, Young G, Lucas AB, Ach R, Bruhn L, Yang X, Amit I, Meissner A, Regev A, Rinn JL, Root DE, Lander ES. lincRNAs act in the circuitry controlling pluripotency and differentiation. Nature. 2011;477:295–300. https://doi.org/10.1038/nature10398.
60. Kretz M, Siprashvili Z, Chu C, Webster DE, Zehnder A, Qu K, Lee CS, Flockhart RJ, Groff AF, Chow J, Johnston D, Kim GE, Spitale RC, Flynn RA, Zheng GXY, Aiyer S, Raj A, Rinn JL, Chang HY, Khavari PA. Control of somatic tissue differentiation by the long non-coding RNA TINCR. Nature. 2013;493:231–5. https://doi.org/10.1038/nature11661.
61. Lopez-Pajares V, Qu K, Zhang J, Webster DE, Barajas BC, Siprashvili Z, Zarnegar BJ, Boxer LD, Rios EJ, Tao S, Kretz M, Khavari PA. A LncRNA-MAF:MAFB transcription factor network regulates epidermal differentiation. Dev Cell. 2015;32:693–706. https://doi.org/10.1016/j.devcel.2015.01.028.
62. Yap KL, Li S, Muñoz-Cabello AM, Raguz S, Zeng L, Mujtaba S, Gil J, Walsh MJ, Zhou M-M. Molecular interplay of the noncoding RNA ANRIL and methylated histone H3 lysine 27 by polycomb CBX7 in transcriptional silencing of INK4a. Mol Cell. 2010;38:662–74. https://doi.org/10.1016/j.molcel.2010.03.021.
63. Pasmant E, Sabbagh A, Vidaud M, Bièche I. ANRIL, a long, noncoding RNA, is an unexpected major hotspot in GWAS. FASEB J. 2011;25:444–8. https://doi.org/10.1096/fj.10-172452.
64. Khaitan D, Dinger ME, Mazar J, Crawford J, Smith MA, Mattick JS, Perera RJ. The melanoma-upregulated long noncoding RNA SPRY4-IT1 modulates apoptosis and invasion. Cancer Res. 2011;71:3852–62. https://doi.org/10.1158/0008-5472.CAN-10-4460.
65. Flockhart RJ, Webster DE, Qu K, Mascarenhas N, Kovalski J, Kretz M, Khavari PA. BRAFV600E remodels the melanocyte transcriptome and induces BANCR to regulate melanoma cell migration. Genome Res. 2012;22:1006–14. https://doi.org/10.1101/gr.140061.112.
66. Lee CS, Ungewickell A, Bhaduri A, Qu K, Webster DE, Armstrong R, Weng WK, Aros CJ, Mah A, Chen RO, Lin M, Sundram U, Chang HY, Kretz M, Kim YH, Khavari PA. Transcriptome sequencing in Sezary syndrome identifies Sezary cell and mycosis fungoides-associated LncRNAs and novel transcripts. Blood. 2012;120:3288. https://doi.org/10.1182/blood-2012-04-423061.
67. Sonkoly E, Bata-Csörgő Z, Pivarcsi A, Polyanka H, Kenderessy-Szabo A, Molnar G, Szentpali K, Bari L, Megyeri K, Mandi Y, Dobozy A, Kemény L, Széll M. Identification and characterization of a novel, psoriasis susceptibility-related noncoding RNA gene, PRINS. J Biol Chem. 2005;280:24159–67. https://doi.org/10.1074/jbc.M501704200.

68. Tsoi LC, Iyer MK, Stuart PE, Swindell WR, Gudjonsson JE, Tejasvi T, Sarkar MK, Li B, Ding J, Voorhees JJ, Kang HM, Nair RP, Chinnaiyan AM, Abecasis GR, Elder JT. Analysis of long non-coding RNAs highlights tissue-specific expression patterns and epigenetic profiles in normal and psoriatic skin. Genome Biol. 2015;16:24. https://doi.org/10.1186/s13059-014-0570-4.
69. Bernard JJ, Cowing-Zitron C, Nakatsuji T, Muehleisen B, Muto J, Borkowski AW, Martinez L, Greidinger EL, Yu BD, Gallo RL. Ultraviolet radiation damages self noncoding RNA and is detected by TLR3. Nat Med. 2012;18:1286–90. https://doi.org/10.1038/nm.2861.
70. Hung T, Wang Y, Lin MF, Koegel AK, Kotake Y, Grant GD, Horlings HM, Shah N, Umbricht C, Wang P, Wang Y, Kong B, Langerød A, Børresen-Dale A-L, Kim SK, van de Vijver M, Sukumar S, Whitfield ML, Kellis M, Xiong Y, Wong DJ, Chang HY. Extensive and coordinated transcription of noncoding RNAs within cell-cycle promoters. Nat Genet. 2011;43:621. https://doi.org/10.1038/ng.848.
71. Schmitt AM, Garcia JT, Hung T, Flynn RA, Shen Y, Qu K, Payumo AY, Peres-da-Silva A, Broz DK, Baum R, Guo S, Chen JK, Attardi LD, Chang HY. An inducible long noncoding RNA amplifies DNA damage signaling. Nat Genet. 2016;48:1370–6. https://doi.org/10.1038/ng.3673.
72. Chang ALS, Bitter PH, Qu K, Lin M, Rapicavoli NA, Chang HY. Rejuvenation of gene expression pattern of aged human skin by broadband light treatment: a pilot study. J Invest Dermatol. 2013;133:394–402. https://doi.org/10.1038/jid.2012.287.
73. Sand M, Bechara FG, Sand D, Gambichler T, Hahn SA, Bromba M, Stockfleth E, Hessam S. Long-noncoding RNAs in basal cell carcinoma. Tumour Biol. 2016;37:10595–608. https://doi.org/10.1007/s13277-016-4927-z.
74. Wang S, Fan W, Wan B, Tu M, Jin F, Liu F, Xu H, Han P. Characterization of long noncoding RNA and messenger RNA signatures in melanoma tumorigenesis and metastasis. PLoS One. 2017;12:e0172498. https://doi.org/10.1371/journal.pone.0172498.
75. Piipponen M, Nissinen L, Farshchian M, Riihilä P, Kivisaari A, Kallajoki M, Peltonen J, Peltonen S, Kähäri V-M. Long noncoding RNA PICSAR promotes growth of cutaneous squamous cell carcinoma by regulating ERK1/2 activity. J Invest Dermatol. 2016;136:1701–10. https://doi.org/10.1016/j.jid.2016.03.028.
76. Ponzio G, Rezzonico R, Bourget I, Allan R, Nottet N, Popa A, Magnone V, Rios G, Mari B, Barbry P. A new long noncoding RNA (LncRNA) is induced in cutaneous squamous cell carcinoma and downregulates several anticancer and cell-differentiation genes in mouse. J Biol Chem. 2017. https://doi.org/10.1074/jbc.M117.776260.
77. Wang Z, Jinnin M, Nakamura K, Harada M, Kudo H, Nakayama W, Inoue K, Nakashima T, Honda N, Fukushima S, Ihn H. Long non-coding RNA TSIX is upregulated in scleroderma dermal fibroblasts and controls collagen mRNA stabilization. Exp Dermatol. 2016;25:131–6. https://doi.org/10.1111/exd.12900.
78. Sun X-J, Wang Q, Guo B, Liu X-Y, Wang B. Identification of skin-related lncRNAs as potential biomarkers that involved in Wnt pathways in keloids. Oncotarget. 2017;8:34236–44. https://doi.org/10.18632/oncotarget.15880.
79. Gupta R, Ahn R, Lai K, Mullins E, Debbaneh M, Dimon M, Arron S, Liao W. Landscape of long noncoding RNAs in psoriatic and healthy skin. J Invest Dermatol. 2016;136:603–9. https://doi.org/10.1016/j.jid.2015.12.009.
80. Chu CC, Qu KK, Zhong FLF, Artandi SES, Chang HYH. Genomic maps of long noncoding RNA occupancy reveal principles of RNA-chromatin interactions. Mol Cell. 2011;44:667–78. https://doi.org/10.1016/j.molcel.2011.08.027.
81. Siprashvili Z, Webster DE, Kretz M, Johnston D, Rinn JL, Chang HY, Khavari PA. Identification of proteins binding coding and non-coding human RNAs using protein microarrays. BMC Genomics. 2012;13:633. https://doi.org/10.1186/1471-2164-13-633.

Chapter 9
RNA Methylation in the Control of Stem Cell Activity and Epidermal Differentiation

Abdulrahim A. Sajini and Michaela Frye

9.1 Introduction

Homeostasis in skin is achieved by stem cells that continuously maintain their population (self-renewal) while generating numerous differentiated progeny. The balance between self-renewal and differentiation is tightly controlled, and aberrant epidermal differentiation leads to more than 100 human skin diseases [1]. The unique properties of stem cells are controlled by a dynamic interplay between extrinsic and intrinsic regulatory mechanisms. Extrinsic mechanisms include signalling from the neighbouring niche cells [2]; and intrinsic mechanisms involve epigenetic, and transcriptional pathways [3]. While the functional role of transcriptional regulators in regulating epidermal stem cell fate is now increasingly understood, far less is known about the roles of post-transcriptional mechanisms in regulating stem cell maintenance and differentiation.

Gene expression is dynamically controlled through reversible chemical modifications on both DNA and histone proteins [4, 5]. RNA modifications are more prevalent and diverse in their chemical nature than DNA modifications [6, 7], yet their importance in regulating gene expression has only recently begun to be explored. All transcripts are subjected to processing, quality control and surveillance pathways, and at each step RNA is dynamically associated with RNA-binding proteins. Post-transcriptional modifications have emerged as important determinants for the dynamic and temporally accurate binding of proteins to their targeted RNA

A. A. Sajini
University of Cambridge, Department of Genetics, Cambridge, UK

Department of Biomedical Engineering, Khalifa University of Science and Technology, Abu Dhabi, United Arab Emirates

M. Frye (✉)
University of Cambridge, Department of Genetics, Cambridge, UK
e-mail: mf346@cam.ac.uk

© Springer International Publishing AG, part of Springer Nature 2018

V. A. Botchkarev, S. E. Millar (eds.), *Epigenetic Regulation of Skin Development and Regeneration*, Stem Cell Biology and Regenerative Medicine,
https://doi.org/10.1007/978-3-319-16769-5_9

molecules [8]. RNA modifications are crucial for development, and aberrant deposition of RNA modifications can lead to complex human diseases, including neuro-developmental disorders and cancer [9–11].

9.2 Detection of Adenosine and Cytosine Methylation in the Transcriptome

There are around 150 known ribonucleoside modifications [6]. Every nucleoside can be chemically modified, yet the importance of many of the reported modifications, and the enzymes responsible for these, are unknown in mammals [12]. Methylation is one of the most common enzyme-catalyzed modifications [6, 7]. The majority of methyl-based modifications are conserved from bacteria to mammals and plants, and their functions include structural and metabolic stabilization as well as functional roles in regulating protein translation [13–16].

The recent development of new techniques and advances in next generation sequencing (NGS) have enabled the identification of methylated adenosines and cytosines globally and in a substrate-specific manner [17–21]. The presence of N6-methyladenosine (m^6A) in polyadenylated mRNA was first described in the 1970s [22, 23], and was estimated to be present at an average level of three m^6A residues per mRNA [24–27]. The presence of cytosine-5 methylation (m^5C) in RNA was also first described in the 1970s, and its detection involved digestion of highly purified RNA followed by separation techniques such as mass spectrometry, which only allowed the identification of m^5C in stable and highly abundant RNAs such as transfer RNAs (tRNAs) and ribosomal RNAs (rRNAs) [28–30].

Bisulfite conversion of cytosine residues to uracil is one method to detect m^5C in both DNA and RNA (Fig. 9.1a). The chemical reaction leaves 5-methylcytosine residues unaffected and sequencing can then reveal the methylation status of a segment of DNA or RNA at single- nucleotide resolution (Fig. 9.1a). Due to the high turnover and instability of RNA, bisulfite sequencing was long thought to be unsuitable for RNA. However, over the last few years RNA bisulfite sequencing has been adapted to RNA and reproducibly and quantitatively detects m^5C in tRNA, rRNA and other abundant non-coding RNA molecules [20, 21, 31–33]. However, RNA bisulfite conversion degrades RNA, which makes the detection of m^5C in rare RNAs challenging and highly susceptible to false positive results [34].

Alternative methods to detect methylation sites are based on RNA-immunoprecipitation (RIP) approaches followed by deep sequencing (Fig. 9.1b). Antibody-mediated capture of m^6A sites combined with parallel sequencing revealed that m^6A preferentially clustered around 3'UTRs, stop codons, and within internal long exons in humans and mice [17, 18]. A large proportion of binding sites also occur within pre-microRNA regions and introns, hence indicating a potential role for the methyl mark in RNA processing and splicing [35–38]. Deposition of m^6A takes place within a DRACH recognition site where D is U, G or A, and R is A or G, and H is C, U or A, and A is the modified base [17, 18]. Although antibody-

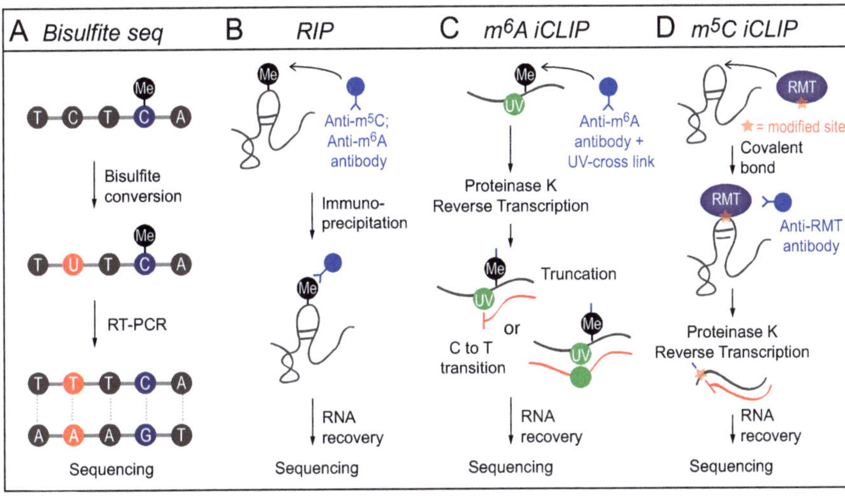

Fig. 9.1 Detection of m⁵C and m⁶A using sequencing based approaches. (**a**) Bisulfite sequencing of RNA that underwent chemical conversion of un-methylated cytosines to uracils detects methylated cytosines at nucleotide resolution. (**b**) RNA immunoprecipitation (RIP) based approaches allow mapping RNA–protein interactions by immunoprecipitating the modified RNA with its methylating enzyme. (**c**, **d**) Modified forms of individual-nucleotide resolution UV cross-linking and immunoprecipitation method (iCLIP) followed by high throughput sequencing to detect m^6A (**c**) or m^5C (**d**) at nucleotide resolution. The cross-linking of the antibodies or covalent binding of the RNA methylase (RMT) causes mis-incorporation or truncations during the reverse transcriptase step that can be detected by sequencing

mediated sequencing for m^6A is specific, it lacks single nucleotide resolution. Detection of the m^6A signature has been achieved by different methods: The first method to map m^6A at single nucleotides resolution was called m^6A-iCLIP (Fig. 9.1c) [39]. m^6A-iCLIP exploits recurrent mutations or cDNA truncations after UV cross-linking m^6A antibodies to RNA targets. After UV cross-linking, specific antibodies against m^6A either substituted the RAC cytosine into thymine, where A is the methylated base or induced pre-mature transcript truncations at the +1 position of m^6A during cDNA synthesis [39]. By sequencing cDNA libraries prepared from RNA fragments containing both antibody-induced signatures, m^6A residues were successfully identified throughout the transcriptome at single nucleotide resolution [39]. The second method detecting m^6A at single nucleotide resolution includes site-specific cleavage and radioactive labelling followed by ligation-assisted extraction and thin layer chromatography (SCARLET) [40]. This method is based on guided RNase H cleavage of candidate 5′ end sites mediated by sequence specific 2′-OMe/2′-H chimeric oligonucleotide for subsequent radiolabelling, and TLC detection. Although this method is laborious, it is able to identify m^6A successfully at single nucleotide resolution in mRNAs and lncRNAs. Interestingly, both single nucleotide resolution methods reported that m^6A occurrence at its consensus motif DRACH is less frequent than predicted in earlier studies [39, 40].

To detect m^5C in RNA, two similar but technically distinct approaches have recently been developed: 5-Azacytidine-mediated RNA Immunoprecipitation (Aza-IP) [21], and methylation-individual nucleotide resolution crosslinking immmunoprecipitation (miCLIP) (Fig. 9.1d) [19]. Both methods rely on covalent bond formation between the RNA methylase (RMT) and its substrate, but differ in the way in which the formation of the stable covalent bond is achieved. Both methods require the over-expression of the respective methylases but in principle work for all m^5C:RNA methylases [14, 41]. However, only miCLIP detects the methylated sites at nucleotide resolution [19, 42]. RNA immunoprecipitation with antibodies followed by high throughput sequencing has been also used to detect m^5C (m^5C-RIP) in tRNA, rRNA and a few coding RNAs in prokaryotes (Fig. 9.1b) [33].

9.3 Regulatory Functions of m^5C in RNA

The chemistry of the deposition of a methyl mark on cytosines is well understood and mediated by a large family of highly conserved m^5C RNA methylases. All RNA:m^5C methyltransferases contain a catalytic domain with a common structural core and the S-Adenosyl methionine (AdoMet)-binding site [14]. Two conserved cysteines, both located within a sequence with similarity to a methyltransferase active site are required for the transfer of the methyl group [41]. To initiate methylation, the catalytic cysteine residue forms a covalent bond with the cytosine pyrimidine ring [43]. The second conserved cysteine residue is then required to break the covalent adduct and releases the methylated RNA and the enzyme [41, 44].

To date, eight enzymes have been confirmed to methylate carbon-5 of cytosines (m^5C) in RNA, NSun1-7 and Dnmt2 (Fig. 9.2) [45]. All m^5C RNA methylases are thought to methylate cytosines in a non-redundant manner, and additional target specificity is given by their distinct subcellular localization (Fig. 9.2). The DNA methyltransferase homolog Dnmt2, NSun2, NSun3 and NSun6 all target transfer RNA (tRNA), yet in a non-overlapping and strict site-specific manner [32, 46–49]. NSun2 is currently the only m5C:RNA methylase with confirmed broader substrate specificity. In addition to tRNA, NSun2 also methylates other non-coding RNAs and a small subset of coding RNAs [19, 21, 50, 51].

The functional consequences of Dnmt2- and NSun2-mediated methylation of tRNAs are now well-understood. Deposition of m^5C by both enzymes protects tRNAs from endonucleolytic cleavage [32, 47, 52]. Lack of NSun2-mediated deposition of m^5C at the variable loop increases the affinity of the tRNA to angiogenin. Angiogenin is an endonuclease that cleaves tRNAs into two small non-coding RNA fragments (Fig. 9.3) [53]. In NSun2-depleted cells, often only the 5′ tRNA-derived small non-coding RNA fragments accumulate, whereas the 3′ tRNA fragments are not detectable [32]. The molecular function of 5′ tRNA fragments is to repress cap-dependent translation by displacing translation initiation and elongation factors from mRNAs or by interfering with efficient transpeptidation [54–57]. Accordingly, global protein synthesis is reduced in the absence of NSun2 *in vitro* and *in vivo* (Fig. 9.3) [32, 50].

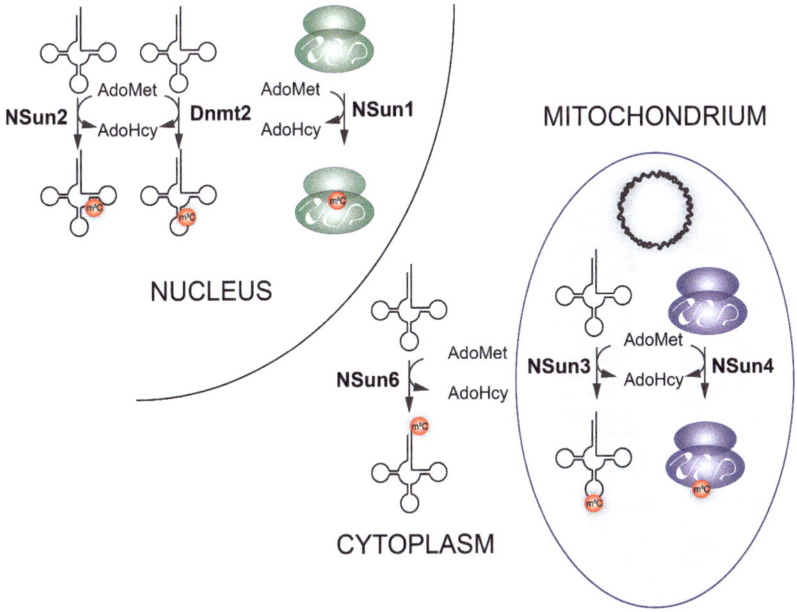

Fig. 9.2 Subcellular localization of the m^{5}C RNA methylases Dnmt2, NSun1-4 and 6. NSun1, 2, and Dnmt2 localize to the nucleus and methylate tRNA or rRNA respectively. NSun6 is found in the cytoplasm and NSun3 and 4 are mitochondrial RNA methylases targeting tRNA and rRNA respectively

In contrast, loss of Dnmt2-mediated methylation of tRNA Asp^{GTC}, Gly^{GCC}, and Val^{AAC} at the anticodon loop causes tRNA-specific fragmentation patterns which leads to specific codon mistranslation [58]. Cytosine-5 methylation of vault non-coding RNAs by NSun2 alters their processing into Argonaute-associated small RNA fragments that can function as microRNAs and regulate expression of targeted mRNAs [19]. The functional roles of the deposition of m^{5}C into coding RNAs remain unclear [20]. Synthetic cytosine-5 methylated mRNAs exhibit increased stability, and loss of methylation in the 3′UTR of p16 has been reported to reduce its stability [59, 60]. However, other studies found no correlation of potentially methylated mRNAs with turnover or stability, indicating that the methyl mark might rather determine cellular localisation or influence protein translation rates [19].

Less is known about the other NSun proteins. Human NSun3 localizes to mitochondria and methylates mitochondrial methionine tRNA (mt-tRNA Met) at the wobble base (C34) [49, 61, 62]. NSun3 methylation of mt-tRNA Met is crucial for efficient mitochondrial protein synthesis and respiratory functions. Loss-of-function mutations in human NSun3 are linked to combined mitochondrial respiratory chain complex deficiency [49].

NSun4-dependent methylation of position C911 in 12S ribosomal RNA in mammalian is essential because NSun4-deletion is embryonic lethal in mice [63]. NSun4 and mitochondrial transcription termination factor 4 (MTERF4) form a complex

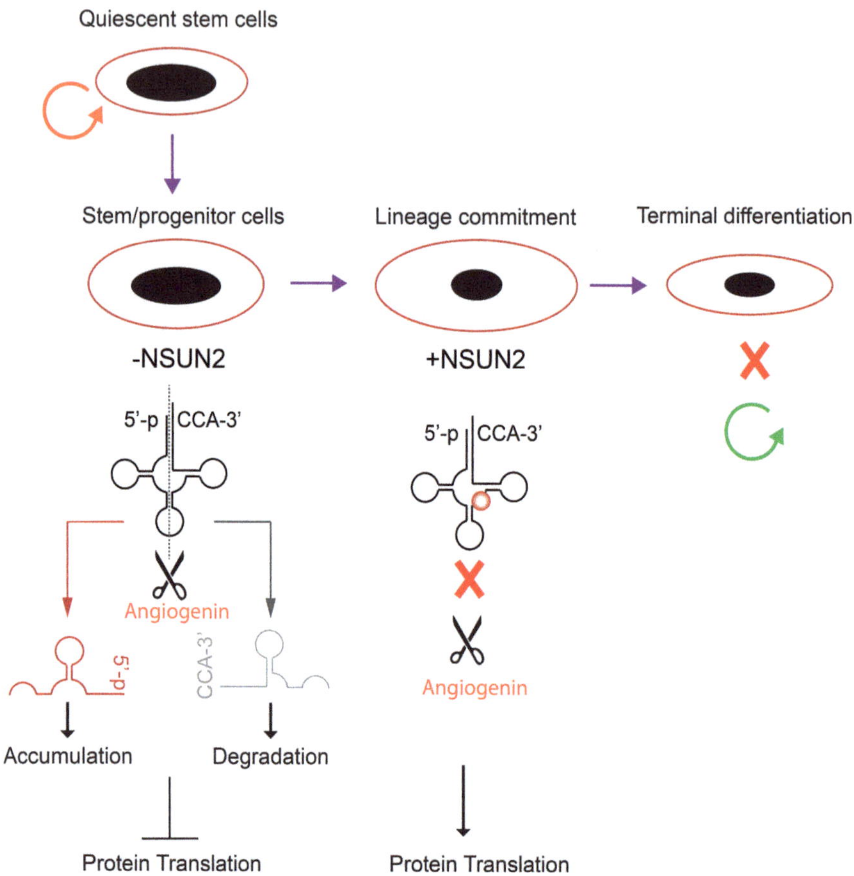

Fig. 9.3 Expression and function of NSun2 in stem, progenitor and differentiated epidermal cells. Stem and progenitor cells express no or low levels of NSun2 leading to hypo-methylation of tRNAs and their cleavage by angiogenin. The 5′ tRNA-derived non-coding RNAs accumulate and inhibit global protein synthesis. External stimuli triggering lineage commitment up-regulate NSun2 and methylated tRNAs are protected from cleavage leading to enhanced protein translation rates. Terminal differentiated cells lack NSun2

necessary for the full maturation of mitochondrial ribosomes by assembling the small, and large ribosomal subunits into monosomes [64, 65]. However, MTERF4 is dispensable for the NSun4 methyl-transferase activity [63].

The yeast homologue of NSun5 (RCM1) targets 25S rRNA at position C2278 for methylation. Methylation of 25S rRNA alters its structure causing global translational changes, which modulate lifespan, and impair stress responses [66]. NSun1 (also known as NOP2 and p120) methylates 25S rRNA at position C2870 [67, 68]. NSun1 regulates the cell cycle and is up-regulated in cancer cells [69]. Recent evidence suggests that NSun7 regulates enhancer RNA function [70]. Loss-of-function mutations in NSun7 are linked to male infertility in humans [71, 72].

9.4 Regulatory Functions of m^6A in RNA

The most abundant internal modification in mRNA is m^6A [23, 28, 73]. Global mapping using m^6A-specific antibodies revealed that mRNA methylation is conserved, widespread and dynamic in eukaryotes [37]. The m^6A modification is installed by a multi-enzyme complex, which contains methyltransferase-like 3 (Mettl3, also known as MT-A70) and methyltransferase-like 14 (Mettl14) at its core, and is associated with additional regulatory factors such as WTAP (Wilm's tumour 1 associating protein) [74, 75]. Transcriptome-wide analyses indicate that METLL3, METTL14 and WTAP share around 36% of binding sites and overlap with the m^6A consensus motif [76]. A large proportion of binding sites also occur in intergenic regions and introns, indicating a role for the methyl mark in pre-mRNA processing and splicing [36, 37]. Dicer-dependent microRNAs were also found to modulate Mettl3 binding to mRNAs with shared sequences, and therefore act as an auxiliary mechanism for m^6A disposition [77]. Silencing of the methyltransferase complex leads to enhanced abundance of m^6A target transcripts, supporting a role of m^6A as a negative regulator of gene expression [37]. However, negative regulation of expression may be at least in part indirect, since m^6A also marks primary microRNAs for processing, thereby promoting initiation of miRNA biogenesis [35].

In contrast to mammalian m^5C, demethylases for the m^6A modification are well-defined. FTO (fat mass and obesity-associated protein) and ALKBH5 (Alpha-Ketoglutarate-Dependent Dioxygenase AlkB Homolog 5) are both able to erase m^6A [78, 79]. Homozygous FTO deficient mice are sub-viable and loss of FTO leads to postnatal growth retardation, significant reduction in adipose tissue, and lean body mass [80]. FTO enzymatic activity is essential for normal human development of the central nervous and cardiovascular systems, and cultured skin fibroblasts from affected subjects show impaired proliferation and accelerated senescence [81]. Similar to m^5C, several proteins have been described to detect m^6A in RNA. YTH domain-containing proteins and heterogenous nuclear ribonucleoprotein (hnRNPs) have been described as "*readers*" of m^6A [82–85]. YTHDF2 mediates mRNA decay, thereby suggesting a role for m^6A RNA as a negative regulator of gene expression [86]. On the other hand, YTHDF1 binding to m^6A-decorated mRNAs promotes translation efficiency by interacting with the translation machinery [87]. Thus, m^6A facilitates the dynamic binding of several regulatory proteins that modulate gene expression, maturation and translation [37]. Recently, a novel nuclear m^6A reader called YTHDC1 was found to promote exon inclusion on modified mRNA [88]. Mechanistically, YTHDC1 binds to m^6A sites within exons and recruit the pre-mRNA processing protein SRSF3, but inhibit SRSF10 binding in order to maintain exons. Lack of YTHDC1 paves the way for SRSF10 binding and as a result exon exclusion of previous included exons [88]. Altogether, m^6A dynamic pattern plays crucial roles in regulating gene expression by facilitating RNA binding proteins, which modulate RNA splicing, RNA metabolism, translation, and micorRNA processing.

9.5 The Importance of m^5C and m^6A in Stem Cell Function and Epidermal Differentiation

The biological consequence of the presence of m^6A in mRNA has been described for mouse and human embryonic stem cells (ESC) and modulates the transition of the pluripotent state towards differentiation. ESC are derived from the inner cell mass of pre-implantation embryos and are pluripotent as they self-renew and retain their ability differentiate into multiple cell types *in vitro* [89–91]. The methyltransferases METTL3 and METTL14 are expressed in ESC, and transcriptome-wide m^6A profiling in mouse and human ESC showed that many core pluripotent genes and developmental regulators carry the modification in their transcripts [36, 75, 92]. METTL3 knockout in ESC promotes self-renewal, but impairs differentiation into mature cardiomyocytes and neurons [92]. METTL3 and 14 have also been associated with the self-renewal of cancer stem cells in human glioblastoma [93]. Molecularly, knockdown of both METTL3 and 14 in glioblastoma increases self-renewal and overall tumorigenicity. In contrast, overexpression of METTL3 or inhibition of m^6A demethylase FTO, suppress initial glioblastoma growth [93].

Studies on the effects of deletions and mutations of m^6A writers, readers and erasers indicate that m^6A methylation also plays important roles in development and tissue differentiation processes. For instance, ablation of METTL3 is embryonic lethal. METTL3 post-implantation embryos retain expression of pluripotent markers, but fail to up-regulate early differentiation markers [36, 92]. Although the role of METTL14 in development has not been described, WTAP, another component of the methylase complex, is an essential factor for early embryonic development. WTAP stabilizes mRNAs involved in cell cycle progression, and its deletion in mice is embryonic lethal due to failure of proliferation in the early [94]. The functions of m^6A "readers" in embryonic development or stem cell fate regulation are not yet known; however a polymorphism in the YTHDF2 gene is associated with human longevity, suggesting a possible role in stem cells and ageing control [95]. Importantly, loss-of-function mutations of the demethylase FTO in humans lead to an autosomal-recessive lethal syndrome characterized by severe growth retardation, microcephaly, psychomotor delay, cardiac deficits, and multiple congenital malformations. At least some of these effects may be due to impaired proliferation and accelerated senescence [81, 96]. Similarly, FTO deficiency in mice leads to postnatal lethality, growth retardation, and multiple malformations [80]. Deletion of the other known demethylase AlkBH5 is not embryonic lethal but causes male infertility in mice due to failure of spermatocytes to initiate and proceed through differentiation [79].

While the precise underlying mechanisms are still not fully understood, it is notable that post-transcriptional pathways are often not required for stem cell self-renewal but are linked to impaired tissue and organ differentiation. Thus, RNA modification pathways are required for the accurate temporal and spatial activation of cellular differentiation programs.

The only post-transcriptional modification that has been functionally linked to epidermal stem cell differentiation and skin homeostasis is the NSun2-mediated formation of m^5C [32, 50, 97, 98]. The expression pattern of NSun2 during mouse development is dynamic and can be detected as early as E3.5 in the inner cell mass of the blastocyst [98]. During development, high levels of NSun2 become gradually restricted to skin, brain and testis. Total deletion of NSun2 delays cell differentiation during the development of all three of these tissues [10, 47, 98, 99]. In the brain, loss-of-function of NSun2 causes neurodevelopmental disorders (i.e. microcephaly) and cognitive deficits in mice and humans [32, 100–104].

Expression of NSun2 is predominantly found in the suprabasal layer of the interfollicular epidermis (IFE) at E15.5 of mouse embryonic development [98]. In adult mouse skin, high expression of NSun2 is confined to the hair follicle and changes dynamically with the hair cycle. The hair follicle offers an excellent model to study regeneration and stem cell fate, as it undergoes cyclic phases of growth (anagen), apoptosis-mediated regression (catagen) and rest (telogen) [105]. In resting skin, NSun2 expression is low or absent in all skin epithelial cells. At the onset of hair growth, when the progeny of bulge stem cells migrate out of their niche and undergo proliferation at the hair germ as progenitor cells, NSun2 expression appears in the lower bulge and at the hair germ. During anagen, NSun2 is most highly expressed in the hair matrix, in committed progenitor cells before they differentiate into the hair shaft [98].

Ablation of NSun2 in skin causes accumulation of quiescent hair follicle stem cells in the hair follicle bulge at the onset of hair growth (anagen), leading to a delay in differentiation of hair follicle lineages [98]. Increased quiescence of NSun2-negative bulge cells is maintained *in vitro*, and cultured NSun2-depleted bulge stem cells fail to differentiate as efficiently as wild type bulge stem cells (Fig. 9.3) [98]. The delay in stem cell activation from the quiescent state may underlie the development of alopecia observed in aged mutant mice.

The delay of differentiation in the absence of NSUN2 can be explained by the finding that NSun2-mediated m^5C is needed for both neural and epidermal cell migration [50, 51]. During brain development, ablation of NSun2 results in a decreased number of differentiated neurons in the upper-layer of the brain cortex at E16.5. It is postulated that lack of NSun2 mediated m^5C drives the accumulation of 5′-derived tRNA fragments, which renders neural stem cells incapable of reacting to differentiation signals [51].

NSun2 methylates around 80% of all expressed tRNAs, and hypo-methylated tRNAs formed in the absence of NSun2 are cleaved by angiogenin, causing the intracellular accumulation of cleaved 5′ tRNA fragments [32]. Cleavage of tRNAs is a conserved response to several stress stimuli in eukaryotes and the resulting tRNA fragments repress protein translation [54, 57, 106]. Reduction of global protein synthesis is a well-known and integral part of stress responses to allow cells to alleviate cellular injury or alternatively induce apoptosis [107]. Low protein synthesis is also a hallmark of stem cells in skin, brain and the hematopoietic system [50, 108, 109], supporting the hypothesis that RNA methylation pathways are functionally involved in lineage commitment but not self-renewal.

In line with the functional role of cleaved tRNAs in response to stress, NSun2−/− cells and tissues exhibit both reduction of global protein synthesis and an activated stress response [32, 50]. In skin, the accumulation of 5'tRNA fragments activates oxidative stress pathways, and NSun2−/− skin is hyper-sensitive to UV light induced stress [32]. Importantly, 5′ tRNA fragments are sufficient and required to induce cellular stress pathways, and brain developmental deficits in NSun2 mutants can be rescued when tRNA cleavage is inhibited through the administration of an angiogenin inhibitor during development [32]. Thus, NSun2-mediated RNA methylation modulates an appropriate and rapid cellular response to a changing microenvironment in the response to external stress stimuli, and down-stream of stem cell activating factors.

Together, deletion of NSun2 causes tRNA cleavage and the tRNA-derived small non-coding RNAs inhibit global protein synthesis. The repression of protein translation rate renders epidermal cells less responsive to external stimuli and thereby cause the accumulation of stem and progenitor cells. Tipping the balance between self-renewal and differentiation pathways towards stem cell self-renewal impacts cellular sensitivity to tumour development. Accordingly, NSun2-knockout mice are more susceptible to skin tumour formation, and NSun2 is up-regulated in human epithelial tumours [50, 97, 110]. Tumour-initiating cells lack NSun2 and are low-translating when compared to non-tumourigenic cells [50]. However, in line with the finding that accumulation of tRNA-derived non-coding RNAs not only inhibits protein synthesis but also activates the stress response, NSun2-negative tumours fail to regenerate after chemotherapeutic drug treatments [50]. Thus, NSun2 is required for cell survival in response to stress stimuli, and the modulation of RNA methylation pathways may represent a novel strategy for cancer treatments.

9.6 Conclusions

Post-transcriptional modifications such as cytosine-5 methylation and N6-methyladenosine are dynamically regulated and are found in coding and non-coding RNAs. The precise mechanisms and functional roles of post-transcriptional methylation in regulating RNA functions are complex but appear to converge in modulating protein synthesis in response to external stimuli, such as stress or differentiation-inducing signals. Distinct post-transcriptional modifications are often dispensable for stem cell self-renewal, but are implicated in the accurate temporal and spatial activation of cellular differentiation programs. Aberrant deposition of methyl marks on RNAs impairs adaptive cellular responses and differentiation pathways, and is linked to complex human diseases and cancer.

Acknowledgments The authors thank members of the Frye Laboratory for helpful discussion. M.F. is supported by grants from Cancer Research UK (CR-UK), the European Research Council (ERC), the Medical Research Council (MRC), and Worldwide Cancer Research (WCR). A.S. was funded by a scholarship from the University of Tabuk, Saudi Arabia.

References

1. Lopez-Pajares V, et al. Genetic pathways in disorders of epidermal differentiation. Trends Genet. 2013;29(1):31–40.
2. Hsu YC, Li L, Fuchs E. Emerging interactions between skin stem cells and their niches. Nat Med. 2014;20(8):847–56.
3. Frye M, Benitah SA. Chromatin regulators in mammalian epidermis. Semin Cell Dev Biol. 2012;23(8):897–905.
4. Botchkarev VA. Epigenetic regulation of epidermal development and keratinocyte differentiation. J Investig Dermatol Symp Proc. 2015;17(1):18–9.
5. Perdigoto CN, et al. Epigenetic regulation of epidermal differentiation. Cold Spring Harb Perspect Med. 2014;4(2):1–19.
6. Machnicka MA, et al. MODOMICS: a database of RNA modification pathways--2013 update. Nucleic Acids Res. 2013;41(Database issue):D262–7.
7. Cantara WA, et al. The RNA modification database, RNAMDB: 2011 update. Nucleic Acids Res. 2011;39(Database issue):D195–201.
8. Helm M, Alfonzo JD. Posttranscriptional RNA modifications: playing metabolic games in a cell's chemical Legoland. Chem Biol. 2014;21(2):174–85.
9. Torres AG, Batlle E, Ribas de Pouplana L. Role of tRNA modifications in human diseases. Trends Mol Med. 2014;20:306.
10. Blanco S, Frye M. Role of RNA methyltransferases in tissue renewal and pathology. Curr Opin Cell Biol. 2014;31C:1–7.
11. Powell CA, Nicholls TJ, Minczuk M. Nuclear-encoded factors involved in post-transcriptional processing and modification of mitochondrial tRNAs in human disease. Front Genet. 2015;6:79.
12. Li S, Mason CE. The pivotal regulatory landscape of RNA modifications. Annu Rev Genomics Hum Genet. 2014;15:127–50.
13. Towns WL, Begley TJ. Transfer RNA methytransferases and their corresponding modifications in budding yeast and humans: activities, predications, and potential roles in human health. DNA Cell Biol. 2012;31(4):434–54.
14. Bujnicki JM, et al. Sequence-structure-function studies of tRNA:m5C methyltransferase Trm4p and its relationship to DNA:m5C and RNA:m5U methyltransferases. Nucleic Acids Res. 2004;32(8):2453–63.
15. Motorin Y, Lyko F, Helm M. 5-methylcytosine in RNA: detection, enzymatic formation and biological functions. Nucleic Acids Res. 2010;38(5):1415–30.
16. Burgess AL, David R, Searle IR. Conservation of tRNA and rRNA 5-methylcytosine in the kingdom Plantae. BMC Plant Biol. 2015;15:199.
17. Dominissini D, et al. Topology of the human and mouse m6A RNA methylomes revealed by m6A-seq. Nature. 2012;485(7397):201–6.
18. Meyer KD, et al. Comprehensive analysis of mRNA methylation reveals enrichment in 3′ UTRs and near stop codons. Cell. 2012;149(7):1635–46.
19. Hussain S, et al. NSun2-mediated cytosine-5 methylation of vault noncoding RNA determines its processing into regulatory small RNAs. Cell Rep. 2013;4(2):255–61.
20. Squires JE, et al. Widespread occurrence of 5-methylcytosine in human coding and noncoding RNA. Nucleic Acids Res. 2012;40(11):5023–33.
21. Khoddami V, Cairns BR. Identification of direct targets and modified bases of RNA cytosine methyltransferases. Nat Biotechnol. 2013;31(5):458–64.
22. Desrosiers R, Friderici K, Rottman F. Identification of methylated nucleosides in messenger RNA from Novikoff hepatoma cells. Proc Natl Acad Sci U S A. 1974;71(10):3971–5.
23. Perry RP, et al. The methylated constituents of L cell messenger RNA: evidence for an unusual cluster at the 5′ terminus. Cell. 1975;4(4):387–94.
24. Rottman F, Shatkin AJ, Perry RP. Sequences containing methylated nucleotides at the 5′ termini of messenger RNAs: possible implications for processing. Cell. 1974;3(3):197–9.

25. Narayan P, Rottman FM. An in vitro system for accurate methylation of internal adenosine residues in messenger RNA. Science. 1988;242(4882):1159–62.
26. Csepany T, et al. Sequence specificity of mRNA N6-adenosine methyltransferase. J Biol Chem. 1990;265(33):20117–22.
27. Rottman FM, et al. N6-adenosine methylation in mRNA: substrate specificity and enzyme complexity. Biochimie. 1994;76(12):1109–14.
28. Dubin DT, Taylor RH. The methylation state of poly A-containing messenger RNA from cultured hamster cells. Nucleic Acids Res. 1975;2(10):1653–68.
29. Maden BE. Locations of methyl groups in 28 S rRNA of Xenopus laevis and man. Clustering in the conserved core of molecule. J Mol Biol. 1988;201(2):289–314.
30. Gambaryan AS, Venkstern TV, Bayev AA. On the mechanism of tRNA methylase-tRNA recognition. Nucleic Acids Res. 1976;3(8):2079–87.
31. Schaefer M, et al. RNA cytosine methylation analysis by bisulfite sequencing. Nucleic Acids Res. 2009;37(2):e12.
32. Blanco S, et al. Aberrant methylation of tRNAs links cellular stress to neuro-developmental disorders. EMBO J. 2014;33(18):2020–39.
33. Edelheit S, et al. Transcriptome-wide mapping of 5-methylcytidine RNA modifications in bacteria, archaea, and yeast reveals m(5)C within archaeal mRNAs. PLoS Genet. 2013;9(6):e1003602.
34. Hussain S, et al. Characterizing 5-methylcytosine in the mammalian epitranscriptome. Genome Biol. 2013;14(11):215.
35. Alarcon CR, et al. N6-methyladenosine marks primary microRNAs for processing. Nature. 2015;519(7544):482–5.
36. Geula S, et al. Stem cells. m6A mRNA methylation facilitates resolution of naive pluripotency toward differentiation. Science. 2015;347(6225):1002–6.
37. Yue Y, Liu J, He C. RNA N6-methyladenosine methylation in post-transcriptional gene expression regulation. Genes Dev. 2015;29(13):1343–55.
38. Ke S, et al. m6A mRNA modifications are deposited in nascent pre-mRNA and are not required for splicing but do specify cytoplasmic turnover. Genes Dev. 2017;31(10):990–1006.
39. Linder B, et al. Single-nucleotide-resolution mapping of m6A and m6Am throughout the transcriptome. Nat Methods. 2015;12(8):767–72.
40. Liu N, Pan T. Probing N(6)-methyladenosine (m(6)a) RNA modification in total RNA with SCARLET. Methods Mol Biol. 2016;1358:285–92.
41. King MY, Redman KL. RNA methyltransferases utilize two cysteine residues in the formation of 5-methylcytosine. Biochemistry. 2002;41(37):11218–25.
42. Sugimoto Y, et al. Analysis of CLIP and iCLIP methods for nucleotide-resolution studies of protein-RNA interactions. Genome Biol. 2012;13(8):R67.
43. Liu Y, Santi DV. m5C RNA and m5C DNA methyl transferases use different cysteine residues as catalysts. Proc Natl Acad Sci U S A. 2000;97(15):8263–5.
44. Redman KL. Assembly of protein-RNA complexes using natural RNA and mutant forms of an RNA cytosine methyltransferase. Biomacromolecules. 2006;7(12):3321–6.
45. Frye M, Blanco S. Post-transcriptional modifications in development and stem cells. Development. 2016;143(21):3871–81.
46. Haag S, et al. NSUN6 is a human RNA methyltransferase that catalyzes formation of m5C72 in specific tRNAs. RNA. 2015;21:1532.
47. Tuorto F, et al. RNA cytosine methylation by Dnmt2 and NSun2 promotes tRNA stability and protein synthesis. Nat Struct Mol Biol. 2012;19(9):900–5.
48. Goll MG, et al. Methylation of tRNAAsp by the DNA methyltransferase homolog Dnmt2. Science. 2006;311(5759):395–8.
49. Van Haute L, et al. Deficient methylation and formylation of mt-tRNA(met) wobble cytosine in a patient carrying mutations in NSUN3. Nat Commun. 2016;7:12039.
50. Blanco S, et al. Stem cell function and stress response are controlled by protein synthesis. Nature. 2016;534(7607):335–40.

51. Flores JV, et al. Cytosine-5 RNA methylation regulates neural stem cell differentiation and motility. Stem Cell Reports. 2017;8(1):112–24.
52. Schaefer M, et al. RNA methylation by Dnmt2 protects transfer RNAs against stress-induced cleavage. Genes Dev. 2010;24(15):1590–5.
53. Fu H, et al. Stress induces tRNA cleavage by angiogenin in mammalian cells. FEBS Lett. 2009;583(2):437–42.
54. Ivanov P, et al. Angiogenin-induced tRNA fragments inhibit translation initiation. Mol Cell. 2011;43(4):613–23.
55. Spriggs KA, Bushell M, Willis AE. Translational regulation of gene expression during conditions of cell stress. Mol Cell. 2010;40(2):228–37.
56. Sobala A, Hutvagner G. Small RNAs derived from the 5′ end of tRNA can inhibit protein translation in human cells. RNA Biol. 2013;10(4):553–63.
57. Gebetsberger J, et al. tRNA-derived fragments target the ribosome and function as regulatory non-coding RNA in Haloferax volcanii. Archaea. 2012;2012:260909.
58. Tuorto F, et al. The tRNA methyltransferase Dnmt2 is required for accurate polypeptide synthesis during haematopoiesis. EMBO J. 2015;34:2350.
59. Warren L, et al. Highly efficient reprogramming to pluripotency and directed differentiation of human cells with synthetic modified mRNA. Cell Stem Cell. 2010;7(5):618–30.
60. Zhang X, et al. The tRNA methyltransferase NSun2 stabilizes p16INK(4) mRNA by methylating the 3′-untranslated region of p16. Nat Commun. 2012;3:712.
61. Haag S, et al. NSUN3 and ABH1 modify the wobble position of mt-tRNAMet to expand codon recognition in mitochondrial translation. EMBO J. 2016;35:2104.
62. Nakano S, et al. NSUN3 methylase initiates 5-formylcytidine biogenesis in human mitochondrial tRNA(Met). Nat Chem Biol. 2016;12(7):546–51.
63. Metodiev MD, et al. NSUN4 is a dual function mitochondrial protein required for both methylation of 12S rRNA and coordination of mitoribosomal assembly. PLoS Genet. 2014;10(2):e1004110.
64. Spahr H, et al. Structure of the human MTERF4-NSUN4 protein complex that regulates mitochondrial ribosome biogenesis. Proc Natl Acad Sci U S A. 2012;109(38):15253–8.
65. Yakubovskaya E, et al. Structure of the essential MTERF4:NSUN4 protein complex reveals how an MTERF protein collaborates to facilitate rRNA modification. Structure. 2012;20(11):1940–7.
66. Schosserer M, et al. Methylation of ribosomal RNA by NSUN5 is a conserved mechanism modulating organismal lifespan. Nat Commun. 2015;6:6158.
67. Sharma S, et al. Identification of a novel methyltransferase, Bmt2, responsible for the N-1-methyl-adenosine base modification of 25S rRNA in Saccharomyces cerevisiae. Nucleic Acids Res. 2013;41(10):5428–43.
68. Bourgeois G, et al. Eukaryotic rRNA modification by yeast 5-Methylcytosine-Methyltransferases and human proliferation-associated antigen p120. PLoS One. 2015;10(7):e0133321.
69. Jhiang SM, Yaneva M, Busch H. Expression of human proliferation-associated nucleolar antigen p120. Cell Growth Differ. 1990;1(7):319–24.
70. Aguilo F, et al. Deposition of 5-Methylcytosine on enhancer RNAs enables the Coactivator function of PGC-1alpha. Cell Rep. 2016;14(3):479–92.
71. Harris T, et al. Sperm motility defects and infertility in male mice with a mutation in Nsun7, a member of the Sun domain-containing family of putative RNA methyltransferases. Biol Reprod. 2007;77(2):376–82.
72. Khosronezhad N, Colagar AH, Jorsarayi SG. T26248G-transversion mutation in exon7 of the putative methyltransferase Nsun7 gene causes a change in protein folding associated with reduced sperm motility in asthenospermic men. Reprod Fertil Dev. 2015;27:471–80.
73. Wei CM, Gershowitz A, Moss B. Methylated nucleotides block 5′ terminus of HeLa cell messenger RNA. Cell. 1975;4(4):379–86.

74. Ping XL, et al. Mammalian WTAP is a regulatory subunit of the RNA N6-methyladenosine methyltransferase. Cell Res. 2014;24(2):177–89.
75. Wang Y, et al. N6-methyladenosine modification destabilizes developmental regulators in embryonic stem cells. Nat Cell Biol. 2014;16(2):191–8.
76. Liu J, et al. A METTL3-METTL14 complex mediates mammalian nuclear RNA N6-adenosine methylation. Nat Chem Biol. 2014;10(2):93–5.
77. Chen T, et al. m(6)A RNA methylation is regulated by microRNAs and promotes reprogramming to pluripotency. Cell Stem Cell. 2015;16(3):289–301.
78. Jia G, et al. N6-methyladenosine in nuclear RNA is a major substrate of the obesity-associated FTO. Nat Chem Biol. 2011;7(12):885–7.
79. Zheng G, et al. ALKBH5 is a mammalian RNA demethylase that impacts RNA metabolism and mouse fertility. Mol Cell. 2013;49(1):18–29.
80. Fischer J, et al. Inactivation of the Fto gene protects from obesity. Nature. 2009;458(7240):894–8.
81. Boissel S, et al. Loss-of-function mutation in the dioxygenase-encoding FTO gene causes severe growth retardation and multiple malformations. Am J Hum Genet. 2009;85(1):106–11.
82. Wang X, et al. N6-methyladenosine-dependent regulation of messenger RNA stability. Nature. 2014;505(7481):117–20.
83. Xu C, et al. Structural basis for selective binding of m6A RNA by the YTHDC1 YTH domain. Nat Chem Biol. 2014;10(11):927–9.
84. Schwartz S, et al. High-resolution mapping reveals a conserved, widespread, dynamic mRNA methylation program in yeast meiosis. Cell. 2013;155(6):1409–21.
85. Liu N, et al. N(6)-methyladenosine-dependent RNA structural switches regulate RNA-protein interactions. Nature. 2015;518(7540):560–4.
86. Wang X, He C. Reading RNA methylation codes through methyl-specific binding proteins. RNA Biol. 2014;11(6):669–72.
87. Wang X, et al. N(6)-methyladenosine modulates messenger RNA translation efficiency. Cell. 2015;161(6):1388–99.
88. Xiao W, et al. Nuclear m(6)A reader YTHDC1 regulates mRNA splicing. Mol Cell. 2016;61(4):507–19.
89. Evans MJ, Kaufman MH. Establishment in culture of pluripotential cells from mouse embryos. Nature. 1981;292(5819):154–6.
90. Martin GR. Isolation of a pluripotent cell line from early mouse embryos cultured in medium conditioned by teratocarcinoma stem cells. Proc Natl Acad Sci U S A. 1981;78(12):7634–8.
91. Thomson JA, et al. Embryonic stem cell lines derived from human blastocysts. Science. 1998;282(5391):1145–7.
92. Batista PJ, et al. m(6)A RNA modification controls cell fate transition in mammalian embryonic stem cells. Cell Stem Cell. 2014;15(6):707–19.
93. Cui Q, et al. m6A RNA methylation regulates the self-renewal and tumorigenesis of glioblastoma stem cells. Cell Rep. 2017;18(11):2622–34.
94. Horiuchi K, et al. Wilms' tumor 1-associating protein regulates G2/M transition through stabilization of cyclin A2 mRNA. Proc Natl Acad Sci U S A. 2006;103(46):17278–83.
95. Cardelli M, et al. A polymorphism of the YTHDF2 gene (1p35) located in an Alu-rich genomic domain is associated with human longevity. J Gerontol A Biol Sci Med Sci. 2006;61(6):547–56.
96. Daoud H, et al. Identification of a pathogenic FTO mutation by next-generation sequencing in a newborn with growth retardation and developmental delay. J Med Genet. 2016;53:200–7.
97. Frye M, Watt FM. The RNA methyltransferase Misu (NSun2) mediates Myc-induced proliferation and is upregulated in tumors. Curr Biol. 2006;16(10):971–81.
98. Blanco S, et al. The RNA-methyltransferase Misu (NSun2) poises epidermal stem cells to differentiate. PLoS Genet. 2011;7(12):e1002403.
99. Hussain S, et al. The mouse Cytosine-5 RNA Methyltransferase NSun2 is a component of the Chromatoid body and required for testis differentiation. Mol Cell Biol. 2013;33(8):1561–70.

100. Khan MA, et al. Mutation in NSUN2, which encodes an RNA methyltransferase, causes autosomal-recessive intellectual disability. Am J Hum Genet. 2012;90(5):856–63.
101. Martinez FJ, et al. Whole exome sequencing identifies a splicing mutation in NSUN2 as a cause of a Dubowitz-like syndrome. J Med Genet. 2012;49(6):380–5.
102. Abbasi-Moheb L, et al. Mutations in NSUN2 cause autosomal-recessive intellectual disability. Am J Hum Genet. 2012;90(5):847–55.
103. Komara M, et al. A novel single-nucleotide deletion (c.1020delA) in NSUN2 causes intellectual disability in an emirati child. J Mol Neurosci. 2015;57:393.
104. Ghadami S, et al. Frequencies of six (five novel) STR markers linked to TUSC3 (MRT7) or NSUN2 (MRT5) genes used for homozygosity mapping of recessive intellectual disability. Clin Lab. 2015;61(8):925–32.
105. Schepeler T, Page ME, Jensen KB. Heterogeneity and plasticity of epidermal stem cells. Development. 2014;141(13):2559–67.
106. Thompson DM, et al. tRNA cleavage is a conserved response to oxidative stress in eukaryotes. RNA. 2008;14(10):2095–103.
107. Wek RC, Jiang HY, Anthony TG. Coping with stress: eIF2 kinases and translational control. Biochem Soc Trans. 2006;34(Pt 1):7–11.
108. Signer RA, et al. Haematopoietic stem cells require a highly regulated protein synthesis rate. Nature. 2014;509(7498):49–54.
109. Llorens-Bobadilla E, et al. Single-cell transcriptomics reveals a population of dormant neural stem cells that become activated upon brain injury. Cell Stem Cell. 2015;17(3):329–40.
110. Frye M, et al. Genomic gain of 5p15 leads to over-expression of Misu (NSUN2) in breast cancer. Cancer Lett. 2010;289:71–80.

Chapter 10
Enhancer-Promoter Interactions and Their Role in the Control of Epidermal Differentiation

Inez Y. Oh and Cristina de Guzman Strong

Abbreviations

3C	Chromosome conformation capture
ChIP	Chromatin immunoprecipitation
CNE	Conserved noncoding element
CNS	Conserved noncoding sequence
CT	Chromosome Territory
CTCF	CCCTC-binding factor
DRR	Distal regulatory region
EDC	Epidermal Differentiation Complex
FLG	Filaggrin
HAT	Histone acetyltransferase
HDAC	Histone deactylase
IVL	Involucrin
LCE	Late cornified envelope
LCR	Locus control region
LOR	Loricrin
PAD	Peptidylarginine deaminase
SPRR	Small proline-rich region
TAD	Topologically associated domains

I. Y. Oh · C. de Guzman Strong (✉)
Division of Dermatology, Department of Medicine, Washington University in St. Louis School of Medicine, St. Louis, MO, USA

Center for Pharmacogenomics, Department of Medicine, Washington University in St. Louis School of Medicine, St. Louis, MO, USA

Center for the Study of Itch, Washington University in St. Louis School of Medicine, St. Louis, MO, USA
e-mail: cristinastrong@wustl.edu

© Springer International Publishing AG, part of Springer Nature 2018
V. A. Botchkarev, S. E. Millar (eds.), *Epigenetic Regulation of Skin Development and Regeneration*, Stem Cell Biology and Regenerative Medicine,
https://doi.org/10.1007/978-3-319-16769-5_10

10.1 Introduction

The epidermis at the surface of the skin provides a tractable and spatially hierarchical model to investigate the development of committed cells. The architecture of the epidermis provides the key structure for the physical barrier of the skin. Epidermal cells, or keratinocytes, in the most internal basal layer of the epidermis must strike a critical balance between self-renewal and differentiation in order to build a functional barrier across the entire body [1]. Keratinocyte self-renewal is characterized by parallel cell division within the basal layer. Expression of Keratin 5 (*K5*) and Keratin 14 (*K14*) marks these basal proliferating keratinocytes.

During differentiation, basal keratinocytes divide asymmetrically, giving rise to suprabasal keratinocytes that migrate outwards, entering the spinous layer [1–3]. Spinous and granular keratinocytes activate Keratin 1 (*K1*) and Keratin 10 (*K10)* expression concomitant with *K5*/*K14* downregulation. During late terminal differentiation, the keratinocytes coordinately express many Epidermal Differentiation Complex (EDC) genes, including filaggrin (*FLG*) and *FLG*-like, late cornified envelope (*LCE*), small proline-rich region (*SPRR*), and *S100* [4–7]. As the keratinocytes reach the outermost stratum corneum, they enucleate and are surrounded by a cornified envelope, the basic structural unit of the skin barrier, which is formed by transglutaminase-1-mediated cross-linking of scaffold proteins. The cornified envelopes are sealed together by keratinocyte-derived extruded lipids to form a semi-permeable barrier [2]. In mice, the pattern of functional barrier acquisition corresponds with maturation of the cornified envelopes, and proceeds from specific dorsal initiation sites at embryonic day (E)16, spreading to converge at the dorsal and ventral midline so that the whole embryo is impermeable by E17 [8].

Epidermal differentiation can be recapitulated *in vitro* by exposure of proliferating keratinocytes to high calcium levels [9, 10]. This process, called calcium switching, stimulates calcium receptor (CaR) and downstream phosphokinase C (PKC) signaling, which activates the Fos/Jun family of transcription factors that play important roles in keratinocyte differentiation (reviewed in [11]). Fos and Jun proteins form homo- or heterodimers that compose the AP-1 transcription factor complex [12]. In normal epidermis as well as in organotypic epidermal cultures, the expression pattern of AP-1 proteins is tightly regulated even within the differentiated layers [13]. Fos proteins are found in the nuclei of both basal and suprabasal keratinocytes. JunB and JunD are expressed in all layers of normal epidermis. Interestingly, c-Jun is expressed in the spinous layer, then disappears and reemerges in the outermost granular layer directly at the transition zone to the stratum corneum. As will be discussed in this chapter, many of the genes expressed in keratinocytes, in either proliferative or differentiated layers of the epidermis, contain AP-1 binding sites in their promoter regions. This suggests that specific combinations of AP-1 protein complexes bind to the enhancers for genes expressed at successive stages of epidermal differentiation.

Transcriptional regulation clearly plays a major role in development of the epidermis. So how does the keratinocyte decipher the genome to activate a transcriptional program specific to epidermal differentiation? Early studies of gene regulation focused on single gene promoters and the transcription factor binding sites contained within [14]. However, not all transcriptional activation is attributable to biochemical activity at the gene promoter, thus suggesting the contribution of other loci. Complete sequencing of the genomes of humans and model organisms has revolutionized our ability to further define, and ascertain the functions of, noncoding sequences in the regulation of gene expression [14–16]. Here we discuss the history of the conceptual advances in our understanding of the roles of enhancer elements in epidermal development. We discuss known functions for enhancer regulation of key genes that define the stages of epidermal development; the genetic, genomic and epigenetic features that allow us to identify enhancers; and the approaches that will enable a thorough understanding of the dynamic role of enhancers in gene activation.

10.2 What Is an Enhancer?

The concept of an "enhancer" emerged in 1981. In that year, Pierre Chambon and George Khoury independently discovered a non-coding 72 bp tandem repeat sequence, upstream of the SV40 early gene promoters that was required for transcription [17, 18]. A subsequent study from Walter Schaffner identified the ability of the SV40 DNA sequence to "enhance" the expression of rabbit β-globin even when the SV40 sequence was placed thousands of base pairs away from the β-globin gene promoter in an expression vector [19]. Chambon observed similar results using the gene for conalbumin [20]. Further experiments performed by Paul Berg and Michael Fromm showed that the SV40 sequence was able to "enhance" transcription independent of its location (upstream or downstream of its target gene) and orientation (forward or reverse) [21]. This established the SV40 sequence as the prototype of a novel genetic element, an enhancer, and established the definition of a classical enhancer as a non-coding sequence that can modulate gene expression in a position- and orientation-independent manner.

The discovery of the SV40 enhancer paved the way for the identification of enhancers in other tissue types [15]. Often, searches for enhancers were prioritized and interrogated in the sequences surrounding the target genes that included upstream or downstream sequences, including 5′ and 3′ untranslated regions (UTRs), introns, and intergenic regions (reviewed in [22]).

In vivo studies of putative enhancers identified the spatiotemporal specificity of these sequences to ensure biologically relevant cell- and tissue-specific gene expression (reviewed in [23, 24]). Biochemical studies further identified clusters or arrays of transcription factor binding sites that act as "building blocks" of cis-regulatory modules that also show enhancer activity (reviewed in [15, 16, 25]). The enrichment of multiple transcription factor binding sites within an enhancer facilitates cell-specific expression largely attributable to combinatorial and differential binding of transcription factor family members in the context of different microenvironments.

Based on these later studies, we more loosely define an enhancer as a non-coding sequence, containing clusters of transcription factor binding sites, that drives cell-, tissue-, or developmental stage-specific gene expression. Subsequent findings, described below, paved the way for the application of comparative genomics to enhancer discovery.

10.3 Transcriptional Regulation in Epidermal Development by Non-Promoter Sequences

Prior to the availability of whole genome sequences, biological insights into the transcriptional activation of key epidermal differentiation target genes initially focused on sequences immediately upstream of the transcription start site, relying on conservation of these sequences between mouse and human to identify putative regulatory regions [14]. These regions were cloned into reporter constructs and tested *in vitro* and *in vivo* to determine their ability to drive gene expression in the expected spatio-temporal pattern. Subsequent genetic deletion studies enabled the discovery of minimal promoter and regulatory core elements for activity and tissue specificity within segments of the cloned fragments. While the discovery of these promoter sequences identified key molecular players in gene activation and facilitated *in vivo* epidermal-specific genetic studies, it also highlighted the paucity of data to explain every nuance in gene expression for epidermal terminal differentiation.

10.3.1 Transcriptional Regulation of the Basal Keratin Genes (K5 and K14)

K5 and *K14* are specifically expressed in mitotically active basal keratinocytes in most types of stratified epithelia. K5 and K14 proteins form stable heterodimers that are further cross-linked to provide structural support for cells [26, 27]. Their co-expression suggests that similar transcriptional mechanisms permit coordinated regulation of the two genes [28, 29].

Earlier studies identified marked similarities between the upstream sequences of *K14* to viral and immunoglobulin enhancer elements [30]. Although 2.5 kb of 5′ upstream and 70 bp of 3′ downstream noncoding sequences of a cloned spliced human *K14* gene are capable of directing gene expression in various cell lines, *in vivo* these regions drive epidermal-specific expression, demonstrated by reporter gene activity in transgenic mice that coincides with endogenous *K14* expression [31]. An additional study of a 2 kb upstream sequence of *K14* in transgenic mice also identified epidermal and outer root sheath hair follicle-specific reporter gene expression, conferred by a 700 bp sequence conserved between mouse and human, that exhibited keratinocyte-specific open chromatin as measured by DNaseI hypersensitivity [32]. The upstream cis-regulatory elements and promoter region of *K14* contained binding sites for the transcription factor ETS, as well as AP-2, AP-1, and SP1 sites shared by the *K5* regulatory region [32–36] (see below).

A similar study of the human *K5* upstream sequence identified a 6 kb region that controls cell type-specific transcriptional activity [37]. A 90 bp sequence within this region is sufficient to activate expression specifically in the epidermis, hair follicles, and tongue. However, the pattern of expression was aberrantly switched from the basal layer to differentiated suprabasal cells. This suggested minimal promoter activity in the 90 bp upstream sequence for keratinocyte-specific expression and a requirement for other enhancers within the 6 kb upstream region for directing expression to the correct layers within stratified epithelia. The link between AP-2, SP1 and other unknown transcription factors for *K5*-tissue-specific activation was later determined based on the discovery of these transcription factor binding sites in the 6 kb sequence upstream of the human *K5* transcription start site and corroborated with comparative studies of the bovine *K5* promoter sequence [35, 37].

The mechanisms controlling coordinate expression of K5 and K14 were further elucidated by the discovery that mice lacking the transcription factor p63 fail to express K5 or K14 and do not develop a stratified epidermis [38, 39]. Biochemical studies revealed the presence of p63-responsive regulatory elements in an epithelial-specific enhancer 1.4 kb upstream of the *K14* promoter [40–43] and in a transcriptionally active region upstream of *K5* [44]. The shared and unique transcription factor binding sites harbored by the *K5* and *K14* enhancers suggest that both common and unique molecular mechanisms are involved in regulating their expression.

10.3.2 *Transcriptional Regulation of the Suprabasal Genes*

10.3.2.1 Early Differentiation: Regulation of *K1* and *K10* Expression

K1 and K10 are expressed in post-mitotic differentiating keratinocytes, and like K14 and K5, form heterodimers that are important for tissue integrity [45]. A 10.8 kb region surrounding the human *K1* gene is sufficient to direct tissue-specific expression in transgenic mice during development, and is responsive to calcium induced differentiation *in vitro* [46]. Two calcium responsive elements were identified within a 1.7 kb enhancer in the 3′ flanking sequences of *K1* [47, 48] including one element at the proximal 5′ end of the enhancer that is activated by AP-1 binding, and another that either suppresses or promotes the calcium response depending on the presence of vitamin D or retinoic acid, respectively [48]. Epidermal specific expression of *K1* is directed by 207 bp of sequence at the distal 3′ end of this enhancer; however the transcription factor responsible for driving this expression has not been identified [47].

Insights into the transcriptional regulation of *K10* expression were provided by analyses of mice deficient in the CCAAT/enhancer binding protein family member

C/EBPβ, which exhibit decreased expression of *K10* and *K1* and mild epidermal hyperplasia [49]. This work was motivated by the observed increases in C/EBPα, C/EBPβ and C/EBPδ mRNA levels, as well as C/EBPα and C/EBPβ protein induction, upon calcium-induced differentiation of mouse primary keratinocytes [50]. DNaseI footprinting and gel-shift assays performed on the human *K10* promoter, which had been identified via homology with the bovine sequence, revealed that C/EBP binding to three distinct elements was required for *K10* expression [51]. While C/EBPα and C/EBPβ both activate *K10*, they are differentially expressed in stratified epidermis. C/EBPβ acts early in the basal and spinous layers of the epidermis and is later superseded by C/EBPα during keratinocyte differentiation and upward migration into the granular layer. Besides C/EBP binding to three intact C/EBP binding sites within the *K10* promoter, full activation of *K10* during differentiation also requires specific binding of AP-2 [51]. Consistent with co-expression of K1 and K10, consensus C/EBP binding sites are also observed in the upstream promoter regions of the *K1* gene [52, 53].

10.3.2.2 Regulation of EDC Genes

Terminal differentiation in the epidermis is marked by the expression of four gene families that lie within the EDC and are coordinately activated at the transcriptional level: small proline-rich region (*SPRR*), late cornified envelope (*LCE*), filaggrin (*FLG*) and *FLG*-like (*FLG*-like), and *S100* genes [4–7]. Pioneer studies to elucidate gene expression in relation to epidermal differentiation targeted the *SPRR* gene family members involucrin (*IVL*) and loricrin (*LOR*). These two important marker genes are distinctively expressed in terminally differentiated keratinocytes, and encode structural proteins that are cross-linked with many of the other proteins encoded by EDC genes to form the cornified envelope, the basic structural unit of the stratum corneum [54]. IVL is cross-linked early in the formation of the cornified envelope [55] and LOR is in turn cross-linked to the existing scaffolding containing IVL [56]. In the developing mouse embryo, *Ivl* and *Lor* transcripts are upregulated as early as E15.5 [57], and protein expression can be observed by E16.5, corresponding to the onset of skin barrier formation [6, 8]. The tight correlation of *IVL* and *LOR* expression with keratinocyte terminal differentiation makes them ideal candidates for studying the mechanisms that underlie the switch from a proliferating to a differentiating program.

10.3.2.3 Involucrin

A 3.7 kb upstream sequence of *IVL* directs keratinocyte-specific expression of a β-galactosidase reporter gene in transgenic mice [58]. Deletion analysis showed that this sequence comprises distal- and proximal-regulatory regions (DRR and PRR) or enhancers [59, 60], and contains an AP-1 binding site that is required for expression and synergizes with an adjacent SP1 binding site [61]. The DRR AP-1 site binds FRA-1, JunB, JunD, and p300, a histone acetyltransferase often associated with enhancers [62, 63], while the SP1 site interacts with SP1, SP3, and KLF4 transcription factors [59, 61, 64–66]. Because FRA-1 and KLF4 are both known to interact with p300 [65, 67], these data indicate that a complex of transcription factors forms on the DRR to drive *IVL* expression during keratinocyte differentiation.

10.3.2.4 Loricrin

Transcriptional activation of mouse loricrin expression was first localized to a 6.5 kb region spanning the loricrin gene [68]. Transgenic reporter mice in which *LOR* coding sequences were replaced by a β-galactosidase gene revealed that the remaining 1.5 kb of 5′-flanking sequence, a small noncoding exon, a 1.1 kb intron, a single coding exon, and 2.2 kb of 3′-flanking sequence from the mouse loricrin gene drive epidermal-specific, but not differentiation-specific expression. Minimal promoter activity, dependent on an AP1 site conserved between mouse and human, was mapped to a 60 bp sequence upstream of the transcription start site. In the case of the human *LOR* gene, enhancers located within 1.5 kb of 5′-flanking sequence and 9 kb of 3′-sequence are responsible for tissue- and differentiation-specific expression of a human *LOR* transgene in transgenic mice [69]. As few as 154 bp of 5′ sequence upstream from the cap site can direct expression specifically in cultured NHEK and HaCAT keratinocytes, in an SP1/c-Jun and p300/CREB-dependent manner [70]. Differential occupation of this site by SP3/CREB-1/CREMα/ATF-1/Jun B, and an AP-2-like protein (named keratinocyte-specific repressor-1 (KSR-1)) in a repressive state, and SP1/c-Jun/p300/CBP in an active state during differentiation, enabled *LOR* transcriptional resolution in stratified layers [70].

10.4 Identification of Enhancers by Comparative Genomics in the Post-Human Genome Era

The early studies described above identified the transcription factors responsible for transcriptional regulation of major epidermal differentiation genes. However, the molecular mechanisms underlying transcription factor-driven activation of these genes were less clear. Furthermore, these studies were limited to the analysis of proximal promoter and enhancer regions. The availability of complete genome sequences for a wide range of model organisms and animal species greatly facilitated the identification of putative enhancer and other regulatory sequences. Specifically, it shifted the discovery of enhancers toward a more systematic procedure – high-throughput and on a genome-wide scale. The discovery of putative enhancers was predicated on the identification of noncoding sequence conservation across multiple species and facilitated by multiple-sequence alignments [71].

Indeed, a study of conserved noncoding sequences present between human and pufferfish or ultraconserved between human and mouse found that many of these conserved noncoding sequences exhibit enhancer activities that are developmental- and tissue-specific in transgenic mouse assays [72]. Below we discuss the discovery of enhancers for relevant skin biology genes using comparative genomics.

10.4.1 SPRR *Genes*

The availability of whole genome data enabled the discovery of conserved noncoding sequences (CNSs and based on sequence alignments) in the *SPRR* locus that is clustered within the EDC [73]. Small proline-rich (SPRR) proteins are the primary constituents of the cornified envelope [74]. Many of the SPRRs are coordinately upregulated under stress conditions to rapidly build a temporary barrier [73]. Of the DNaseI hypersensitive sites (HSs) residing within CNSs, one demonstrated enhancer activity under conditions when the *SPRR* genes are coordinately upregulated, suggesting its potential as an enhancer region that coordinates *SPRR* gene expression.

10.4.2 PADI3

The family of peptidylarginine deaminases (PADs) is encoded by a cluster of 5 *PADI* genes on human chromosome 1p35–36 [75]. Peptidylarginine deaminase 3 (*PADI3*) encodes PAD3, which is involved in filaggrin metabolism, releasing individual filaggrin monomers that contribute to the natural moisturizing functions of the skin barrier. Using a comparative genomic approach, an enhancer located 86 kb from the *PADI3* gene promoter was identified and determined to be calcium sensitive [76]. This enhancer was found to trigger expression of *PADI3* upon epidermal keratinocyte differentiation, and links *PADI3* expression to AP-1 transcription factors through chromatin opening and looping (see Sect. 10.5).

10.4.3 The Role of the CNE 923 Enhancer in Coordinate Regulation of the EDC

The coordinate regulation of EDC genes, as well as the synteny and linearity of the EDC locus across a wide range of mammalian species, suggest a molecular mechanism originating at the proximal genomic level. To delineate this, investigators screened for enhancer elements within the EDC [7]. In 2010, 48 conserved noncoding elements (CNEs) within the human EDC were identified from sequence alignments of orthologous EDC loci across eutherian (human, chimpanzee, macaque, mouse, rat, dog) and metatherian (opossum) mammals. Approximately 50% of these CNEs exhibited dynamic regulatory activity, and were thus identified as potential cis-regulatory elements or enhancers that might synergistically or independently coordinate EDC gene expression during skin barrier formation. Among these, human CNE 923, located approximately 923 kb from the transcriptional start site of the most 5' EDC gene, *S100A10*, induced the highest luciferase reporter activity in proliferating and differentiated keratinocytes. CNE 923 exhibited DNaseI hypersensitivity in primary human keratinocytes, and was sufficient to drive reporter gene expression in the developing E16.5 epidermis in transgenic reporter mice. The activity of CNE 923 was monitored in an independent transgenic mouse line and was sufficient to drive β-galactosidase activity in the same dorsal to ventral pattern of barrier acquisition that coincided with EDC gene activation [57]. These studies provided compelling evidence for CNE 923 as an epidermal-specific enhancer and a potential LCR. CNE 923 was also noted to form dynamic chromatin interactions with a number of EDC genes, and this was sensitive to keratinocyte differentiation and dependent on the AP-1 transcription factor. The role of CNE 923 for mediating EDC chromatin architecture is discussed further in Sect. 10.7.

10.4.4 p63

A long-range cis-regulatory enhancer of p63 (p63LRE) spanning a 12 kb region in mice was recently identified by comparative genomics [77]. p63LRE comprises two evolutionarily conserved modules acting in concert to control tissue- and layer-specific expression of the *p63* gene. Both modules are in an accessible and active chromatin state in human and mouse keratinocytes and in embryonic epidermis, and are strongly bound by p63. p63LRE activity is dependent on p63 expression in embryonic skin and also in the commitment of human induced pluripotent stem cells toward an epithelial cell fate. C/EBPα, C/EBPβ, and the POU domain-containing protein Pou3f1 repress p63 expression during keratinocyte differentiation by binding the p63LRE enhancer. The availability of these transcription factors in the outermost layers of the epidermis accounts for increased repression of p63, thereby relieving p63-mediated repression of EDC and keratinocyte differentiation genes in these layers, and limiting p63 activity to the basal layers. We discuss p63-bound sites and enhancer regions on a genome-wide scale in Sect. 10.7.4.

10.5 Epigenetics and Chromatin Remodeling

So far, we have discussed studies that make compelling arguments for transcriptional regulation in which the key determinants are transcription factor binding to individual nucleotide motifs. However, this view has been challenged by our increased understanding of the non-random packaging of linear DNA into histones, which alters the accessibility of DNA segments to transcription factor binding, and by the discovery of specific post-translational modifications to histone components that dynamically regulate DNA accessibility. Specifically, the identification of histone deacetylase (HDAC) and histone acetyltransferase (HAT) enzymes and their targets furnished the first direct evidence linking histone modification states to transcriptional regulation [78, 79]. These findings paved the way for investigations into higher order chromatin structure and genome compartmentalization. HATs catalyze the transfer of an acetyl group from acetyl CoA to the ε-amino group of lysine residues on histones. This mark is generally associated with active genes. Conversely, HDACs remove the acetyl group from acetyl-lysine (Ac-Lys) to regenerate the free ε-amino group, causing chromatin compaction and a transcriptionally repressive environment (reviewed in [80, 81]). The identification and characterization of other classes of histone modifying enzymes soon followed, implicating kinases [82, 83], lysine and arginine-specific methyltransferases [84–86], arginine deiminases [87, 88], ubiquitinases [89], deubiquitinases [90–92], and lysine- and arginine-specific demethylases (HDMs) in transcription regulation [93–95]. Collectively, these discoveries highlighted the role of chromatin modifications in governing eukaryotic gene expression and other DNA-dependent functions and ushered in a new era of chromatin-based epigenetic studies (reviewed in [96, 97]).

Within the context of the skin, p300 and CBP, two enhancer-associated HATs, have been implicated in the regulation of *IVL* and *LOR* (discussed earlier) [65, 70]. Epidermal-specific deletion of *Actl6a*, an essential component of HATs, resulted in de-repression of KLF4 and Brg1/Brm binding, thereby aberrantly activating epidermal differentiation genes and abolishing epidermal progenitor function [98]. HDACs also play important roles in epidermal development, specifically in preventing senescence of basal progenitor cells [99, 100]. Epidermal-specific deletion of both HDAC1 and 2 in embryonic epidermis resulted in a phenotype resembling the effects of loss of *p63*. This phenotype was associated with de-repression of ΔNp63-repressed target genes including the senescence gene *p16* [101]. HDAC1/2 localizes to the promoter regions of ΔNp63-repressed targets in cultured human keratinocytes, and histones in these regions are hyper-acetylated following HDAC inhibition, indicating a requirement for HDAC1/2 in ΔNp63-mediated repression. Together, these data reveal essential roles for histone modifying enzymes in controlling the activities of key regulators of epidermal development (reviewed in [102]).

Ezh2, an essential component of the polycomb repressor complex 2 (PRC2) is also important for epidermal development [103]. Epidermal-specific loss of Ezh2 resulted in early epidermal differentiation owing to precocious recruitment of AP-1 transcription factor to the EDC. This indicated a role for Ezh2 in gene repression in proliferating keratinocytes by promoting histone H3K27 trimethylation (H3K27me3). However, loss of Ezh2, while decreasing H3K27me3 marks, was not sufficient to alter transcriptional status unless AP-1 was also recruited to the affected region, emphasizing an important and direct role for AP-1 in keratinocyte differentiation.

Recent cell biology studies have also highlighted the importance of chromatin remodeling in permitting efficient and coordinate regulation of clusters of genes in epidermal development. During epidermal development, the EDC locus moves away from the nuclear periphery and towards the nuclear interior prior to activation of EDC gene expression [104]. Ablation of either p63, a master regulator of epidermal development [38, 39, 105], or Satb1, a higher-order genome organizer that binds to the EDC in epidermal progenitor cells, caused altered chromatin conformation of the EDC, and loss of expression of genes that lie in the central domain of the EDC [106]. These findings identified Satb1 as an important downstream target of p63 required for proper establishment of higher-order EDC chromatin structure and coordinated gene expression [106]. Similarly, p63 and its direct target Brg1 are essential in remodeling the higher-order chromatin structure of the EDC and positioning the locus within the 3D chromatin landscape to allow efficient expression of EDC genes in epidermal progenitor cells during skin development [107, 108].

10.6 Methods to Identify Enhancers by ENCODE in the Post-Human Genome Era

Increased understanding of the chromatin state of the genome has forced us to re-examine initial models for the control of gene expression that focused entirely on the role of cis-regulatory elements (Sect. 10.3). Following completion of the human genome sequence, the National Human Genome Research Institute, recognizing the need to more fully understand the regulation of gene expression, launched the Encyclopedia of Non-Coding Elements (ENCODE), a collaborative public research project to identify and characterize the function of noncoding elements in the genome, and develop the tools and technology to achieve this goal [63]. Next-generation sequencing was instrumental in producing these genome datasets in a cost-effective manner. The ENCODE studies as well as work by others have greatly facilitated our ability to identify enhancers on a genome-wide scale based on chromatin modifications that are unique to these regulatory elements, such as DNaseI hypersensitivity (open chromatin), histone modification epigenetic marks (H3K27Ac, H3K4me1), and transcription factor binding (p300, activating TFs) associated with functional enhancers (reviewed in [15, 25, 109]). High-throughput chromatin immunoprecipitation linked with deep sequencing (ChIP-seq) for specific histone modifications has enabled the discovery of new enhancers and paved the way for further downstream functional analyses (reviewed in [15]). Table 10.1 lists signature genomic marks and methods to identify enhancers.

Chromatin looping and tracking have been proposed as models to explain how distant enhancers are able to regulate their target genes [143, 144]. The first experimental demonstration of direct interactions between distantly located enhancers and target genes was made possible by the development of chromosome conformation capture (3C) techniques [125]. 3C was used to demonstrate that loop formation between the β-globin LCR enhancer and gene accompanied transcriptional activation [145–147], and established a paradigm that was later validated for numerous

Table 10.1 Genomic features of enhancers and methods for prediction and functional validation

	Genomic landmarks	Method for prediction	Features associated with enhancers
Enhancer prediction	Histone modifications	ChIP-seq [110]	Enriched: H3K27Ac, H3K4me1, H3K4me2, H3K4me3, H3K79me2, H3K9Ac
			Depleted: H3K27me3
	Chromatin accessibility	FAIRE-seq [111], ATAC-seq [112], DNaseI hypersensitivity mapping (DNase-seq) [113], Repli-seq [114], MNase-seq [115]	Open chromatin
	Transcriptional activity	ChIP-seq [110], RNA-seq [116], cap analysis gene expression (CAGE) [117], chromatin interaction analysis by paired end tag sequencing (ChIA-PET) [118], RNA annotation and mapping of promoters for analysis of gene expression (RAMPAGE) [119], RNA-PET [120]	TFBS clusters, RNA pol II, enhancer RNA (eRNA)
	Enhancer associated proteins	Protein sequencing by tandem mass spectrometry assay [121], SELEX [122123124]	CTCF, mediator, Cohesin, EP300
	Long range DNA looping	3C-based methods (3C [125], 4C [126], 5C [127], Hi-Seq [128]), ChIA-PET [118]	Enhancers associate with promoters and other enhancers
	DNA methylation	Whole genome shotgun bisulfite sequencing [129], reduced representation bisulfite sequencing (RRBS) [130], MeDIP [131], MRE-seq [132]	Lower methylation
	Conservation	Multiple sequence alignment [14]	Highly conserved

	Scale	Method for validation
Functional enhancer validation/identification	Low throughput	Transgenic models [133]
		Enhancer trap [134]
		Cell-based reporter assays [135, 136]
	Medium throughput	Site-specific integration FACS-sequencing (SIF-seq) [137]
	High throughput	Functional identification of regulatory elements within accessible chromatin (FIREWACh) [138]
		Self-transcribing active regulatory region sequencing (STARR-seq) [139]
		Massively parallel reporter assays (MPRAs) [140]
		Cis-regulatory element sequencing (CRE-seq) [141]
		Thousands of reporters integrated in parallel (TRIP) [142]

other loci, including the α-globin gene cluster, T_H2, IFNG, MHC class II and IgH loci [148]. Transcription factor ChIP-chip studies also revealed that enhancers could be located even further from their target genes than previously thought, as far as 10–20 kbs to several Mbs away [149]. Often, these proximal and distal enhancers interact to co-regulate a target gene.

Recent improvements in chromosome conformation capture methods have allowed us to examine the chromatin interactions of genomic regions at varying levels of depth and resolution. 4C (circular chromosome conformation capture) detects all interacting sequences with a sequence of interest using a bait such as an enhancer [126]. 5C (chromosome conformation capture carbon-copy) is designed to detect many known interactions with numerous baits and typically within a gene locus [127], while the Hi-C approach is aimed at detecting all chromatin interactions [128]. Methods such as ChIA-PET (Chromatin interaction analysis with paired-end tag sequencing) combine 5C and Hi-C methods to simultaneously identify genome-wide chromatin interactions and the proteins that bind interacting sequences [150]. Evidence for epigenetic modifications, chromatin looping, and the interplay between the two has been obtained relatively recently, and has provided new insights to our understanding of the biochemical aspects of enhancer-mediated transcriptional regulation.

10.7 Emerging Concepts for Understanding Enhancer-Promoter Interactions

10.7.1 Mechanisms Underlying Enhancer-Promoter Interactions and Topological Association of Chromatin Domains in the Regulation of Gene Expression

Apart from the aforementioned roles of p63, Satb1, Brg1, and AP-1 in transcriptional regulation during epidermal differentiation, additional molecular mechanisms underlying transcriptional regulation by enhancers are less clear. Here we discuss studies that have elucidated the mechanisms of enhancer-promoter interactions in epidermal biology, and lessons we can learn from other tissue models.

10.7.1.1 The Formation and Biology of Enhancer-Promoter Chromatin Loops

Formation of enhancer-promoter chromatin loops as a mechanism to drive gene activation has emerged as a major concept in the exploration of enhancer-promoter interactions. One of the best studied loci is the evolutionarily conserved β-globin locus that plays key roles in hematopoiesis (reviewed in [151]). The 5 globin genes (ε, Gγ, Aγ, δ and β) form a cluster, and are expressed in a developmental-stage- and tissue-specific manner, controlled by a LCR. The GATA-1 transcription factor and Ldb1 are required to form a chromatin interaction between the β-globin LCR and the β-globin promoter for transcriptional activation in erythroid cells [152, 153]. The requirement for formation of a chromatin loop for gene transcription was demonstrated using artificial zinc fingers (ZF) [154] to force chromatin loop formation by tethering Ldb1 to the β-globin locus control region in GATA-1 null erythroblasts. This was found to be sufficient to activate β-globin gene expression. This work was the first to demonstrate the causality of chromatin spatial interactions in promoting gene transcription.

10.7.1.2 Regulation of the EDC by CNE 923 Via AP-1 Mediated Chromatin Interactions

The organization of the EDC as a conserved cluster of genes with related functions [7] is reminiscent of the organization of the β-globin locus [151]. Moreover, the identification of the EDC enhancer CNE 923 as an epidermal specific enhancer (described in Sect. 10.4.3), suggests its potential function as an LCR that drives coordinate and concomitant EDC gene expression in a manner similar to the β-globin LCR. In line with this, 3C experiments performed with respect to the mouse orthologous CNE 923 sequence in proliferating primary mouse keratinocytes demonstrated that this sequence interacts with nine EDC gene promoters (*Sprr2a1*, *Sprr2d*, *Sprr2f*, *Sprr1b*, *Sprr3*, *Ivl*, *Lce1b*, *Lce1a2*, and *Crct1* gene promoters) that lie as far as 500 kb from CNE 923, despite the lack of EDC gene expression in these cells, suggesting a poised and enhancer-mediated chromatin state [57]. After calcium-induced differentiation, CNE 923 interacted with the promoters of 11 EDC gene, including *Lce3b*, *S100a6*, *Sprr2a1*, *Sprr2b*, *Sprr3*, *Sprr4*, *Ivl*, *Lce6a*, *Lce1b*, *Lce1e*, and *Crct1*. The interaction between CNE 923 and *S100a6*, located 2 Mb apart, suggested that the EDC chromatin domain is compacted during differentiation to bring more linearly distal genes into close proximity with the enhancer. These results highlight dynamic chromatin looping interactions with 923 that are associated with concomitant EDC gene expression.

Bioinformatics analysis of human CNE 923 identified two highly conserved sequence blocks (PhastCons) that are required for enhancer activity. A consensus AP-1 transcription factor binding site within the most 5′ block is required for maximal enhancer activity, and pharmacological inhibition of AP-1 binding in calcium-induced keratinocytes represses EDC gene expression, and causes loss of c-Jun/AP-1 binding to 923 and aberrant chromatin remodeling. These observations identify a link between an epidermal-specific EDC enhancer and c-Jun/AP-1 transcription factor binding, and together with other studies [104, 106, 107], suggest that further analysis of the 3D structure of chromatin would aid our understanding of EDC regulation.

10.7.2 Higher Level Chromatin Architecture: Topologically Associated Domains (TAD) and Chromosome Territories (CT)

Topologically associated domains (TADs) were first identified by Hi-C [155, 156], and are defined as distinct clusters of enhancer-promoter interactions [157, 158]. At the highest order of chromosome organization, spatially proximal TADs compose a chromosome territory (CT), a compartment within the nucleus that is often segregated in a chromosome-specific manner (reviewed in [159]). Actively transcribed gene-rich loci that are in an open conformation are more likely to loop out of their CTs, suggesting that the space between CTs is important for genomic loci to access the transcription machinery (reviewed in [159]).

The importance of long range enhancer-promoter interactions in the context of a CT was demonstrated in studies of the developing limb bud [160], where differential expression of the Sonic hedgehog (*Shh*) gene is mediated by specific interactions between the *Shh* promoter and a long-range enhancer MFCS1. In the intermediate portion of the limb bud, which lacks *Shh* expression, the long-range enhancer is spatially and linearly distant from the *Shh* coding region. In anterior limb bud cells, the long-range enhancer interacts with the *Shh* coding region, but the interactors remain in a poised state within their CT. However, in cells of the zone of polarizing activity (ZPA) where *Shh* is actively expressed, 3D-FISH showed that the interacting regions relocate outside the CT.

A similar mechanism has been observed in the control of EDC gene expression in differentiating keratinocytes, where Satb1 binds to several sites across the EDC locus, compacting the EDC chromatin architecture into a densely looped structure that, upon relocalization of the locus into the nuclear interior by Brg1, enables efficient and coordinate activation of EDC genes [106, 107]. Similar mechanisms may be employed to control genes such as *K1* and *K10* which are coordinately activated as keratinocytes transition from basal to suprabasal layers of the epidermis.

10.7.3 Involvement of Cohesin and CTCF in Forming Active Chromatin Hubs

Enhancer activity can also be modulated by insulators that function as physical barriers to optimal enhancer-promoter formation for transcriptional activation. Here we discuss newly recognized attributes of enhancers and new direct roles for CCCTC-binding factor (CTCF) bound insulators in enhancer–promoter interactions and in broadly configuring the genome (Reviewed in [143, 161]).

Cohesin is a complex of proteins that holds sister chromatids together after DNA replication, until the sister chromatids separate at anaphase (reviewed in [162]).

Analysis of the effects of mutations in cohesin subunits identified a role for cohesin in regulating enhancer-promoter interactions and gene expression (reviewed in [162]). Subsequently, it was discovered that mammalian cohesin complexes can be recruited to DNaseI hypersensitive sites and conserved noncoding sequences by the CTCF DNA binding protein [163–165]. CTCF often binds at insulators and at boundary elements to demarcate active chromatin hubs and limit the effect of enhancers [166], and cohesin contributes to CTCF's enhancer blocking activity [163, 165]. Studies of the apolipoprotein gene cluster [167], the globin locus [168], and the T-cell receptor (*Tcra*) locus [169] demonstrated cooperation of CTCF and cohesin to mediate insulators corresponding to TAD boundaries, thereby maintaining proper chromatin loop formation and localization of transcriptional apparatus at the gene promoters to control gene expression.

These mechanisms are relevant to human disease, as chromosomal rearrangements of the conserved TAD-spanning *WNT6/IHH/EPHA4/PAX3* locus that disrupt a CTCF-associated boundary domain within a TAD, cause limb malformations in humans. Mice harboring the equivalent disease-relevant rearrangements were generated using CRISPR/Cas9-genome editing, and displayed ectopic limb expression of a gene that lies within the locus but is not normally expressed in limb development, due to misplacement of a cluster of limb enhancers relative to TAD boundaries [158]. This finding demonstrates the functional importance of TADs for orchestrating gene expression via genome architecture and suggests the utility of analyzing disease-associated large-scale chromosomal rearrangements in delineating TAD boundaries.

Genomic studies in other tissue types and loci have also greatly advanced our understanding of the molecular mechanisms of CTCF-mediated chromatin looping events. Chromatin conformation Hi-C capture data with parallel CTCF ChIP-seq identified a nonrandom pattern of forward and reverse orientation for a given pair of CTCF binding sites involved within a chromatin loop [170–172]. Subsequent studies later identified a functional role for the directionality of CTCF binding sites to influence chromatin topology and enhancer-promoter function [173, 174]. CRISPR/Cas9-generated inversion of a genomic region spanning CTCF boundary elements in the P-cadherin enhancer altered chromatin topology [173]. Deletions of individual CTCF binding sites by CRISPR/Cas9 led to loss of CTCF, reduced cohesin binding, and reduced or abolished chromatin looping, while inversions restored CTCF and cohesin binding but reduced chromatin looping [174]. However, the impact of the presence or directionality of CTCF binding sites on proximal gene expression varied, suggesting the influence of additional factors. This was further evidenced by deletions of 4 CTCF binding sites in the casein locus that did not affect gene expression but instead caused more distal *de novo Sultd1* activation by way of other nearby enhancers [175]. These studies demonstrated a novel governing principle for chromatin architecture in controlling gene expression, and holds great potential for more accurate and contextual prediction of the functionality of enhancers in the skin and other tissues.

10.7.4 The Role of Mediator and Super-Enhancers

More recently, a new class of enhancers called "super-enhancers" has been identified [176]. Super-enhancers are marked by high levels of Mediator coactivator complex occupation as determined by ChIP-seq and span much larger distances than typical enhancers (approximately 8.7 kb versus 703 bp). Mediator is a major component of the transcription pre-initiation complex (PIC) machinery with RNA polymerase II (RNA pol II) and is required for activator-dependent transcription *in vitro* and *in vivo* (reviewed in [177]). Reduced levels of Mediator specifically affect gene expression near super-enhancers [176]. This was convincingly demonstrated by the loss of enhancer-promoter loops of select genes upon deletion of Mediator [178, 179]. Mediator-occupied super-enhancers also exhibited enriched binding of transcription factors that are master regulators involved in cell-identity in ESCs, pro-B cells, T helper cells, myotubes, and macrophages, among other cell types. However, one can argue that Mediator and even cohesin binding may not be entirely necessary for gene activation as it was recently determined that chromatin looping between the globin LCR and the β globin locus via Ldb1 was established in the absence of Mediator and cohesion binding [180]. More recent work identified an emergent paradigm for super-enhancers in the direct biogenesis of master regulator miRNAs for tissue specificity [181].

A role for super-enhancers was recently identified in epidermal stem cells [182]. The target genes associated with epidermal stem cell-specific super-enhancers identified by H3K27Ac and Mediator ChIP-seq methods contain a high frequency of binding motifs for the transcription factors Sox9, Lhx2, Nfatc1 and Nfib, that are important for maintaining hair follicle stem cells [183–186]. ChIP-seq experiments showed that these transcription factors bind at high frequency to super-enhancers relative to typical enhancers. Lineage tracing during mouse epidermal development, wound-healing, and in cell culture showed that super-enhancers are remodeled according to their cellular environment, supporting the idea that enhancers are activated or silenced in a lineage-specific fashion. The dynamic behavior of these regulatory elements in human keratinocyte progenitors as well as in their differentiating progeny was further supported by the discovery of genomic profiles for superenhancers (H3K27ac) and typical enhancers (H3K4me1, H3K27ac, H3K4me3) [187] that were different across these distinct keratinocyte states. Binding of the p63 master regulator for keratinocyte differentiation was observed in a core set of super-enhancers and enhancers (shared between mouse and human) associated with keratinocyte-specific gene expression, but also notably in mouse-specific regions that underlie species-specific transcriptional differences [188].

Analysis of sites bound by Mediator thus enables identification of key transcription factors and enhancer sequences in a variety of cell types and their sensitivity to changing conditions, highlighting the potential of this approach as a tool to pinpoint important regulatory sequences involved in cell and tissue homeostasis, even without prior knowledge of the transcription factors or genes involved.

10.7.5 Non-coding RNAs

As discussed above, the establishment and maintenance of chromosome territories or TADs bring enhancer and promoter interactions into close proximity to facilitate gene activation (reviewed in [157]). In addition to the roles of cohesin and CTCF occupation in enhancer-promoter interactions, long non-coding RNAs (or lncRNAs) may also serve as a scaffold for the assembly of transcription factors and chromatin remodeling enzymes at the promoter [189]. The lncRNA *HOTAIR*, first discovered in adult skin [190], targets members of the polycomb repressive complex 2 (PRC2) to specific genomic loci, including Ezh2 [189] which, as discussed earlier, regulates epidermal differentiation [103]. *HOTAIR* also directly interacts with LSD1 [189], a protein complex that is associated with maintenance of epidermal stem cells in an undifferentiated state [191]. Moreover, PRC2 and LSD1 interact in a *HOTAIR* dependent manner [189], suggesting a potential role for *HOTAIR* in epidermal development.

More recently, evidence of transcription of functional enhancer RNA (eRNA) from non-coding enhancer sequences has emerged [192]. eRNAs are considered a separate class of non-coding RNA. Unlike lncRNAs, which are marked by H3K4me3 at their promoters and are frequently spliced and polyadenylated, eRNAs are transcribed from enhancer regions marked by H3K4me1, an absence of H3K4me3 histone modifications, and to a lesser extent, post-transcriptional modifications. During transcriptional activation, enhancer RNAs participate in a bi-directional enhancer-promoter activation feedback loop, whereby chromatin looping between the enhancer and promoter brings the bi-directionally transcribed eRNA near the target gene to drive gene expression, at the same time allowing the eRNA to stabilize the enhancer-promoter loop with the help of the Mediator complex (reviewed in [87]). This mechanism for eRNA mediated gene activation has been identified in the expression of various lineage-specific genes, including macrophage-specific genes [193], estrogen-regulated genes in breast cancer cells [194], and p53-regulated genes that induce cell cycle arrest [195]. To date, a role for eRNAs in epidermal development has yet to be identified, but should be kept in mind as a possible mode of gene regulation.

10.8 Conclusion

In this Chapter, we have outlined the history of scientific discoveries related to the transcriptional regulation of key genes involved in skin biology leading up to the concept of enhancers and the molecular mechanisms that orchestrate this process at the chromatin level in the post-human genome and post-ENCODE eras. Our current knowledge of the mechanisms regulating expression of genes involved in determining the proliferative or differentiated state of keratinocytes identifies common patterns in modes of regulation. These observations are consistent with mechanisms identified in other tissue systems. However, regulation of keratinocyte-specific genes appears to heavily utilize p63, AP-1, SP1, and C/EBP transcription factors in particular.

Earlier studies of proximal regions of epidermal genes including keratin genes, *IVL,* and *LOR,* and functional genetic approaches to delineate the sequences that drive transcriptional activities, underscored the role of the complex interplay of AP-1 proteins, SP1 and C/EBP transcription factors in epidermal gene regulation. However, we have yet to understand the larger regulatory landscape of the keratin gene clusters.

By contrast, studies of the EDC locus have benefited from the availability of bioinformatics tools and high-throughput methods, revealing that proper control of gene expression is an intricate hierarchy of events that depends firstly on appropriate post-translational regulation of histones to designate chromatin regions as accessible or inaccessible, followed by further organization and remodeling of the EDC together with proper nuclear spatial positioning [106, 107]. The establishment of regulatory landscapes and chromosome territories or topologically associated domains (TADs) brings regulatory enhancers and promoters into close proximity, allowing AP-1, together with other as yet unknown factors, to be poised for efficient coordinate activation of the EDC genes upon induction of epidermal differentiation [57].

The varied roles of AP-1 transcription factors provide an example of how different combinations of common transcription factors are brought together to form complexes to modulate or alter gene expression under different conditions such as spatial context and developmental stage. Distinct complexes may be formed by altering the occupancy of DNA binding proteins at arrays of enhancer elements in close proximity to gene promoters. Complex assembly may also be driven in part by long-range chromatin interactions that bring distal enhancers close to their target gene promoters, and this can be mediated in a myriad of ways. Within the field of skin biology, we have only recently begun to define the mechanisms by which enhancer-promoter interactions initially occur.

10.9 Future Studies

Further dissection of the regulatory principles underlying the gene expression patterns that accompany and drive epidermal development will require a two-pronged approach. The availability of whole genome sequences from increasing numbers of species [196], high-throughput techniques such as ChIP-seq, RNA-seq, ATAC-seq, and the development of bioinformatics tools to allow the integration of such data [197], have made it possible to approach the analyses of gene regulation at the genomics level. Specifically, these advances have enabled researchers to identify enhancer landscapes across the genome and further elucidate the principles that govern enhancer-promoter interactions and genome organization relevant to gene expression. Such an approach identified a surprising role for the DNA methyltransferases,

Dnmt3a and Dnmt3b, in occupying enhancers and functionally linked them to epidermal stem cell function [198]. Furthermore, the discovery of skin disease sequence variants enriched in superenhancers highlights the clinical impact of this research and future studies in this area [187].

While several key mechanisms, such as the roles of chromatin looping, 3D genomic architecture, and non-coding RNAs were first demonstrated in non-epidermal tissues and cell-types, they are crucial in constructing a framework for understanding epidermal-specific enhancer-promoter interactions. For example, cohesin and CTCF have been shown to play a role in maintaining enhancer-promoter interactions in a multitude of cell-types and tissues [167–169, 199]. This strongly suggests that they are also likely to be major players in mediating the formation of chromatin loops in differentiating keratinocytes. This could be tested by identifying changes in CTCF and cohesin occupation in the different epidermal layers, as well as through functional genetic studies.

In order to completely understand the mechanisms driving enhancer regulated gene expression, we must continue to incorporate multi-disciplinary approaches and novel methods to approach the problem from genetic, molecular and cellular perspectives. Drawing from the expertise of the evolutionary biology field, we are now able to identify candidate regions with regulatory potential faster than ever before [71, 72, 200]. Additional enhancers that are not evolutionarily conserved can be identified using sequencing technologies (see Table 10.1 and [201]). Current advances in genome editing (CRISPR/Cas9, TALENs) have also made it easier to test the functions of endogenous enhancers [158, 202–204] and provide an improvement over the use of artificial transgenes that remove enhancers from their appropriate genomic context. The precision of these genome-editing methods enables us to directly test hypotheses regarding enhancer functions at specific locus/loci within the regulatory landscape. For example, CRISPR/Cas9 genome editing of several enhancers has demonstrated their requirement for gene expression [158, 202–204]. A high-throughput genetic screen for a targeted set of regulatory elements in the *POU5F1* locus using CRISPR/Cas9 editing has also confirmed the functional role of enhancers for *POU5F1* expression, while simultaneously revealing a new class of "TEMP" enhancers that are characterized by temporary loss of gene expression and weak reporter activity [205]. CRISPR/Cas9-mediated recombination of orthologous yet divergent enhancer sequences in mice has also provided a comparative functional assay to further assess the importance of transcription factor binding sites during development [204]. For instance, in vivo replacement of a mouse Sonic hedgehog enhancer with the orthologous snake-specific Sonic hedgehog enhancer led to a limb defect that was rescued by introduction of an ETS binding site that had been lost in the snake. In addition to modern genetics and genomics approaches, molecular tools, such as live-imaging and high-resolution microscopy and biomechanics studies will extend our understanding of the dynamics of enhancer-promoter interactions.

Elucidation of the molecular biology and biochemistry of enhancer-promoter interactions has set the stage for a new era of investigation into the mechanisms of transcriptional regulation. Armed with new methodologies for genome sequencing and editing and protein engineering to both discover enhancers and to rapidly test their functions, we are well placed to achieve a more comprehensive understanding of the principles of genome architecture that modulate cellular transcriptomes.

References

1. Hsu YC, Li L, Fuchs E. Emerging interactions between skin stem cells and their niches. Nat Med. 2014;20(8):847–56.
2. Lechler T, Fuchs E. Asymmetric cell divisions promote stratification and differentiation of mammalian skin. Nature. 2005;437(7056):275–80.
3. Blanpain C, Fuchs E. Epidermal homeostasis: a balancing act of stem cells in the skin. Nat Rev Mol Cell Biol. 2009;10(3):207–17.
4. Mischke D, Korge BP, Marenholz I, Volz A, Ziegler A. Genes encoding structural proteins of epidermal cornification and S100 calcium-binding proteins form a gene complex ("epidermal differentiation complex") on human chromosome 1q21. J Invest Dermatol. 1996;106(5):989–92.
5. Zhao XP, Elder JT. Positional cloning of novel skin-specific genes from the human epidermal differentiation complex. Genomics. 1997;45(2):250–8.
6. Marshall D, Hardman MJ, Nield KM, Byrne C. Differentially expressed late constituents of the epidermal cornified envelope. Proc Natl Acad Sci U S A. 2001;98(23):13031–6.
7. de Guzman Strong C, Conlan S, Deming CB, Cheng J, Sears KE, Segre JA. A milieu of regulatory elements in the epidermal differentiation complex syntenic block: implications for atopic dermatitis and psoriasis. Hum Mol Genet. 2010;19(8):1453–60.
8. Hardman MJ, Sisi P, Banbury DN, Byrne C. Patterned acquisition of skin barrier function during development. Development. 1998;125(8):1541–52.
9. Yuspa SH, Kilkenny AE, Steinert PM, Roop DR. Expression of murine epidermal differentiation markers is tightly regulated by restricted extracellular calcium concentrations in vitro. J Cell Biol. 1989;109(3):1207–17.
10. Pillai S, Bikle DD, Mancianti ML, Cline P, Hincenbergs M. Calcium regulation of growth and differentiation of normal human keratinocytes: modulation of differentiation competence by stages of growth and extracellular calcium. J Cell Physiol. 1990;143(2):294–302.
11. Bikle DD, Xie Z, Tu CL. Calcium regulation of keratinocyte differentiation. Expert Rev Endocrinol Metab. 2012;7(4):461–72.
12. Shaulian E, Karin M. AP-1 in cell proliferation and survival. Oncogene. 2001;20(19):2390–400.
13. Mehic D, Bakiri L, Ghannadan M, Wagner EF, Tschachler E. Fos and jun proteins are specifically expressed during differentiation of human keratinocytes. J Invest Dermatol. 2005;124(1):212–20.
14. Visel A, Bristow J, Pennacchio LA. Enhancer identification through comparative genomics. Semin Cell Dev Biol. 2007;18(1):140–52.
15. Shlyueva D, Stampfel G, Stark A. Transcriptional enhancers: from properties to genome-wide predictions. Nat Rev Genet. 2014;15(4):272–86.
16. Levo M, Segal E. In pursuit of design principles of regulatory sequences. Nat Rev Genet. 2014;15(7):453–68.
17. Benoist C, Chambon P. In vivo sequence requirements of the SV40 early promotor region. Nature. 1981;290(5804):304–10.
18. Gruss P, Dhar R, Khoury G. Simian virus 40 tandem repeated sequences as an element of the early promoter. Proc Natl Acad Sci U S A. 1981;78(2):943–7.

19. Banerji J, Rusconi S, Schaffner W. Expression of a beta-globin gene is enhanced by remote SV40 DNA sequences. Cell. 1981;27(2 Pt 1):299–308.
20. Moreau P, Hen R, Wasylyk B, Everett R, Gaub MP, Chambon P. The SV40 72 base repair repeat has a striking effect on gene expression both in SV40 and other chimeric recombinants. Nucleic Acids Res. 1981;9(22):6047–68.
21. Fromm M, Berg P. Simian virus 40 early- and late-region promoter functions are enhanced by the 72-base-pair repeat inserted at distant locations and inverted orientations. Mol Cell Biol. 1983;3(6):991–9.
22. Kleinjan DA, van Heyningen V. Long-range control of gene expression: emerging mechanisms and disruption in disease. Am J Hum Genet. 2005;76(1):8–32.
23. Levine M. Transcriptional enhancers in animal development and evolution. Curr Biol. 2010;20(17):R754–63.
24. Williamson I, Hill RE, Bickmore WA. Enhancers: from developmental genetics to the genetics of common human disease. Dev Cell. 2011;21(1):17–9.
25. Hardison RC, Taylor J. Genomic approaches towards finding cis-regulatory modules in animals. Nat Rev Genet. 2012;13(7):469–83.
26. Coulombe PA, Fuchs E. Elucidating the early stages of keratin filament assembly. J Cell Biol. 1990;111(1):153–69.
27. Lee CH, Coulombe PA. Self-organization of keratin intermediate filaments into cross-linked networks. J Cell Biol. 2009;186(3):409–21.
28. Moll R, Franke WW, Schiller DL, Geiger B, Krepler R. The catalog of human cytokeratins: patterns of expression in normal epithelia, tumors and cultured cells. Cell. 1982;31(1):11–24.
29. Lersch R, Stellmach V, Stocks C, Giudice G, Fuchs E. Isolation, sequence, and expression of a human keratin K5 gene: transcriptional regulation of keratins and insights into pairwise control. Mol Cell Biol. 1989;9(9):3685–97.
30. Marchuk D, McCrohon S, Fuchs E. Complete sequence of a gene encoding a human type I keratin: sequences homologous to enhancer elements in the regulatory region of the gene. Proc Natl Acad Sci U S A. 1985;82(6):1609–13.
31. Vassar R, Rosenberg M, Ross S, Tyner A, Fuchs E. Tissue-specific and differentiation-specific expression of a human K14 keratin gene in transgenic mice. Proc Natl Acad Sci U S A. 1989;86(5):1563–7.
32. Sinha S, Degenstein L, Copenhaver C, Fuchs E. Defining the regulatory factors required for epidermal gene expression. Mol Cell Biol. 2000;20(7):2543–55.
33. Leask A, Rosenberg M, Vassar R, Fuchs E. Regulation of a human epidermal keratin gene: sequences and nuclear factors involved in keratinocyte-specific transcription. Genes Dev. 1990;4(11):1985–98.
34. Leask A, Byrne C, Fuchs E. Transcription factor AP2 and its role in epidermal-specific gene expression. Proc Natl Acad Sci U S A. 1991;88(18):7948–52.
35. Ohtsuki M, Flanagan S, Freedberg IM, Blumenberg M. A cluster of five nuclear proteins regulates keratin gene transcription. Gene Expr. 1993;3(2):201–13.
36. Sinha S, Fuchs E. Identification and dissection of an enhancer controlling epithelial gene expression in skin. Proc Natl Acad Sci U S A. 2001;98(5):2455–60.
37. Byrne C, Fuchs E. Probing keratinocyte and differentiation specificity of the human K5 promoter in vitro and in transgenic mice. Mol Cell Biol. 1993;13(6):3176–90.
38. Mills AA, Zheng B, Wang XJ, Vogel H, Roop DR, Bradley A. p63 is a p53 homologue required for limb and epidermal morphogenesis. Nature. 1999;398(6729):708–13.
39. Yang A, Schweitzer R, Sun D, Kaghad M, Walker N, Bronson RT, Tabin C, Sharpe A, Caput D, Crum C, McKeon F. p63 is essential for regenerative proliferation in limb, craniofacial and epithelial development. Nature. 1999;398(6729):714–8.
40. Candi E, Rufini A, Terrinoni A, Dinsdale D, Ranalli M, Paradisi A, De Laurenzi V, Spagnoli LG, Catani MV, Ramadan S, Knight RA, Melino G. Differential roles of p63 isoforms in epidermal development: selective genetic complementation in p63 null mice. Cell Death Differ. 2006;13(6):1037–47.
41. Romano RA, Birkaya B, Sinha S. A functional enhancer of keratin14 is a direct transcriptional target of deltaNp63. J Invest Dermatol. 2007;127(5):1175–86.

42. Boldrup L, Coates PJ, Gu X, Nylander K. DeltaNp63 isoforms regulate CD44 and keratins 4, 6, 14 and 19 in squamous cell carcinoma of head and neck. J Pathol. 2007;213(4):384–91.
43. Medawar A, Virolle T, Rostagno P, de la Forest-Divonne S, Gambaro K, Rouleau M, Aberdam D. DeltaNp63 is essential for epidermal commitment of embryonic stem cells. PLoS One. 2008;3(10):e3441.
44. Romano RA, Ortt K, Birkaya B, Smalley K, Sinha S. An active role of the DeltaN isoform of p63 in regulating basal keratin genes K5 and K14 and directing epidermal cell fate. PLoS One. 2009;4(5):e5623.
45. Magin TM, Vijayaraj P, Leube RE. Structural and regulatory functions of keratins. Exp Cell Res. 2007;313(10):2021–32.
46. Rosenthal DS, Steinert PM, Chung S, Huff CA, Johnson J, Yuspa SH, Roop DR. A human epidermal differentiation-specific keratin gene is regulated by calcium but not negative modulators of differentiation in transgenic mouse keratinocytes. Cell Growth Differ. 1991;2(2):107–13.
47. Huff CA, Yuspa SH, Rosenthal D. Identification of control elements 3′ to the human keratin 1 gene that regulate cell type and differentiation-specific expression. J Biol Chem. 1993;268(1):377–84.
48. Lu B, Rothnagel JA, Longley MA, Tsai SY, Roop DR. Differentiation-specific expression of human keratin 1 is mediated by a composite AP-1/steroid hormone element. J Biol Chem. 1994;269(10):7443–9.
49. Zhu S, Oh HS, Shim M, Sterneck E, Johnson PF, Smart RC. C/EBPbeta modulates the early events of keratinocyte differentiation involving growth arrest and keratin 1 and keratin 10 expression. Mol Cell Biol. 1999;19(10):7181–90.
50. Oh HS, Smart RC. Expression of CCAAT/enhancer binding proteins (C/EBP) is associated with squamous differentiation in epidermis and isolated primary keratinocytes and is altered in skin neoplasms. J Invest Dermatol. 1998;110(6):939–45.
51. Maytin EV, Lin JC, Krishnamurthy R, Batchvarova N, Ron D, Mitchell PJ, Habener JF. Keratin 10 gene expression during differentiation of mouse epidermis requires transcription factors C/EBP and AP-2. Dev Biol. 1999;216(1):164–81.
52. Johnson LD, Idler WW, Zhou XM, Roop DR, Steinert PM. Structure of a gene for the human epidermal 67-kDa keratin. Proc Natl Acad Sci U S A. 1985;82(7):1896–900.
53. Rieger M, Franke WW. Identification of an orthologous mammalian cytokeratin gene. High degree of intron sequence conservation during evolution of human cytokeratin 10. J Mol Biol. 1988;204(4):841–56.
54. Segre JA. Epidermal barrier formation and recovery in skin disorders. J Clin Invest. 2006;116(5):1150–8.
55. Eckert RL, Crish JF, Efimova T, Dashti SR, Deucher A, Bone F, Adhikary G, Huang G, Gopalakrishnan R, Balasubramanian S. Regulation of involucrin gene expression. J Invest Dermatol. 2004;123(1):13–22.
56. Nithya S, Radhika T, Jeddy N. Loricrin – an overview. J Oral Maxillofac Pathol. 2015;19(1):64–8.
57. Oh IY, Albea DA, Goodwin ZA, Quiggle AM, Baker BP, Guggisberg AM, Geahlen JH, Kroner GM, de Guzman Strong C. Regulation of the dynamic chromatin architecture of the epidermal differentiation complex is mediated by a c-Jun/AP-1-modulated enhancer. J Invest Dermatol. 2014;134:2371–80.
58. Carroll JM, Taichman LB. Characterization of the human involucrin promoter using a transient beta-galactosidase assay. J Cell Sci. 1992;103(Pt 4):925–30.
59. Welter JF, Crish JF, Agarwal C, Eckert RL. Fos-related antigen (Fra-1), junB, and junD activate human involucrin promoter transcription by binding to proximal and distal AP1 sites to mediate phorbol ester effects on promoter activity. J Biol Chem. 1995;270(21):12614–22.
60. Banks EB, Crish JF, Eckert RL. Transcription factor Sp1 activates involucrin promoter activity in non-epithelial cell types. Biochem J. 1999;337(Pt 3):507–12.
61. Banks EB, Crish JF, Welter JF, Eckert RL. Characterization of human involucrin promoter distal regulatory region transcriptional activator elements-a role for Sp1 and AP1 binding sites. Biochem J. 1998;331(Pt 1):61–8.

62. Ogryzko VV, Schiltz RL, Russanova V, Howard BH, Nakatani Y. The transcriptional coactivators p300 and CBP are histone acetyltransferases. Cell. 1996;87(5):953–9.
63. ENCODE Project Consortium, Birney E, Stamatoyannopoulos JA, Dutta A, Guigó R, Gingeras TR, Margulies EH, Weng Z, Snyder M, Dermitzakis ET, Thurman RE, Kuehn MS, Taylor CM, Neph S, Koch CM, Asthana S, Malhotra A, Adzhubei I, Greenbaum JA, Andrews RM, Flicek P, Boyle PJ, Cao H, Carter NP, Clelland GK, Davis S, Day N, Dhami P, Dillon SC, Dorschner MO, Fiegler H, Giresi PG, Goldy J, Hawrylycz M, Haydock A, Humbert R, James KD, Johnson BE, Johnson EM, Frum TT, Rosenzweig ER, Karnani N, Lee K, Lefebvre GC, Navas PA, Neri F, Parker SC, Sabo PJ, Sandstrom R, Shafer A, Vetrie D, Weaver M, Wilcox S, Yu M, Collins FS, Dekker J, Lieb JD, Tullius TD, Crawford GE, Sunyaev S, Noble WS, Dunham I, Denoeud F, Reymond A, Kapranov P, Rozowsky J, Zheng D, Castelo R, Frankish A, Harrow J, Ghosh S, Sandelin A, Hofacker IL, Baertsch R, Keefe D, Dike S, Cheng J, Hirsch HA, Sekinger EA, Lagarde J, Abril JF, Shahab A, Flamm C, Fried C, Hackermüller J, Hertel J, Lindemeyer M, Missal K, Tanzer A, Washietl S, Korbel J, Emanuelsson O, Pedersen JS, Holroyd N, Taylor R, Swarbreck D, Matthews N, Dickson MC, Thomas DJ, Weirauch MT, Gilbert J, Drenkow J, Bell I, Zhao X, Srinivasan KG, Sung WK, Ooi HS, Chiu KP, Foissac S, Alioto T, Brent M, Pachter L, Tress ML, Valencia A, Choo SW, Choo CY, Ucla C, Manzano C, Wyss C, Cheung E, Clark TG, Brown JB, Ganesh M, Patel S, Tammana H, Chrast J, Henrichsen CN, Kai C, Kawai J, Nagalakshmi U, Wu J, Lian Z, Lian J, Newburger P, Zhang X, Bickel P, Mattick JS, Carninci P, Hayashizaki Y, Weissman S, Hubbard T, Myers RM, Rogers J, Stadler PF, Lowe TM, Wei CL, Ruan Y, Struhl K, Gerstein M, Antonarakis SE, Fu Y, Green ED, Karaöz U, Siepel A, Taylor J, Liefer LA, Wetterstrand KA, Good PJ, Feingold EA, Guyer MS, Cooper GM, Asimenos G, Dewey CN, Hou M, Nikolaev S, Montoya-Burgos JI, Löytynoja A, Whelan S, Pardi F, Massingham T, Huang H, Zhang NR, Holmes I, Mullikin JC, Ureta-Vidal A, Paten B, Seringhaus M, Church D, Rosenbloom K, Kent WJ, Stone EA; NISC Comparative Sequencing Program; Baylor College of Medicine Human Genome Sequencing Center; Washington University Genome Sequencing Center; Broad Institute; Children's Hospital Oakland Research Institute, Batzoglou S, Goldman N, Hardison RC, Haussler D, Miller W, Sidow A, Trinklein ND, Zhang ZD, Barrera L, Stuart R, King DC, Ameur A, Enroth S, Bieda MC, Kim J, Bhinge AA, Jiang N, Liu J, Yao F, Vega VB, Lee CW, Ng P, Shahab A, Yang A, Moqtaderi Z, Zhu Z, Xu X, Squazzo S, Oberley MJ, Inman D, Singer MA, Richmond TA, Munn KJ, Rada-Iglesias A, Wallerman O, Komorowski J, Fowler JC, Couttet P, Bruce AW, Dovey OM, Ellis PD, Langford CF, Nix DA, Euskirchen G, Hartman S, Urban AE, Kraus P, Van Calcar S, Heintzman N, Kim TH, Wang K, Qu C, Hon G, Luna R, Glass CK, Rosenfeld MG, Aldred SF, Cooper SJ, Halees A, Lin JM, Shulha HP, Zhang X, Xu M, Haidar JN, Yu Y, Ruan Y, Iyer VR, Green RD, Wadelius C, Farnham PJ, Ren B, Harte RA, Hinrichs AS, Trumbower H, Clawson H, Hillman-Jackson J, Zweig AS, Smith K, Thakkapallayil A, Barber G, Kuhn RM, Karolchik D, Armengol L, Bird CP, de Bakker PI, Kern AD, Lopez-Bigas N, Martin JD, Stranger BE, Woodroffe A, Davydov E, Dimas A, Eyras E, Hallgrímsdóttir IB, Huppert J, Zody MC, Abecasis GR, Estivill X, Bouffard GG, Guan X, Hansen NF, Idol JR, Maduro VV, Maskeri B, McDowell JC, Park M, Thomas PJ, Young AC, Blakesley RW, Muzny DM, Sodergren E, Wheeler DA, Worley KC, Jiang H, Weinstock GM, Gibbs RA, Graves T, Fulton R, Mardis ER, Wilson RK, Clamp M, Cuff J, Gnerre S, Jaffe DB, Chang JL, Lindblad-Toh K, Lander ES, Koriabine M, Nefedov M, Osoegawa K, Yoshinaga Y, Zhu B, de Jong PJ. Identification and analysis of functional elements in 1% of the human genome by the ENCODE pilot project. Nature. 2007;447(7146):799–816.
64. Balasubramanian S, Sturniolo MT, Dubyak GR, Eckert RL. Human epidermal keratinocytes undergo (−)-epigallocatechin-3-gallate-dependent differentiation but not apoptosis. Carcinogenesis. 2005;26(6):1100–8.
65. Crish JF, Eckert RL. Synergistic activation of human involucrin gene expression by Fra-1 and p300 – evidence for the presence of a multiprotein complex. J Invest Dermatol. 2008;128(3):530–41.
66. Chew YC, Adhikary G, Xu W, Wilson GM, Eckert RL. Protein kinase C delta increases Kruppel-like factor 4 protein, which drives involucrin gene transcription in differentiating keratinocytes. J Biol Chem. 2013;288(24):17759–68.

67. Kaczynski J, Cook T, Urrutia R. Sp1- and Kruppel-like transcription factors. Genome Biol. 2003;4(2):206.
68. DiSepio D, Jones A, Longley MA, Bundman D, Rothnagel JA, Roop DR. The proximal promoter of the mouse loricrin gene contains a functional AP-1 element and directs keratinocyte-specific but not differentiation-specific expression. J Biol Chem. 1995;270(18):10792–9.
69. Yoneda K, Steinert PM. Overexpression of human loricrin in transgenic mice produces a normal phenotype. Proc Natl Acad Sci U S A. 1993;90(22):10754–8.
70. Jang SI, Steinert PM. Loricrin expression in cultured human keratinocytes is controlled by a complex interplay between transcription factors of the Sp1, CREB, AP1, and AP2 families. J Biol Chem. 2002;277(44):42268–79.
71. Siepel A, Bejerano G, Pedersen JS, Hinrichs AS, Hou M, Rosenbloom K, Clawson H, Spieth J, Hillier LW, Richards S, Weinstock GM, Wilson RK, Gibbs RA, Kent WJ, Miller W, Haussler D. Evolutionarily conserved elements in vertebrate, insect, worm, and yeast genomes. Genome Res. 2005;15(8):1034–50.
72. Pennacchio LA, Ahituv N, Moses AM, Prabhakar S, Nobrega MA, Shoukry M, Minovitsky S, Dubchak I, Holt A, Lewis KD, Plajzer-Frick I, Akiyama J, De Val S, Afzal V, Black BL, Couronne O, Eisen MB, Visel A, Rubin EM. In vivo enhancer analysis of human conserved non-coding sequences. Nature. 2006;444(7118):499–502.
73. Martin N, Patel S, Segre JA. Long-range comparison of human and mouse Sprr loci to identify conserved noncoding sequences involved in coordinate regulation. Genome Res. 2004;14(12):2430–8.
74. Cabral A, Voskamp P, Cleton-Jansen AM, South A, Nizetic D, Backendorf C. Structural organization and regulation of the small proline-rich family of cornified envelope precursors suggest a role in adaptive barrier function. J Biol Chem. 2001;276(22):19231–7.
75. Chavanas S, Mechin MC, Takahara H, Kawada A, Nachat R, Serre G, Simon M. Comparative analysis of the mouse and human peptidylarginine deiminase gene clusters reveals highly conserved non-coding segments and a new human gene, PADI6. Gene. 2004;330:19–27.
76. Chavanas S, Adoue V, Mechin MC, Ying S, Dong S, Duplan H, Charveron M, Takahara H, Serre G, Simon M. Long-range enhancer associated with chromatin looping allows AP-1 regulation of the peptidylarginine deiminase 3 gene in differentiated keratinocyte. PLoS One. 2008;3(10):e3408.
77. Antonini D, Sirico A, Aberdam E, Ambrosio R, Campanile C, Fagoonee S, Altruda F, Aberdam D, Brissette JL, Missero C. A composite enhancer regulates p63 gene expression in epidermal morphogenesis and in keratinocyte differentiation by multiple mechanisms. Nucleic Acids Res. 2015;43(2):862–74.
78. Taunton J, Hassig CA, Schreiber SL. A mammalian histone deacetylase related to the yeast transcriptional regulator Rpd3p. Science. 1996;272(5260):408–11.
79. Brownell JE, Zhou J, Ranalli T, Kobayashi R, Edmondson DG, Roth SY, Allis CD. Tetrahymena histone acetyltransferase A: a homolog to yeast Gcn5p linking histone acetylation to gene activation. Cell. 1996;84(6):843–51.
80. Yang XJ, Seto E. HATs and HDACs: from structure, function and regulation to novel strategies for therapy and prevention. Oncogene. 2007;26(37):5310–8.
81. Wang Z, Zang C, Cui K, Schones DE, Barski A, Peng W, Zhao K. Genome-wide mapping of HATs and HDACs reveals distinct functions in active and inactive genes. Cell. 2009;138(5):1019–31.
82. Sassone-Corsi P, Mizzen CA, Cheung P, Crosio C, Monaco L, Jacquot S, Hanauer A, Allis CD. Requirement of Rsk-2 for epidermal growth factor-activated phosphorylation of histone H3. Science. 1999;285(5429):886–91.
83. Thomson S, Clayton AL, Hazzalin CA, Rose S, Barratt MJ, Mahadevan LC. The nucleosomal response associated with immediate-early gene induction is mediated via alternative MAP kinase cascades: MSK1 as a potential histone H3/HMG-14 kinase. EMBO J. 1999;18(17):4779–93.
84. Chen D, Ma H, Hong H, Koh SS, Huang SM, Schurter BT, Aswad DW, Stallcup MR. Regulation of transcription by a protein methyltransferase. Science. 1999;284(5423):2174–7.

85. Gary JD, Lin WJ, Yang MC, Herschman HR, Clarke S. The predominant protein-arginine methyltransferase from Saccharomyces cerevisiae. J Biol Chem. 1996;271(21):12585–94.
86. Rea S, Eisenhaber F, O'Carroll D, Strahl BD, Sun ZW, Schmid M, Opravil S, Mechtler K, Ponting CP, Allis CD, Jenuwein T. Regulation of chromatin structure by site-specific histone H3 methyltransferases. Nature. 2000;406(6796):593–9.
87. Wang Y, Wysocka J, Sayegh J, Lee YH, Perlin JR, Leonelli L, Sonbuchner LS, McDonald CH, Cook RG, Dou Y, Roeder RG, Clarke S, Stallcup MR, Allis CD, Coonrod SA. Human PAD4 regulates histone arginine methylation levels via demethylimination. Science. 2004;306(5694):279–83.
88. Cuthbert GL, Daujat S, Snowden AW, Erdjument-Bromage H, Hagiwara T, Yamada M, Schneider R, Gregory PD, Tempst P, Bannister AJ, Kouzarides T. Histone deimination antagonizes arginine methylation. Cell. 2004;118(5):545–53.
89. Robzyk K, Recht J, Osley MA. Rad6-dependent ubiquitination of histone H2B in yeast. Science. 2000;287(5452):501–4.
90. Emre NC, Ingvarsdottir K, Wyce A, Wood A, Krogan NJ, Henry KW, Li K, Marmorstein R, Greenblatt JF, Shilatifard A, Berger SL. Maintenance of low histone ubiquitylation by Ubp10 correlates with telomere-proximal Sir2 association and gene silencing. Mol Cell. 2005;17(4):585–94.
91. Gardner RG, Nelson ZW, Gottschling DE. Ubp10/Dot4p regulates the persistence of ubiquitinated histone H2B: distinct roles in telomeric silencing and general chromatin. Mol Cell Biol. 2005;25(14):6123–39.
92. Henry KW, Wyce A, Lo WS, Duggan LJ, Emre NC, Kao CF, Pillus L, Shilatifard A, Osley MA, Berger SL. Transcriptional activation via sequential histone H2B ubiquitylation and deubiquitylation, mediated by SAGA-associated Ubp8. Genes Dev. 2003;17(21):2648–63.
93. Chang B, Chen Y, Zhao Y, Bruick RK. JMJD6 is a histone arginine demethylase. Science. 2007;318(5849):444–7.
94. Shi Y, Lan F, Matson C, Mulligan P, Whetstine JR, Cole PA, Casero RA, Shi Y. Histone demethylation mediated by the nuclear amine oxidase homolog LSD1. Cell. 2004;119(7):941–53.
95. Tsukada Y, Fang J, Erdjument-Bromage H, Warren ME, Borchers CH, Tempst P, Zhang Y. Histone demethylation by a family of JmjC domain-containing proteins. Nature. 2006;439(7078):811–6.
96. Marmorstein R, Trievel RC. Histone modifying enzymes: structures, mechanisms, and specificities. Biochim Biophys Acta. 2009;1789(1):58–68.
97. Berger SL, Kouzarides T, Shiekhattar R, Shilatifard A. An operational definition of epigenetics. Genes Dev. 2009;23(7):781–3.
98. Bao X, Tang J, Lopez-Pajares V, Tao S, Qu K, Crabtree GR, Khavari PA. ACTL6a enforces the epidermal progenitor state by suppressing SWI/SNF-dependent induction of KLF4. Cell Stem Cell. 2013;12(2):193–203.
99. Frye M, Fisher AG, Watt FM. Epidermal stem cells are defined by global histone modifications that are altered by Myc-induced differentiation. PLoS One. 2007;2(8):e763.
100. Kashiwagi M, Morgan BA, Georgopoulos K. The chromatin remodeler Mi-2beta is required for establishment of the basal epidermis and normal differentiation of its progeny. Development. 2007;134(8):1571–82.
101. LeBoeuf M, Terrell A, Trivedi S, Sinha S, Epstein JA, Olson EN, Morrisey EE, Millar SE. Hdac1 and Hdac2 act redundantly to control p63 and p53 functions in epidermal progenitor cells. Dev Cell. 2010;19(6):807–18.
102. Bannister AJ, Kouzarides T. Regulation of chromatin by histone modifications. Cell Res. 2011;21(3):381–95.
103. Ezhkova E, Pasolli HA, Parker JS, Stokes N, Su IH, Hannon G, Tarakhovsky A, Fuchs E. Ezh2 orchestrates gene expression for the stepwise differentiation of tissue-specific stem cells. Cell. 2009;136(6):1122–35.

104. Gdula MR, Poterlowicz K, Mardaryev AN, Sharov AA, Peng Y, Fessing MY, Botchkarev VA. Remodeling of three-dimensional organization of the nucleus during terminal keratinocyte differentiation in the epidermis. J Invest Dermatol. 2013;133(9):2191–201.
105. Vigano MA, Lamartine J, Testoni B, Merico D, Alotto D, Castagnoli C, Robert A, Candi E, Melino G, Gidrol X, Mantovani R. New p63 targets in keratinocytes identified by a genome-wide approach. EMBO J. 2006;25(21):5105–16.
106. Fessing MY, Mardaryev AN, Gdula MR, Sharov AA, Sharova TY, Rapisarda V, Gordon KB, Smorodchenko AD, Poterlowicz K, Ferone G, Kohwi Y, Missero C, Kohwi-Shigematsu T, Botchkarev VA. p63 regulates Satb1 to control tissue-specific chromatin remodeling during development of the epidermis. J Cell Biol. 2011;194(6):825–39.
107. Mardaryev AN, Gdula MR, Yarker JL, Emelianov VU, Poterlowicz K, Sharov AA, Sharova TY, Scarpa JA, Joffe B, Solovei I, Chambon P, Botchkarev VA, Fessing MY. p63 and Brg1 control developmentally regulated higher-order chromatin remodelling at the epidermal differentiation complex locus in epidermal progenitor cells. Development. 2014;141(1):101–11.
108. Sethi I, Sinha S, Buck MJ. Role of chromatin and transcriptional co-regulators in mediating p63-genome interactions in keratinocytes. BMC Genomics. 2014;15:1042.
109. Ong CT, Corces VG. Enhancer function: new insights into the regulation of tissue-specific gene expression. Nat Rev Genet. 2011;12(4):283–93.
110. Landt SG, Marinov GK, Kundaje A, Kheradpour P, Pauli F, Batzoglou S, Bernstein BE, Bickel P, Brown JB, Cayting P, Chen Y, DeSalvo G, Epstein C, Fisher-Aylor KI, Euskirchen G, Gerstein M, Gertz J, Hartemink AJ, Hoffman MM, Iyer VR, Jung YL, Karmakar S, Kellis M, Kharchenko PV, Li Q, Liu T, Liu XS, Ma L, Milosavljevic A, Myers RM, Park PJ, Pazin MJ, Perry MD, Raha D, Reddy TE, Rozowsky J, Shoresh N, Sidow A, Slattery M, Stamatoyannopoulos JA, Tolstorukov MY, White KP, Xi S, Farnham PJ, Lieb JD, Wold BJ, Snyder M. ChIP-seq guidelines and practices of the ENCODE and modENCODE consortia. Genome Res. 2012;22(9):1813–31.
111. Giresi PG, Kim J, McDaniell RM, Iyer VR, Lieb JD. FAIRE (Formaldehyde-Assisted Isolation of Regulatory Elements) isolates active regulatory elements from human chromatin. Genome Res. 2007;17(6):877–85.
112. Buenrostro JD, Giresi PG, Zaba LC, Chang HY, Greenleaf WJ. Transposition of native chromatin for fast and sensitive epigenomic profiling of open chromatin, DNA-binding proteins and nucleosome position. Nat Methods. 2013;10(12):1213–8.
113. Song L, Crawford GE. DNase-seq: a high-resolution technique for mapping active gene regulatory elements across the genome from mammalian cells. Cold Spring Harb Protoc. 2010;2010(2):pdb prot5384.
114. Hansen RS, Thomas S, Sandstrom R, Canfield TK, Thurman RE, Weaver M, Dorschner MO, Gartler SM, Stamatoyannopoulos JA. Sequencing newly replicated DNA reveals widespread plasticity in human replication timing. Proc Natl Acad Sci U S A. 2010;107(1):139–44.
115. Zaret K. Micrococcal nuclease analysis of chromatin structure. Curr Protoc Mol Biol., Chapter 21:Unit. 2005;21:21.
116. Mortazavi A, Williams BA, McCue K, Schaeffer L, Wold B. Mapping and quantifying mammalian transcriptomes by RNA-Seq. Nat Methods. 2008;5(7):621–8.
117. Carninci P, Kvam C, Kitamura A, Ohsumi T, Okazaki Y, Itoh M, Kamiya M, Shibata K, Sasaki N, Izawa M, Muramatsu M, Hayashizaki Y, Schneider C. High-efficiency full-length cDNA cloning by biotinylated CAP trapper. Genomics. 1996;37(3):327–36.
118. Fullwood MJ, Liu MH, Pan YF, Liu J, Xu H, Mohamed YB, Orlov YL, Velkov S, Ho A, Mei PH, Chew EG, Huang PY, Welboren WJ, Han Y, Ooi HS, Ariyaratne PN, Vega VB, Luo Y, Tan PY, Choy PY, Wansa KD, Zhao B, Lim KS, Leow SC, Yow JS, Joseph R, Li H, Desai KV, Thomsen JS, Lee YK, Karuturi RK, Herve T, Bourque G, Stunnenberg HG, Ruan X, Cacheux-Rataboul V, Sung WK, Liu ET, Wei CL, Cheung E, Ruan Y. An oestrogen-receptor-alpha-bound human chromatin interactome. Nature. 2009;462(7269):58–64.
119. Batut P, Gingeras TR. RAMPAGE: promoter activity profiling by paired-end sequencing of 5'-complete cDNAs. Curr Protoc Mol Biol, 104:Unit. 2013;25B:11.

120. Ruan X, Ruan Y. Genome wide full-length transcript analysis using 5' and 3' paired-end-tag next generation sequencing (RNA-PET). Methods Mol Biol. 2012;809:535–62.
121. Aebersold R, Mann M. Mass spectrometry-based proteomics. Nature. 2003;422(6928): 198–207.
122. Jolma A, Kivioja T, Toivonen J, Cheng L, Wei G, Enge M, Taipale M, Vaquerizas JM, Yan J, Sillanpaa MJ, Bonke M, Palin K, Talukder S, Hughes TR, Luscombe NM, Ukkonen E, Taipale J. Multiplexed massively parallel SELEX for characterization of human transcription factor binding specificities. Genome Res. 2010;20(6):861–73.
123. Zhao Y, Granas D, Stormo GD. Inferring binding energies from selected binding sites. PLoS Comput Biol. 2009;5(12):e1000590.
124. Zykovich A, Korf I, Segal DJ. Bind-n-Seq: high-throughput analysis of in vitro protein-DNA interactions using massively parallel sequencing. Nucleic Acids Res. 2009;37(22):e151.
125. Dekker J, Rippe K, Dekker M, Kleckner N. Capturing chromosome conformation. Science. 2002;295(5558):1306–11.
126. Zhao Z, Tavoosidana G, Sjolinder M, Gondor A, Mariano P, Wang S, Kanduri C, Lezcano M, Sandhu KS, Singh U, Pant V, Tiwari V, Kurukuti S, Ohlsson R. Circular chromosome conformation capture (4C) uncovers extensive networks of epigenetically regulated intra- and interchromosomal interactions. Nat Genet. 2006;38(11):1341–7.
127. Dostie J, Richmond TA, Arnaout RA, Selzer RR, Lee WL, Honan TA, Rubio ED, Krumm A, Lamb J, Nusbaum C, Green RD, Dekker J. Chromosome Conformation Capture Carbon Copy (5C): a massively parallel solution for mapping interactions between genomic elements. Genome Res. 2006;16(10):1299–309.
128. Lieberman-Aiden E, van Berkum NL, Williams L, Imakaev M, Ragoczy T, Telling A, Amit I, Lajoie BR, Sabo PJ, Dorschner MO, Sandstrom R, Bernstein B, Bender MA, Groudine M, Gnirke A, Stamatoyannopoulos J, Mirny LA, Lander ES, Dekker J. Comprehensive mapping of long-range interactions reveals folding principles of the human genome. Science. 2009;326(5950):289–93.
129. Cokus SJ, Feng S, Zhang X, Chen Z, Merriman B, Haudenschild CD, Pradhan S, Nelson SF, Pellegrini M, Jacobsen SE. Shotgun bisulphite sequencing of the Arabidopsis genome reveals DNA methylation patterning. Nature. 2008;452(7184):215–9.
130. Meissner A, Gnirke A, Bell GW, Ramsahoye B, Lander ES, Jaenisch R. Reduced representation bisulfite sequencing for comparative high-resolution DNA methylation analysis. Nucleic Acids Res. 2005;33(18):5868–77.
131. Jacinto FV, Ballestar E, Esteller M. Methyl-DNA immunoprecipitation (MeDIP): hunting down the DNA methylome. Biotechniques. 2008;44:1. 35, 37, 39 passim
132. Maunakea AK, Nagarajan RP, Bilenky M, Ballinger TJ, D'Souza C, Fouse SD, Johnson BE, Hong C, Nielsen C, Zhao Y, Turecki G, Delaney A, Varhol R, Thiessen N, Shchors K, Heine VM, Rowitch DH, Xing X, Fiore C, Schillebeeckx M, Jones SJ, Haussler D, Marra MA, Hirst M, Wang T, Costello JF. Conserved role of intragenic DNA methylation in regulating alternative promoters. Nature. 2010;466(7303):253–7.
133. Kothary R, Clapoff S, Darling S, Perry MD, Moran LA, Rossant J. Inducible expression of an hsp68-lacZ hybrid gene in transgenic mice. Development. 1989;105(4):707–14.
134. le T, A Fraser SE. Enhancer and gene traps for molecular imaging and genetic analysis in zebrafish. Dev Growth Differ. 2013;55(4):434–45.
135. Smale ST. Luciferase assay. Cold Spring Harb Protoc. 2010;2010(5):pdb prot5421.
136. Smale ST. Beta-galactosidase assay. Cold Spring Harb Protoc. 2010;2010(5):pdb prot5423.
137. Dickel DE, Zhu Y, Nord AS, Wylie JN, Akiyama JA, Afzal V, Plajzer-Frick I, Kirkpatrick A, Gottgens B, Bruneau BG, Visel A, Pennacchio LA. Function-based identification of mammalian enhancers using site-specific integration. Nat Methods. 2014;11(5):566–71.
138. Murtha M, Tokcaer-Keskin Z, Tang Z, Strino F, Chen X, Wang Y, Xi X, Basilico C, Brown S, Bonneau R, Kluger Y, Dailey L. FIREWACh: high-throughput functional detection of transcriptional regulatory modules in mammalian cells. Nat Methods. 2014;11(5):559–65.

139. Vockley CM, Guo C, Majoros WH, Nodzenski M, Scholtens DM, Hayes MG, Lowe WL, Reddy TE Jr. Massively parallel quantification of the regulatory effects of noncoding genetic variation in a human cohort. Genome Res. 2015;25(8):1206–14.
140. Inoue F, Ahituv N. Decoding enhancers using massively parallel reporter assays. Genomics. 2015;106(3):159–64.
141. Kwasnieski JC, Mogno I, Myers CA, Corbo JC, Cohen BA. Complex effects of nucleotide variants in a mammalian cis-regulatory element. Proc Natl Acad Sci U S A. 2012;109(47):19498–503.
142. Akhtar W, de Jong J, Pindyurin AV, Pagie L, Meuleman W, de Ridder J, Berns A, Wessels LF, Lohuizen v, M van Steensel B. Chromatin position effects assayed by thousands of reporters integrated in parallel. Cell. 2013;154(4):914–27.
143. Krivega I, Dean A. Enhancer and promoter interactions-long distance calls. Curr Opin Genet Dev. 2012;22(2):79–85.
144. Dean A. On a chromosome far, far away: LCRs and gene expression. Trends Genet. 2006;22(1):38–45.
145. Carter D, Chakalova L, Osborne CS, Dai YF, Fraser P. Long-range chromatin regulatory interactions in vivo. Nat Genet. 2002;32(4):623–6.
146. Tolhuis B, Palstra RJ, Splinter E, Grosveld F, de Laat W. Looping and interaction between hypersensitive sites in the active beta-globin locus. Mol Cell. 2002;10(6):1453–65.
147. Palstra RJ, Tolhuis B, Splinter E, Nijmeijer R, Grosveld F, de Laat W. The beta-globin nuclear compartment in development and erythroid differentiation. Nat Genet. 2003;35(2):190–4.
148. Kadauke S, Blobel GA. Chromatin loops in gene regulation. Biochim Biophys Acta. 2009;1789(1):17–25.
149. Hong JW, Hendrix DA, Levine MS. Shadow enhancers as a source of evolutionary novelty. Science. 2008;321(5894):1314.
150. Li G, Fullwood MJ, Xu H, Mulawadi FH, Velkov S, Vega V, Ariyaratne PN, Mohamed YB, Ooi HS, Tennakoon C, Wei CL, Ruan Y, Sung WK. ChIA-PET tool for comprehensive chromatin interaction analysis with paired-end tag sequencing. Genome Biol. 2010;11(2):R22.
151. Kiefer CM, Hou C, Little JA, Dean A. Epigenetics of beta-globin gene regulation. Mutat Res. 2008;647(1–2):68–76.
152. Song SH, Hou C, Dean A. A positive role for NLI/Ldb1 in long-range beta-globin locus control region function. Mol Cell. 2007;28(5):810–22.
153. Tripic T, Deng W, Cheng Y, Zhang Y, Vakoc CR, Gregory GD, Hardison RC, Blobel GA. SCL and associated proteins distinguish active from repressive GATA transcription factor complexes. Blood. 2009;113(10):2191–201.
154. Deng W, Lee J, Wang H, Miller J, Reik A, Gregory PD, Dean A, Blobel GA. Controlling long-range genomic interactions at a native locus by targeted tethering of a looping factor. Cell. 2012;149(6):1233–44.
155. Dixon JR, Selvaraj S, Yue F, Kim A, Li Y, Shen Y, Hu M, Liu JS, Ren B. Topological domains in mammalian genomes identified by analysis of chromatin interactions. Nature. 2012;485(7398):376–80.
156. Sexton T, Yaffe E, Kenigsberg E, Bantignies F, Leblanc B, Hoichman M, Parrinello H, Tanay A, Cavalli G. Three-dimensional folding and functional organization principles of the Drosophila genome. Cell. 2012;148(3):458–72.
157. Symmons O, Uslu VV, Tsujimura T, Ruf S, Nassari S, Schwarzer W, Ettwiller L, Spitz F. Functional and topological characteristics of mammalian regulatory domains. Genome Res. 2014;24(3):390–400.
158. Lupianez DG, Kraft K, Heinrich V, Krawitz P, Brancati F, Klopocki E, Horn D, Kayserili H, Opitz JM, Laxova R, Santos-Simarro F, Gilbert-Dussardier B, Wittler L, Borschiwer M, Haas SA, Osterwalder M, Franke M, Timmermann B, Hecht J, Spielmann M, Visel A, Mundlos S. Disruptions of topological chromatin domains cause pathogenic rewiring of gene-enhancer interactions. Cell. 2015;161(5):1012–25.
159. Fraser J, Williamson I, Bickmore WA, Dostie J. An overview of genome organization and how we got there: from FISH to Hi-C. Microbiol Mol Biol Rev. 2015;79(3):347–72.

160. Amano T, Sagai T, Tanabe H, Mizushina Y, Nakazawa H, Shiroishi T. Chromosomal dynamics at the Shh locus: limb bud-specific differential regulation of competence and active transcription. Dev Cell. 2009;16(1):47–57.
161. Buffry AD, Mendes CC, McGregor AP. The functionality and evolution of eukaryotic transcriptional enhancers. Adv Genet. 2016;96:143–206.
162. Hagstrom KA, Meyer BJ. Condensin and cohesin: more than chromosome compactor and glue. Nat Rev Genet. 2003;4(7):520–34.
163. Parelho V, Hadjur S, Spivakov M, Leleu M, Sauer S, Gregson HC, Jarmuz A, Canzonetta C, Webster Z, Nesterova T, Cobb BS, Yokomori K, Dillon N, Aragon L, Fisher AG, Merkenschlager M. Cohesins functionally associate with CTCF on mammalian chromosome arms. Cell. 2008;132(3):422–33.
164. Wendt KS, Yoshida K, Itoh T, Bando M, Koch B, Schirghuber E, Tsutsumi S, Nagae G, Ishihara K, Mishiro T, Yahata K, Imamoto F, Aburatani H, Nakao M, Imamoto N, Maeshima K, Shirahige K, Peters JM. Cohesin mediates transcriptional insulation by CCCTC-binding factor. Nature. 2008;451(7180):796–801.
165. Rubio ED, Reiss DJ, Welcsh PL, Disteche CM, Filippova GN, Baliga NS, Aebersold R, Ranish JA, Krumm A. CTCF physically links cohesin to chromatin. Proc Natl Acad Sci U S A. 2008;105(24):8309–14.
166. Wallace JA, Felsenfeld G. We gather together: insulators and genome organization. Curr Opin Genet Dev. 2007;17(5):400–7.
167. Mishiro T, Ishihara K, Hino S, Tsutsumi S, Aburatani H, Shirahige K, Kinoshita Y, Nakao M. Architectural roles of multiple chromatin insulators at the human apolipoprotein gene cluster. EMBO J. 2009;28(9):1234–45.
168. Hou C, Dale R, Dean A. Cell type specificity of chromatin organization mediated by CTCF and cohesin. Proc Natl Acad Sci U S A. 2010;107(8):3651–6.
169. Seitan VC, Hao B, Tachibana-Konwalski K, Lavagnolli T, Mira-Bontenbal H, Brown KE, Teng G, Carroll T, Terry A, Horan K, Marks H, Adams DJ, Schatz DG, Aragon L, Fisher AG, Krangel MS, Nasmyth K, Merkenschlager M. A role for cohesin in T-cell-receptor rearrangement and thymocyte differentiation. Nature. 2011;476(7361):467–71.
170. Rao SS, Huntley MH, Durand NC, Stamenova EK, Bochkov ID, Robinson JT, Sanborn AL, Machol I, Omer AD, Lander ES, Aiden EL. A 3D map of the human genome at kilobase resolution reveals principles of chromatin looping. Cell. 2014;159(7):1665–80.
171. Jabbari S, Fitzmaurice T, Munoz F, Lafuente C, Zentner P, Bustamante JG. Cross-border collaboration in oncology: a model for United States-Mexico border health. Int J Radiat Oncol Biol Phys. 2015;92(3):509–11.
172. Vietri Rudan M, Barrington C, Henderson S, Ernst C, Odom DT, Tanay A, Hadjur S. Comparative Hi-C reveals that CTCF underlies evolution of chromosomal domain architecture. Cell Rep. 2015;10(8):1297–309.
173. Guo Y, Xu Q, Canzio D, Shou J, Li J, Gorkin DU, Jung I, Wu H, Zhai Y, Tang Y, Lu Y, Wu Y, Jia Z, Li W, Zhang MQ, Ren B, Krainer AR, Maniatis T, Wu Q. CRISPR inversion of CTCF sites alters genome topology and enhancer/promoter function. Cell. 2015;162(4):900–10.
174. de Wit E, Vos ES, Holwerda SJ, Valdes-Quezada C, Verstegen MJ, Teunissen H, Splinter E, Wijchers PJ, Krijger PH, de Laat W. CTCF binding polarity determines chromatin looping. Mol Cell. 2015;60(4):676–84.
175. Lee HK, Willi M, Wang C, Yang CM, Smith HE, Liu C, Hennighausen L. Functional assessment of CTCF sites at cytokine-sensing mammary enhancers using CRISPR/Cas9 gene editing in mice. Nucleic Acids Res. 2017;45(8):4606–18.
176. Whyte WA, Orlando DA, Hnisz D, Abraham BJ, Lin CY, Kagey MH, Rahl PB, Lee TI, Young RA. Master transcription factors and mediator establish super-enhancers at key cell identity genes. Cell. 2013;153(2):307–19.
177. Poss ZC, Ebmeier CC, Taatjes DJ. The mediator complex and transcription regulation. Crit Rev Biochem Mol Biol. 2013;48(6):575–608.

178. Park SW, Li G, Lin YP, Barrero MJ, Ge K, Roeder RG, Wei LN. Thyroid hormone-induced juxtaposition of regulatory elements/factors and chromatin remodeling of Crabp1 dependent on MED1/TRAP220. Mol Cell. 2005;19(5):643–53.
179. Wang Q, Carroll JS, Brown M. Spatial and temporal recruitment of androgen receptor and its coactivators involves chromosomal looping and polymerase tracking. Mol Cell. 2005;19(5):631–42.
180. Krivega I, Dean A. LDB1-mediated enhancer looping can be established independent of mediator and cohesin. Nucleic Acids Res. 2017;45:8255.
181. Suzuki HI, Young RA, Sharp PA. Super-enhancer-mediated RNA processing revealed by integrative microRNA network analysis. Cell. 2017;168(6):1000–14.
182. Adam RC, Yang H, Rockowitz S, Larsen SB, Nikolova M, Oristian DS, Polak L, Kadaja M, Asare A, Zheng D, Fuchs E. Pioneer factors govern super-enhancer dynamics in stem cell plasticity and lineage choice. Nature. 2015;521(7552):366–70.
183. Folgueras AR, Guo X, Pasolli HA, Stokes N, Polak L, Zheng D, Fuchs E. Architectural niche organization by LHX2 is linked to hair follicle stem cell function. Cell Stem Cell. 2013;13(3):314–27.
184. Kadaja M, Keyes BE, Lin M, Pasolli HA, Genander M, Polak L, Stokes N, Zheng D, Fuchs E. SOX9: a stem cell transcriptional regulator of secreted niche signaling factors. Genes Dev. 2014;28(4):328–41.
185. Keyes BE, Segal JP, Heller E, Lien WH, Chang CY, Guo X, Oristian DS, Zheng D, Fuchs E. Nfatc1 orchestrates aging in hair follicle stem cells. Proc Natl Acad Sci U S A. 2013;110(51):E4950–9.
186. Chang CY, Pasolli HA, Giannopoulou EG, Guasch G, Gronostajski RM, Elemento O, Fuchs E. NFIB is a governor of epithelial-melanocyte stem cell behaviour in a shared niche. Nature. 2013;495(7439):98–102.
187. Klein RH, Lin Z, Hopkin AS, Gordon W, Tsoi LC, Liang Y, Gudjonsson JE, Andersen B. GRHL3 binding and enhancers rearrange as epidermal keratinocytes transition between functional states. PLoS Genet. 2017;13(4):e1006745.
188. Sethi I, Gluck C, Zhou H, Buck MJ, Sinha S. Evolutionary re-wiring of p63 and the epigenomic regulatory landscape in keratinocytes and its potential implications on species-specific gene expression and phenotypes. Nucleic Acids Res. 2017;45:8208.
189. Tsai MC, Manor O, Wan Y, Mosammaparast N, Wang JK, Lan F, Shi Y, Segal E, Chang HY. Long noncoding RNA as modular scaffold of histone modification complexes. Science. 2010;329(5992):689–93.
190. Rinn JL, Kertesz M, Wang JK, Squazzo SL, Xu X, Brugmann SA, Goodnough LH, Helms JA, Farnham PJ, Segal E, Chang HY. Functional demarcation of active and silent chromatin domains in human HOX loci by noncoding RNAs. Cell. 2007;129(7):1311–23.
191. Mulder KW, Wang X, Escriu C, Ito Y, Schwarz RF, Gillis J, Sirokmany G, Donati G, Uribe-Lewis S, Pavlidis P, Murrell A, Markowetz F, Watt FM. Diverse epigenetic strategies interact to control epidermal differentiation. Nat Cell Biol. 2012;14(7):753–63.
192. Kim TK, Hemberg M, Gray JM. Enhancer RNAs: a class of long noncoding RNAs synthesized at enhancers. Cold Spring Harb Perspect Biol. 2015;7(1):a018622.
193. Lam MT, Cho H, Lesch HP, Gosselin D, Heinz S, Tanaka-Oishi Y, Benner C, Kaikkonen MU, Kim AS, Kosaka M, Lee CY, Watt A, Grossman TR, Rosenfeld MG, Evans RM, Glass CK. Rev-Erbs repress macrophage gene expression by inhibiting enhancer-directed transcription. Nature. 2013;498(7455):511–5.
194. Li W, Notani D, Ma Q, Tanasa B, Nunez E, Chen AY, Merkurjev D, Zhang J, Ohgi K, Song X, Oh S, Kim HS, Glass CK, Rosenfeld MG. Functional roles of enhancer RNAs for oestrogen-dependent transcriptional activation. Nature. 2013;498(7455):516–20.
195. Melo CA, Drost J, Wijchers PJ, van de Werken H, de Wit E, Oude Vrielink JA, Elkon R, Melo SA, Leveille N, Kalluri R, de Laat W, Agami R. eRNAs are required for p53-dependent enhancer activity and gene transcription. Mol Cell. 2013;49(3):524–35.
196. Lindblad-Toh K, Garber M, Zuk O, Lin MF, Parker BJ, Washietl S, Kheradpour P, Ernst J, Jordan G, Mauceli E, Ward LD, Lowe CB, Holloway AK, Clamp M, Gnerre S, Alföldi J,

Beal K, Chang J, Clawson H, Cuff J, Di Palma F, Fitzgerald S, Flicek P, Guttman M, Hubisz MJ, Jaffe DB, Jungreis I, Kent WJ, Kostka D, Lara M, Martins AL, Massingham T, Moltke I, Raney BJ, Rasmussen MD, Robinson J, Stark A, Vilella AJ, Wen J, Xie X, Zody MC; Broad Institute Sequencing Platform and Whole Genome Assembly Team, Baldwin J, Bloom T, Chin CW, Heiman D, Nicol R, Nusbaum C, Young S, Wilkinson J, Worley KC, Kovar CL, Muzny DM, Gibbs RA; Baylor College of Medicine Human Genome Sequencing Center Sequencing Team, Cree A, Dihn HH, Fowler G, Jhangiani S, Joshi V, Lee S, Lewis LR, Nazareth LV, Okwuonu G, Santibanez J, Warren WC, Mardis ER, Weinstock GM, Wilson RK; Genome Institute at Washington University, Delehaunty K, Dooling D, Fronik C, Fulton L, Fulton B, Graves T, Minx P, Sodergren E, Birney E, Margulies EH, Herrero J, Green ED, Haussler D, Siepel A, Goldman N, Pollard KS, Pedersen JS, Lander ES, Kellis M. A high-resolution map of human evolutionary constraint using 29 mammals. Nature. 2011;478(7370):476–82.

197. Hoffman MM, Ernst J, Wilder SP, Kundaje A, Harris RS, Libbrecht M, Giardine B, Ellenbogen PM, Bilmes JA, Birney E, Hardison RC, Dunham I, Kellis M, Noble WS. Integrative annotation of chromatin elements from ENCODE data. Nucleic Acids Res. 2013;41(2):827–41.
198. Rinaldi L, Datta D, Serrat J, Morey L, Solanas G, Avgustinova A, Blanco E, Pons JI, Matallanas D, Von Kriegsheim A, Di C, L Benitah SA. Dnmt3a and Dnmt3b associate with enhancers to regulate human epidermal stem cell homeostasis. Cell Stem Cell. 2016;19(4):491–501.
199. Hadjur S, Williams LM, Ryan NK, Cobb BS, Sexton T, Fraser P, Fisher AG, Merkenschlager M. Cohesins form chromosomal cis-interactions at the developmentally regulated IFNG locus. Nature. 2009;460(7253):410–3.
200. Hemberg M, Gray JM, Cloonan N, Kuersten S, Grimmond S, Greenberg ME, Kreiman G. Integrated genome analysis suggests that most conserved non-coding sequences are regulatory factor binding sites. Nucleic Acids Res. 2012;40(16):7858–69.
201. Fisher S, Grice EA, Vinton RM, Bessling SL, McCallion AS. Conservation of RET regulatory function from human to zebrafish without sequence similarity. Science. 2006;312(5771):276–9.
202. Fanucchi S, Shibayama Y, Burd S, Weinberg MS, Mhlanga MM. Chromosomal contact permits transcription between coregulated genes. Cell. 2013;155(3):606–20.
203. Sander JD, Joung JK. CRISPR-Cas systems for editing, regulating and targeting genomes. Nat Biotechnol. 2014;32(4):347–55.
204. Kvon EZ, Kamneva OK, Melo US, Barozzi I, Osterwalder M, Mannion BJ, Tissieres V, Pickle CS, Plajzer-Frick I, Lee EA, Kato M, Garvin TH, Akiyama JA, Afzal V, Lopez-Rios J, Rubin EM, Dickel DE, Pennacchio LA, Visel A. Progressive loss of function in a limb enhancer during snake evolution. Cell. 2016;167(3):633–642.e611.
205. Diao YR, Li B, Meng ZP, Jung I, Lee AY, Dixon J, Maliskova L, Guan KL, Shen Y, Ren B. A new class of temporarily phenotypic enhancers identified by CRISPR/Cas9-mediated genetic screening. Genome Res. 2016;26(3):397–405.

Chapter 11
Integration of Biochemical and Mechanical Signals at the Nuclear Periphery: Impacts on Skin Development and Disease

Rachel M. Stewart, Megan C. King, and Valerie Horsley

Abbreviations

AJs	Adherens junctions
APC	Adenomatous polyposis coli
BAF	Barrier-to-autointegration factor
ChIP-seq	Chromatin immunoprecipitation sequencing
DDR	DNA damage response
DSB	Double-strand break
ECM	Extracellular Matrix
EDC	Epidermal differentiation complex
GA repeat	Guanine-adenine repeat
H2K9me2/3	Di- or trimethylation of lysine 9 on histone H2
H3K27me3	Trimethylation of lysine 27 on histone H3
HDAC	Histone deacetylase
HGPS	Hutchinson-Gilford Progeria Syndrome
Hi-C	High-throughput chromosome capture
IFE	Interfollicular epidermis
INM	Inner nuclear membrane
LEM	LAP2, emerin and MAN1
ONM	Outer nuclear membrane
LAD	Lamina associated domain

R. M. Stewart · M. C. King (✉)
Department of Cell Biology, Yale School of Medicine, New Haven, CT, USA
e-mail: megan.king@yale.edu

V. Horsley (✉)
Department of Dermatology, Yale School of Medicine, New Haven, CT, USA

Department of Molecular, Cell and Developmental Biology, Yale University, New Haven, CT, USA
e-mail: valerie.horsley@yale.edu

© Springer International Publishing AG, part of Springer Nature 2018

V. A. Botchkarev, S. E. Millar (eds.), *Epigenetic Regulation of Skin Development and Regeneration*, Stem Cell Biology and Regenerative Medicine,
https://doi.org/10.1007/978-3-319-16769-5_11

LAP1	Lamina-associated polypeptide 1
LAP2	Lamina-associated polypeptide 2
LBR	Lamin B receptor
LINC	Linker of Nucleoskeleton and Cytoskeleton
MEF	Mouse embryonic fibroblast
RGD	Arginylglycylaspartic acid
TAN lines	Transmembrane actin-associated nuclear lines

11.1 Introduction: Skin Epithelial Keratinocytes and the Nucleus

The skin and its appendages, including hair follicles, sweat glands, and sebaceous glands, provide essential functions for animal survival, such as protection from water loss and environmental insults, as well as the capacity for tactile sense. The interfollicular epidermis (IFE), which composes the outermost layer of the skin, consists of a series of functionally and structurally distinct stratified layers. The basal layer contains proliferative progenitor keratinocytes that tightly bind to extracellular matrix (ECM) molecules in the basement membrane, which separates the epidermis from the underlying dermis. During embryonic development, basal keratinocytes divide asymmetrically, giving rise to daughter cells that move upwards to form the suprabasal layers (spinous, granular, and stratum corneum); the resulting cells of the stratified epithelium generate a cornified envelope essential for the barrier function of the skin [1]. A subset of basal cells also gives rise to hair follicles and other appendages [2–4].

Keratinocyte differentiation during epidermal development requires coordinated changes in the cytoskeleton that support the generation and mechanical stability of cell-cell adhesions, including adherens junctions (AJs), desmosomes, and tight junctions. Keratin intermediate filament proteins, the major structural component of the IFE, interact with integrin-based hemidesmosome adhesions that connect the basal IFE to the dermis (Fig. 11.1a). E-cadherin-mediated cell-cell junctions, located throughout the epidermis and its appendages, connect to actin and microtubule networks through catenin molecules [5]. As basal cells move upward during differentiation, another class of intercellular junction, desmosomes, which contain desmosomal cadherins and are connected to keratins, becomes abundant (Fig. 11.1a). Desmosome density and desmosomal cadherin composition markedly change as cells differentiate [6]. Finally, tight junctions assemble in the granular layer [7–9]. These junctions provide mechanical stability to the epidermis and are essential for the barrier function of the skin.

During the process of terminal differentiation in the IFE and during hair follicle stem cell activation and differentiation, the cell nucleus goes through remarkable changes in shape and size [10]. In the most differentiated cells in the IFE and hair follicle, a poorly understood process leads to beneficial loss of the nucleus. When intact, the nucleus is bounded by the nuclear envelope that serves to physically and functionally segregate the genome from the cytoplasmic milieu. Research over the last several decades has highlighted that, in addition to this role, the nucleus additionally serves to integrate transcriptional, biochemical, and mechanical networks within cells and tissues. In this chapter, we will discuss how the structure and organization of the nucleus impacts gene expression, genome integrity, cell and tissue level mechanics, and disease in the context of skin homeostasis and regeneration.

11.2 The Nuclear Lamina and Nuclear Organization

The nucleus is organized into distinct but dynamic subnuclear compartments, including chromosome territories and nuclear bodies (nucleoli, Cajal bodies, etc). One critical nuclear landmark is the nuclear envelope, which is contiguous with the endoplasmic reticulum and is comprised of both an inner and outer nuclear membrane (INM and ONM, respectively). The interface between the nuclear interior and the INM is characterized by the nuclear lamina, a highly complex proteinaceous structure that consists of a meshwork of the nuclear lamins, an array of lamin-interacting integral INM proteins, and the associated chromatin, which is typically heterochromatic in nature [11]. In addition, the nuclear lamina is physically connected across the nuclear envelope lumen to integral ONM proteins that mechanically attach the cytoskeleton to the chromatin in the nuclear interior [12].

In vitro studies have provided some insight into the potential structure of lamins, which are members of the type V intermediate filament protein family [13]. Much like other intermediate filament proteins, the lamins exist as parallel coiled-coil dimers *in vitro*, and interact in a head-to-tail fashion to generate long polymers [14]. These polymers also associate with one another in parallel to give rise to large networks [15–17]. *In vivo* studies using super-resolution microscopy suggest that the A-type and B-type lamins are organized into distinct meshworks associated with the INM [18, 19]. The lamins are classified as A-type or B-type depending on their chemical and structural qualities. The B-type lamins, lamin B1 and B2, are transcribed from the separate *LMNB1* and *LMNB2* genes, while A-type lamins, lamins A and C (lamin A/C), are derived from alternative splicing of the single *LMNA* gene. While B-type lamins are expressed in all metazoan cell types throughout development, including embryonic stem cells, A-type lamins are only found in differentiated cells [20–25]. Most A- and B-type lamins are farnesylated during biosynthesis, the exception being lamin C. While farnesylation of B-type lamins appears to be constitutive and promotes association with the INM, A-type lamins undergo proteolytic cleavage by the zinc metalloproteinase Zmpste24 to remove the farnesylated C-terminal region [26]. This processing is critical, as a mutation that disrupts cleavage of the lamin A farnesylated tail gives rise to the pathogenic "progerin" form of lamin A and is the cause of the accelerated aging disorder progeria [27].

Interestingly, lamin composition varies with both differentiation state and tissue type [22, 23]. Although the broad requirements for lamin A/C in skin biology have

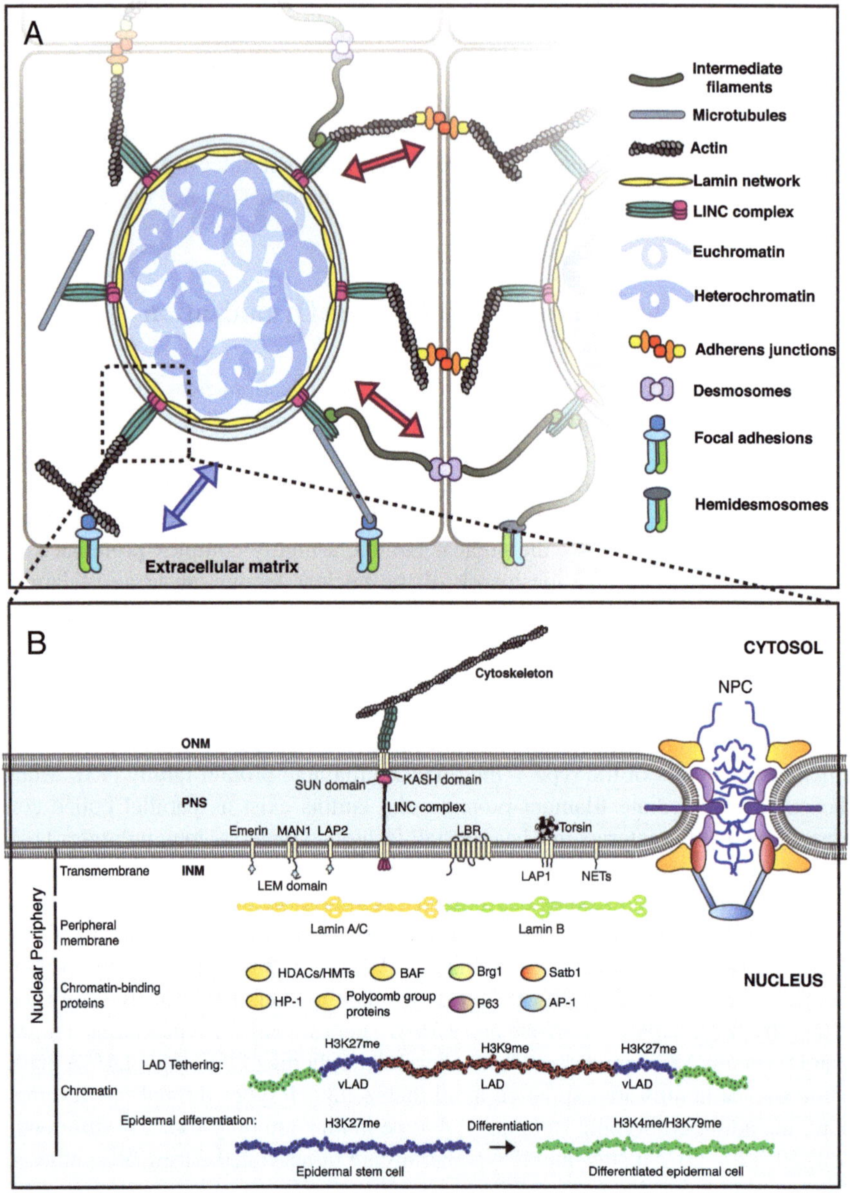

Fig. 11.1 (**a**) The nuclear lamina allows crosstalk between the nuclear envelope and plasma membrane. The nuclear periphery is a richly complicated cellular compartment that functions in chromatin organization, transcriptional regulation, mechanosensing, integration of the nucleus into the cytoskeleton, and DNA damage repair, among other functions. In epidermal keratinocytes, the nuclear lamina network is mechanically integrated into the cytoskeleton via nuclear envelope-spanning LINC complexes. Actin, microtubules, and intermediate filaments interact with the LINC complex through either direct or indirect interactions. Through these cytoskeletal connections, forces exerted on plasma membrane adhesions from either the extracellular matrix, mediated by

not yet been examined in great depth, lamins have been shown to play roles in epidermal differentiation and development. Interestingly, lamin A/C and LBR appear to have temporally distinct functions during cellular differentiation in various mammalian cell types, although at least one must be present for proper tethering of heterochromatin to the INM [28]. Using the hair follicle as a model tissue that contains both highly differentiated and undifferentiated cells, the authors of this study found that transit amplifying matrix progenitors and bulge stem cells in the hair follicle express only LBR, while the more differentiated cells of the follicular dermal papilla, epidermis, and dermis express lamin A/C [28]. Emerin and other LEM domain proteins may be particularly important in dermal papilla cells [28]. Further, mice expressing the progerin form of lamin A that is associated with premature aging exhibit aberrant LBR expression in suprabasal layers, in many cases concomitant with aberrant compaction of DNA at the nuclear periphery [29]. Epidermal differentiation appears sensitive to these defects, as these mice also display epidermal thickening and mislocalization of keratin 5 to suprabasal layers [29]. Other work has shown that simultaneous ablation of lamins A/C, lamin B1, and lamin B2 in mice produces a thickened epidermis with abnormal development of the stratum corneum and hypotrophic hair follicles, coincident with incursion of endoplasmic reticulum components into the chromatin of epidermal keratinocytes [30].

While lamin B1 and B2 are uniformly distributed in human epidermal cells, lamin A/C may be expressed at lower levels in human basal layer cells of the IFE [24]. However, a similar analysis of lamin expression in murine skin found high levels of lamin A/C in the basal layer, with decreased expression in suprabasal

Fig. 11.1 (continued) integrin-based focal adhesions and hemidesmosomes, or from adjacent cells, mediated by adherens junctions and desmosomes, can be transmitted to the nuclear interior to alter cellular function. Likewise, recent work indicates that changes at the nuclear envelope can alter the function of the cytoskeleton and plasma membrane adhesions. Thus, crosstalk exists between plasma membrane adhesions, the cytoskeleton, and components of the nuclear periphery. (**b**) Schematic of nuclear periphery components involved in epidermal homeostasis. The nuclear envelope is composed of both inner and outer nuclear membranes (INM and ONM), as well as conduits for macromolecular transport known as nuclear pore complexes (NPCs). Constituents of the nuclear periphery, including integral INM proteins, peripheral membrane proteins, and chromatin-binding proteins have all been shown to influence epidermal development and function. The LINC complex, which consists of SUN and KASH domain proteins, spans the INM, perinuclear space, and ONM to connect the nuclear periphery to the cytoskeleton. A number of other significant integral INM proteins function at the nuclear periphery, including the LAP2-Emerin-MAN1 (LEM) domain proteins, lamin B receptor (LBR), LAP1/Torsin, and a multitude of additional nuclear envelope transmembrane proteins (NETs) with currently unknown functions. Lamins A/C and lamin B1 and B2 form a meshwork underlying the INM. A number of chromatin-binding proteins, which also exhibit crosstalk with other nuclear periphery components, modulate chromatin organization and function, including the chromatin remodeling factor Brg1, the chromatin organizer Satb1, polycomb group proteins, heterochromatin-binding protein HP-1, histone methyltransferases (HMTs), histone deacetylases (HDACs), and a number of transcription factors including p63 and the AP-1 complex. In general, lamina-associated domains (LADs) are rich in the histone mark H3K9me, while H3K27me is commonly found at the LAD borders, known as variable LADs (vLAD). In the context of epidermal differentiation, a switch from H3K27me to H3K4me and H3K79me marks has been shown to mediate the transcriptional activation of lineage commitment genes

layers, suggesting either differences in the expression pattern of these proteins in mice and humans, or differences in antibody accessibility [25, 31].

The disassembly of lamin A/C during keratinocyte terminal differentiation is required for epidermal function [32]. Normally, lamin A/C are highly phosphorylated by the serine/threonine kinase AKT1 in the granular and cornified layers of the IFE, and loss of this activity results in parakeratosis, the retention of undegraded nuclei [32]. Deletion of AKT1 also produces increased BMP2/SMAD1 signaling and altered terminal differentiation in the IFE, including decreased keratin 1/10 and loricrin expression [32]. Interestingly, mice lacking keratins 1/10 exhibit premature nuclear degradation, with concomitant reductions in the levels of lamin A/C and the INM proteins emerin and SUN1 in suprabasal keratinocytes [33]. These results suggest that a feedback mechanism exists in the IFE between the nuclear lamina and cellular signaling, although the mechanistic basis for this signaling circuit remains unclear.

In addition to farnesylation, multiple lamin-binding integral INM proteins reinforce interactions between the lamins and the INM. The largest class of these proteins is comprised of members of the LAP2, emerin, and MAN1 (LEM) domain family [34]. These proteins share a common 40 amino acid motif known as the LEM domain that facilitates their interaction with chromatin through the DNA-binding Barrier-to-Autointegration Factor (BAF or BANF1). Additional proteins such as LAP1 (lamina-associated polypeptide 1), LAP2 (lamina-associated polypeptide 2) and LBR (lamin B receptor) also interact with lamins, as well as carrying out distinct biochemical functions [35–37]. Of particular interest are the lamin-binding SUN (Sad2/UNC-84) domain proteins that make up the nuclear aspect of the LInker of Nucleoskeleton and Cytoskeleton (LINC) complex. SUN domain proteins, found in the INM, bind to the ONM-resident KASH (Klarsicht/ANC-1/SYNE homology) domain proteins (also called Nesprins) in the nuclear envelope lumen, allowing LINC complexes to bridge both membranes of the nuclear envelope [12]. KASH domain proteins facilitate either direct or indirect interactions with actin, microtubules, and intermediate filaments in the cytoplasm, thereby facilitating a continuous connection between the nuclear interior and cytoplasmic cytoskeleton [12]. Interestingly, mutations in many of the lamin-associated INM proteins drive diseases (so-called "nuclear envelopathies") that phenocopy those associated with mutations in the lamins themselves [38], suggesting both complex and integrated functions for the nuclear lamina.

11.3 The Nuclear Lamina Is Mechanically Integrated into the Cell by the LINC Complex

As in most tissues, the cells of the epidermis and dermis contain an interconnected cytoskeletal network that can propagate forces from externally applied mechanical stimuli. Forces from both ECM adhesions and cell-cell junctions can propagate to the nuclear interior [39, 40] (Fig. 11.1a) and reorganize nuclear proteins [22, 41, 42].

One of the earliest demonstrations of this continuous connection was performed by applying pulling forces to plasma membrane-localized integrins on single cells using RGD-coated microbeads and monitoring nuclear distention in the direction of pulling [40]. Magnetic twisting cytometry induces dissociation of the nuclear-localized Cajal body components coilin and SMN within seconds in HeLa cells, indicating that mechanical stimulation is sufficient to induce changes to nuclear architecture [42]. Intranuclear movements have also been observed in response to compressive and shear forces [43].

How are mechanical signals communicated between the cell surface and the nuclear interior? The LINC complexes at the nuclear envelope provide a conduit for forces to be transmitted from the ECM or adjacent cells, through the cytoskeleton, to the nuclear lamina (Fig. 11.1a). Indeed, the LINC complex is essential for cytoskeletal-nuclear force transmission, as its disruption eliminates the nuclear response to mechanical stimuli [44–47] and abrogates cytoskeleton-dependent nuclear positioning in a broad range of contexts (reviewed in [48]). An emerging area of research also suggests that alterations in the composition of the nuclear lamina can likewise impact cell-ECM adhesions [49–54] (Fig. 11.1a).

While the nuclear response to ECM-derived forces is well established, cell-cell adhesion can also deliver forces on the nucleus to alter nuclear position [55, 56], suggesting a link between the nucleus and intercellular adhesions. Several components of the LINC complex are expressed in the IFE and its appendages, including multiple KASH domain-containing proteins and the widely expressed SUN domain proteins SUN1 and SUN2. Surprisingly, SUN1 is absent from the hair follicle after morphogenesis, and mice lacking SUN2 display transient alopecia during the first hair growth cycle [56]. Ultrastructural and cell culture experiments revealed alterations in the integrity of cell-cell adhesion, possibly due to a loss of LINC complex-dependent organization and/or maintenance of the cytoskeletal network and desmosome based adhesions [56]. Thus, the nucleus plays an essential role in allowing epidermal keratinocytes to generate mechanically stable cell-cell adhesions.

In addition to changes in nuclear position and the organization of intranuclear components, forces propagated to the nuclear interior can also influence the nuclear proteome and nuclear mechanics (Fig. 11.2). The expression level of lamin A/C scales with tissue stiffness *in vivo* and with substrate stiffness *in vitro*: lamin A/C levels are high in stiff tissues like bone and lower in soft tissues like brain [22]. As a result, nuclear stiffness scales with ECM stiffness, giving rise to the concept of "mechanical reciprocity" between the nucleus and the substrate that cells are grown on. Interestingly, the correlation of lamin expression level with stiffness appears to be a specific function of lamin A/C, as lamin B levels do not vary as drastically between tissue or substrate types [22]. Additionally, B-type lamins do not seem to play the same role as lamin A/C in influencing nuclear mechanics [57]. Importantly, while low substrate stiffness is known to decrease keratinocyte-ECM adhesion to induce differentiation [58], the effect of cell-ECM or intercellular adhesion-derived forces on lamin A/C levels and/or their polymerization state in keratinocytes has not yet been investigated.

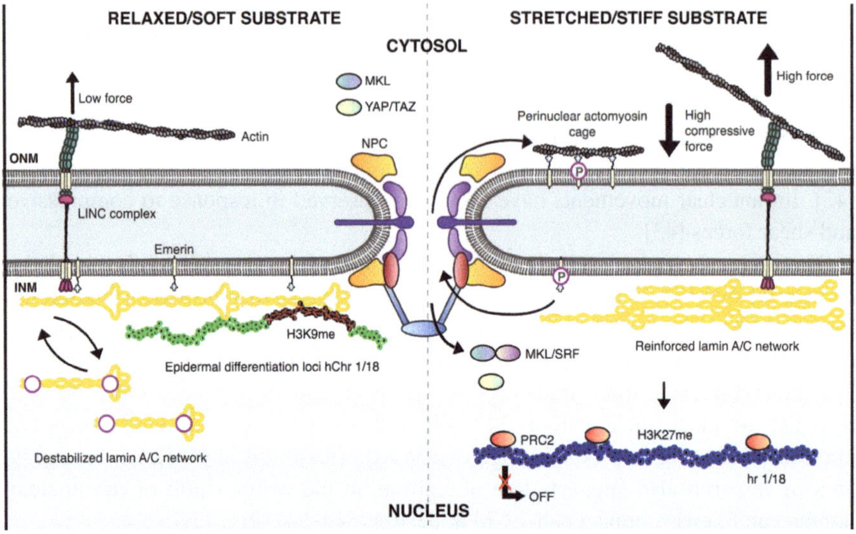

Fig. 11.2 The mechanosensitive function of the nuclear periphery modulates epidermal function. On softer substrates/tissues or when cells are not exposed to stretch, cytoskeletal tension is low. Signaling proteins involved in the sensing of cytoskeletal tension, like MKL and YAP/TAZ, remain inactive in the cytoplasm. Low levels of force are exerted on the LINC complex via the cytoskeleton. Emerin is localized to the inner nuclear membrane, where it associates with lamins A/C. Under these conditions, epitopes in lamins A/C are available for phosphorylation, leading to rapid turnover of the lamin network. In human epidermal stem cells, epidermal differentiation loci on chromosomes 1 and 18 associate with the lamina, most likely mediated by H3K9me chromatin marks. On stiffer substrates/tissues or upon cell stretching, higher cytoskeletal forces are exerted on the LINC complex and transmitted to the nuclear interior; cellular compression forces can also be transmitted to the nuclear interior without the LINC complex. These changes in cytoskeletal organization promote nuclear localization of MKL and YAP/TAZ, and the transcriptional activation of their target genes; importantly, the nuclear periphery plays a role in modulating the activity of these factors. Lamins A/C undergo reinforcement in response to elevated cytoskeletal tension, leading to multimerization and stabilization of the lamin network. This process may involve the burial of force-sensitive epitopes in lamins A/C, preventing their phosphorylation and turnover. When under tension, the LINC complex exhibits stronger association with lamins A/C, while emerin is phosphorylated, potentially weakening its interaction with the nuclear lamina. In human epidermal stem cells, this altered association with lamins A/C could drive the relocation of emerin from the inner nuclear membrane to the outer nuclear membrane, allowing for the emerin-dependent formation of an actomyosin cage around the nucleus. Reduced emerin localization to the inner nuclear membrane facilitates loss of H3K9me and accumulation of H3K27me marks on the promoters of epidermal differentiation loci, resulting in global chromatin rearrangements and dissociation from the periphery. Combined with PRC2 association and reduced RNAPII accumulation at promoters, these changes drive transcriptional silencing of lineage commitment genes. The reinforcement of the lamin network, the formation of a perinuclear actomyosin cage, and the release of chromatin from the nuclear periphery may protect the genome from force-induced damage under conditions of high cytoskeletal strain

While the structural integrity provided by the lamin A/C polymer network is proposed to impart nuclear stiffness directly, the associated chromatin may also influence the mechanical properties of the nucleus. Mechanical measurements made with optical tweezers on isolated nuclei derived from a fission yeast model, *Schizosaccharomyces pombe*, which lacks lamins, have shown that chromatin association with the nuclear periphery can impart nuclear stiffness in the absence of lamins, and can influence the viscoelastic properties of nuclei [59]. Thus, the peripherally localized heterochromatin that associates with the lamina may also play a role in defining the mechanics of nuclei. Indeed, evidence suggests that heterochromatin may also buffer forces exerted on the nucleus, acting in concert with the lamins [60–62]. Cell migration, which is a mechanically taxing process for the nucleus, leads to an increase in facultative heterochromatin at the leading edge of the nucleus in a mouse melanoma cell line, further suggesting that chromatin condensation and nuclear mechanical challenge can be linked [63]. The lamin network may also influence local chromatin mobility, which is limited by the association of heterochromatin with the nuclear periphery [64]. Indeed, the loss of lamin A in mammalian cells [65] or INM proteins in yeast [59] results in increased chromatin mobility, which can strongly influence nuclear mechanics, particularly nuclear viscosity, and therefore nuclear function.

In addition to changes in expression level, a central theme in the response of the nuclear lamina to mechanical input appears to be force-dependent changes in lamin A/C multimerization [31], which may serve several functions (Fig. 11.2). First, reinforcement of the lamin A/C network and subsequent nuclear stiffening may improve the efficiency of force transmission from the cytoskeleton through the LINC complex to the lamin network. This change in nuclear stiffness would be important for processes where the nucleus must be actively positioned or moved, such as during polarization prior to cell migration [66]. Second, the differential exposure of binding sites for interaction partners, including chromatin, histones, and INM proteins, may serve as a regulatory mechanism for guiding signal transduction and/or transcription. Finally, modulation of the lamin network may influence how susceptible the genome is to mechanically-induced DNA damage. Defects in the remodeling of the lamina, including chromatin, in response to force could result in increased damage; this may explain why the ablation of SUN proteins or lamin A/C can produce increases in baseline DNA damage [67] and apoptosis in response to mechanical challenge [68].

11.4 The Nuclear Lamina and Chromatin Organization

In addition to its roles in defining nuclear mechanics, the nuclear lamina, and lamin A/C and LBR in particular, sequester heterochromatic regions of the genome at the nuclear periphery [28, 69, 70]. Lamina associated domains (LADs), which account for 40% of the mammalian genome [71], are recruited to lamins through the coordinated activity of transcription factors, distinctly localized heterochromatic marks, and sequence specificity (Fig. 11.1b). Lamina-associated sequences, sometimes enriched in extended GA-rich sequence motifs, in combination with the facultative heterochromatin marks H3K27me3 and H3K9me2/3, are capable of driving peripheral association of ectopic LADs [72, 73]. These marks have been identified in other studies as important facilitators of lamina tethering [71, 74, 75]. However, GA repeats are not found in all LADs [71], and indeed non-GA sequences in LADs appear to be sufficient for lamina-targeting [72]. The transcription factor YY1, as well as histone deacetylase 3 (HDAC3), LAP2β, and lamin A/C appear to be required for peripheral recruitment [72], although additional transcription factors such as Zbtb7b may also be involved in particular cases, such as in recruiting the *IgH* locus to the lamina in fibroblasts [73]. Interestingly, either lamin C alone, or an optimum ratio of lamin A to lamin C, may specifically be involved in this process, suggesting distinct roles for lamin A and lamin C in LAD tethering [72].

Efforts to understand whether the tethering of individual genes to the nuclear periphery is sufficient to repress gene transcription have given mixed results, depending on the experimental system analyzed [76, 77]. Importantly, lamins appear dispensable for repression at the nuclear periphery, just as transcriptional repression is not an essential requirement for peripheral recruitment. Indeed, genomic loci with transcriptionally active histone marks are also found in this subnuclear compartment (reviewed in [78]). Further, the release of LADs from the nuclear periphery is not sufficient to induce their transcriptional activation [79]. Thus, lamin A/C does not seem to directly influence heterochromatinization, and instead merely sequesters these regions at the nuclear periphery. However, several INM proteins have an established connection with the regulation of chromatin state. For example, emerin interacts with HDAC3, an HDAC found in the NCoR complex that represses transcription, recruiting it to the nuclear lamina, and promoting its activity [80].

While techniques such as ChIP-seq and Hi-C have allowed static "images" of genome-nuclear lamina interactions to be identified, dynamic alterations in genome organization have also been observed using microscopy-based techniques [75]. Using these methods, it has been observed that LADs in mammalian cells undergo stochastic rearrangements in sub-nuclear localization, from peripheral to nucleolar regions, following mitosis [75]. These results suggest that the peripheral localization of LADs is not an essential regulator of gene expression and cell function in differentiated cells. Further, Hi-C analysis of chromosome territories in single mammalian cells suggests that chromosome organization is highly dynamic and variable between individual cells [81]. Similar analyses of LAD dynamics in differentiating epidermal cells will be an essential avenue of experimental inquiry.

While association with the lamina is not generally sufficient to predict transcriptional state, several studies suggest that the nuclear lamina helps tune gene expression in different tissue types and at different stages of cellular differentiation and development [82]. The characteristic changes in nuclear shape and size that occur during epidermal terminal differentiation occur concomitantly with changes in the 3D architecture, and therefore the "transcriptional microenvironment", of epidermal keratinocytes [10]. Indeed, while basal progenitors exhibit markers of active transcription, increasingly suprabasal cells show global decreases in active transcriptional markers and increases in pericentromeric heterochromatin [10]. Further, a chromosomal region termed the epidermal differentiation complex (EDC), that consists of sixty consecutive genes necessary for epidermal stratification and the production of a cornified envelope (reviewed in [83]), has been observed to associate more frequently with pericentromeric heterochromatin regions in suprabasal cells, and relocates away from the nuclear periphery towards the nuclear interior [10, 84, 85]. These results suggest that functional reorganization of chromatin within the nucleus occurs during epidermal differentiation. Interestingly, relocation of the EDC locus is accompanied by the increased transcription of EDC genes, and is dependent on the chromatin remodeling factor Brg1 and the chromatin organizer Satb1; both of these chromatin interactors are regulated by p63, a transcription factor essential for epidermal specification and progenitor activity [85, 86]. In addition, the differentiation process is associated with the loss of H3K27me3 chromatin marks on differentiation-specific genes, and their replacement with H3K79me2 and H3K4me3 marks, although the full complexity of this transition is still under investigation [87] (Fig. 11.1b).

More generally, regulated histone deacetylation is important for both epidermal development and hair follicle specification. The chromatin remodeling factor Brg1, which interacts with HDAC proteins, is required for epidermal differentiation [88]. Conditional deletion of Mi-2b, a component of the HDAC1 and HDAC2-containing NURD complex, results in failure of hair follicle specification and atrophic epidermis due to loss of epidermal progenitor cells [89]. Similarly, deletion of both HDAC1 and HDAC2 in murine epidermis results in thin skin due to the upregulation of the specific target genes normally repressed by p63, as well as increases in acetylated p53 [90]. Furthermore, these mutant mice do not form hair follicles, teeth, or tongue papillae, all of which derive from basal epidermal keratinocytes. Additionally, treatment of adult skin with the HDAC inhibitor trichostatin A can induce hair follicle stem cell activity [91], while deletion of HDAC1 and HDAC2 postnatally in the epidermis results in alopecia, claw dystrophy, and hyperkeratosis [92]. Whether these functions of HDACs are related to the regulation of transcriptional repression at the nuclear periphery is unknown. Interestingly, mice lacking p63 display aberrant nuclear morphology, as well as alterations in the expression of LINC complex components, including the increased expression of SUN2 and decreased expression of SUN1, Nesprin 3, lamin A/C, lamin B1, and plectin [85, 93]. These changes are further linked to relocation of repressive heterochromatin marks H3K27me3, H3K9me3, and HP1α away from the nuclear periphery, as well as reorganization of the keratin-encoding loci KtyI and KtyII towards repressive chromocenters [93]. Thus, p63 may carry out aspects of its influence on transcriptional control of epidermal differentiation in coordination with the nuclear lamina.

In addition to histone deacetylation and chromatin remodelers, the nuclear lamina can also serve as a platform for the concentration of transcription factors and components of signaling pathways, thereby influencing gene expression. For example, the lamin A/C scaffold is postulated to regulate the transcription factor c-Fos, which heterodimerizes with c-Jun to form the AP-1 complex. The sequestration of c-Fos at the nuclear periphery by lamin A/C prevents c-Fos/c-Jun heterodimerization, attenuating AP-1 DNA binding and transcription [94]. The interaction of c-Fos with lamin A/C appears to be regulated by ERK1/2-dependent c-Fos phosphorylation, which allows it to be released from the periphery and activate transcription [95]. Interestingly, AP-1 regulates EDC gene expression in both proliferating and differentiating keratinocytes *in vitro* [96] and coordinates with Ezh2 in the polycomb complex to regulate epidermal differentiation [97]. Thus, the nuclear lamina may support epidermal differentiation by regulating AP-1 activity.

Several additional nuclear envelope components can influence specific signal transduction cascades, including signaling through the Wnt pathway, itself an important regulator of stem cell renewal and differentiation in the epidermis [98]. Emerin interacts with β-catenin via its adenomatous polyposis coli (APC) homology domain and antagonizes Wnt signaling, potentially by promoting β-catenin export from the nucleus [99–101]. Conversely, the LINC complex component Nesprin-2 may positively regulate the nuclear localization of β-catenin [102]. Thus, because the level of β-catenin signaling influences stem cell lineage choice (reviewed in [98, 103]), the nuclear lamina-dependent tuning of nuclear β-catenin levels may impact lineage selection.

TGF-β is another important modulator of epidermal stem cell activity and wound healing that is regulated by nuclear lamina components. MAN1 antagonizes TGF-β/BMP signaling by binding Smad2 and Smad3, upstream regulators of TGF-β signaling, although the mechanism of inhibition is unclear [104, 105]. Disruption of this activity by a loss-of-function mutation in MAN1 is linked to Buschke-Ollendorff syndrome, characterized by skeletal dysplasia and skin abnormalities [106]. Interestingly, increased TGF-β signaling is often seen in fibrotic skin disorders, such as those that characterize the early aging disorder progeria, associated with expression of the progerin form of lamin A [107, 108]. Once again, Nesprin-2 may act conversely to promote Smad activity, as mice expressing a mutant actin-binding domain-null Nesprin-2G exhibit delayed nuclear accumulation of Smad2/3, as well as delayed wound healing *in vivo* [109]. These changes have also been linked to aberrant fibroblast differentiation and keratinocyte proliferation in response to wounding [109]. Thus, in addition to the lamins, other nuclear envelope components likely act to modulate signaling cascades leading to proper stimulation and repression of signaling circuits in the epidermis.

11.5 Maintenance of Genome Integrity

In addition to regulating chromatin dynamics, activation state, and organization, increasing data suggest that the lamina also plays a role in maintaining genome integrity. Although this topic has been less well studied to date, there are likely two primary mechanisms that underlie the functions of the lamina in genome integrity: one in the prevention of DNA double-strand breaks (DSBs), and the other in modulating the efficiency and fidelity of repairing the DSBs that do occur. Mechanical stresses have been shown to induce DNA damage, leading to apoptosis, in vascular smooth muscle cells [110]. A role for the nuclear lamina in supporting mechanical stability when high force is exerted on the nucleus may protect the genome from damage. For example, mechanical stresses in the form of nuclear deformation during cellular migration through pores have been shown to induce apoptosis in lung carcinoma A549 cells with partial knockdown of lamin A [68]. While this study did not directly examine whether DNA damage was the driver of the apoptotic response, increases in genome instability, including DNA lesions and telomere dysfunction, have been observed in response to mutations in nuclear lamina components. For example, mouse embryonic fibroblasts (MEFs) from *Sun1/Sun2−/−* mice exhibit elevated basal DNA damage and increased sensitivity to DNA damaging agents [67], as do *C. elegans* deficient for the SUN domain-containing protein UNC-84 [111].

In addition, the nuclear lamina likely contributes to DSB repair pathway choice. Components of the nuclear lamina have been found to directly interact with DNA damage signaling or repair-associated proteins. Nesprin-1 physically interacts with MSH2 and MSH6, components of the mismatch repair pathway [112]. A Nesprin-2 isoform lacking a KASH domain may also influence the DNA damage response (DDR), specifically influencing ATM localization to sites of damage through its interaction with ERK1/2 [113]. Further, SUN1 and SUN2 have been implicated in modulating DDR in MEFs [67], and the budding yeast SUN protein Mps3p recruits DSBs to the nuclear periphery [114, 115]. Interestingly, the bridging of persistent DSBs through the LINC complex to dynamic cytoplasmic microtubules promotes homology-directed repair in fission yeast [116], while a similar LINC complex-mediated process may inhibit non-homologous end joining in favor of homology-directed repair in the *C. elegans* germline [111]. Lamin A/C may also influence DDR by directly interacting with DDR components, such as Ku70 and γ-H2AX [117], and influencing the formation of repair foci (reviewed in [118–122]). Indeed, MEFs null for Zmpste24 or expressing unprocessed prelamin A exhibit delayed 53BP1 and Rad51 recruitment to sites of damage, leading to defective repair and irreparable DSBs [121]. Further, defects in DNA damage repair were identified in a restrictive dermopathy-like disease – characterized by severe epidermal defects – which arose through mutation of lamin A and subsequent postnatal loss of mature lamin A expression [123]. This lamin A mutation resulted in the accumulation of DSBs *in vivo*, as well as decreased 53BP1 localization at breaks and impaired DNA repair in human fibroblasts [123].

While it is not yet mechanistically clear if or how defects in DNA repair contribute to progeria and other diseases of the lamina, fibroblasts from progeria patients, *Zmpste24−/−* mice (defective in lamin A processing), or MEFs overexpressing unprocessed prelamin A all show increased susceptibility to DNA damage [121]. Interestingly, SUN1 has been shown to accumulate in progeroid-expressing and lamin A-null mouse fibroblasts [124]. Given the potential role of SUN proteins in the DDR, Lei et al. proposed a mechanism for progerin-toxicity where excessive accumulation of SUN1 at the nuclear envelope could lead to hyperactive DDR signaling [67]. Indeed, the loss of SUN1 is capable of rescuing many of the defects in a mouse model of progeria [124]. The loss of SUN1 in this context would theoretically reduce DDR signaling in lamin A mutant mice, and minimize the associated defects. However, it remains possible that the amelioration of the progeria phenotype may instead be due to a concomitant decrease in the number of LINC complexes, and therefore the force exerted on the nucleus. The emerging connection between lamins A/C, lamin B1, and autophagy [125–130] may additionally play a role in modulating cellular senescence and aging in progeria patients. For example, epidermal keratinocytes deficient for Atg7 exhibit decreased lamin B1 expression coupled with increased DNA damage foci, altered lipid metabolism, and cellular senescence [129]. As an alternative model, progerin itself may interfere with the activity of repair factors, thereby promoting irreparable DSBs [119, 131].

As genome integrity is essential for skin homeostasis and regeneration (reviewed in [132]), a potential role for the nuclear lamina in protecting the genome and even repairing DNA has significant consequences for epidermal function. The homeostasis of different components of the epidermis, including the stratified epithelium, hair follicles, and the sebaceous glands, is maintained by resident stem cell niches; in the case of the hair follicle, stem cells located in the bulge region promote hair follicle regeneration throughout adulthood [133]. Adult bulge stem cells may utilize distinct DNA repair strategies at different developmental stages, as these cells rely on non-homologous end joining and expression of the anti-apoptotic protein Bcl2 during the resting phase of the hair cycle [134]. Interestingly, progerin has been implicated in suppressing 53BP1-mediated non-homologous end joining in human keratinocytes following UVA radiation [135]. Further, ablation of BRCA1, an important mediator of homologous recombination, in the murine epidermis results in DNA damage accumulation, apoptosis, and loss of transit amplifying cells and bulge stem cells, preventing adult regeneration of hair follicles [136]. Additional work must be carried out to determine if the nuclear lamina's regulation of DNA repair has implications for skin homeostasis and tumorigenesis.

11.6 Mechanosensing: The Lamina and the LINC Complex

In addition to biochemical signaling, mechanical signals are also known to influence chromatin structure, gene expression, and differentiation (reviewed in [137, 138] (Fig. 11.2). Cell geometry can influence gene expression downstream of MKL1/SRF signaling and HDAC3 localization [139], while modulation of the cytoskeleton and/or LINC complexes can modify nuclear morphology, chromatin organization, and gene expression [140–142]. Similarly, cell geometry, substrate stiffness, and intracellular tension can influence stem cell lineage selection [58, 143, 144].

Interaction of the nucleus with the cytoskeleton through LINC complexes provides a mechanism for external forces to be propagated from the ECM or adjacent cells to the nuclear interior directly. Thus, in addition to the modulation of nuclear mechanics in response to extracellular biochemical cues, mechanosensing at the nuclear lamina is also likely critical for the cell to integrate and respond to mechanical cues from adhesions and the cytoskeleton. An attractive model is the possibility that the nuclear lamina responds to mechanical inputs in a fashion similar to that described for E-cadherin-based AJs and integrin-based focal adhesions, which increase in size in response to exogenous force [145–147]. The acute application of force on Nesprin-1 molecules on isolated nuclei using magnetic tweezers has been shown to produce nuclear stiffening within seconds, an effect dependent on the LINC complex, emerin, and lamin A/C [46]. Further, force application to nuclei in this context increases the association of lamin A/C with LINC complexes [46]. Nuclear stiffening in response to stress has been observed in other contexts as well: endothelial cell nuclei stiffen in response to shear stress [148], while HeLa cells exposed to shear exhibit increased peripheral recruitment of lamin A/C [149]. Interestingly, acute force application to isolated nuclei also results in emerin phosphorylation [46]. This process is required for force-induced nuclear stiffening, as the expression of a phosphomutant emerin prevents nuclear stiffening and abrogates an increased association of LINC complexes with lamin A/C [46]. As both SUN2 and emerin bind to the same region of lamin A/C, the force-induced phosphorylation of emerin in this context may decrease lamin A/C-emerin affinity and promote a lamin A/C-SUN2 interaction. In support of this notion, *emerin*-null nuclei exhibit increased nuclear stiffness, suggesting that the association of lamin A/C with SUN proteins is even greater than in control nuclei [46] (Fig. 11.2).

Interestingly, uniaxial stretch applied to human epidermal progenitor cells drives the redistribution of emerin from the INM to the ONM, leading to the formation of a perinuclear non-muscle myosin II-actin scaffold, nuclear actin depletion and inactivation of RNAPII, and the subsequent H3K27me3- and PRC2-dependent transcriptional repression of genes associated with epidermal differentiation [150] (Fig. 11.2). While a role for emerin phosphorylation in this process has not yet been identified, reduced association between lamin A/C and emerin could facilitate the redistribution of emerin to the ONM in this context. Additional nuclear lamina components may have the capacity to swap between the INM and ONM, such as short nesprin isoforms [151–154] and LUMA [155].

Like emerin, lamin A/C has also been shown to undergo force-induced phosphorylation, perhaps in response to unfolding of its immunoglobulin-like domain [22, 41]. During mitosis, lamin A is phosphorylated on Ser22, prompting disassembly of the lamin network in preparation for nuclear envelope breakdown [156]. This same modification appears to be sensitive to mechanical input: increased intracellular cytoskeletal tension, induced by growing mesenchymal stem cells on stiff substrates, decreases phosphorylation of Ser22, while reduced tension, either through plating on soft substrates or inhibition of myosin II, increases phosphorylation [22, 41] (Fig. 11.2). These changes in phosphorylation level are seen within tens of minutes, and correspond to changes in nuclear stiffness as measured by micropipette aspiration [41].

While a definitive role for forced unfolding has not yet been identified in lamin A/C function, the discovery of two mechanically-regulated epitopes of lamin A/C hints at such a mechanism [31]. The first epitope is located within the first 50 residues of the N-terminus, while the second is a conformational epitope consisting of sequences in the Ig domain and the Ig-proximal unstructured linker. These regions become inaccessible to antibodies at the basal side of nuclei in response to compressive forces, including those applied by an apical actin cap [31], an actin-rich meshwork found in close association with the nucleus in various cell types [157]. The observed differential lamin A epitope accessibility is regulated by cytoskeletal tension, and requires intact LINC complexes [31]. Further, steered molecular dynamics simulations show that the loss of the epitope can be linked to lamin A/C multimerization, and therefore reinforcement of the lamin network [31]. Several previously identified stress-sensitive phosphorylation sites [22], including Ser22, are present in the two epitopes. These results suggest that cytoskeletal tension could regulate the phosphorylation state of lamin A/C by influencing the exposure of phosphorylation sites to kinases and/or phosphatases, potentially through partial unfolding of the Ig domain and concomitant multimerization of lamins (Fig. 11.2).

This force-dependent exposure of lamin A/C epitopes may play an important role in stem cell lineage selection. Fascinatingly, human mesenchymal stem cells that are shunted down osteogenic or adipogenic lineages exhibit different levels of basal epitope exposure, with the lamin epitope more polarized in osteoblasts than in adipocytes [31]. Previous work has also identified actin cap and LINC complex-dependent polarization of lamin A/C to the apical side of nuclei, as well as the apical polarization of histone marks H3K12ac and H4K5ac, which are associated with active transcription [158]. Interestingly, the conformational epitope in lamin A/C overlaps with binding sites for DNA, histones, emerin, and SUN proteins [31]. Thus, the force-dependent differential exposure of these binding sites may influence the interaction of lamin A/C with the genome, either directly or through modulation of the activity of its binding partners. As described earlier, emerin is capable of binding to DNA, and can also interact with and activate HDAC3 [80]. Again, these responses harken back to the force-dependent modulation of adhesion morphology and function at the cell surface.

Interestingly, Ihalainen et al. showed that integrin-based connectivity to the nucleus might play a crucial role in determining the polarity of lamin A/C epitope exposure, suggesting this activity of lamins may be important for cell types that are exposed to both cell-cell and cell-ECM adhesions. Thus, the polarized, force-dependent exposure of lamin A/C epitopes may be particularly relevant for the skin, where polarized basal layer cells – which contact both the basal lamina and other keratinocytes – must differentiate in a highly regulated manner to correctly form the layers of the IFE. Further, other work has shown that the geometry of exogenous force application to cells may influence nuclear lamina-dependent regulation of gene expression by modulating the extent of chromatin stretch [159].

While mechanosensing at the nuclear interior by lamin A requires the LINC complex, likely for force transduction across the nuclear envelope, it is also possible that the LINC complex is itself mechanoresponsive. The majority of the extranuclear domains of Nesprin proteins are composed of spectrin repeats, known to undergo force-induced unfolding, which could serve multiple functions [160, 161].

As with other mechanosensitive proteins, such as talin [162], this unfolding could either abrogate or reveal cryptic interacting sites, potentially influencing signaling pathways or LINC complex oligomerization [153, 163]. Increased affinity between Nesprins under tension could allow increased force to be transmitted to the nuclear lamina and could therefore mediate changes in lamina-regulated signaling, chromatin interactions, and transcription. Alternatively, unfolding of the spectrin repeats could instead release tension on LINC complexes, preventing excessive and potentially dangerous levels of force from being exerted on the nuclear interior. Indeed, using an optical tweezers assay to apply force to LINC complexes on isolated nuclei, Balikov et al. showed that nesprin spectrin repeats may undergo unfolding in response to physiological levels of force [164]. Interestingly, the LINC complex is not only capable of adapting to the sustained application of high forces, such as during the formation of TAN lines for nuclear positioning, but also to transient (10s of milliseconds) [42, 45] or low magnitude [165] mechanical signals. This suggests that the LINC complex may be able to coordinate differential responses tuned to the magnitude and geometries of forces exerted on the nucleus, which could in turn differentially influence lamin A/C activity.

Thus far, studies of nuclear lamina mechanosensitivity have focused on isolated cells in which mechanical cues are driven exclusively by cell-ECM adhesions. Indeed, little is known about the potential role intercellular adhesions, and the balance between forces generated at cell-cell and cell-ECM adhesions, may have on the lamina, particularly in tissues. In the current model, mechanical signals originating at intercellular adhesions would be relayed to the nucleus in the same way as signals from cell-ECM adhesions, resulting in changes in lamin A/C phosphorylation and structure. However, the types of mechanical inputs studied to date represent unnaturally high levels of force or asymmetrical compressive forces (due to apical actin cap formation). In an epithelial sheet, the force distribution at intercellular adhesions reaches a homeostatic equilibrium, suggesting that these cells may respond differently to changes in tension *in vivo* [166].

Many of the mechanosensing functions of cells are dependent upon YAP/TAZ and/or MKL/SRF signaling [167]. YAP1 regulates epidermal stem cell proliferation and tissue expansion [168, 169] and is downstream of α-catenin, which links YAP signaling to cadherin junctions [170]. MKL/SRF signaling is an essential regulator of epidermal differentiation [171–173] and hair follicle morphogenesis [174]. SRF further regulates epidermal differentiation by modulating actomyosin-guided mitotic spindle orientation during asymmetric divisions of basal layer cells to promote epidermal stratification [171], as well as controlling the expression and localization of AP-1 family members, and the C/EBPα transcription factor [174]. Postnatal deletion of SRF in mice yields alterations in cell-cell and cell-ECM adhesion as well as altered differentiation and actin organization, leading to an inflammatory hyperproliferative phenotype similar to psoriasis [172].

Interestingly, YAP/TAZ and MKL/SRF are regulated by the nuclear lamina (Fig. 11.2). In MEFs, the transcription of the mechanically sensitive immediate early genes *egr-1* and *iex-1* in response to cyclic strain requires lamin A/C and emerin [175, 176]. Further, YAP/TAZ and MKL/SRF signaling are perturbed upon loss of lamin A/C [22, 177, 178] and a nonphosphorylatable mutant lamin A/C has been shown to increase gene expression compared to a phosphomimetic, suggesting that the structure of the lamin network can influence mechanosensitive transcription [41]. Further, emerin, through its mechanosensing capacity, is specifically required for MKL/SRF-dependent gene expression on stiff versus soft substrates, implicating it in the regulation of fibrosis and other pathologies involving tissue stiffening [179].

Thus, a model is emerging in which a mechanical stimulus can be rapidly transmitted from the cytoskeleton to the nuclear interior through LINC complexes, leading to force-induced changes in posttranslational modifications of nuclear lamina components, subsequent lamina reinforcement, and transcriptional regulation. This model is consistent with recent work showing that YAP/TAZ works with AP-1 to activate target genes via chromatin looping [180]. Phosphorylation of emerin may link nuclear lamina changes to these signaling pathways by promoting MKL/SRF signaling, potentially by promoting actin polymerization and/or organization in the cytoplasm and nucleoplasm [177, 181]. However, mutant emerin protein that cannot be phosphorylated does not influence YAP/TAZ transcription [46], suggesting that modulation of these two pathways may occur through distinct mechanisms. These changes may promote increased stiffness of the nuclear lamina through reinforcement of the lamin network [22, 41, 46]. How the nuclear lamina functions in mechanosensing and through these signaling pathways during epidermal differentiation and homeostasis will be an interesting area of future investigation.

11.7 The Nuclear Lamina and Human Disease

Disruption of the nuclear lamina leads to a host of rare diseases, known as laminopathies and nuclear envelopathies, which include muscle dystrophies, premature aging, lipodystrophies, peripheral nerve disorders, and bone diseases (reviewed in [182]). These disorders are often associated with hallmark defects in epidermal structure and function, which hints at the special role the nuclear lamina plays in epidermal homeostasis. While we are just beginning to dissect the etiology of human diseases associated with the nuclear lamina, the burgeoning field of epidermal nuclear biology highlight the importance of this domain in regulating epidermal development and integrity.

In particular, mutations in lamin A, associated with either Hutchinson-Gilford Progeria Syndrome (HGPS) or other laminopathies, have been implicated in altering interactions between LADs and the nuclear lamina. Numerous studies have shown that loss of lamin A/C, or expression of lamin A/C mutants, can result in global changes in 3D genome organization and gene expression. Further, global changes in H3K27me3 marks, dissociation of chromatin from lamin A/C and the lamina, and gene expression changes have been identified in epidermal fibroblasts from HGPS patients [183]. However, in line with the fact that lamin association does not necessarily influence transcriptional state, these changes do not always induce altered gene expression. Other work suggests that progerin expression is not sufficient to induce such vast perturbation of chromatin organization and transcription [79]. While progerin preferentially interacts with a subset of genes distinct from lamin A/C, these altered associations are not associated with global expression changes [79]. A complex array of interactions between nuclear lamina proteins and chromatin-interactors or –modifiers has been identified. These interactions may be disrupted in diseases where the nuclear lamina is altered, such as in HGPS and restrictive dermopathy [184]. Further work will be required to understand the exact changes that occur upon lamin A/C disruption, and the mechanisms by which these occur. The use of sophisticated tools to map the dynamic interactions of chromatin with the nuclear periphery over time [75], including in embryonic development, will prove an essential component in understanding the etiology of laminopathies and nuclear envelopathies.

11.8 Summary

Skin is subject to constant mechanical challenge through external insults and as a result of the cellular changes that occur during development and homeostasis. Recent data suggest that tissue homeostasis requires crosstalk between the cell surface (in the form of cell-ECM and cell-cell adhesions) and the nucleus to coordinate biochemical and mechanical cues. The LINC complex, which spans the nuclear envelope, plays a critical role in communicating mechanical information from the cytoskeleton to the nuclear interior. The nuclear lamina responds to forces transduced by the LINC complex by coordinating changes in chromatin organization and transcriptional output; the lamina also helps to maintain genome integrity, although the mechanisms at play will require further investigation. Understanding how the lamina coordinates these integrated functions will be critical to defining how lamin dysfunction contributes to defects in epidermal structure and function.

References

1. Candi E, Schmidt R, Melino G. The cornified envelope: a model of cell death in the skin. Nat Rev Mol Cell Biol. 2005;6:328–40. https://doi.org/10.1038/nrm1619.
2. Blanpain C, Fuchs E. Epidermal homeostasis: a balancing act of stem cells in the skin. Nat Rev Mol Cell Biol. 2009;10:207–17. https://doi.org/10.1038/nrm2636.
3. Lechler T, Fuchs E. Asymmetric cell divisions promote stratification and differentiation of mammalian skin. Nature. 2005;437:275–80. https://doi.org/10.1038/nature03922.
4. Koster MI, Roop DR. Asymmetric cell division in skin development: a new look at an old observation. Dev Cell. 2005;9:444–6. https://doi.org/10.1016/j.devcel.2005.09.009.
5. Simpson CL, Patel DM, Green KJ. Deconstructing the skin: cytoarchitectural determinants of epidermal morphogenesis. Nat Rev Mol Cell Biol. 2011;12:565–80. https://doi.org/10.1038/nrm3175.
6. Delva E, Tucker DK, Kowalczyk AP. The desmosome. Cold Spring Harb Perspect Biol. 2009;1:a002543. https://doi.org/10.1101/cshperspect.a002543.

7. Furuse M, Hata M, Furuse K, Yoshida Y, Haratake A, Sugitani Y, Noda T, Kubo A, Tsukita S. Claudin-based tight junctions are crucial for the mammalian epidermal barrier: a lesson from claudin-1-deficient mice. J Cell Biol. 2002;156:1099–111. https://doi.org/10.1083/jcb.200110122.
8. Morita K, Itoh M, Saitou M, Ando-Akatsuka Y, Furuse M, Yoneda K, Imamura S, Fujimoto K, Tsukita S. Subcellular distribution of tight junction-associated proteins (occludin, ZO-1, ZO-2) in rodent skin. J Invest Dermatol. 1998;110:862–6. https://doi.org/10.1046/j.1523-1747.1998.00209.x.
9. Schlüter H, Wepf R, Moll I, Franke WW. Sealing the live part of the skin: the integrated meshwork of desmosomes, tight junctions and curvilinear ridge structures in the cells of the uppermost granular layer of the human epidermis. Eur J Cell Biol. 2004;83:655–65. https://doi.org/10.1078/0171-9335-00434.
10. Gdula MR, Poterlowicz K, Mardaryev AN, Sharov AA, Peng Y, Fessing MY, Botchkarev VA. Remodeling of three-dimensional organization of the nucleus during terminal keratinocyte differentiation in the epidermis. J Invest Dermatol. 2013;133:2191–201. https://doi.org/10.1038/jid.2013.66.
11. Burke B, Stewart CL. The nuclear lamins: flexibility in function. Nat Rev Mol Cell Biol. 2013;14:13–24. https://doi.org/10.1038/nrm3488.
12. Chang W, Worman HJ, Gundersen GG. Accessorizing and anchoring the LINC complex for multifunctionality. J Cell Biol. 2015;208:11–22. https://doi.org/10.1083/jcb.201409047.
13. Peter A, Stick R. Evolutionary aspects in intermediate filament proteins. Curr Opin Cell Biol. 2015;32:48–55. https://doi.org/10.1016/j.ceb.2014.12.009.
14. Heitlinger E, Peter M, Lustig A, Villiger W, Nigg EA, Aebi U. The role of the head and tail domain in lamin structure and assembly: analysis of bacterially expressed chicken lamin A and truncated B2 lamins. J Struct Biol. 1992;108:74–89.
15. Herrmann H, Aebi U. Intermediate filaments: molecular structure, assembly mechanism, and integration into functionally distinct intracellular Scaffolds. Annu Rev Biochem. 2004;73:749–89. https://doi.org/10.1146/annurev.biochem.73.011303.073823.
16. Gruenbaum Y, Medalia O. Lamins: the structure and protein complexes. Curr Opin Cell Biol. 2015;32:7–12. https://doi.org/10.1016/j.ceb.2014.09.009.
17. Turgay Y, Eibauer M, Goldman AE, Shimi T, Khayat M, Ben-Harush K, Dubrovsky-Gaupp A, Sapra KT, Goldman RD, Medalia O. The molecular architecture of lamins in somatic cells. Nature. 2017;543:261–4. https://doi.org/10.1038/nature21382.
18. Shimi T, Kittisopikul M, Tran J, Goldman AE, Adam SA, Zheng Y, Jaqaman K, Goldman RD. Structural organization of nuclear lamins A, C, B1 and B2 revealed by super-resolution microscopy. Mol Biol Cell. 2015;26(22):4075–86. https://doi.org/10.1091/mbc.E15-07-0461.
19. Xie W, Chojnowski A, Boudier T, Lim JSY, Ahmed S, Ser Z, Stewart C, Burke B. A-type Lamins form distinct filamentous networks with differential nuclear pore complex associations. Curr Biol. 2016;26:2651–8. https://doi.org/10.1016/j.cub.2016.07.049.
20. Worman HJ, Lazaridis I, Georgatos SD. Nuclear lamina heterogeneity in mammalian cells. Differential expression of the major lamins and variations in lamin B phosphorylation. J Biol Chem. 1988;263:12135–41.
21. Constantinescu D, Gray HL, Sammak PJ, Schatten GP, Csoka AB. Lamin A/C expression is a marker of mouse and human embryonic stem cell differentiation. Stem Cells. 2006;24:177–85. https://doi.org/10.1634/stemcells.2004-0159.
22. Swift J, Ivanovska IL, Buxboim A, Harada T, Dingal PCDP, Pinter J, Pajerowski JD, Spinler KR, Shin J-W, Tewari M, Rehfeldt F, Speicher DW, Discher DE. Nuclear lamin-A scales with tissue stiffness and enhances matrix-directed differentiation. Science. 2013;341:1240104. https://doi.org/10.1126/science.1240104.
23. Shin J-W, Spinler KR, Swift J, Chasis JA, Mohandas N, Discher DE. Lamins regulate cell trafficking and lineage maturation of adult human hematopoietic cells. Proc Natl Acad Sci USA. 2013;110:18892–7. https://doi.org/10.1073/pnas.1304996110.
24. Venables RS, McLean S, Luny D, Moteleb E, Morley S, Quinlan RA, Lane EB, Hutchison CJ. Expression of individual lamins in basal cell carcinomas of the skin. Br J Cancer. 2001;84:512–9. https://doi.org/10.1054/bjoc.2000.1632.

25. Hanif M, Rosengardten Y, Sagelius H, Rozell B, Eriksson M. Differential expression of A-type and B-type lamins during hair cycling. PLoS One. 2009;4:e4114. https://doi.org/10.1371/journal.pone.0004114.
26. Rusiñol AE, Sinensky MS. Farnesylated lamins, progeroid syndromes and farnesyl transferase inhibitors. J Cell Sci. 2006;119:3265–72. https://doi.org/10.1242/jcs.03156.
27. De Sandre-Giovannoli A, Bernard R, Cau P, Navarro C, Amiel J, Boccaccio I, Lyonnet S, Stewart CL, Munnich A, Le Merrer M, Lévy N. Lamin a truncation in Hutchinson-Gilford progeria. Science. 2003;300:2055. https://doi.org/10.1126/science.1084125.
28. Solovei II, Wang ASA, Thanisch KK, Schmidt CSC, Krebs SS, Zwerger MM, Cohen TVT, Devys DD, Foisner RR, Peichl LL, Herrmann HH, Blum HH, Engelkamp DD, Stewart CLC, Leonhardt HH, Joffe BB. LBR and lamin A/C sequentially tether peripheral heterochromatin and inversely regulate differentiation. Cell. 2013;152:584–98.
29. McKenna T, Rosengardten Y, Viceconte N, Baek J-H, Grochová D, Eriksson M. Embryonic expression of the common progeroid lamin A splice mutation arrests postnatal skin development. Aging Cell. 2014;13:292–302. https://doi.org/10.1111/acel.12173.
30. Jung H-J, Tatar A, Tu Y, Nobumori C, Yang SH, Goulbourne CN, Herrmann H, Fong LG, Young SG. An absence of nuclear lamins in keratinocytes leads to ichthyosis, defective epidermal barrier function, and intrusion of nuclear membranes and endoplasmic reticulum into the nuclear chromatin. Mol Cell Biol. 2014;34:4534–44. https://doi.org/10.1128/MCB.00997-14.
31. Ihalainen TO, Aires L, Herzog FA, Schwartlander R, Moeller J, Vogel V. Differential basal-to-apical accessibility of lamin A/C epitopes in the nuclear lamina regulated by changes in cytoskeletal tension. Nat Mater. 2015. https://doi.org/10.1038/nmat4389.
32. Naeem AS, Zhu Y, Di WL, Marmiroli S, O'Shaughnessy RFL. AKT1-mediated Lamin A/C degradation is required for nuclear degradation and normal epidermal terminal differentiation. Cell Death Differ. 2015. https://doi.org/10.1038/cdd.2015.62.
33. Wallace L, Roberts-Thompson L, Reichelt J. Deletion of K1/K10 does not impair epidermal stratification but affects desmosomal structure and nuclear integrity. J Cell Sci. 2012;125:1750–8. https://doi.org/10.1242/jcs.097139.
34. Barton LJ, Soshnev AA, Geyer PK. Networking in the nucleus: a spotlight on LEM-domain proteins. Curr Opin Cell Biol. 2015;34:1–8. https://doi.org/10.1016/j.ceb.2015.03.005.
35. Shin J-Y, Dauer WT, Worman HJ. Lamina-associated polypeptide 1: protein interactions and tissue-selective functions. Semin Cell Dev Biol. 2014;29:164–8. https://doi.org/10.1016/j.semcdb.2014.01.010.
36. Gesson K, Vidak S, Foisner R. Lamina-associated polypeptide (LAP)2α and nucleoplasmic lamins in adult stem cell regulation and disease. Semin Cell Dev Biol. 2014;29:116–24. https://doi.org/10.1016/j.semcdb.2013.12.009.
37. Olins AL, Rhodes G, Welch DBM, Zwerger M, Olins DE. Lamin B receptor: multi-tasking at the nuclear envelope. Nucleus. 2010;1:53–70. https://doi.org/10.4161/nucl.1.1.10515.
38. Dauer WT, Worman HJ. The nuclear envelope as a signaling node in development and disease. Dev Cell. 2009;17:626–38. https://doi.org/10.1016/j.devcel.2009.10.016.
39. Wang N, Tytell JD, Ingber DE. Mechanotransduction at a distance: mechanically coupling the extracellular matrix with the nucleus. Nat Rev Mol Cell Biol. 2009;10:75–82. https://doi.org/10.1038/nrm2594.
40. Maniotis AJ, Ingber DE, Chen CS. Demonstration of mechanical connections between integrins, cytoskeletal filaments, and nucleoplasm that stabilize nuclear structure. Proc Natl Acad Sci USA. 1997;94:849–54.
41. Buxboim A, Swift J, Irianto J, Spinler KR, Dingal PCDP, Athirasala A, Kao Y-RC, Cho S, Harada T, Shin J-W, Discher DE. Matrix elasticity regulates lamin-A,C phosphorylation and turnover with feedback to actomyosin. Curr Biol. 2014;24:1909–17. https://doi.org/10.1016/j.cub.2014.07.001.
42. Poh Y-C, Shevtsov SP, Chowdhury F, Wu DC, Na S, Dundr M, Wang N. Dynamic force-induced direct dissociation of protein complexes in a nuclear body in living cells. Nat Commun. 2012;3:866. https://doi.org/10.1038/ncomms1873.

43. Booth-Gauthier EA, Alcoser TA, Yang G, Dahl KN. Force-induced changes in subnuclear movement and rheology. Biophys J. 2012;103:2423–31. https://doi.org/10.1016/j.bpj.2012.10.039.
44. Lombardi ML, Jaalouk DE, Shanahan CM, Burke B, Roux KJ, Lammerding J. The interaction between nesprins and sun proteins at the nuclear envelope is critical for force transmission between the nucleus and cytoskeleton. J Biol Chem. 2011;286:26743–53. https://doi.org/10.1074/jbc.M111.233700.
45. Chambliss AB, Khatau SB, Erdenberger N, Robinson DK, Hodzic D, Longmore GD, Wirtz D. The LINC-anchored actin cap connects the extracellular milieu to the nucleus for ultrafast mechanotransduction. Sci Rep. 2013;3:1087. https://doi.org/10.1038/srep01087.
46. Guilluy C, Osborne LD, Van Landeghem L, Sharek L, Superfine R, Garcia-Mata R, Burridge K. Isolated nuclei adapt to force and reveal a mechanotransduction pathway in the nucleus. Nat Cell Biol. 2014;16:376–81. https://doi.org/10.1038/ncb2927.
47. Morgan JT, Pfeiffer ER, Thirkill TL, Kumar P, Peng G, Fridolfsson HN, Douglas GC, Starr DA, Barakat AI. Nesprin-3 regulates endothelial cell morphology, perinuclear cytoskeletal architecture, and flow-induced polarization. Mol Biol Cell. 2011;22:4324–34. https://doi.org/10.1091/mbc.E11-04-0287.
48. Gundersen GG, Worman HJ. Nuclear positioning. Cell. 2013;152:1376–89. https://doi.org/10.1016/j.cell.2013.02.031.
49. Thakar K, May CK, Rogers A, Carroll CW. Opposing roles for distinct LINC complexes in regulation of the small GTPase RhoA. Mol Biol Cell. 2017;28:182–91. https://doi.org/10.1091/mbc.E16-06-0467.
50. Mounkes LC, Kozlov SV, Rottman JN, Stewart CL. Expression of an LMNA-N195K variant of A-type lamins results in cardiac conduction defects and death in mice. Hum Mol Genet. 2005;14:2167–80. https://doi.org/10.1093/hmg/ddi221.
51. Frock RL, Chen SC, Da D-F, Frett E, Lau C, Brown C, Pak DN, Wang Y, Muchir A, Worman HJ, Santana LF, Ladiges WC, Rabinovitch PS, Kennedy BK. Cardiomyocyte-specific expression of lamin a improves cardiac function in Lmna-/- mice. PLoS One. 2012;7:e42918. https://doi.org/10.1371/journal.pone.0042918.
52. Hale CM, Shrestha AL, Khatau SB, Stewart-Hutchinson PJ, Hernandez L, Stewart CL, Hodzic D, Wirtz D. Dysfunctional connections between the nucleus and the actin and microtubule networks in laminopathic models. Biophys J. 2008;95:5462–75. https://doi.org/10.1529/biophysj.108.139428.
53. Lee JSH, Hale CM, Panorchan P, Khatau SB, George JP, Tseng Y, Stewart CL, Hodzic D, Wirtz D. Nuclear lamin A/C deficiency induces defects in cell mechanics, polarization, and migration. Biophys J. 2007;93:2542–52. https://doi.org/10.1529/biophysj.106.102426.
54. Kim D-H, Khatau SB, Feng Y, Walcott S, Sun SX, Longmore GD, Wirtz D. Actin cap associated focal adhesions and their distinct role in cellular mechanosensing. Sci Rep. 2012;2:555. https://doi.org/10.1038/srep00555.
55. Dupin I, Camand E, Etienne-Manneville S. Classical cadherins control nucleus and centrosome position and cell polarity. J Cell Biol. 2009;185:779–86. https://doi.org/10.1083/jcb.200812034.
56. Stewart RM, Zubek AE, Rosowski KA, Schreiner SM, Horsley V, King MC. Nuclear-cytoskeletal linkages facilitate cross talk between the nucleus and intercellular adhesions. J Cell Biol. 2015;209:403–18. https://doi.org/10.1083/jcb.201502024.
57. Lammerding J, Fong LG, Ji JY, Reue K, Stewart CL, Young SG, Lee RT. Lamins A and C but not lamin B1 regulate nuclear mechanics. J Biol Chem. 2006;281:25768–80. https://doi.org/10.1074/jbc.M513511200.
58. Trappmann B, Gautrot JE, Connelly JT, Strange DGT, Li Y, Oyen ML, Cohen Stuart MA, Boehm H, Li B, Vogel V, Spatz JP, Watt FM, Huck WTS. Extracellular-matrix tethering regulates stem-cell fate. Nat Mater. 2012;11:642–9. https://doi.org/10.1038/nmat3339.
59. Schreiner SM, Koo PK, Zhao Y, Mochrie SGJ, King MC. The tethering of chromatin to the nuclear envelope supports nuclear mechanics. Nat Commun. 2015;6:7159. https://doi.org/10.1038/ncomms8159.

60. Dahl KN, Kahn SM, Wilson KL, Discher DE. The nuclear envelope lamina network has elasticity and a compressibility limit suggestive of a molecular shock absorber. J Cell Sci. 2004;117:4779–86. https://doi.org/10.1242/jcs.01357.
61. Stephens AD, Banigan EJ, Adam SA, Goldman RD, Marko JF. Chromatin and lamin A determine two different mechanical response regimes of the cell nucleus. Mol Biol Cell. 2017;28:1984–96. https://doi.org/10.1091/mbc.E16-09-0653.
62. Furusawa T, Rochman M, Taher L, Dimitriadis EK, Nagashima K, Anderson S, Bustin M. Chromatin decompaction by the nucleosomal binding protein HMGN5 impairs nuclear sturdiness. Nat Commun. 2015;6:6138. https://doi.org/10.1038/ncomms7138.
63. Gerlitz G, Bustin M. Efficient cell migration requires global chromatin condensation. J Cell Sci. 2010;123:2207–17. https://doi.org/10.1242/jcs.058271.
64. Chubb JR, Boyle S, Perry P, Bickmore WA. Chromatin motion is constrained by association with nuclear compartments in human cells. Curr Biol. 2002;12:439–45.
65. Bronshtein I, Kepten E, Kanter I, Berezin S, Lindner M, Redwood AB, Mai S, Gonzalo S, Foisner R, Shav-Tal Y, Garini Y. Loss of lamin A function increases chromatin dynamics in the nuclear interior. Nat Commun. 2015;6:8044. https://doi.org/10.1038/ncomms9044.
66. Luxton GWG, Gomes ER, Folker ES, Vintinner E, Gundersen GG. Linear arrays of nuclear envelope proteins harness retrograde actin flow for nuclear movement. Science. 2010;329:956–9. https://doi.org/10.1126/science.1189072.
67. Lei K, Zhu X, Xu R, Shao C, Xu T, Zhuang Y, Han M. Inner nuclear envelope proteins SUN1 and SUN2 play a prominent role in the DNA damage response. Curr Biol. 2012;22:1609–15. https://doi.org/10.1016/j.cub.2012.06.043.
68. Harada T, Swift J, Irianto J, Shin J-W, Spinler KR, Athirasala A, Diegmiller R, Dingal PCDP, Ivanovska IL, Discher DE. Nuclear lamin stiffness is a barrier to 3D migration, but softness can limit survival. J Cell Biol. 2014;204:669–82. https://doi.org/10.1083/jcb.201308029.
69. Mattout A, Pike BL, Towbin BD, Bank EM, Gonzalez-Sandoval A, Stadler MB, Meister P, Gruenbaum Y, Gasser SM. An EDMD mutation in C. elegans lamin blocks muscle-specific gene relocation and compromises muscle integrity. Curr Biol. 2011;21:1603–14. https://doi.org/10.1016/j.cub.2011.08.030.
70. Towbin BD, Meister P, Pike BL, Gasser SM. Repetitive transgenes in C. elegans accumulate heterochromatic marks and are sequestered at the nuclear envelope in a copy-number- and lamin-dependent manner. Cold Spring Harb Symp Quant Biol. 2010;75:555–65. https://doi.org/10.1101/sqb.2010.75.041.
71. Guelen L, Pagie L, Brasset E, Meuleman W, Faza MB, Talhout W, Eussen BH, de Klein A, Wessels L, de Laat W, van Steensel B. Domain organization of human chromosomes revealed by mapping of nuclear lamina interactions. Nature. 2008;453:948–51. https://doi.org/10.1038/nature06947.
72. Harr JC, Luperchio TR, Wong X, Cohen E, Wheelan SJ, Reddy KL. Directed targeting of chromatin to the nuclear lamina is mediated by chromatin state and A-type lamins. J Cell Biol. 2015;208:33–52. https://doi.org/10.1083/jcb.201405110.
73. Zullo JM, Demarco IA, Piqué-Regi R, Gaffney DJ, Epstein CB, Spooner CJ, Luperchio TR, Bernstein BE, Pritchard JK, Reddy KL, Singh H. DNA sequence-dependent compartmentalization and silencing of chromatin at the nuclear lamina. Cell. 2012;149:1474–87. https://doi.org/10.1016/j.cell.2012.04.035.
74. Towbin BD, González-Aguilera C, Sack R, Gaidatzis D, Kalck V, Meister P, Askjaer P, Gasser SM. Step-wise methylation of histone H3K9 positions heterochromatin at the nuclear periphery. Cell. 2012;150:934–47. https://doi.org/10.1016/j.cell.2012.06.051.
75. Kind J, Pagie L, Ortabozkoyun H, Boyle S, de Vries SS, Janssen H, Amendola M, Nolen LD, Bickmore WA, van Steensel B. Single-cell dynamics of genome-nuclear lamina interactions. Cell. 2013;153:178–92. https://doi.org/10.1016/j.cell.2013.02.028.
76. Kumaran RI, Spector DL. A genetic locus targeted to the nuclear periphery in living cells maintains its transcriptional competence. J Cell Biol. 2008;180:51–65. https://doi.org/10.1083/jcb.200706060.

77. Reddy KL, Zullo JM, Bertolino E, Singh H. Transcriptional repression mediated by repositioning of genes to the nuclear lamina. Nature. 2008;452:243–7. https://doi.org/10.1038/nature06727.
78. Van de Vosse DW, Wan Y, Wozniak RW, Aitchison JD. Role of the nuclear envelope in genome organization and gene expression. Wiley Interdiscip Rev Syst Biol Med. 2011;3:147–66. https://doi.org/10.1002/wsbm.101.
79. Kubben N, Adriaens M, Meuleman W, Voncken JW, van Steensel B, Misteli T. Mapping of lamin A- and progerin-interacting genome regions. Chromosoma. 2012;121:447–64. https://doi.org/10.1007/s00412-012-0376-7.
80. Demmerle J, Koch AJ, Holaska JM. The nuclear envelope protein emerin binds directly to histone deacetylase 3 (HDAC3) and activates HDAC3 activity. J Biol Chem. 2012;287:22080–8. https://doi.org/10.1074/jbc.M111.325308.
81. Nagano T, Lubling Y, Stevens TJ, Schoenfelder S, Yaffe E, Dean W, Laue ED, Tanay A, Fraser P. Single-cell Hi-C reveals cell-to-cell variability in chromosome structure. Nature. 2013;502:59–64. https://doi.org/10.1038/nature12593.
82. Andrés V, González JM. Role of A-type lamins in signaling, transcription, and chromatin organization. J Cell Biol. 2009;187:945–57. https://doi.org/10.1083/jcb.200904124.
83. Kypriotou M, Huber M, Hohl D. The human epidermal differentiation complex: cornified envelope precursors, S100 proteins and the "fused genes" family. Exp Dermatol. 2012;21:643–9. https://doi.org/10.1111/j.1600-0625.2012.01472.x.
84. Williams RRE, Broad S, Sheer D, Ragoussis J. Subchromosomal positioning of the epidermal differentiation complex (EDC) in keratinocyte and lymphoblast interphase nuclei. Exp Cell Res. 2002;272:163–75. https://doi.org/10.1006/excr.2001.5400.
85. Mardaryev AN, Gdula MR, Yarker JL, Emelianov VU, Emelianov VN, Poterlowicz K, Sharov AA, Sharova TY, Scarpa JA, Joffe B, Solovei I, Chambon P, Botchkarev VA, Fessing MY. p63 and Brg1 control developmentally regulated higher-order chromatin remodelling at the epidermal differentiation complex locus in epidermal progenitor cells. Development. 2014;141:101–11. https://doi.org/10.1242/dev.103200.
86. Fessing MY, Mardaryev AN, Gdula MR, Sharov AA, Sharova TY, Rapisarda V, Gordon KB, Smorodchenko AD, Poterlowicz K, Ferone G, Kohwi Y, Missero C, Kohwi-Shigematsu T, Botchkarev VA. p63 regulates Satb1 to control tissue-specific chromatin remodeling during development of the epidermis. J Cell Biol. 2011;194:825–39. https://doi.org/10.1083/jcb.201101148.
87. Lien W-H, Guo X, Polak L, Lawton LN, Young RA, Zheng D, Fuchs E. Genome-wide maps of histone modifications unwind in vivo chromatin states of the hair follicle lineage. Cell Stem Cell. 2011;9:219–32. https://doi.org/10.1016/j.stem.2011.07.015.
88. Indra AK, Dupé V, Bornert J-M, Messaddeq N, Yaniv M, Mark M, Chambon P, Metzger D. Temporally controlled targeted somatic mutagenesis in embryonic surface ectoderm and fetal epidermal keratinocytes unveils two distinct developmental functions of BRG1 in limb morphogenesis and skin barrier formation. Development. 2005;132:4533–44. https://doi.org/10.1242/dev.02019.
89. Kashiwagi M, Morgan BA, Georgopoulos K. The chromatin remodeler Mi-2beta is required for establishment of the basal epidermis and normal differentiation of its progeny. Development. 2007;134:1571–82. https://doi.org/10.1242/dev.001750.
90. LeBoeuf M, Terrell A, Trivedi S, Sinha S, Epstein JA, Olson EN, Morrisey EE, Millar SE. Hdac1 and Hdac2 act redundantly to control p63 and p53 functions in epidermal progenitor cells. Dev Cell. 2010;19:807–18. https://doi.org/10.1016/j.devcel.2010.10.015.
91. Frye M, Fisher AG, Watt FM. Epidermal stem cells are defined by global histone modifications that are altered by Myc-induced differentiation. PLoS One. 2007;2:e763. https://doi.org/10.1371/journal.pone.0000763.
92. Hughes MW, Jiang T-X, Lin S-J, Leung Y, Kobielak K, Widelitz RB, Chuong CM. Disrupted ectodermal organ morphogenesis in mice with a conditional histone deacetylase 1, 2 deletion in the epidermis. J Invest Dermatol. 2014;134:24–32. https://doi.org/10.1038/jid.2013.283.

93. Rapisarda V, Malashchuk I, Asamaowei IE, Poterlowicz K, Fessing MY, Sharov AA, Karakesisoglou I, Botchkarev VA, Mardaryev A. p63 transcription factor regulates nuclear shape and expression of nuclear envelope-associated genes in epidermal keratinocytes. J Invest Dermatol. 2017. https://doi.org/10.1016/j.jid.2017.05.013.
94. Ivorra C, Kubicek M, González JM, Sanz-González SM, Alvarez-Barrientos A, O'Connor J-E, Burke B, Andrés V. A mechanism of AP-1 suppression through interaction of c-Fos with lamin A/C. Genes Dev. 2006;20:307–20. https://doi.org/10.1101/gad.349506.
95. González JM, Navarro-Puche A, Casar B, Crespo P, Andrés V. Fast regulation of AP-1 activity through interaction of lamin A/C, ERK1/2, and c-Fos at the nuclear envelope. J Cell Biol. 2008;183:653–66. https://doi.org/10.1083/jcb.200805049.
96. Oh IY, Albea DM, Goodwin ZA, Quiggle AM, Baker BP, Guggisberg AM, Geahlen JH, Kroner GM, de Guzman Strong C. Regulation of the dynamic chromatin architecture of the epidermal differentiation complex is mediated by a c-Jun/AP-1-modulated enhancer. J Invest Dermatol. 2014;134:2371–80. https://doi.org/10.1038/jid.2014.44.
97. Ezhkova E, Pasolli HA, Parker JS, Stokes N, Su I-H, Hannon G, Tarakhovsky A, Fuchs E. Ezh2 orchestrates gene expression for the stepwise differentiation of tissue-specific stem cells. Cell. 2009;136:1122–35. https://doi.org/10.1016/j.cell.2008.12.043.
98. Blanpain C, Fuchs E. Epidermal stem cells of the skin. Annu Rev Cell Dev Biol. 2006;22:339–73. https://doi.org/10.1146/annurev.cellbio.22.010305.104357.
99. Markiewicz E, Tilgner K, Barker N, van de Wetering M, Clevers H, Dorobek M, Hausmanowa-Petrusewicz I, Ramaekers FCS, Broers JLV, Blankesteijn WM, Salpingidou G, Wilson RG, Ellis JA, Hutchison CJ. The inner nuclear membrane protein emerin regulates beta-catenin activity by restricting its accumulation in the nucleus. EMBO J. 2006;25:3275–85. https://doi.org/10.1038/sj.emboj.7601230.
100. Tilgner K, Wojciechowicz K, Jahoda C, Hutchison C, Markiewicz E. Dynamic complexes of A-type lamins and emerin influence adipogenic capacity of the cell via nucleocytoplasmic distribution of beta-catenin. J Cell Sci. 2009;122:401–13. https://doi.org/10.1242/jcs.026179.
101. Stubenvoll A, Rice M, Wietelmann A, Wheeler M, Braun T. Attenuation of Wnt/β-catenin activity reverses enhanced generation of cardiomyocytes and cardiac defects caused by the loss of emerin. Hum Mol Genet. 2015;24:802–13. https://doi.org/10.1093/hmg/ddu498.
102. Neumann S, Schneider M, Daugherty RL, Gottardi CJ, Eming SA, Beijer A, Noegel AA, Karakesisoglou I. Nesprin-2 interacts with {alpha}-catenin and regulates Wnt signaling at the nuclear envelope. J Biol Chem. 2010;285:34932–8. https://doi.org/10.1074/jbc.M110.119651.
103. Lim X, Nusse R. Wnt signaling in skin development, homeostasis, and disease. Cold Spring Harb Perspect Biol. 2013;5:a008029. https://doi.org/10.1101/cshperspect.a008029.
104. Lin F, Morrison JM, Wu W, Worman HJ. MAN1, an integral protein of the inner nuclear membrane, binds Smad2 and Smad3 and antagonizes transforming growth factor-beta signaling. Hum Mol Genet. 2005;14:437–45. https://doi.org/10.1093/hmg/ddi040.
105. Bengtsson L. What MAN1 does to the Smads. TGFbeta/BMP signaling and the nuclear envelope. FEBS J. 2007;274:1374–82.
106. Hellemans J, Preobrazhenska O, Willaert A, Debeer P, Verdonk PCM, Costa T, Janssens K, Menten B, Van Roy N, Vermeulen SJT, Savarirayan R, Van Hul W, Vanhoenacker F, Huylebroeck D, De Paepe A, Naeyaert J-M, Vandesompele J, Speleman F, Verschueren K, Coucke PJ, Mortier GR. Loss-of-function mutations in LEMD3 result in osteopoikilosis, Buschke-Ollendorff syndrome and melorheostosis. Nat Genet. 2004. Published online: 10 December 2001; doi:101038/ng789 36:1213–18. doi: https://doi.org/10.1038/ng1453.
107. Asano Y, Ihn H, Yamane K, Kubo M, Tamaki K. Impaired Smad7-Smurf-mediated negative regulation of TGF-beta signaling in scleroderma fibroblasts. J Clin Invest. 2004;113:253–64. https://doi.org/10.1172/JCI16269.
108. Mori Y, Chen S-J, Varga J. Expression and regulation of intracellular SMAD signaling in scleroderma skin fibroblasts. Arthritis Rheum. 2003;48:1964–78. https://doi.org/10.1002/art.11157.

109. Rashmi RN, Eckes B, Glöckner G, Groth M, Neumann S, Gloy J, Sellin L, Walz G, Schneider M, Karakesisoglou I, Eichinger L, Noegel AA. The nuclear envelope protein Nesprin-2 has roles in cell proliferation and differentiation during wound healing. Nucleus. 2012;3:172–86. https://doi.org/10.4161/nucl.19090.
110. Mayr M, Hu Y, Hainaut H, Xu Q. Mechanical stress-induced DNA damage and rac-p38MAPK signal pathways mediate p53-dependent apoptosis in vascular smooth muscle cells. FASEB J. 2002;16:1423–5. https://doi.org/10.1096/fj.02-0042fje.
111. Lawrence KS, Tapley EC, Cruz VE, Li Q, Aung K, Hart KC, Schwartz TU, Starr DA, Engebrecht J. LINC complexes promote homologous recombination in part through inhibition of nonhomologous end joining. J Cell Biol. 2016;215:801–21. https://doi.org/10.1083/jcb.201604112.
112. Sur I, Neumann S, Noegel AA. Nesprin-1 role in DNA damage response. Nucleus. 2014;5:173–91. https://doi.org/10.4161/nucl.29023.
113. Warren DT, Tajsic T, Porter LJ, Minaisah RM, Cobb A, Jacob A, Rajgor D, Zhang QP, Shanahan CM. Nesprin-2-dependent ERK1/2 compartmentalisation regulates the DNA damage response in vascular smooth muscle cell ageing. Cell Death Differ. 2015;22:1540–50. https://doi.org/10.1038/cdd.2015.12.
114. Oza P, Jaspersen SL, Miele A, Dekker J, Peterson CL. Mechanisms that regulate localization of a DNA double-strand break to the nuclear periphery. Genes Dev. 2009;23:912–27. https://doi.org/10.1101/gad.1782209.
115. Kalocsay M, Hiller NJ, Jentsch S. Chromosome-wide Rad51 spreading and SUMO-H2A.Z--dependent chromosome fixation in response to a persistent DNA double-strand break. Mol Cell. 2009;33:335–43. https://doi.org/10.1016/j.molcel.2009.01.016.
116. Swartz RK, Rodriguez EC, King MC. A role for nuclear envelope-bridging complexes in homology-directed repair. Mol Biol Cell. 2014;25:2461–71. https://doi.org/10.1091/mbc.E13-10-0569.
117. Kubben N, Voncken JW, Demmers J, Calis C, van Almen G, Pinto Y, Misteli T. Identification of differential protein interactors of lamin A and progerin. Nucleus. 2010;1:513–25. https://doi.org/10.4161/nucl.1.6.13512.
118. Gonzalez-Suarez I, Gonzalo S. Nurturing the genome: A-type lamins preserve genomic stability. Nucleus. 2010;1:129–35. https://doi.org/10.4161/nucl.1.2.10797.
119. Gonzalo S. DNA damage and lamins. Adv Exp Med Biol. 2014;773:377–99. https://doi.org/10.1007/978-1-4899-8032-8_17.
120. Manju K, Muralikrishna B, Parnaik VK. Expression of disease-causing lamin A mutants impairs the formation of DNA repair foci. J Cell Sci. 2006;119:2704–14. https://doi.org/10.1242/jcs.03009.
121. Liu B, Wang J, Chan KM, Tjia WM, Deng W, Guan X, Huang J-D, Li KM, Chau PY, Chen DJ, Pei D, Pendas AM, Cadiñanos J, López-Otín C, Tse HF, Hutchison C, Chen J, Cao Y, Cheah KSE, Tryggvason K, Zhou Z. Genomic instability in laminopathy-based premature aging. Nat Med. 2005;11:780–5. https://doi.org/10.1038/nm1266.
122. Mahen R, Hattori H, Lee M, Sharma P, Jeyasekharan AD, Venkitaraman AR. A-type lamins maintain the positional stability of DNA damage repair foci in mammalian nuclei. PLoS One. 2013;8:e61893. https://doi.org/10.1371/journal.pone.0061893.
123. Starke S, Meinke P, Camozzi D, Mattioli E, Pfaeffle R, Siekmeyer M, Hirsch W, Horn LC, Paasch U, Mitter D, Lattanzi G, Wehnert M, Kiess W. Progeroid laminopathy with restrictive dermopathy-like features caused by an isodisomic LMNA mutation p.R435C. Aging (Albany NY). 2013;5:445–59.
124. Chen C-Y, Chi Y-H, Mutalif RA, Starost MF, Myers TG, Anderson SA, Stewart CL, Jeang K-T. Accumulation of the inner nuclear envelope protein Sun1 is pathogenic in progeric and dystrophic laminopathies. Cell. 2012;149:565–77. https://doi.org/10.1016/j.cell.2012.01.059.
125. Choi JC, Worman HJ. Reactivation of autophagy ameliorates LMNA cardiomyopathy. Autophagy. 2013;9:110–1. https://doi.org/10.4161/auto.22403.

126. Choi JC, Muchir A, Wu W, Iwata S, Homma S, Morrow JP, Worman HJ. Temsirolimus activates autophagy and ameliorates cardiomyopathy caused by lamin A/C gene mutation. Sci Transl Med. 2012;4:144ra102. https://doi.org/10.1126/scitranslmed.3003875.
127. Lenain C, Gusyatiner O, Douma S, van den Broek B, Peeper DS. Autophagy-mediated degradation of nuclear envelope proteins during oncogene-induced senescence. Carcinogenesis. 2015;36:1263–74. https://doi.org/10.1093/carcin/bgv124.
128. Dou Z, Xu C, Donahue G, Shimi T, Pan J-A, Zhu J, Ivanov A, Capell BC, Drake AM, Shah PP, Catanzaro JM, Ricketts MD, Lamark T, Adam SA, Marmorstein R, Zong W-X, Johansen T, Goldman RD, Adams PD, Berger SL. Autophagy mediates degradation of nuclear lamina. Nature. 2015;527:105–9. https://doi.org/10.1038/nature15548.
129. Song X, Narzt MS, Nagelreiter IM, Hohensinner P, Terlecki-Zaniewicz L, Tschachler E, Grillari J, Gruber F. Autophagy deficient keratinocytes display increased DNA damage, senescence and aberrant lipid composition after oxidative stress in vitro and in vivo. Redox Biol. 2017;11:219–30. https://doi.org/10.1016/j.redox.2016.12.015.
130. Akinduro O, Sully K, Patel A, Robinson DJ, Chikh A, McPhail G, Braun KM, Philpott MP, Harwood CA, Byrne C, O'Shaughnessy RFL, Bergamaschi D. Constitutive autophagy and Nucleophagy during epidermal differentiation. J Invest Dermatol. 2016;136:1460–70. https://doi.org/10.1016/j.jid.2016.03.016.
131. Musich PR, Zou Y. DNA-damage accumulation and replicative arrest in Hutchinson-Gilford progeria syndrome. Biochem Soc Trans. 2011;39:1764–9. https://doi.org/10.1042/BST20110687.
132. Sotiropoulou PA, Blanpain C. Development and homeostasis of the skin epidermis. Cold Spring Harb Perspect Biol. 2012;4:a008383. https://doi.org/10.1101/cshperspect.a008383.
133. Goldstein J, Horsley V. Home sweet home: skin stem cell niches. CMLS, Cell Mol Life Sci. 2012;69:2573–82. https://doi.org/10.1007/s00018-012-0943-3.
134. Sotiropoulou PA, Candi A, Mascré G, De Clercq S, Youssef KK, Lapouge G, Dahl E, Semeraro C, Denecker G, Marine J-C, Blanpain C. Bcl-2 and accelerated DNA repair mediates resistance of hair follicle bulge stem cells to DNA-damage-induced cell death. Nat Cell Biol. 2010;12:572–82. https://doi.org/10.1038/ncb2059.
135. Huang X, Pan Y, Cao D, Fang S, Huang K, Chen J, Chen A. UVA-induced upregulation of progerin suppresses 53BP1-mediated NHEJ DSB repair in human keratinocytes via progerin-lamin A complex formation. Oncol Rep. 2017;37:3617–24. https://doi.org/10.3892/or.2017.5603.
136. Sotiropoulou PA, Karambelas AE, Debaugnies M, Candi A, Bouwman P, Moers V, Revenco T, Rocha AS, Sekiguchi K, Jonkers J, Blanpain C. BRCA1 deficiency in skin epidermis leads to selective loss of hair follicle stem cells and their progeny. Genes Dev. 2013;27:39–51. https://doi.org/10.1101/gad.206573.112.
137. Vogel V. Mechanotransduction involving multimodular proteins: converting force into biochemical signals. Annu Rev Biophys Biomol Struct. 2006;35:459–88. https://doi.org/10.1146/annurev.biophys.35.040405.102013.
138. Ivanovska IL, Shin J-W, Swift J, Discher DE. Stem cell mechanobiology: diverse lessons from bone marrow. Trends Cell Biol. 2015;25:523–32. https://doi.org/10.1016/j.tcb.2015.04.003.
139. Jain N, Iyer KV, Kumar A, Shivashankar GV. Cell geometric constraints induce modular gene-expression patterns via redistribution of HDAC3 regulated by actomyosin contractility. Proc Natl Acad Sci USA. 2013;110:11349–54. https://doi.org/10.1073/pnas.1300801110.
140. Ramdas NM, Shivashankar GV. Cytoskeletal control of nuclear morphology and chromatin organization. J Mol Biol. 2015;427:695–706. https://doi.org/10.1016/j.jmb.2014.09.008.
141. Versaevel M, Braquenier J-B, Riaz M, Grevesse T, Lantoine J, Gabriele S. Super-resolution microscopy reveals LINC complex recruitment at nuclear indentation sites. Sci Rep. 2014;4:7362. https://doi.org/10.1038/srep07362.
142. Alam SG, Zhang Q, Prasad N, Li Y, Chamala S, Kuchibhotla R, Kc B, Aggarwal V, Shrestha S, Jones AL, Levy SE, Roux KJ, Nickerson JA, Lele TP. The mammalian LINC complex reg-

ulates genome transcriptional responses to substrate rigidity. Sci Rep. 2016;6:38063. https://doi.org/10.1038/srep38063.
143. McBeath R, Pirone DM, Nelson CM, Bhadriraju K, Chen CS. Cell shape, cytoskeletal tension, and RhoA regulate stem cell lineage commitment. Dev Cell. 2004;6:483–95. https://doi.org/10.1038/nrm1413.
144. Engler AJ, Sen S, Sweeney HL, Discher DE. Matrix elasticity directs stem cell lineage specification. Cell. 2006;126:677–89. https://doi.org/10.1016/j.cell.2006.06.044.
145. le Duc Q, Shi Q, Blonk I, Sonnenberg A, Wang N, Leckband D, de Rooij J. Vinculin potentiates E-cadherin mechanosensing and is recruited to actin-anchored sites within adherens junctions in a myosin II-dependent manner. J Cell Biol. 2010;189:1107–15. https://doi.org/10.1083/jcb.201001149.
146. Yonemura S, Wada Y, Watanabe T, Nagafuchi A, Shibata M. Alpha-catenin as a tension transducer that induces adherens junction development. Nat Cell Biol. 2010;12:533–42. https://doi.org/10.1038/ncb2055.
147. Galbraith CG, Yamada KM, Sheetz MP. The relationship between force and focal complex development. J Cell Biol. 2002;159:695–705. https://doi.org/10.1083/jcb.200204153.
148. Deguchi S, Maeda K, Ohashi T, Sato M. Flow-induced hardening of endothelial nucleus as an intracellular stress-bearing organelle. J Biomech. 2005;38:1751–9. https://doi.org/10.1016/j.jbiomech.2005.06.003.
149. Philip JT, Dahl KN. Nuclear mechanotransduction: response of the lamina to extracellular stress with implications in aging. J Biomech. 2008;41:3164–70. https://doi.org/10.1016/j.jbiomech.2008.08.024.
150. Le HQ, Ghatak S, Yeung C-YC, Tellkamp F, Günschmann C, Dieterich C, Yeroslaviz A, Habermann B, Pombo A, Niessen CM, Wickström SA. Mechanical regulation of transcription controls Polycomb-mediated gene silencing during lineage commitment. Nat Cell Biol. 2016;18:864–75. https://doi.org/10.1038/ncb3387.
151. Morris GE, Randles KN. Nesprin isoforms: are they inside or outside the nucleus? Biochem Soc Trans. 2010;38:278–80. https://doi.org/10.1042/BST0380278.
152. Wheeler MA, Davies JD, Zhang Q, Emerson LJ, Hunt J, Shanahan CM, Ellis JA. Distinct functional domains in nesprin-1alpha and nesprin-2beta bind directly to emerin and both interactions are disrupted in X-linked Emery-Dreifuss muscular dystrophy. Exp Cell Res. 2007;313:2845–57. https://doi.org/10.1016/j.yexcr.2007.03.025.
153. Mislow JMK, Holaska JM, Kim MS, Lee KK, Segura-Totten M, Wilson KL, McNally EM. Nesprin-1alpha self-associates and binds directly to emerin and lamin A in vitro. FEBS Lett. 2002;525:135–40.
154. Haque F, Lloyd DJ, Smallwood DT, Dent CL, Shanahan CM, Fry AM, Trembath RC, Shackleton S. SUN1 interacts with nuclear lamin A and cytoplasmic nesprins to provide a physical connection between the nuclear lamina and the cytoskeleton. Mol Cell Biol. 2006;26:3738–51. https://doi.org/10.1128/MCB.26.10.3738-3751.2006.
155. Bengtsson L, Otto H. LUMA interacts with emerin and influences its distribution at the inner nuclear membrane. J Cell Sci. 2008;121:536–48. https://doi.org/10.1242/jcs.019281.
156. Heald R, McKeon F. Mutations of phosphorylation sites in lamin A that prevent nuclear lamina disassembly in mitosis. Cell. 1990;61:579–89.
157. Khatau SB, Kim D-H, Hale CM, Bloom RJ, Wirtz D. The perinuclear actin cap in health and disease. Nucleus. 2010;1:337–42. https://doi.org/10.4161/nucl.1.4.12331.
158. Kim D-H, Wirtz D. Cytoskeletal tension induces the polarized architecture of the nucleus. Biomaterials. 2015;48:161–72. https://doi.org/10.1016/j.biomaterials.2015.01.023.
159. Tajik A, Zhang Y, Wei F, Sun J, Jia Q, Zhou W, Singh R, Khanna N, Belmont AS, Wang N. Transcription upregulation via force-induced direct stretching of chromatin. Nat Mater. 2016;15:1287–96. https://doi.org/10.1038/nmat4729.
160. Autore F, Pfuhl M, Quan X, Williams A, Roberts RG, Shanahan CM, Fraternali F. Large-scale modelling of the divergent spectrin repeats in nesprins: giant modular proteins. PLoS One. 2013;8:e63633. https://doi.org/10.1371/journal.pone.0063633.

161. Johnson CP, Tang H-Y, Carag C, Speicher DW, Discher DE. Forced unfolding of proteins within cells. Science. 2007;317:663–6. https://doi.org/10.1126/science.1139857.
162. Fillingham I, Gingras AR, Papagrigoriou E, Patel B, Emsley J, Critchley DR, Roberts GCK, Barsukov IL. A vinculin binding domain from the talin rod unfolds to form a complex with the vinculin head. Structure. 2005;13:65–74. https://doi.org/10.1016/j.str.2004.11.006.
163. Lu W, Schneider M, Neumann S, Jaeger V-M, Taranum S, Munck M, Cartwright S, Richardson C, Carthew J, Noh K, Goldberg M, Noegel AA, Karakesisoglou I. Nesprin interchain associations control nuclear size. CMLS, Cell Mol Life Sci. 2012;69:3493–509. https://doi.org/10.1007/s00018-012-1034-1.
164. Balikov DA, Brady SK, Ko UH, Shin JH, de Pereda JM, Sonnenberg A, Sung H-J, Lang MJ. The nesprin-cytoskeleton interface probed directly on single nuclei is a mechanically rich system. Nucleus. 2017;5:1–14. https://doi.org/10.1080/19491034.2017.1322237.
165. Uzer G, Thompson WR, Sen B, Xie Z, Yen SS, Miller S, Bas G, Styner M, Rubin CT, Judex S, Burridge K, Rubin J. Cell Mechanosensitivity to extremely low-magnitude signals is enabled by a LINCed nucleus. Stem Cells. 2015;33:2063–76. https://doi.org/10.1002/stem.2004.
166. Tambe DT, Hardin CC, Angelini TE, Rajendran K, Park CY, Serra-Picamal X, Zhou EH, Zaman MH, Butler JP, Weitz DA, Fredberg JJ, Trepat X. Collective cell guidance by cooperative intercellular forces. Nat Mater. 2011;10:469–75. https://doi.org/10.1038/nmat3025.
167. Dupont S, Morsut L, Aragona M, Enzo E, Giulitti S, Cordenonsi M, Zanconato F, Le Digabel J, Forcato M, Bicciato S, Elvassore N, Piccolo S. Role of YAP/TAZ in mechanotransduction. Nature. 2011;474:179–83. https://doi.org/10.1038/nature10137.
168. Zhang H, Pasolli HA, Fuchs E. Yes-associated protein (YAP) transcriptional coactivator functions in balancing growth and differentiation in skin. Proc Natl Acad Sci USA. 2011;108:2270–5. https://doi.org/10.1073/pnas.1019603108.
169. Schlegelmilch K, Mohseni M, Kirak O, Pruszak J, Rodriguez JR, Zhou D, Kreger BT, Vasioukhin V, Avruch J, Brummelkamp TR, Camargo FD. Yap1 acts downstream of α-catenin to control epidermal proliferation. Cell. 2011;144:782–95. https://doi.org/10.1016/j.cell.2011.02.031.
170. Silvis MR, Kreger BT, Lien W-H, Klezovitch O, Rudakova GM, Camargo FD, Lantz DM, Seykora JT, Vasioukhin V. α-catenin is a tumor suppressor that controls cell accumulation by regulating the localization and activity of the transcriptional coactivator Yap1. Sci Signal. 2011;4:ra33. https://doi.org/10.1126/scisignal.2001823.
171. Luxenburg C, Pasolli HA, Williams SE, Fuchs E. Developmental roles for Srf, cortical cytoskeleton and cell shape in epidermal spindle orientation. Nat Cell Biol. 2011;13:203–14. https://doi.org/10.1038/ncb2163.
172. Koegel H, Tobel von L, Schäfer M, Alberti S, Kremmer E, Mauch C, Hohl D, Wang X-J, Beer H-D, Bloch W, Nordheim A, Werner S. Loss of serum response factor in keratinocytes results in hyperproliferative skin disease in mice. J Clin Invest. 2009;119:899–910. https://doi.org/10.1172/JCI37771.
173. Connelly JT, Gautrot JE, Trappmann B, Tan DW-M, Donati G, Huck WTS, Watt FM. Actin and serum response factor transduce physical cues from the microenvironment to regulate epidermal stem cell fate decisions. Nat Cell Biol. 2010;12:711–8. https://doi.org/10.1038/ncb2074.
174. Lin C, Hindes A, Burns CJ, Koppel AC, Kiss A, Yin Y, Ma L, Blumenberg M, Khnykin D, Jahnsen FL, Crosby SD, Ramanan N, Efimova T. Serum response factor controls transcriptional network regulating epidermal function and hair follicle morphogenesis. J Invest Dermatol. 2013;133:608–17. https://doi.org/10.1038/jid.2012.378.
175. Lammerding J, Schulze PC, Takahashi T, Kozlov S, Sullivan T, Kamm RD, Stewart CL, Lee RT. Lamin A/C deficiency causes defective nuclear mechanics and mechanotransduction. J Clin Invest. 2004;113:370–8. https://doi.org/10.1172/JCI19670.
176. Lammerding J, Hsiao J, Schulze PC, Kozlov S, Stewart CL, Lee RT. Abnormal nuclear shape and impaired mechanotransduction in emerin-deficient cells. J Cell Biol. 2005;170:781–91. https://doi.org/10.1083/jcb.200502148.

177. Ho CY, Jaalouk DE, Vartiainen MK, Lammerding J. Lamin A/C and emerin regulate MKL1-SRF activity by modulating actin dynamics. Nature. 2013;497:507–11. https://doi.org/10.1038/nature12105.
178. Bertrand AT, Ziaei S, Ehret C, Duchemin H, Mamchaoui K, Bigot A, Mayer M, Quijano-Roy S, Desguerre I, Lainé J, Ben Yaou R, Bonne G, Coirault C. Cellular microenvironments reveal defective mechanosensing responses and elevated YAP signaling in LMNA-mutated muscle precursors. J Cell Sci. 2014;127:2873–84. https://doi.org/10.1242/jcs.144907.
179. Willer MK, Carroll CW. Substrate stiffness-dependent regulation of the SRF-Mkl1 co-activator complex requires the inner nuclear membrane protein Emerin. J Cell Sci. 2017;130:2111–8. https://doi.org/10.1242/jcs.197517.
180. Zanconato F, Forcato M, Battilana G, Azzolin L, Quaranta E, Bodega B, Rosato A, Bicciato S, Cordenonsi M, Piccolo S. Genome-wide association between YAP/TAZ/TEAD and AP-1 at enhancers drives oncogenic growth. Nat Cell Biol. 2015;17:1218–27. https://doi.org/10.1038/ncb3216.
181. Holaska JM, Kowalski AK, Wilson KL. Emerin caps the pointed end of actin filaments: evidence for an actin cortical network at the nuclear inner membrane. PLoS Biol. 2004;2:E231. https://doi.org/10.1371/journal.pbio.0020231.
182. Schreiber KH, Kennedy BK. When lamins go bad: nuclear structure and disease. Cell. 2013;152:1365–75. https://doi.org/10.1016/j.cell.2013.02.015.
183. McCord RP, Nazario-Toole A, Zhang H, Chines PS, Zhan Y, Erdos MR, Collins FS, Dekker J, Cao K. Correlated alterations in genome organization, histone methylation, and DNA-lamin A/C interactions in Hutchinson-Gilford progeria syndrome. Genome Res. 2013;23:260–9. https://doi.org/10.1101/gr.138032.112.
184. Gonzalo S, Kreienkamp R. DNA repair defects and genome instability in Hutchinson-Gilford progeria syndrome. Curr Opin Cell Biol. 2015;34:75–83. https://doi.org/10.1016/j.ceb.2015.05.007.

Chapter 12
Epigenetic Regulation of Skin Wound Healing

Andrei N. Mardaryev

List of Abbreviations

DNMT	DNA methyltransferase
ECM	extracellular matrix
EMT	epithelial–mesenchymal transition
HAT	histone acetyltransferase
HDAC	histone deacetylase
INFγ	interferon γ
LPS	lipopolysaccharide
MMP	matrix metalloproteinase
PCAF	P300/CBP-associated factor
PcG	polycomb group protein
PRC2	Polycomb Repressive Complex 2
T2D	type II diabetes
TET	ten-eleven translocation
TLR	Toll-like receptor
TSA	trichostatin A
VEGF	vascular endothelial growth factor

12.1 Introduction

The skin evolved as a vital barrier protecting animals from damaging external factors, such as mechanical and chemical insults and microorganisms, as well as preventing water loss. The skin barrier must be restored expeditiously after injury to maintain homeostasis and survival. To achieve this, a complex and dynamic process

A. N. Mardaryev (✉)
Centre for Skin Sciences, School of Chemistry and Biosciences, Faculty of Life Science, University of Bradford, Bradford, UK
e-mail: a.mardaryev@bradford.ac.uk

© Springer International Publishing AG, part of Springer Nature 2018

V. A. Botchkarev, S. E. Millar (eds.), *Epigenetic Regulation of Skin Development and Regeneration*, Stem Cell Biology and Regenerative Medicine,
https://doi.org/10.1007/978-3-319-16769-5_12

has evolved that involves coordinated multicellular events to repair the structure and function of damaged tissue.

At the wound site, coordinated production and influx of growth factors, extracellular matrix components, and various cell types mediate the healing process and its transition through distinct yet overlapping phases of inflammation, proliferation, and remodeling [1, 2]. Multiple cell lineages in the skin respond to injury by transiently activating or repressing up to 1000 different genes to achieve skin closure [3]. During the last two decades, significant progress has been made in understanding the complex regulatory mechanisms that control gene expression during wound healing [1, 4–6]. While the roles of signaling pathways and transcription factors have been studied extensively, the role of epigenetic regulators in skin repair is an emerging area of interest.

The molecular basis of epigenetic processes is complex and involves modulation of chromatin structure via covalent DNA and histone modifications, ATP-dependent chromatin remodelling and positioning of histone variants, regulation of non-coding RNAs, as well as higher-order chromatin remodelling to bring distal regulatory elements to genes in three-dimensional (3D) nuclear space [7, 8]. Epigenetic changes in chromatin structure are dynamic and potentially reversible; thus, a thorough understanding of these processes may help to uncover new therapeutic targets for many pathological conditions, including poor wound healing.

Recent progress in the field of cutaneous epigenetics revealed that chromatin regulators are essential for epidermal homeostasis and contribute to the pathogenesis of several skin diseases, including skin cancer and psoriasis [9–13]. Recent studies have revealed that certain epigenetic factors regulate aspects of the wound healing process such as cell migration, cell proliferation, angiogenesis and myofibroblast differentiation; however, there is still a need for more research to clarify their exact roles during cutaneous wound healing further. Gaining a complete picture of the biological functions of chromatin regulators, their genomic targets, and their mechanisms of action will enable appropriate manipulation of epigenetic mechanisms to promote wound closure and functional restoration of the skin barrier after injury.

The roles of noncoding/microRNAs are covered elsewhere in this volume and in several recent reviews and original publications [14–17]. In this Chapter, I will summarize recent advances in our understanding of the roles of epigenetic factors in controlling gene expression and behavior of multiple cell lineages at successive stages of wound healing. I will also discuss alterations in epigenetic mechanisms that are linked to poor wound healing and excessive pathological scarring, and will suggest the potential utility of targeting epigenetic factors for the treatment of these conditions.

12.2 Dynamic Expression of Epigenetic Regulators in Skin Injury

The importance of epigenetic modulators in regulation of skin repair is highlighted by their contrasting spatial and temporal expression patterns in the intact and healing skin. Research to date has focused in particular on selected polycomb group proteins (PcG) and histone deacetylases (HDACs) that display dynamic expression in human or mouse skin after injury. Shaw and Martin [18] found that expression levels of mRNAs for the Polycomb Repressive Complex 2 (PRC2) genes *Ezh2*, *Suz12*, and *Eed* that mediate trimethylation of lysine 27 on histone H3 (H3K27me3) are transiently decreased in wounded epithelium. Immunohistochemical analysis revealed reduced levels of Eed and Ezh2 proteins at the wound margin, while their expression was abundant at sites distant from the wound [18]. In contrast, the expression levels of H3K27 histone demethylases Jmjd3 (also known as Kdm6b) and Utx, which generally act in opposition to PcG factors, are transiently increased at the wound edge immediately after wounding [18, 19]. Levels of both Eed and Ezh2 are, however, restored once the regenerated epidermis has fully matured, suggesting that activation of repair genes via loss of PcG- mediated silencing occurs transiently [18, 19].

The expression levels of histone deacetylases also change dynamically during wound healing. For example, HDAC2 is specifically up-regulated at wound margins along with an abundant presence of HDAC2 high-expressing dermal cells in the wound bed at later stages of healing in both mouse and human skin wounds [20]. By contrast, expression levels of HDAC1, 4, or 7 are unchanged in wounded epidermis and dermis after injury [20].

Further research is required to delineate the expression patterns of other epigenetic regulators in wound healing. Systematic screening for differentially expressed chromatin modifiers would help to create a clearer picture of the involvement of the epigenetic machinery in regulation of repair genes in the skin, and to lay the groundwork for functional studies. The identification of cell type-specific and/or wound stage-specific epigenetic factors would be of particular interest. This could be achieved by isolating pure cell populations from the distinct lineages involved in wound healing, including epithelial, dermal, adipose, endothelial and immune cells, using fluorescent flow cytometry-based cell sorting, followed by high-throughput transcriptome analysis (RNA-seq) and genome-wide mapping of selected epigenetic factors (ChIP-seq) at different phases of the wound healing process.

12.3 Epigenetic Regulation of the Inflammatory Response After Skin Injury

Immediately after skin injury, activation of the innate immune system leads to a local inflammatory reaction. This rapid response is initiated by activated platelets that release the contents of their granules, as well as injury-induced degranulation of local mast cells at the site of injury. Inflammatory cytokines produced by activated platelets and mast cells attract phagocytic immune cells leading to initial infiltration of the wound bed by polymorphonuclear leukocytes, and followed by recruitment of monocytes and their differentiation into macrophages [21]. Both types of inflammatory cells provide non-specific defense against pathogens by clearing necrotic tissue at the site of injury and producing pro-inflammatory cytokines. Inflammatory cells also secrete growth factors, which activate cell migration, proliferation, angiogenesis, and extracellular matrix (ECM) synthesis at later stages of wound healing [1, 6, 22, 23]. Therefore, the inflammatory response to the injury is a critical step that is ubiquitously employed for proper wound healing in multiple tissues; however, it must be tightly controlled to allow the reparative process to take place [24, 25].

Macrophages are key effector cells that play pleotropic roles in both the induction and resolution of inflammation after injury. Upon recruitment to the site of injury, bone marrow-derived monocytes are stimulated to differentiate into M1 pro-inflammatory macrophages by pro-inflammatory cytokine interferon γ (INFγ) in cooperation with bacterial-derived Toll-like receptor (TLR) ligands such as lipopolysaccharide (LPS) [26, 27]. Activated M1 macrophages produce a number of inflammatory cytokines and reactive oxygen and nitrogen species (i.e. NO, O_2-) that promote the inflammatory response and help to clear pathogens and damaged tissue [28, 29].

Inflammatory stimuli, including TNF-α, IL-1, and exposure to bacterial LPS, activate the canonical NFκB, IRF and STAT signaling pathways, which in turn regulate expression of key inflammatory genes in M1 macrophages [30–33]. Importantly, NFκB-controlled cytokines, such as IL-1, IL-6, IL-8, IL-12 and TNF-α mediate the inflammatory response during wound healing [1, 6, 23, 34].

Recent studies highlighted an important contribution of epigenetic mechanisms to NFκB-mediated expression of pro-inflammatory cytokines in activated M1 macrophages.

Balanced activity of histone acetyltransferases (HATs) and HDACs, that respectively regulate histone acetylation and deacetylation at gene promoter regions, is important for controlling inflammatory gene expression in macrophages. Bacterial LPS regulates several members of the HDAC family at the mRNA level in murine bone marrow-derived macrophages. LPS transiently represses and then induces expression of a number of HDACs (HDACs 1, 4, 5, 7, 8), coinciding with up-regulation and repression of several pro-inflammatory genes [35]. Furthermore, in alveolar macrophages, reduction of HDAC activity induces rapid CBP/p300-mediated acetylation at the promoter regions of several pro-inflammatory cytokines such as CXCL8/IL8, leading to their transcriptional activation and sustained chronic inflammation in airways [36]. These studies suggest that HDACs act as potent and selective negative regulators of pro-inflammatory gene expression and prevent excessive inflammatory responses in macrophages [35]. However, the roles of HDACs in the regulation of inflammation are complex as these factors can repress both pro- and anti-inflammatory genes [37]. Furthermore, small molecule inhibitors of HDACs (HDACi) are found to trigger both pro- and anti-inflammatory effects in a range of inflammation-relevant cell types [38, 39].

Jmjd3/Kdm6b, a selective H3K27me3 demethylase, is another direct NFκB target that is induced in macrophages in response to bacterial products and inflammatory cytokines [40]. Upon stimulation with LPS, Jmjd3 de-represses PcG target genes by reducing the levels of H3K27me3 [40]. Surprisingly, however, Jmjd3 also plays roles in inflammation independent of H3K27me3 demethylation. At most LPS-induced inflammatory genes, Jmjd3 is preferentially recruited to transcription start sites that are characterized by high levels of H3K4me3, a marker of gene activity, and RNA polymerase II, but are not marked with H3K27me3 [41]. This observation suggests that many inflammatory genes are not PcG targets in macrophages.

By contrast with the findings of De Santa et al. [41], Gallagher et al. [42] showed that induction of IL-12 expression in M1 macrophages at wound sites is linked to reduction of H3K27me3 histone marks in the *IL-12* gene promoter region. Removal of H3K27me3 is associated with up-regulation of Jmjd3/Kdm6b expression in bone marrow-derived monocyte progenitor/stem cells and macrophages that are programmed toward a proinflammatory M1 phenotype [42].

In order to limit inflammation and allow the healing cascade to progress, macrophages must switch in a timely fashion from an M1 pro-inflammatory to an M2 anti-inflammatory phenotype [21, 43, 44]. This switch is stimulated by Th2 cytokines such as IL-4 and IL-13 [45, 46]. Recent data suggest that the switching process is influenced by epigenetic modifications, including histone acetylation. Mullican et al. [47] showed that in unstimulated macrophages, deletion of HDAC3 activates expression of immune and inflammatory genes that are positively regulated by IL-4 exposure in wild-type macrophages. These data indicate that HDAC3 represses a subset of IL-4-regulated genes and restricts activation of M2 anti-inflammatory macrophages, while supporting the M1 pro-inflammatory phenotype [47]. Chromatin remodeling is also mechanistically important in the acquisition of the M2-macrophage phenotype, as Ishii et al. [48] demonstrated that IL-4 stimulation leads to STAT6-mediated Jmjd3 expression. Increased Jmjd3 levels contribute to a decrease in H3K27me2/3 deposition and transcriptional activation of specific M2 macrophage marker genes [48].

Collectively, these data indicate that Jmjd3 plays crucial roles in the acquisition of both macrophage phenotypes by up-regulating the expression of M1 or M2 specific genes [40–42, 48]. By doing so, Jmjd3 can enhance both pro-inflammatory and anti-inflammatory responses in concert with cell-type-specific transcription factors that target it to gene promoters in a context-dependent manner [49]. Importantly, JMJD3 expression can be induced by many inflammatory mediators as well as stress inducers, including metabolic, hypoxic and oncogenic stress. Thus, an inflammatory milieu and cellular stress can alter immune responses and impair wound healing by affecting the levels of JMJD3 expression and activity. As an example, persistence of M1 macrophages and over-expression of IL-12 lead to chronic inflammation and delayed wound healing in type II diabetes (T2D) [42, 50, 51]. Importantly, IL-12 production in M1 macrophages can be modulated by inhibiting Jmjd3 in a diet-induced obese (DIO) mouse model, which closely mimics the T2D in humans. This observation suggests a novel avenue for treatment of chronic diabetic wounds using histone demethylase inhibitors [42]. Together, these studies highlight Jmjd3/Kdm6b as a key chromatin regulator that is involved in the epigenetic control of inflammatory genes and in controlling the differentiation and cell identity of macrophages in tissue repair after injury.

12.4 Epigenetic Regulation of Keratinocyte Proliferation and Migration in Wound Healing

Within few hours after injury, keratinocytes at the edges of wounded epidermis, and adjacent hair follicle and sweat gland cells begin to proliferate and migrate, forming an epidermal "migrating tongue" [52–54]. This process is initiated by signaling pathways in epithelial, dermal and immune cells at the wound edges, which release a large number of different cytokines and growth factors, including TGFα/β, EGF, KGF, IGF-1, and NGF [6]. For re-epithelization to take place, keratinocytes at the wound margin have to profoundly change their morphology and function in a process that resembles certain aspects of the epithelial–mesenchymal transition (EMT) [55–57]. As keratinocytes migrate, they modify their terminal differentiation program by reducing expression of differentiation-associated proteins such as keratins 1 and 10, and by de novo production of the injury associated keratins 6, 16, and 17 [58, 59]. Once the newly formed epidermis covers the wound, keratinocytes re-differentiate into stratified squamous epithelium to restore important barrier functions.

A number of epigenetic mechanisms have been shown to be involved in regulating keratinocyte proliferation. In developing epidermis and postnatal homeostatic conditions, activity of DNA methyltransferase 1 (DNMT1), the histone methyltransferase Setd8, the histone demethylase Jmdj3, HDAC1/2, the PRC1 subunit Cbx4, the PRC2 subunits Ezh1/2, the ATP-dependent chromatin remodeler Chd4, and the genome organizer Special AT-rich sequence binding protein 1 (SATB1) maintain proliferation activity and control differentiation of epidermal progenitor cells via regulation of proliferation-associated and differentiation genes [60–68]. However, the role of these and other epigenetic regulators in keratinocyte proliferation and differentiation during skin repair has not been fully investigated.

12.4.1 The Roles of Histone Methylation and Demethylation in Keratinocyte Proliferation and Migration After Skin Injury

In response to injury, keratinocytes at the wound edge rapidly induce expression of a large number of genes involved in multiple cellular events, including cell cycle regulation, migration, adhesion and ECM degradation [1, 2, 69]. This process is paralleled by a transient decrease in repressive H3K27me3 histone modification and reduced expression of the PRC2 subunits Ezh2, Suz12 and Eed in wounded mouse epidermis [18]. Interestingly in intact skin, H3K27me3 primarily marks histones in epidermal cells with virtually no expression in dermal cells [18], suggesting that epidermal keratinocytes are the main cellular targets for PcG-mediated epigenetic gene silencing in unwounded skin. This observation is further supported by the fact that deletion of any of the three PRC2 core subunits Ezh2, Suz12 and Eed inhibits epidermal and hair follicle stem cell proliferation via up-regulation of senescence-associated genes in the Ink4b-Arf-Ink4a locus, leading to defective epidermal regeneration in vivo [61, 70].

Wound-induced H3K27me3 reduction is inversely linked to marked up-regulation of the H3K27me3 demethylases Jmjd3 and Utx and induction of the expression of several genes such as Egfr and Myc that contribute to repair of injured epidermis by stimulating keratinocyte proliferation and migration [18, 71–73]. The role of Jmjd3 in wound re-epithelization has been further studied by Na et al. [19]. Similar to the *in vivo* context, human keratinocytes upregulate Jmjd3 expression in *in vitro* wound healing models [19]. Upon scratch-wounding of cell monolayers, Jmjd3 induces NFκB-dependent expression of genes encoding several inflammatory cytokines, including IL-1b, IL-6, IL-8, IL-15, IL-23a, CCL20 and TNF-α matrix metalloproteinases such as MMP1/2/9/10/13/14/26, and the growth factors HB-EGF and HGH, that are known to be up-regulated in keratinocytes during the early phase of wound healing [19]. Furthermore, Jmdj3-depleted keratinocytes demonstrated delayed cell migration and reduced proliferation. Crucially, the observed functions of Jmdj3 required its histone demethylase activity, as inhibition of this activity with a small molecule inhibitor GSK-14 significantly attenuated the up-regulation of inflammatory, MMP, and growth factor genes in wounded keratinocytes, resulting in delayed wound closure and keratinocyte migration [19].

PRC2-mediated gene repression is also counterbalanced by Trithorax group proteins of transcriptional activators, such as Ash1l. *Ash1l* encodes a SET domain–containing protein with H3K36 methyltransferase activity, which has been shown to regulate a number of genes by antagonizing H3K27me3 [74–76]. In *Ash1l* mutant mice, wound healing is impaired due to delayed re-epithelialization, increased proliferation and altered expression of keratinocyte markers. This chronic wound phenotype in *Ash1l* mutants is linked to elevated c-Myc expression in the epidermis adjacent to wounds [77]. Together, these studies indicate that tissue repair genes are tightly controlled by Polycomb and Trithorax group proteins, whereas their transient activation via loss of PRC2-mediated gene silencing is required to permit epithelial closure [18, 19].

The functions of PRC1 proteins in wound healing remain unclear. However, recent studies point towards a possible involvement of PRC1 proteins in re-epithelialization via regulation of keratinocyte proliferation and migration. During mouse skin development and in cultured human keratinocytes, the PRC1 subunit Cbx4 maintains proliferative activity and long-term survival of epidermal progenitor cells via suppression of Ink4b-Arf-Ink4a locus genes, and concomitantly prevents their premature terminal differentiation [66, 78]. Similar to Cbx4, another PRC1 subunit Bmi1 (also known as Pcgf4) promotes epidermal keratinocyte proliferation and survival; furthermore its overexpression leads to keratinocyte transformation and is linked to skin carcinogenesis [79–83]. Interestingly in hair-free areas of mouse skin, Bmi1 is expressed in long-term multipotent sweat gland stem cells that maintain the entire eccrine unit and the inter-adnexal epidermis. Upon irradiation-induced injury, Bmi1-positive stem cells rapidly proliferate and regenerate injured epithelial tissue [84]. Up-regulation of Bmi1 expression is associated with enhanced EMT in keratinocytes, as well as increased motility and metastatic activity of BMI1-positive epithelial tumor cells [85–87]. The latter observation is also relevant for wound healing as keratinocytes at the wound margin undergo partial and reversible EMT characterized by loss of intercellular adhesion and polarity, increased secretion of extracellular matrix-degrading proteinases, a shift from production of keratin to vimentin intermediate filaments, and increased cell motility [88].

12.4.2 The Roles of Histone Acetylation and Deacetylation in Wound Re-Epithelialization

Several studies have provided evidence for the importance of histone acetylation in regulating gene expression and skin re-epithelialization after wounding. Spallotta et al. [89] reported that application of the class I-IIa histone deacetylase inhibitor trichostatin A (TSA) accelerated skin regeneration, while a class III HDAC inhibitor Sirtinol inhibited the stimulatory effect of TSA. Class III HDAC activators significantly accelerated skin regeneration by increasing keratinocyte proliferation and migration via endothelial NO synthase phosphorylation and NO production [89]. The latter was associated with S-nitrosylation of HDAC2, which lead to loss of HDAC2 activity and its decreased binding to the promoter regions of several genes, including *Igf1*, *Fgf10*, and *Egf*, that regulate keratinocyte proliferation during wound healing [89]. These data suggest opposing effects of class I-IIa and class III HDACs in skin regeneration, whereby decreased class I-IIa HDAC activity and activation of Class III HDAC sirtuins (SIRTs) promotes wound repair [89]. Importantly, increased activity of HATs and a global increase in acetylation of lysine residues in histones and other proteins in keratinocytes at the wound edge could be mechanistically linked to wound re-epithelialization via activation of repair genes. Indeed, augmentation of the HAT activity of P300/CBP-associated factor (PCAF) using the specific chemical activator SPV-106 significantly increased histone acetylation in wounded epidermis and enhanced wound repair in mice [90].

Similar effects of class I-IIa HDAC inhibition on wound repair were observed following the administration of the HDAC inhibitor TSA in other wound healing models. Intraperitoneal injection of TSA in mice following digital amputation increased proliferation, collagen deposition, and the number of Rex1+ and Sca1+ stem/progenitor cells at the amputation site, resulting in enhanced digit regeneration compared with controls [91]. HDAC inhibition is also beneficial in promoting regeneration of other tissues, such as the spinal cord, after injury. For instance, treatment with the HDAC inhibitor valporic acid was found to reduce apoptosis and improve functional recovery of spinal cord injuries in mice [92].

12.4.3 The Roles of DNA Methylation and Demethylation in Skin Re-Epithelialization After Injury

DNA methylation is one of the key epigenetic mechanisms controlling transcription factor binding and gene expression. DNA methyltransferase 1 (DNMT1) maintains DNA methylation patterns after cellular replication and is required to retain proliferative activity of epidermal and hair follicle stem/progenitor cells in the skin under homeostatic conditions [67, 93]. Another group of DNA methyltransferases, DNMT3a and DNMT3b, perform de novo methylation of previously unmethylated cytosine residues, resulting in altered gene expression. Recent studies highlited importance of DNMT3a and DNMT3b in epidermal progenitor cells (EpPCs). DNMT3a and DNMT3b are critical for the self-renewal of human EPCs, whereas DNMT3a is also required for their proper differentiation [94]. Both enzymes associate with the most active enhancers in human EPCs and maintain high transcriptionsl activity of genes necessary for epidermal stem cell identity and maintenance [94]. However, in mice, Dnmt3a and Dnmt3b are dispensable for steady-state homeostasis, but prevent carcinogen-induced tumor development in the epidermis [95].

The roles of DNMTs in skin regeneration after woinding are still not fully understood. Sun et al. [96] reported increased expression of Dnmt3b in incisional wounds in mice. The latter was associated with increased global DNA methylation in skin tissue at early time points after incision [96]. Increased DNMT3a expression has also been reported in wound epithelium during limb regeneration in the axolotl [97]. Using a model in which skin wounds normally do not regenerate, Aguilar et al. [97] demonstrated that treatment with the DNA methyltransferase inhibitor decitabine (also known as 5-aza2'-deoxycitidine, or 5-aza-dC) stimulated regeneration, while untreated wounds failed to regenerate. A similar effect of DNMT inhibition on wound healing was shown in the mouse model of digit regeneration after amputation [91]: treatment with 5-aza-dC alone or in combination with the HDAC inhibitor

TSA significantly enhanced digit regeneration compared with controls [91]. Together these data suggest that DNA methylation plays an important role in wound healing by regulating the expression of genes that control cell behavior during regeneration.

To date, there are no known DNA demethylating enzymes that directly catalyze removal of the methyl group from 5-methylcytosine (5-mC). Instead, the ten-eleven translocation (TET) enzymes cause step-wise oxidization of 5-mC to produce 5-hydroxymethylcytosine (5-hmC), which is then oxidized to 5-formylcytosine (5fC) and 5-carboxylcytosine (5caC) [98–100]. 5fC and 5caC can be converted to unmodified cytosine by Terminal deoxynucleotidyl transferase (TdT) or Thymine DNA glycosylase (TDG) via base excision repair [101]. The role of TET enzymes in skin homeostasis is still a mystery; however, recent studies suggest crucial roles for these factors in controlling the expression of genes that regulate reparative processes in the skin. Recently, Zhang et al. [102] reported that overexpression of TET2 decreased cell migration and proliferation in cultured primary human keratinocytes, whereas TET2 inhibition had an opposite effect, leading to significantly increased cell proliferation and migration. Furthermore, TET2 is upregulated in the epidermis of T2D patients; this results in hypomethylation of the *MMP9* gene promoter and significantly greater expression of *MMP9* in chronic diabetic wounds compared to wounds from non-diabetic patients [102, 103]. MMP9 is critical for proper wound repair as its deletion in mice leads to impaired wound healing associated with reduced keratinocyte migration, delayed reepithelization, and disordered collagen synthesis [104]. However, excessive production of MMP9 is linked to poor wound healing in diabetic wound ulcers due to an imbalance in the synthesis and degradation of ECM [105–107]. Collectively, these data demonstrate that tight epigenetic control at the level of DNA methylation is critical for regulation of MMP9 expression in wounded epithelium for effective repair. It is conceivable that similar mechanisms control the expression of other wound repair genes that are induced in keratinocytes in response to injury.

These studies highlight the dramatic effects that epigenetic modulations have on keratinocyte proliferation and migration in injured skin. For complete restoration of the epidermal barrier, regenerated epidermis must first undergo re-differentiation, and subsequently needs to restore expression of terminal differentiation genes following injury. Epigenetic mechanisms are vital for keratinocyte differentiation and stratification of the epidermis, both during development and in postnatal homeostasis. Numerous studies have demonstrated that deletion of epigenetic factors such as DNMT1 and Cbx4 leads to premature epidermal differentiation, while JMJD3 overexpression accelerates differentiation [66–68]. The roles of these factors in epidermal re-differentiation following injury, however, remain to be fully elucidated.

12.5 Epigenetic Control of Fibroblast Function and Neo-Angiogenesis After Injury

The dermis undergoes major regeneration and restructuring following skin injury and is re-established in a process known as fibroplasia; this involves fibroblast proliferation and deposition of type III collagen and other matrix proteins [1]. Subpopulations of fibroblasts transform into myofibroblasts to aid wound contraction and produce new immature collagen fibrils, which will later be remodeled to become mature type IV collagen.

As soon as the wound is closed and inflammation subsides, remodeling of the granulation tissue and reorganization of the ECM begins by continuous synthesis, degradation, and maturation of collagen fibers in the dermis. The new blood vessels within the scar regress and mature to form a functional network [108]. Most myofibroblasts, which are responsible for wound contraction, undergo apoptosis at this stage. The resulting scar tissue is either resolved or permanently remains at the wound site. However, the scar tissue never restores the original mechanical strength and functional properties of the *skin*, and lacks hair follicles, sebaceous glands, and sweat glands [109, 110].

12.5.1 Roles of Epigenetic Regulators in the Control of Fibroblast Activity in Skin Repair

12.5.1.1 Histone Acetylation

Epigenetic mechanisms alter the behavior of fibroblasts during wound healing and collagen production. Fitzgerald O'Connor et al. [20] reported overexpression of HDAC2 in the skin fibroblasts in response to wounding in mice, as well as in human normal and keloid scar tissues. Interestingly, treatment with transforming growth factor beta-1 (TGFβ1) increased HDAC2 expression in both human and mouse fibroblasts [20]. Furthermore, HDAC4 was found to be required for TGFβ1-mediated myofibroblastic differentiation, a critical event that contributes to contraction and closure of wounds [111]; correspondingly, HDAC inhibitors block this transformation [112]. HDACs also control collagen production by myofibroblasts, while HDAC inhibition with TSA blocks collagen gene expression in human fibroblasts [113, 114]. These studies suggest that HDAC inhibitors have potential as anti-fibrotic agents for prevention or treatment of pathological wound scarring, for instance in hypertrophic and keloid scars. Consistent with this, Class I/II HDAC inhibition with TSA has been shown to inhibit collagen synthesis and promote apoptosis of keloid fibroblasts [115].

Sirtuin 1 (SIRT1), a Class III HDAC, is decreased in hypertrophic scar tissue, and its knockdown in hypertrophic scar-derived fibroblasts leads to increased expression of α-smooth muscle actin (α-SMA) and TGFβ1-regulated collagen 1

(Col1) and collagen 3 (Col3) [116]. In contrast, up-regulation of SIRT1 by its agonist resveratrol not only inhibits the expression of α-SMA, Col1 and Col3 but also blocks their activation by TGFβ1 in both hypertrophic and normal skin-derived fibroblasts [116]. Furthermore, Ikeda et al. [117] reported that in keloid-derived fibroblasts, resveratrol significantly downregulates the expression of Col1, α-SMA and TGF-β1, and decreases fibroblasts proliferation [117]. Collectively, these studies reveal that activation of Class III HDACs along with inhibition of Class I/II HDACs may be a promising therapeutic strategy for prevention and treatment of pathological scarring after skin injury.

12.5.1.2 DNA Methylation

Recent studies suggest that aberrant DNA methylation can also contribute to the development of pathological scars. Russell et al. studied the epigenetic status of keloid fibroblasts, finding that their phenotype and propensity to form benign dermal tumors may be due to differential DNA methylation of proliferative genes [118]. DNMT1 is upregulated in keloid fibroblasts [119, 120], and DNMT inhibition with 5-aza-dC restores expression of selected genes that are silenced in keloids [118]. Furthermore, 5-aza-dC treatment decreases TGFβ1 mRNA expression, while increasing transcription of inhibitory Smad7 in keloid fibroblasts [119]. DNMT inhibition also decreases expression of collagen and connective tissue growth factor (CTGF), while causing cell cycle arrest and apoptosis in keloid fibroblasts [118, 119]. Thus, DNMT inhibitors have potential utility in the treatment of pathological wound scarring.

12.5.2 Epigenetic Regulation of Angiogenesis

Neo-angiogenesis, stimulated by vascular endothelial growth factor (VEGF), platelet-derived growth factor, basic fibroblast growth factor, and TGFβ, permits angiogenic capillary buds to grow into the wound bed and form a microvascular network that sustains granulation tissue. Epigenetic factors have been shown to control endothelial cell differentiation, migration and sprouting by regulating the expression of vascular growth factors and their receptors. HDAC6, 7 and 9 positively regulate vascular tube formation and enhance angiogenic gene expression [121, 122], while VEGF signaling, which is critical for vascular morphogenesis and endothelial differentiation, is attenuated by HDAC inhibition [123]. Additionally, human umbilical vein endothelial cells fail to form capillary buds following treatment with the HDAC inhibitors TSA and SAHA due to inhibition of VEGF-induced expression of its receptors VEGFR-1 and VEGFR-2 [123].

DNA methylation also plays an important role in the differentiation of bone marrow-derived progenitor cells into endothelial cells after skin injury. Inhibition of DNMT activity with 5-aza-dC has been shown to induce expression of the endothelial cell marker genes VE-cadherin, vWF, and Flk1 in the progenitor cells and stimulates vascular network formation [124]. DNMTs play similar roles in embryonic stem cells and mesenchymal stem cells. In these cell types, the promoters of genes controlling endothelial lineage differentiation and angiogenesis are suppressed by hypermethylation, and DNMT inhibition leads to their activation [125, 126].

Members of the PcG family of proteins are also involved in angiogenesis. VEGF mediated stimulation of angiogenic activity is accompanied by upregulation of Ezh2 in normal blood vessels [127, 128] and in tumor vasculature, supporting a link between epigenetic mechanisms and growth factor signaling [128]. These data suggest that PcG-mediated gene silencing is critical for activation of endothelial lineage genes and vasculogenesis. Endothelial cells also express high levels of JARID1B, a histone 3 lysine 4 (H3K4) demethylase, which suppresses gene expression via reducing H3K4me2/3 levels at the promoters of actively expressed genes [129]. Fork et al. [129] reported that knockout or inhibition of JARID1B attenuated the angiogenic capacity of human endothelial cells in vitro and murine endothelial cells in vivo. The latter effects were associated with reactivation of *HOXA5*, a JARID1B target gene, which is normally highly expressed in quiescent undifferentiated endothelial progenitor cells and possesses anti-angiogenic activity [129, 130]. Importantly, HOXA5 has been suggested to play a critical role in wound healing as its forced expression in wounds impairs angiogenesis and skin repair in mice [131], and suppresses keratinocyte growth and epidermal reconstruction in organotypic cultures [132]. Whether the *HOXA5* gene is activated in poorly healing or chronic wounds that are deficient in angiogenesis is an interesting topic for further investigation.

Members of the heterochromatin protein 1 (HP1) family recognize methylated histones and are involved in the organization of heterochromatin formation and gene silencing in many organisms. Recently, HP1a has been shown to be a key regulator that drives endothelial progenitor cell (EPC) differentiation to mature endothelial cells. Maeng et al. [133] showed that HPa1 is induced during EPC differentiation and its overexpression promotes EPC angiogenic activity in vitro and in vivo by repressing progenitor cells marker genes and activating angiogenic genes [133]. HP1a is thought to regulate angiogenic gene transcription by interacting with chromatin modifying factors such as the methyl-CpG binding domain, Polycomb group ring finger 2 (PCGF2), and DNA methyltransferases at early stages of EPC differentiation [133].

12.6 Summary

Epigenetic regulators are involved in the control of multiple aspects of the wound healing process by regulating the adaptive behaviors of keratinocytes, fibroblasts, endothelial cells and inflammatory/immune cells in response to tissue injury. Epigenetic factors participate in complex regulatory mechanisms that either

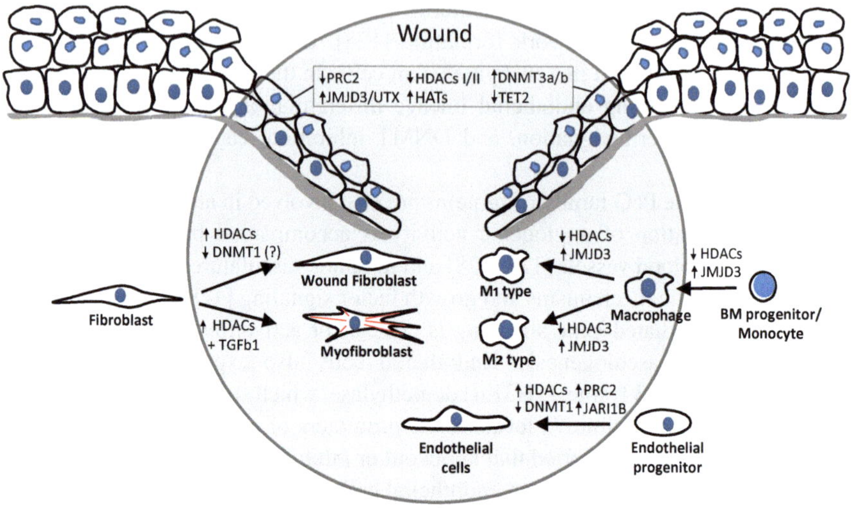

Fig. 12.1 Epigenetic factors regulate multiple cell types in the wounded skin. In response to wounding, epigenetic factors participate in complex regulatory mechanisms that either stimulate or repress gene activation to transiently alter cellular phenotype and behavior, and interfere with growth factor activities in several cell lineages such as keratinocytes, fibroblasts, macrophages, and endothelial cells

stimulate or repress gene activation to transiently alter cellular phenotypes and behavior, and interact with growth factor activity (Fig. 12.1). Transient down-regulation of factors that constitute the repressive epigenetic machinery, such as PcG proteins, is a feature of epigenetic responses to injury in multiple cell lineages. In parallel, histone demethylases are upregulated, leading to enrichment of accessible chromatin that together with transcription factor activity facilitates expression of wound repair genes. Similarly, HDAC and DNMT inhibition can also re-activate expression of genes that stimulate reparative processes after skin injury in both epithelial and mesenchymal cell lineages.

A fine balance of epigenetic regulation is critically important for efficient skin repair, as disruption of these mechanisms results in defective healing, for instance in chronic diabetic wounds and keloid scar formation. Deeper understanding of the epigenetic mechanisms that regulate the healing process is likely to identify novel potential therapeutic targets, ultimately permitting the development of enhanced treatments for chronic and poorly healing wounds and for excessive wound scarring. Systematic studies will be required to characterize the complex epigenetic changes that are induced by skin injury.

References

1. Gurtner GC, Werner S, Barrandon Y, Longaker MT. Wound repair and regeneration. Nature. 2008;453(7193):314–21. https://doi.org/10.1038/nature07039.
2. Schafer M, Werner S. Transcriptional control of wound repair. Annu Rev Cell Dev Biol. 2007;23:69–92. https://doi.org/10.1146/annurev.cellbio.23.090506.123609.
3. Cooper L, Johnson C, Burslem F, Martin P. Wound healing and inflammation genes revealed by array analysis of 'macrophageless' PU.1 null mice. Genome Biol. 2005;6(1):R5. https://doi.org/10.1186/gb-2004-6-1-r5.
4. Eming SA, Martin P, Tomic-Canic M. Wound repair and regeneration: mechanisms, signaling, and translation. Sci Transl Med. 2014;6(265):265sr6. https://doi.org/10.1126/scitranslmed.3009337.
5. MacLeod AS, Mansbridge JN. The innate immune system in acute and chronic wounds. Adv Wound Care (New Rochelle). 2016;5(2):65–78. https://doi.org/10.1089/wound.2014.0608.
6. Werner S, Grose R. Regulation of wound healing by growth factors and cytokines. Physiol Rev. 2003;83(3):835–70. https://doi.org/10.1152/physrev.00031.2002.
7. Bickmore WA, van Steensel B. Genome architecture: domain organization of interphase chromosomes. Cell. 2013;152(6):1270–84. https://doi.org/10.1016/j.cell.2013.02.001.
8. Ho L, Crabtree GR. Chromatin remodelling during development. Nature. 2010;463(7280):474–84. https://doi.org/10.1038/nature08911.
9. Botchkarev VA, Gdula MR, Mardaryev AN, Sharov AA, Fessing MY. Epigenetic regulation of gene expression in keratinocytes. J Invest Dermatol. 2012;132(11):2505–21. https://doi.org/10.1038/jid.2012.182.
10. Liang Y, Chang C, Lu Q. The genetics and epigenetics of atopic dermatitis-filaggrin and other polymorphisms. Clin Rev Allergy Immunol. 2015. https://doi.org/10.1007/s12016-015-8508-5.
11. Makino T, Jinnin M. Genetic and epigenetic abnormalities in systemic sclerosis. J Dermatol. 2016;43(1):10–8. https://doi.org/10.1111/1346-8138.13221.
12. Millington GW. Epigenetics and dermatological disease. Pharmacogenomics. 2008;9(12):1835–50. https://doi.org/10.2217/14622416.9.12.1835.
13. Nguyen CM, Liao W. Genomic imprinting in psoriasis and atopic dermatitis: a review. J Dermatol Sci. 2015;80(2):89–93. https://doi.org/10.1016/j.jdermsci.2015.08.004.
14. Banerjee J, Sen CK. microRNA and wound healing. Adv Exp Med Biol. 2015;888:291–305. https://doi.org/10.1007/978-3-319-22671-2_15.
15. Fahs F, Bi X, Yu FS, Zhou L, Mi QS. New insights into microRNAs in skin wound healing. IUBMB Life. 2015;67(12):889–96. https://doi.org/10.1002/iub.1449.
16. Li D, Wang A, Liu X, Meisgen F, Grunler J, Botusan IR, et al. MicroRNA-132 enhances transition from inflammation to proliferation during wound healing. J Clin Invest. 2015;125(8):3008–26. https://doi.org/10.1172/JCI79052.
17. Liu J, Luo C, Yin Z, Li P, Wang S, Chen J, et al. Downregulation of let7b promotes COL1A1 and COL1A2 expression in dermis and skin fibroblasts during heat wound repair. Mol Med Rep. 2016;13(3):2683–8. https://doi.org/10.3892/mmr.2016.4877.
18. Shaw T, Martin P. Epigenetic reprogramming during wound healing: loss of polycomb-mediated silencing may enable upregulation of repair genes. EMBO Rep. 2009;10(8):881–6. https://doi.org/10.1038/embor.2009.102.
19. Na J, Lee K, Na W, Shin JY, Lee MJ, Yune TY, et al. Histone H3K27 Demethylase JMJD3 in cooperation with NF-kappaB regulates keratinocyte wound healing. J Invest Dermatol. 2016;136(4):847–58. https://doi.org/10.1016/j.jid.2015.11.029.
20. Fitzgerald O'Connor EJ, Badshah II, Addae LY, Kundasamy P, Thanabalasingam S, Abioye D, et al. Histone deacetylase 2 is upregulated in normal and keloid scars. J Invest Dermatol. 2012;132(4):1293–6. https://doi.org/10.1038/jid.2011.432.
21. Italiani P, Boraschi D. From monocytes to M1/M2 macrophages: phenotypical vs. functional differentiation. Front Immunol. 2014;5:514. https://doi.org/10.3389/fimmu.2014.00514.

22. Haertel E, Werner S, Schafer M. Transcriptional regulation of wound inflammation. Semin Immunol. 2014;26(4):321–8. https://doi.org/10.1016/j.smim.2014.01.005.
23. Trinchieri G. Interleukin-12 and the regulation of innate resistance and adaptive immunity. Nat Rev Immunol. 2003;3(2):133–46. https://doi.org/10.1038/nri1001.
24. Cooper PR, Takahashi Y, Graham LW, Simon S, Imazato S, Smith AJ. Inflammation-regeneration interplay in the dentine-pulp complex. J Dent. 2010;38(9):687–97. https://doi.org/10.1016/j.jdent.2010.05.016.
25. Filbin MT. How inflammation promotes regeneration. Nat Neurosci. 2006;9(6):715–7. https://doi.org/10.1038/nn0606-715.
26. Delneste Y, Charbonnier P, Herbault N, Magistrelli G, Caron G, Bonnefoy JY, et al. Interferon-gamma switches monocyte differentiation from dendritic cells to macrophages. Blood. 2003;101(1):143–50. https://doi.org/10.1182/blood-2002-04-1164.
27. Mills CD, Thomas AC, Lenz LL, Munder M. Macrophage: SHIP of immunity. Front Immunol. 2014;5:620. https://doi.org/10.3389/fimmu.2014.00620.
28. Arango Duque G, Descoteaux A. Macrophage cytokines: involvement in immunity and infectious diseases. Front Immunol. 2014;5:491. https://doi.org/10.3389/fimmu.2014.00491.
29. Bosca L, Zeini M, Traves PG, Hortelano S. Nitric oxide and cell viability in inflammatory cells: a role for NO in macrophage function and fate. Toxicology. 2005;208(2):249–58. https://doi.org/10.1016/j.tox.2004.11.035.
30. Hayden MS, West AP, Ghosh S. NF-kappaB and the immune response. Oncogene. 2006;25(51):6758–80. https://doi.org/10.1038/sj.onc.1209943.
31. Lawrence T. The nuclear factor NF-kappaB pathway in inflammation. Cold Spring Harb Perspect Biol. 2009;1(6):a001651. https://doi.org/10.1101/cshperspect.a001651.
32. Mitchell S, Vargas J, Hoffmann A. Signaling via the NFkappaB system. Wiley Interdiscip Rev Syst Biol Med. 2016. https://doi.org/10.1002/wsbm.1331.
33. Sharif O, Bolshakov VN, Raines S, Newham P, Perkins ND. Transcriptional profiling of the LPS induced NF-kappaB response in macrophages. BMC Immunol. 2007;8:1. https://doi.org/10.1186/1471-2172-8-1.
34. Hubner G, Brauchle M, Smola H, Madlener M, Fassler R, Werner S. Differential regulation of pro-inflammatory cytokines during wound healing in normal and glucocorticoid-treated mice. Cytokine. 1996;8(7):548–56. https://doi.org/10.1006/cyto.1996.0074.
35. Aung HT, Schroder K, Himes SR, Brion K, van Zuylen W, Trieu A, et al. LPS regulates proinflammatory gene expression in macrophages by altering histone deacetylase expression. FASEB J. 2006;20(9):1315–27. https://doi.org/10.1096/fj.05-5360com.
36. Ito K, Ito M, Elliott WM, Cosio B, Caramori G, Kon OM, et al. Decreased histone deacetylase activity in chronic obstructive pulmonary disease. N Engl J Med. 2005;352(19):1967–76. https://doi.org/10.1056/NEJMoa041892.
37. Villagra A, Sotomayor EM, Seto E. Histone deacetylases and the immunological network: implications in cancer and inflammation. Oncogene. 2010;29(2):157–73. https://doi.org/10.1038/onc.2009.334.
38. Halili MA, Andrews MR, Sweet MJ, Fairlie DP. Histone deacetylase inhibitors in inflammatory disease. Curr Top Med Chem. 2009;9(3):309–19.
39. Roger T, Lugrin J, Le Roy D, Goy G, Mombelli M, Koessler T, et al. Histone deacetylase inhibitors impair innate immune responses to Toll-like receptor agonists and to infection. Blood. 2011;117(4):1205–17. https://doi.org/10.1182/blood-2010-05-284711.
40. De Santa F, Totaro MG, Prosperini E, Notarbartolo S, Testa G, Natoli G. The histone H3 lysine-27 demethylase Jmjd3 links inflammation to inhibition of polycomb-mediated gene silencing. Cell. 2007;130(6):1083–94. https://doi.org/10.1016/j.cell.2007.08.019.
41. De Santa F, Narang V, Yap ZH, Tusi BK, Burgold T, Austenaa L, et al. Jmjd3 contributes to the control of gene expression in LPS-activated macrophages. EMBO J. 2009;28(21):3341–52. https://doi.org/10.1038/emboj.2009.271.
42. Gallagher KA, Joshi A, Carson WF, Schaller M, Allen R, Mukerjee S, et al. Epigenetic changes in bone marrow progenitor cells influence the inflammatory phenotype and alter

wound healing in type 2 diabetes. Diabetes. 2015;64(4):1420–30. https://doi.org/10.2337/db14-0872.
43. Barrientos S, Stojadinovic O, Golinko MS, Brem H, Tomic-Canic M. Growth factors and cytokines in wound healing. Wound Repair Regen. 2008;16(5):585–601. https://doi.org/10.1111/j.1524-475X.2008.00410.x.
44. Porcheray F, Viaud S, Rimaniol AC, Leone C, Samah B, Dereuddre-Bosquet N, et al. Macrophage activation switching: an asset for the resolution of inflammation. Clin Exp Immunol. 2005;142(3):481–9. https://doi.org/10.1111/j.1365-2249.2005.02934.x.
45. Gordon S, Martinez FO. Alternative activation of macrophages: mechanism and functions. Immunity. 2010;32(5):593–604. https://doi.org/10.1016/j.immuni.2010.05.007.
46. Murray PJ, Allen JE, Biswas SK, Fisher EA, Gilroy DW, Goerdt S, et al. Macrophage activation and polarization: nomenclature and experimental guidelines. Immunity. 2014;41(1):14–20. https://doi.org/10.1016/j.immuni.2014.06.008.
47. Mullican SE, Gaddis CA, Alenghat T, Nair MG, Giacomin PR, Everett LJ, et al. Histone deacetylase 3 is an epigenomic brake in macrophage alternative activation. Genes Dev. 2011;25(23):2480–8. https://doi.org/10.1101/gad.175950.111.
48. Ishii M, Wen H, Corsa CA, Liu T, Coelho AL, Allen RM, et al. Epigenetic regulation of the alternatively activated macrophage phenotype. Blood. 2009;114(15):3244–54. https://doi.org/10.1182/blood-2009-04-217620.
49. Salminen A, Kaarniranta K, Hiltunen M, Kauppinen A. Histone demethylase Jumonji D3 (JMJD3/KDM6B) at the nexus of epigenetic regulation of inflammation and the aging process. J Mol Med (Berl). 2014;92(10):1035–43. https://doi.org/10.1007/s00109-014-1182-x.
50. Falanga V. Wound healing and its impairment in the diabetic foot. Lancet. 2005;366(9498):1736–43. https://doi.org/10.1016/S0140-6736(05)67700-8.
51. Wetzler C, Kampfer H, Stallmeyer B, Pfeilschifter J, Frank S. Large and sustained induction of chemokines during impaired wound healing in the genetically diabetic mouse: prolonged persistence of neutrophils and macrophages during the late phase of repair. J Invest Dermatol. 2000;115(2):245–53. https://doi.org/10.1046/j.1523-1747.2000.00029.x.
52. Ito M, Liu Y, Yang Z, Nguyen J, Liang F, Morris RJ, et al. Stem cells in the hair follicle bulge contribute to wound repair but not to homeostasis of the epidermis. Nat Med. 2005;11(12):1351–4. https://doi.org/10.1038/nm1328.
53. Langton AK, Herrick SE, Headon DJ. An extended epidermal response heals cutaneous wounds in the absence of a hair follicle stem cell contribution. J Invest Dermatol. 2008;128(5):1311–8. https://doi.org/10.1038/sj.jid.5701178.
54. Lu C, Fuchs E. Sweat gland progenitors in development, homeostasis, and wound repair. Cold Spring Harb Perspect Med. 2014;4(2). https://doi.org/10.1101/cshperspect.a015222.
55. Hudson LG, Newkirk KM, Chandler HL, Choi C, Fossey SL, Parent AE, et al. Cutaneous wound reepithelialization is compromised in mice lacking functional Slug (Snai2). J Dermatol Sci. 2009;56(1):19–26. https://doi.org/10.1016/j.jdermsci.2009.06.009.
56. Kusewitt DF, Choi C, Newkirk KM, Leroy P, Li Y, Chavez MG, et al. Slug/Snai2 is a downstream mediator of epidermal growth factor receptor-stimulated reepithelialization. J Invest Dermatol. 2009;129(2):491–5. https://doi.org/10.1038/jid.2008.222.
57. Savagner P, Kusewitt DF, Carver EA, Magnino F, Choi C, Gridley T, et al. Developmental transcription factor slug is required for effective re-epithelialization by adult keratinocytes. J Cell Physiol. 2005;202(3):858–66. https://doi.org/10.1002/jcp.20188.
58. McGowan K, Coulombe PA. The wound repair-associated keratins 6, 16, and 17. Insights into the role of intermediate filaments in specifying keratinocyte cytoarchitecture. Subcell Biochem. 1998;31:173–204.
59. Wong P, Coulombe PA. Loss of keratin 6 (K6) proteins reveals a function for intermediate filaments during wound repair. J Cell Biol. 2003;163(2):327–37. https://doi.org/10.1083/jcb.200305032.

60. Driskell I, Oda H, Blanco S, Nascimento E, Humphreys P, Frye M. The histone methyltransferase Setd8 acts in concert with c-Myc and is required to maintain skin. EMBO J. 2012;31(3):616–29. https://doi.org/10.1038/emboj.2011.421.
61. Ezhkova E, Lien WH, Stokes N, Pasolli HA, Silva JM, Fuchs E. EZH1 and EZH2 cogovern histone H3K27 trimethylation and are essential for hair follicle homeostasis and wound repair. Genes Dev. 2011;25(5):485–98. https://doi.org/10.1101/gad.2019811.
62. Ezhkova E, Pasolli HA, Parker JS, Stokes N, Su IH, Hannon G, et al. Ezh2 orchestrates gene expression for the stepwise differentiation of tissue-specific stem cells. Cell. 2009;136(6):1122–35. https://doi.org/10.1016/j.cell.2008.12.043.
63. Fessing MY, Mardaryev AN, Gdula MR, Sharov AA, Sharova TY, Rapisarda V, et al. p63 regulates Satb1 to control tissue-specific chromatin remodeling during development of the epidermis. J Cell Biol. 2011;194(6):825–39. https://doi.org/10.1083/jcb.201101148.
64. Kashiwagi M, Morgan BA, Georgopoulos K. The chromatin remodeler Mi-2beta is required for establishment of the basal epidermis and normal differentiation of its progeny. Development. 2007;134(8):1571–82. https://doi.org/10.1242/dev.001750.
65. LeBoeuf M, Terrell A, Trivedi S, Sinha S, Epstein JA, Olson EN, et al. Hdac1 and Hdac2 act redundantly to control p63 and p53 functions in epidermal progenitor cells. Dev Cell. 2010;19(6):807–18. https://doi.org/10.1016/j.devcel.2010.10.015.
66. Mardaryev AN, Liu B, Rapisarda V, Poterlowicz K, Malashchuk I, Rudolf J, et al. Cbx4 maintains the epithelial lineage identity and cell proliferation in the developing stratified epithelium. J Cell Biol. 2016;212(1). https://doi.org/10.1083/jcb.201506065.
67. Sen GL, Reuter JA, Webster DE, Zhu L, Khavari PA. DNMT1 maintains progenitor function in self-renewing somatic tissue. Nature. 2010;463(7280):563–7. https://doi.org/10.1038/nature08683.
68. Sen GL, Webster DE, Barragan DI, Chang HY, Khavari PA. Control of differentiation in a self-renewing mammalian tissue by the histone demethylase JMJD3. Genes Dev. 2008;22(14):1865–70. https://doi.org/10.1101/gad.1673508.
69. Roy S, Khanna S, Rink C, Biswas S, Sen CK. Characterization of the acute temporal changes in excisional murine cutaneous wound inflammation by screening of the wound-edge transcriptome. Physiol Genomics. 2008;34(2):162–84. https://doi.org/10.1152/physiolgenomics.00045.2008.
70. Dauber KL, Perdigoto CN, Valdes VJ, Santoriello FJ, Cohen I, Ezhkova E. Dissecting the roles of Polycomb repressive complex 2 subunits in the control of skin development. J Invest Dermatol. 2016. https://doi.org/10.1016/j.jid.2016.02.809.
71. Ching YH, Sutton TL, Pierpont YN, Robson MC, Payne WG. The use of growth factors and other humoral agents to accelerate and enhance burn wound healing. Eplasty. 2011;11:e41.
72. Doma E, Rupp C, Baccarini M. EGFR-ras-raf signaling in epidermal stem cells: roles in hair follicle development, regeneration, tissue remodeling and epidermal cancers. Int J Mol Sci. 2013;14(10):19361–84. https://doi.org/10.3390/ijms141019361.
73. Shi Y, Shu B, Yang R, Xu Y, Xing B, Liu J, et al. Wnt and Notch signaling pathway involved in wound healing by targeting c-Myc and Hes1 separately. Stem Cell Res Ther. 2015;6:120. https://doi.org/10.1186/s13287-015-0103-4.
74. Dorighi KM, Tamkun JW. The trithorax group proteins Kismet and ASH1 promote H3K36 dimethylation to counteract Polycomb group repression in Drosophila. Development. 2013;140(20):4182–92. https://doi.org/10.1242/dev.095786.
75. Miyazaki H, Higashimoto K, Yada Y, Endo TA, Sharif J, Komori T, et al. Ash1l methylates Lys36 of histone H3 independently of transcriptional elongation to counteract polycomb silencing. PLoS Genet. 2013;9(11):e1003897. https://doi.org/10.1371/journal.pgen.1003897.
76. Tanaka Y, Katagiri Z, Kawahashi K, Kioussis D, Kitajima S. Trithorax-group protein ASH1 methylates histone H3 lysine 36. Gene. 2007;397(1–2):161–8. https://doi.org/10.1016/j.gene.2007.04.027.

77. Li G, Ye Z, Shi C, Sun L, Han M, Zhuang Y, et al. The histone Methyltransferase Ash1l is required for epidermal homeostasis in mice. Sci Rep. 2017;7:45401. https://doi.org/10.1038/srep45401.
78. Luis NM, Morey L, Mejetta S, Pascual G, Janich P, Kuebler B, et al. Regulation of human epidermal stem cell proliferation and senescence requires polycomb- dependent and -independent functions of Cbx4. Cell Stem Cell. 2011;9(3):233–46. https://doi.org/10.1016/j.stem.2011.07.013.
79. Balasubramanian S, Adhikary G, Eckert RL. The Bmi-1 polycomb protein antagonizes the (−)-epigallocatechin-3-gallate-dependent suppression of skin cancer cell survival. Carcinogenesis. 2010;31(3):496–503. https://doi.org/10.1093/carcin/bgp314.
80. Cordisco S, Maurelli R, Bondanza S, Stefanini M, Zambruno G, Guerra L, et al. Bmi-1 reduction plays a key role in physiological and premature aging of primary human keratinocytes. J Invest Dermatol. 2010;130(4):1048–62. https://doi.org/10.1038/jid.2009.355.
81. Lee K, Adhikary G, Balasubramanian S, Gopalakrishnan R, McCormick T, Dimri GP, et al. Expression of Bmi-1 in epidermis enhances cell survival by altering cell cycle regulatory protein expression and inhibiting apoptosis. J Invest Dermatol. 2008;128(1):9–17. https://doi.org/10.1038/sj.jid.5700949.
82. Reinisch CM, Uthman A, Erovic BM, Pammer J. Expression of BMI-1 in normal skin and inflammatory and neoplastic skin lesions. J Cutan Pathol. 2007;34(2):174–80. https://doi.org/10.1111/j.1600-0560.2006.00587.x.
83. Wang Q, Li WL, You P, Su J, Zhu MH, Xie DF, et al. Oncoprotein BMI-1 induces the malignant transformation of HaCaT cells. J Cell Biochem. 2009;106(1):16–24. https://doi.org/10.1002/jcb.21969.
84. Ohe S, Tanaka T, Yanai H, Komai Y, Omachi T, Kanno S, et al. Maintenance of sweat glands by stem cells located in the acral epithelium. Biochem Biophys Res Commun. 2015;466(3):333–8. https://doi.org/10.1016/j.bbrc.2015.09.022.
85. Kurihara K, Isobe T, Yamamoto G, Tanaka Y, Katakura A, Tachikawa T. Expression of BMI1 and ZEB1 in epithelial-mesenchymal transition of tongue squamous cell carcinoma. Oncol Rep. 2015;34(2):771–8. https://doi.org/10.3892/or.2015.4032.
86. Qiao B, Chen Z, Hu F, Tao Q, Lam AK. BMI-1 activation is crucial in hTERT-induced epithelial-mesenchymal transition of oral epithelial cells. Exp Mol Pathol. 2013;95(1):57–61. https://doi.org/10.1016/j.yexmp.2013.05.004.
87. Yuan W, Yuan Y, Zhang T, Wu S. Role of Bmi-1 in regulation of ionizing irradiation-induced epithelial-mesenchymal transition and migration of breast cancer cells. PLoS One. 2015;10(3):e0118799. https://doi.org/10.1371/journal.pone.0118799.
88. Arnoux V, Côme C, Kusewitt D, Hudson L, Savagner P. Cutaneous wound reepithelialization. In: Savanger P, editor. Rise and fall of epithelial phenotype: concepts of epithelial-mesenchymal transition. Berlin: Springer; 2005. p. 111–34.
89. Spallotta F, Cencioni C, Straino S, Nanni S, Rosati J, Artuso S, et al. A nitric oxide-dependent cross-talk between class I and III histone deacetylases accelerates skin repair. J Biol Chem. 2013;288(16):11004–12. https://doi.org/10.1074/jbc.M112.441816.
90. Spallotta F, Cencioni C, Straino S, Sbardella G, Castellano S, Capogrossi MC, et al. Enhancement of lysine acetylation accelerates wound repair. Commun Integr Biol. 2013;6(5):e25466. https://doi.org/10.4161/cib.25466.
91. Wang G, Badylak SF, Heber-Katz E, Braunhut SJ, Gudas LJ. The effects of DNA methyltransferase inhibitors and histone deacetylase inhibitors on digit regeneration in mice. Regen Med. 2010;5(2):201–20. https://doi.org/10.2217/rme.09.91.
92. Lv L, Sun Y, Han X, Xu CC, Tang YP, Dong Q. Valproic acid improves outcome after rodent spinal cord injury: potential roles of histone deacetylase inhibition. Brain Res. 2011;1396:60–8. https://doi.org/10.1016/j.brainres.2011.03.040.
93. Li J, Jiang TX, Hughes MW, Wu P, Yu J, Widelitz RB, et al. Progressive alopecia reveals decreasing stem cell activation probability during aging of mice with epidermal deletion of

DNA methyltransferase 1. J Invest Dermatol. 2012;132(12):2681–90. https://doi.org/10.1038/jid.2012.206.
94. Rinaldi L, Datta D, Serrat J, Morey L, Solanas G, Avgustinova A, et al. Dnmt3a and Dnmt3b associate with enhancers to regulate human epidermal stem cell homeostasis. Cell Stem Cell. 2016;19(4):491–501. https://doi.org/10.1016/j.stem.2016.06.020.
95. Rinaldi L, Avgustinova A, Martin M, Datta D, Solanas G, Prats N, et al. Loss of Dnmt3a and Dnmt3b does not affect epidermal homeostasis but promotes squamous transformation through PPAR-gamma. eLife. 2017;6. https://doi.org/10.7554/eLife.21697.
96. Sun Y, Sahbaie P, Liang D, Li W, Shi X, Kingery P, et al. DNA methylation modulates nociceptive sensitization after incision. PLoS One. 2015;10(11):e0142046. https://doi.org/10.1371/journal.pone.0142046.
97. Aguilar C, Gardiner DM. DNA methylation dynamics regulate the formation of a regenerative wound epithelium during axolotl limb regeneration. PLoS One. 2015;10(8):e0134791. https://doi.org/10.1371/journal.pone.0134791.
98. Ito S, D'Alessio AC, Taranova OV, Hong K, Sowers LC, Zhang Y. Role of Tet proteins in 5mC to 5hmC conversion, ES-cell self-renewal and inner cell mass specification. Nature. 2010;466(7310):1129–33. https://doi.org/10.1038/nature09303.
99. Ito S, Shen L, Dai Q, Wu SC, Collins LB, Swenberg JA, et al. Tet proteins can convert 5-methylcytosine to 5-formylcytosine and 5-carboxylcytosine. Science. 2011;333(6047):1300–3. https://doi.org/10.1126/science.1210597.
100. Tahiliani M, Koh KP, Shen Y, Pastor WA, Bandukwala H, Brudno Y, et al. Conversion of 5-methylcytosine to 5-hydroxymethylcytosine in mammalian DNA by MLL partner TET1. Science. 2009;324(5929):930–5. https://doi.org/10.1126/science.1170116.
101. Zhang L, Lu X, Lu J, Liang H, Dai Q, Xu GL, et al. Thymine DNA glycosylase specifically recognizes 5-carboxylcytosine-modified DNA. Nat Chem Biol. 2012;8(4):328–30. https://doi.org/10.1038/nchembio.914.
102. Zhang J, Yang C, Wang C, Liu D, Lao G, Liang Y, et al. AGE-induced keratinocyte MMP-9 expression is linked to TET2-mediated CpG Demethylation. Wound Repair Regen. 2016. https://doi.org/10.1111/wrr.12426.
103. Tan Q, Wang W, Chuan Y, Zhang J, Sun K, Luo HC, et al. Alpha-ketoglutarate is associated with delayed wound healing in diabetes. Clin Endocrinol. 2016. https://doi.org/10.1111/cen.13047.
104. Kyriakides TR, Wulsin D, Skokos EA, Fleckman P, Pirrone A, Shipley JM, et al. Mice that lack matrix metalloproteinase-9 display delayed wound healing associated with delayed reepithelization and disordered collagen fibrillogenesis. Matrix Biol. 2009;28(2):65–73. https://doi.org/10.1016/j.matbio.2009.01.001.
105. Liu Y, Min D, Bolton T, Nube V, Twigg SM, Yue DK, et al. Increased matrix metalloproteinase-9 predicts poor wound healing in diabetic foot ulcers. Diabetes Care. 2009;32(1):117–9. https://doi.org/10.2337/dc08-0763.
106. Lobmann R, Zemlin C, Motzkau M, Reschke K, Lehnert H. Expression of matrix metalloproteinases and growth factors in diabetic foot wounds treated with a protease absorbent dressing. J Diabetes Complicat. 2006;20(5):329–35. https://doi.org/10.1016/j.jdiacomp.2005.08.007.
107. Soo C, Shaw WW, Zhang X, Longaker MT, Howard EW, Ting K. Differential expression of matrix metalloproteinases and their tissue-derived inhibitors in cutaneous wound repair. Plast Reconstr Surg. 2000;105(2):638–47.
108. Adams RH, Alitalo K. Molecular regulation of angiogenesis and lymphangiogenesis. Nat Rev Mol Cell Biol. 2007;8(6):464–78. https://doi.org/10.1038/nrm2183.
109. Ding J, Tredget EE. The role of chemokines in fibrotic wound healing. Adv Wound Care (New Rochelle). 2015;4(11):673–86. https://doi.org/10.1089/wound.2014.0550.
110. Xue M, Jackson CJ. Extracellular matrix reorganization during wound healing and its impact on abnormal scarring. Adv Wound Care (New Rochelle). 2015;4(3):119–36. https://doi.org/10.1089/wound.2013.0485.

111. Glenisson W, Castronovo V, Waltregny D. Histone deacetylase 4 is required for TGFbeta1-induced myofibroblastic differentiation. Biochim Biophys Acta. 2007;1773(10):1572–82. https://doi.org/10.1016/j.bbamcr.2007.05.016.
112. Guo W, Shan B, Klingsberg RC, Qin X, Lasky JA. Abrogation of TGF-beta1-induced fibroblast-myofibroblast differentiation by histone deacetylase inhibition. Am J Physiol Lung Cell Mol Physiol. 2009;297(5):L864–70. https://doi.org/10.1152/ajplung.00128.2009.
113. Ghosh AK, Mori Y, Dowling E, Varga J. Trichostatin A blocks TGF-beta-induced collagen gene expression in skin fibroblasts: involvement of Sp1. Biochem Biophys Res Commun. 2007;354(2):420–6. https://doi.org/10.1016/j.bbrc.2006.12.204.
114. Rombouts K, Niki T, Greenwel P, Vandermonde A, Wielant A, Hellemans K, et al. Trichostatin A, a histone deacetylase inhibitor, suppresses collagen synthesis and prevents TGF-beta(1)-induced fibrogenesis in skin fibroblasts. Exp Cell Res. 2002;278(2):184–97.
115. Diao JS, Xia WS, Yi CG, Wang YM, Li B, Xia W, et al. Trichostatin A inhibits collagen synthesis and induces apoptosis in keloid fibroblasts. Arch Dermatol Res. 2011;303(8):573–80. https://doi.org/10.1007/s00403-011-1140-1.
116. Bai XZ, Liu JQ, Yang LL, Fan L, He T, Su LL, et al. Identification of sirtuin 1 as a promising therapeutic target for hypertrophic scars. Br J Pharmacol. 2016;173(10):1589–601. https://doi.org/10.1111/bph.13460.
117. Ikeda K, Torigoe T, Matsumoto Y, Fujita T, Sato N, Yotsuyanagi T. Resveratrol inhibits fibrogenesis and induces apoptosis in keloid fibroblasts. Wound Repair Regen. 2013;21(4):616–23. https://doi.org/10.1111/wrr.12062.
118. Russell SB, Russell JD, Trupin KM, Gayden AE, Opalenik SR, Nanney LB, et al. Epigenetically altered wound healing in keloid fibroblasts. J Invest Dermatol. 2010;130(10):2489–96. https://doi.org/10.1038/jid.2010.162.
119. Yang E, Qipa Z, Hengshu Z. The expression of DNMT1 in pathologic scar fibroblasts and the effect of 5-aza-2-Deoxycytidine on cytokines of pathologic scar fibroblasts. Wounds. 2014;26(5):139–46.
120. Yang E, Zou QP, Zhang HS. Expression and significance of DNMT1 in human keloid fibroblast. Zhonghua Zheng Xing Wai Ke Za Zhi. 2013;29(2):117–20.
121. Kaluza D, Kroll J, Gesierich S, Manavski Y, Boeckel JN, Doebele C, et al. Histone deacetylase 9 promotes angiogenesis by targeting the antiangiogenic microRNA-17-92 cluster in endothelial cells. Arterioscler Thromb Vasc Biol. 2013;33(3):533–43. https://doi.org/10.1161/ATVBAHA.112.300415.
122. Kaluza D, Kroll J, Gesierich S, Yao TP, Boon RA, Hergenreider E, et al. Class IIb HDAC6 regulates endothelial cell migration and angiogenesis by deacetylation of cortactin. EMBO J. 2011;30(20):4142–56. https://doi.org/10.1038/emboj.2011.298.
123. Deroanne CF, Bonjean K, Servotte S, Devy L, Colige A, Clausse N, et al. Histone deacetylases inhibitors as anti-angiogenic agents altering vascular endothelial growth factor signaling. Oncogene. 2002;21(3):427–36. https://doi.org/10.1038/sj.onc.1205108.
124. Mahpatra S, Firpo MT, Bacanamwo M. Inhibition of DNA methyltransferases and histone deacetylases induces bone marrow-derived multipotent adult progenitor cells to differentiate into endothelial cells. Ethn Dis. 2010;20(1 Suppl 1):S160–4.
125. Banerjee S, Bacanamwo M. DNA methyltransferase inhibition induces mouse embryonic stem cell differentiation into endothelial cells. Exp Cell Res. 2010;316(2):172–80. https://doi.org/10.1016/j.yexcr.2009.08.011.
126. Zhang R, Wang N, Zhang LN, Huang N, Song TF, Li ZZ, et al. Knockdown of Dnmt1 and Dnmt3a promotes the angiogenesis of hMSCs leading to arterial specific differentiation. Stem Cells. 2016. https://doi.org/10.1002/stem.2288.
127. Smits M, Mir SE, Nilsson RJ, van der Stoop PM, Niers JM, Marquez VE, et al. Down-regulation of miR-101 in endothelial cells promotes blood vessel formation through reduced repression of EZH2. PLoS One. 2011;6(1):e16282. https://doi.org/10.1371/journal.pone.0016282.

128. Turunen MP, Yla-Herttuala S. Epigenetic regulation of key vascular genes and growth factors. Cardiovasc Res. 2011;90(3):441–6. https://doi.org/10.1093/cvr/cvr109.
129. Fork C, Gu L, Hitzel J, Josipovic I, Hu J, SzeKa Wong M, et al. Epigenetic regulation of angiogenesis by JARID1B-induced repression of HOXA5. Arterioscler Thromb Vasc Biol. 2015;35(7):1645–52. https://doi.org/10.1161/ATVBAHA.115.305561.
130. Cuevas I, Layman H, Coussens L, Boudreau N. Sustained endothelial expression of HoxA5 in vivo impairs pathological angiogenesis and tumor progression. PLoS One. 2015;10(3):e0121720. https://doi.org/10.1371/journal.pone.0121720.
131. Kachgal S, Mace KA, Boudreau NJ. The dual roles of homeobox genes in vascularization and wound healing. Cell Adhes Migr. 2012;6(6):457–70. https://doi.org/10.4161/cam.22164.
132. Liang Y, Xia L, Du Z, Sheng L, Chen H, Chen G, et al. HOXA5 inhibits keratinocytes growth and epidermal formation in organotypic cultures in vitro and in vivo. J Dermatol Sci. 2012;66(3):197–206. https://doi.org/10.1016/j.jdermsci.2012.02.019.
133. Maeng YS, Kwon JY, Kim EK, Kwon YG. Heterochromatin protein 1 alpha (HP1alpha: CBX5) is a key regulator in differentiation of endothelial progenitor cells to endothelial cells. Stem Cells. 2015;33(5):1512–22. https://doi.org/10.1002/stem.1954.

Index

© Springer International Publishing AG, part of Springer Nature 2018

V. A. Botchkarev, S. E. Millar (eds.), *Epigenetic Regulation of Skin Development and Regeneration*, Stem Cell Biology and Regenerative Medicine,
https://doi.org/10.1007/978-3-319-16769-5

MIX
Papier aus verantwortungsvollen Quellen
Paper from responsible sources
FSC® C105338

If you have any concerns about our products,
you can contact us on
ProductSafety@springernature.com

In case Publisher is established outside the EU,
the EU authorized representative is:
Springer Nature Customer Service Center GmbH
Europaplatz 3, 69115 Heidelberg, Germany

Printed by Libri Plureos GmbH
in Hamburg, Germany